ADVANCE PRAISE

From a senior scholar and historian of physics:
"Fascinating . . . A marvelous book."

—David Topper, author of *Einstein for Anyone* and *How Einstein Created Relativity Out of Physics and Astronomy*

From a quantum chemist and inventor:
"Excellent book. . . . Very articulate."

—Simon West, chief scientific officer at WRS Bioproducts

From a business strategist:
"Crisp, clear and enticing. Friendly and accessible style. Dives deeply into complex matters."

—Doug Davison, founder of Sequoia Energy

From a civil servant:
"An enjoyable read."

—Norm Brandson, former deputy minister of environment

From a mathematician:
"A huge accomplishment."

—David Miller, physicist analyzing *Voyager* space probe space-plasma data for Carmel Research Center

From a movie producer:
"It may well be one of the most important books for this century."

—David Cherniack, All in One Films, producer of *Stephen Hawking's Universe*

From a quantum physicist:

"Monumental. Everyone can find in it something . . . to resonate with. Delicious with a kind of connectivity among apparently different things. It's not just elegant. It interconnects everything. Bringing together . . . concepts at the heart of the biggest mysteries. The intriguing part is its simplicity. [Its] history of scientific thinking . . . goes through many, many, many different thinkers. Irremediably elegant and aesthetic."
—Maria Luisa Chiofalo, University of Pisa

From a chemical engineer:

"Magnificent. A joy to follow the thoughts. I enjoyed the book—the entire experience—immensely."
—Gord Collis, J.R. Simplot Company (retired)

From a particle physicist:

"Astounding. A wonderful book."
—Kyle Shiells, University of Manitoba

From a leading publicist:

"A textbook but on the existence of things. The wide breadth of quotes is breath taking."
—Barbara Monteiro, Monteiro & Company

From a Nobel-winning cosmologist:

"[An] impressive romp through mathematics, physics, philosophy, and a good deal of what seems to me to be bold hypotheses."
—Jim Peebles, Princeton University

The True Nature of Reality

TIME
NOW

COLIN GILLESPIE

RODIN
BOOKS

Copyright © 2025 Colin Gillespie

All rights reserved.

No portion of this book may be reproduced in any fashion, print, facsimile, or electronic, or by any method yet to be developed, without the express written permission of the publisher.

Hardcover ISBN 978-1-957588-34-6
eBook ISBN 978-1-957588-35-3

PUBLISHED BY RODIN BOOKS INC
666 Old Country Road, Suite 510
Garden City, New York 11530

www.rodinbooks.com

Book and cover design by Alexia Garaventa

Manufactured in the United States

BOOK TWO OF THREE

Theme of the Work:

Science is the study of emergence.
Religion is reflection on all else.
The beginning of it all brings them together.

For Rosemarie (Waratah Rose)
and for Lisbeth
in requiem

And for Carole
and for Carolyn

Sine quem nihil

All men by nature desire to know.

Aristotle (c. 340 BC)

When we have unified enough certain knowledge,
we will understand who we are and why we are here.

Edward O. Wilson (1998)

Why is there something rather than nothing,
and why are things as they are?
Questions for the twenty-first century.

Frank Close (2011)

We are searching for the nature of reality,
but what does that mean?

Jim Peebles (2022)

CONTENTS

Preface ... 1
Introduction ... 7

Part I: Some Seminal Ideas ... 13
1. Says Who? ... 15
2. The Cosmos ... 22
3. The Old Ontology ... 25
4. A New Ontology ... 34
5. Distant Relations ... 43
6. The A-tom ... 47
7. The Singular Ingredient ... 51
8. The Multiplicity Affliction ... 53
9. Simply Put ... 57
10. Being One and Many ... 63
11. The Substance of Space ... 66
12. Riemann's Grains of Space ... 74
13. Planck's New Number ... 76
14. Physics Meets the Quantum ... 87
15. Strings and Things ... 95
16. Small Change ... 103
17. Being Begun ... 105
18. Expanding Reality ... 112
19. Tunnel Vision ... 118
20. Learning Curve ... 122
21. Saving Inflation ... 126
22. Einstein's Universe ... 133
23. Alles in Ordnung ... 137
24. The Inside Story ... 141

25. A Remote Entanglement 144
26. One Is All 150
27. No Not Now 154
28. Two-Timing 158
29. Einstein's Best Blunder 165
30. Just Imagine 173
31. Buns in the Physics Oven 176
32. Sufficient Reason 181
33. The Whole Nine Yards 183
34. The Single Solution 186
35. Desperately Seeking Sundance 190

Part II: Getting Real 203
36. Road Closed 205
37. On Not Being Lost 208
38. Search for Sense 213
39. Cherry Picking 220
40. The Power of Two 222
41. Think Links 228
42. Getting Going 236
43. Being Causal, Being Random 243
44. Unpacking the Sundance Story 249
45. One Quantum for All 255
46. In Throes of Threes 263
47. What's the Matter? 266
48. You'll Get a Charge Out of This 268
49. A Light Variation 271
50. Traveling Light 276
51. Moving Parts 279
52. The Cosmic Clock 284
53. Runaway Inflation 291
54. Hitting the Universal Brakes 294
55. Room for Improvement 297
56. Being Now 301

57. Doing Next	307
58. Ultimate Cause	312
59. Universal Disorder	316
60. The Direction of Time	320
61. The Flow of Time	324
62. The Closed World	326
63. The Location Illusion	330
64. On a Whole New Plane	337
65. Across the River and Into the Trees	340
66. Cool Your Jets	343
67. Dark Energy Is Space	348
68. Black Holes Are Dark Matter	351
69. A Production Check	357
70. Escape from a Monster Black Hole	363
71. The Antimatter Matter	367
Part III: Men at Work	**371**
72. In a Few Words	373
73. All in a Good Cause	378
74. Causing Problems	381
75. The Two-Slit Mystery	386
76. All the Same	392
77. Big Business	395
78. Being Pointless	399
79. A Reality Check	402
Part IV: On the Way Out	**409**
80. Desperately Seeking Susy	411
81. A Particular Problem	413
82. Taking Your Time	416
83. Shedding Spacetime	420
84. Fields of Dreams	427
85. A Conservation Conversation	431

86. Mastering Mass	435
87. Elementary My Dear Weinberg	439
88. Fudge It	442
89. A Suitable Space	444
90. That's Nothing	448
91. Something for Nothing	454
92. Higgledy-Piggledy	457
93. A Singular Difference	460

Part V: The Emerging Picture	**465**
94. In the Beginning	467
95. Property Development	471
96. Get with the Metric	475
97. Inlaws and Outlaws	478
98. The Sessile Source	482
99. Relatively Unfounded	485
100. New Light on the Old Quantum	492
101. Leading the Charge	496
102. The Hole Story	499
103. Out of Time	508

Part VI: The Numbers Game	**513**
104. Real Math	515
105. Them Apples	523
106. Getting Set	528
107. Real Imagination	532
108. Got Your Number	535
109. Zero Zeroes	539
110. Infinite Wisdom	543
111. The Upper Crust	547
112. Humbled Pi	551
113. Squaring the Planck Circle	554
114. O To Be Incomplete	559

Part VII: Testing the Waters — 563
115. Off the Scale — 565
116. See No Evil — 568
117. On the Other Hand — 571
118. Nothing Doing — 575
119. Where All the Action Is — 577
120. The Goldilocks Enigma — 580
121. Missing Multiverses — 584
122. This Changes Everything (Almost) — 590
123. Economy Class — 593
124. Passing in the Ontologic Dark — 597
125. The End of Zen — 601
126. Nothing but the Truth — 603

Part VIII: Real Progress — 607
127. The Dream of Reality — 609
128. The Mystery of Motion — 614
129. New-Views News — 617
130. Personal Limits — 624
131. Newfound Realism — 628
132. A Higher Reality — 632
133. A Future for Physics — 639
134. Being — 647
135. Doing — 655
136. You and I and We — 658

Acknowledgments — 665
Image Sources — 667
Endnotes — 671
Subject Index — 783

PREFACE

> I'd like to know
> what this whole show
> is all about
> before it's out.
>
> Piet Hein (1969)[1]

This book is about what's really going on. It seeks a fundamental answer to poet (also mathematician, physicist, freedom fighter, and buddy of Niels Bohr) Piet Hein's question.

And it is the story of a personal journey.

The question *What is real?* has seen several thousand years of human mental striving. It has long lacked a coherent answer.[2] For reasons that will become clear, this is about to change.

We each try to make sense of the world in our own way. But in the end our view of reality is socially constructed.[3] Which is to say, together we create a kind of mental picture of the world. Maybe without thinking overmuch, we then entrench it in our language. In turn, language—historian and philosopher Yuval Harari calls it "the operating system of our civilization"[4]—informs our worldview.[5]

Problem is, our mental picture is plain wrong. Indeed, if we consider how it varies over times and between persons—and how wide is the dispute about the way things work—we know it must be wrong.

This book builds a new view of reality that aims to be true.

Let me—a physicist—confess I found this new view hard to take. It challenges much I was taught. Yet it draws on thoughts of many Nobel laureates and famed philosophers, each in their own way seeking understanding of what's real.

Most of all, it taps into two physicists who each, a century apart, discovered a window into a strange world far smaller than the atom—a world that's unknown to this day because neither knew (or could accept) what he had accomplished.

Change is afoot behind the scenes. For example, in 2009, Carlo Rovelli—a physicist with philosophic inclinations, one of several whose ideas you will see much more of here—wrote,

> [O]ur present understanding of the physical world at the fundamental level is in a state of great confusion.[6]

But the story is about much more than physics: It is an adventure at the last frontier in humankind's long quest for understanding. As Rovelli also noted,

> *What is real? How is one to know?* These are among the most ancient questions not only of philosophical inquiry proper, but of human thought.[7]

The bold promise of this story is it will answer fundamental questions of philosophy; it will solve troubling problems now besetting science; and it will be simple.

For thousands of years, philosophers have sought a true view of reality. We will revisit old insights, choosing those that can, taken together, offer a consistent worldview. Decision theorist Sheena Iyengar studies choices like this; she said innovation is . . .

> . . . a novel, useful combination of old ideas that come together to solve a complex problem.[8]

Wary readers may already harbor doubts. But hark to humankind's maybe most successful chooser of ideas, Albert Einstein, from whose insights we will often draw:[9]

> [T]he grand aim of all science . . . is to cover the greatest possible number of empirical facts by logical deduction from the smallest possible number of hypotheses. . . . The theorist who undertakes such a labor should not be carped at as "fanciful"; on the contrary, he should be granted the right to give free [rein] to his fancy, for there is no other way to the goal. His is no idle daydreaming, but a search for the logically simplest possibilities and their consequences.[10]

Our task is exactly that: *A search for the logically simplest possibilities and their consequences.* And here's the secret sauce: It recently became possible to succeed.

Since the early nineteen hundreds, philosophy and physics have been at odds about *What is real?* and even *Does "real" have a meaning?* Their deepest divide is whether a reality exists; this forms in turn a deep schism in modern physics. Physicist Lee Smolin is another whose thoughts we will often draw upon. He outlined his view of the schism:

> Realist approaches assume we are able to arrive sooner or later at a true representation of the world. . . . [R]ealists are interested in ontology, which is the study of what exists. By contrast, anti-realists believe we cannot know what really exists.[11]

Physics has its test of truth, *What works?*[12] When a theory conflicts with observation, it must be cast aside—or so the myth of scientific method goes.[a] In practice an entrenched theory mutates and so survives. Today's physics consists mostly of highly mutated theories.

By contrast, philosophy is prone to ask, *How do we know what's real?* It's prone to become lost in swamps of words.

Neither physics nor philosophy confers consistent clarity on what *real* means.

A book about what's real needs to be clear about its meaning. We will use the toughest test: For us to accept something as truly real, it must be an accurately described objective aspect of the universe.[b] This may seem obvious. Yet little of philosophy or physics even aims to pass this test.[c]

It is much tougher than the facile "quotes" test of truth logician and mathematician Alfred Tarski famously proposed for his theory of theories,

> "It is raining" is true if and only if it is raining.[13]

---

a See also chapter 133, which may be found at page 639.
b This test itself is—as is unavoidable—a construct of our minds; but it is subject to the universe's oversight.
c See chapter 131 at page 628.

With our test—and a dash of history sauce—we will find a new view of the world, one that explains much more than the view we've been using.

This kind of view is called an *ontology*.[a] So we may say the reason for this book is we want a true ontology.

Even in my teens I wanted to understand. *Why?* was often on my lips or in my mind. And so of course I studied physics.

It took me years to get the message that physics does not do understanding.[14] It does what works. It invents theories that make good predictions but it doesn't ask them to make sense. These days they mostly don't.

It was not always so. For example, Nobel Prize–winning physicist Steven Weinberg said,

> In a list of the ten (or twelve) greatest physicists, we can trace the history of our progress in explaining the world.[15]

However, most on his list made their mark before the nineteen twenties when the search for understanding that drove physics vanished, deep-sixed by the quantum theory[16] with its *Shut-up-and-calculate* reply to anyone who dared ask *Why?*[17]

Some may be surprised to hear that physics turned away from what is real. It turned instead to testing theories with experiments. It got good at this. It also got complex; now, nobody understands it, not even physicists; maybe especially the physicists.[18] Thus, physics has become both useful and meaningless. I like useful; but I do still want to understand.

Our quest for understanding will embrace some seminal ideas; each will take us one step in the right direction. It is our good fortune that today the key ideas are in view; we can at last arrive at a true view of reality—what is real and what is not—and explain *why*.

Like *Time One*, this book aims to be accessible to the interested reader. Unlike *Time One*, in which detective fiction overlaid the concepts, this is plain vanilla fact.

---

a From Greek ov (pronounced "on"), the verb to *be*. The interplay between *be* and *do* will emerge here as a universal motif.

In the end, it answers questions of concern to all who think about their world: *What is all this?* and *What am I doing here?* So, while it is the universe's story, it is our story too. It may take a little while but, trust me, we will get there.

Colin Gillespie
Winnipeg, 2025

INTRODUCTION

[A]bout 13.7 billion years ago, . . . the entire universe was squashed into a single point with zero size. . . .

Stephen Hawking (1988)[19]

The Big Bang's beginning—widely envisioned just as physicist and mathematician Stephen Hawking said—has made its way into our worldview. Few physicists believed it; certainly not Hawking. Yet it is the curtainraiser for the ever-evolving and grossly inconsistent story we will call the *old ontology*.

This book builds a new way to understand our world—a *new ontology*.

Little of it is my work. Rather, it has been my good fortune to be in some of the right places at the right times when key pieces of the puzzle hove into our view. The reader will come to see what I mean. I recall author Frank Herbert's turn of phrase . . .

> . . . endless queues of happenstance meeting at this nexus.[20]

The new ontology will set out for the first time a coherent understanding of our world from beginning to, if not end, at least now. It will embrace science scenes ranging from tiniest (called the Planck scale) to hugest (universe scale).

The Planck size[a] is mind-bendingly tiny. The universe is mind-bogglingly huge. Yet, tiny to huge, it is all one world; we need a way to see it so. We need to discover what Plato called . . .

> . . . the first principle of all that exists.[21]

The need is pressing. Physics, driver of new ideas and economic growth, is bogged down in the mire of its remarkable success.[22] At

---

[a] So called because Max Planck discovered it; see chapter 13, beginning at page 76.

the pinnacle of his career, Einstein warned of the problem and anticipated its solution:

> Concepts that have proven useful in ordering things easily achieve such an authority over us that we forget their earthly origins and accept them as unalterable givens. . . . They will be . . . replaced by others if a new system can be established that we prefer for whatever reason.[23]

That "new system" needs a new foundation, a worldview with consistent answers to our deepest questions, a worldview that, unlike the current one, can lay claim to being real.

In the long struggle to understand reality, some doubted it was possible. For example, in his 1943 book, *Physics and Philosophy*, physicist, mathematician, and philosopher Sir James Jeans said,

> We see that we can never understand the true nature of reality.[24]

Long before him, others had more hopeful views. Some two thousand six hundred years earlier, philosopher Thales of Miletus tried to explain the world, sowing the seeds of what physicist and historian of science Gerald Holton called the "Ionian Enchantment":[25]

> [A] conviction . . . that the world is orderly and can be explained by a small number of laws.[26]

Through the ages, science sought those laws. Physics now espouses rather more than "a small number." It's not even clear how many,[27] but certainly more than Thales would have wanted.

Nonetheless we are about to find we can explain them all with a few simple premises. Indeed, it seems those laws arise from one law of astonishing simplicity.[a] Thales would be overjoyed.

Speaking in 1918 (about physicist Max Planck, whose work looms large in what follows[b]), Einstein said,

> Man tries to make for himself in the fashion that suits him best a simplified and intelligible picture of the world;

---

[a] Found in chapter 19 at page 118.
[b] See chapters 13 and 14, which are at pages 76 and 87.

> he then tries to some extent to substitute this cosmos of his for the world of experience, and thus to overcome it.[28]

Around that time, physics was aggressively abandoning reality. Einstein was himself unwilling instigator of the sea change. His dogged insistence that light *waves* are also *particles*[29] launched quantum physics. To his dismay, its fuzzy worldview grew at the expense of crisp reality.

In 2014, Rovelli said,

> Einstein conceded that the theory was a giant step forward in our understanding of the world, but he remained convinced that things could not be as strange as it proposed— that "behind" it there must be a further, more reasonable explanation. A century later we are at the same point.[30]

Quantum theory soon became the most successful, most precise, and most absurd theory ever. It always works; it's always accurate; it always makes no sense.[31] Quantum theory's PR policy is captured by that Mary Poppins line, "I never explain anything."[32]

Einstein was latterly a realist;[33] the philosophy called *realism* holds that an objective reality exists.[34] Quantum theory seems to say there *is* no such reality.[35]

Even in its own terms it is inexplicable. Thus, we have science writer Giles Sparrow saying,

> Quantum physics is . . . deeply troubling for our understanding of how the universe works.[36]

Explaining why quantum theory works so well—indeed, why it works at all—is another task this book sets out to tackle. Fear not, fair reader; this too will be simple.

Another basic task is, *What is time?* As philosopher Hans Reichenbach said,

> The problem of time has always baffled the human mind.[37]

We will ask other simple questions, such as, Why are there three space dimensions? A century of successful physics has produced no viable responses to them, yet we will see they all have simple and consistent answers.

How will we find those answers? Partly with this simple insight, *Science treats as imaginary some things that are real and vice versa.* Discerning what is (and isn't) real (and why) will be our lodestar.

We will need knowledge science offers. As philosopher Willard Quine said,

> For my part I do, qua lay physicist, believe in physical objects and not in Homer's gods; and I consider it a scientific error to believe otherwise.[38]

So, we will depend on science, but cautiously. We will avoid the quagmires of imagination and escape sinkholes of contradiction that litter its landscape. We will check its concepts down to their foundations before we buy in.

How can it be, the discerning reader may be asking, that fundamental questions could be answerable yet remain so long unanswered?

This question will lead us to a better grasp of what our science is and what it isn't, and of practical dilemmas facing scientists who seek success. Science tends to sit in silos that stay separate since scientists must specialize to be successful.[39] As scientist extraordinaire Edward O. Wilson,[a] leading advocate for searching across silos, said,

> The most productive scientists . . . have no time to think about the big picture, and see little profit in it.[40]

And Rovelli said,

> [T]he "pragmatic" scientist ignores conceptual questions and physical insights, and only cares about developing a theory.[41]

Our searching across silos will bring together concepts from various fields. Mindful of Nobel prize–winning physicist and famous teacher Richard Feynman's caution . . .

> In talking about the impact of ideas in one field on ideas in another field, one is always apt to make a fool of oneself.[42]

---

a He specialized in ants.

Introduction   11

... wherever possible, I will allow those with the ideas to speak for themselves, from original sources.[a]

Some of those sources are in languages other than English, those before the nineteen forties often in German,[43] most significantly those of Einstein, who, as his friend, physicist, and science historian Abraham Pais, said, was ...

... a highly gifted stylist of the German language.[44]

Nuance in such sources is important, especially Einstein's, on whose thoughts this work will heavily depend.[b] Where extant translations appear to miss the mark, I use my own (and provide original text).[c]

The reader will find a degree of warp and weft in the story's unfolding, sometimes returning to old ground with new perspective.[d] I'm sorry if this irritates. It is in the nature of the tale that I am setting out to tell—understanding what is real—which is a weave rather than a thread, one that is indeed more intricately interwoven than even such recursion can discover.[e]

---

a I also include secondary sources selected for accessibility, without relying on their complete accuracy.
b His name appears some seven hundred times.
c Language changes over spans of decades; I had the good fortune that my *Hochdeutsch* teachers, Betty and Charles Howlett, were almost of Einstein's generation.
d Linkages are cross-referenced in the footnotes, supported by a detailed subject index at the end, both of which will be integral to the style of reasoning, see note 143.
e As political scientist Margaret Canovan said of philosopher Hannah Arendt's *The Human Condition* (in her introduction; see my note 102 at her page viii): "There are more intertwined strands of thought than can possibly be followed at first reading, and even repeated readings are liable to bring surprises."

PART I

SOME SEMINAL IDEAS

Science begins with a vision.
Carlo Rovelli (2014)[45]

1
SAYS WHO?

> We need to go back to the insights [about the nature of reality] behind general relativity and quantum field theory, learn to hold them together in our minds, and dare to imagine a world more strange, more beautiful, but ultimately more reasonable than our current theories of it.
>
> John Baez (2001)[46]

Through the ages our question *What is real?* has engaged thinkers of a philosophical persuasion. Of late it has been taken up by a few physicists. One thing they have in common is they want to understand.

It is hard to choose among them. But the views and roles of some will be of special interest here. Their views, right or wrong, help set the scene. They distill for us from scientific silos some essences of physics' insights (and its key confusions) that may belong in (or need to be kept out of) the picture we are seeking.

These, then, are physicists who bridge the divide into philosophy. Their cross-border works support a mini-genre that confronts our theme with many questions. They tend to share a problem Herbert put this way, | Leading physicists seek a new view of reality.

> From the top of the mountain, you cannot see the mountain.[47]

One way or another, though, we often will be building with their words. So, by way of a brief introduction, here's a selective bestiary of sorts.

John Baez is a mathematician and a physicist who has a philosophic bent. With his observation, in the epigram above,[a] he goes on

---
a We will revisit what he said, in chapter 136 at page 658, to see how right he was.

to say that fundamental physics should have room for philosophizing, a view few of his colleagues seem to share. He blogs in a loose zone between math, physics, and philosophy. He sometimes writes in what he maybe thinks is simple language like,

> [I]t implies that the Standard Model gauge group consists of the symmetries of an octonionic qutrit that restrict to symmetries of an octonionic qubit and preserve all the structure arising from a choice of unit imaginary octonion.[48]

As anyone ('cept me and likely thee) can see, he's speaking of constructing Jordan algebras. I don't hold such impenetrable stuff against him: In all seriousness he would be my pick to build new math the physics world will need if it begins to navigate reality.

Adam Becker is an astrophysicist and science writer who wrote a book that asks, *What is real?*[49] He ended, with no answer, saying,

> There is something real, out in the world, that somehow resembles the quantum. We just don't know what that means yet. . . . This is the great enterprise.[50]

(We will find that every word of this was right.[a])

He followed up in 2022 with a popular article on space and time.[51] Here too he left us hanging:

> Will we ever know the real nature of space and time?[52]

This might seem less than helpful but does give us the score so far: universe 3, science 0.

Physicist and science-communicator extraordinaire Brian Greene has a relaxed approach to poking fun at fundamental physics. Neuroscientist Stuart Firestein said,

> [He] is a theoretical physicist who is also well known for his elegantly written descriptions of the hard-to-get-your-head-around concepts that make up the Alice in Wonderland worlds of relativity and quantum mechanics. He is a unifier.[53]

---

a In chapter 45 at page 255.

Greene asks key questions with short words and wraps key points in punchy lines:

> [T]he big bang leaves out the bang. It tells us nothing about what banged, why it banged, how it banged, or, frankly, whether it really banged at all.[54]

He digs into deep problems we will tackle, like "Is Space a Human Abstraction or a Physical Entity?"[55] He gave that one a chapter but—as so often (he's far from alone in this)—ended with a reprise of the question.

He leans a lot on Einstein (I do too); yet, like many physicists, he seemed to not know of (or forgot to mention) Einstein's answer to his *Is space . . . a physical entity?* question.[a]

He makes movies that explore deep issues.[56] Once there is an understanding of reality, Greene would be my pick to tell the world.

Former physicist Fotini Markopoulou gets an honorable mention—even though she's *former*[57]—as, in my view, she may be the deepest fundamental-physics thinker of them all; and she has special roles here. Former colleague, Smolin,[b] said,

> Fotini is extremely original, original to a fault. Most scientists pick up on ideas which are dominant, which come from living figures, and develop them incrementally. She doesn't do that—she works solely on her own ideas.[58]

She was inspired . . .

> . . . to pursue physics as a quest to understand reality from within.[59]

She brought a wide view to the search for quantum gravity, the pot of gold at the far end of the fundamental-physics rainbow. She thought about the implications of the fact that we are studying the universe from *inside*.[c] Twenty years ago, she plugged the kind of math

---

a See chapter 10 at page 63, especially note 261.
b They were once wed.
c For example, observations of systems in general relativity are made from inside the system while those made of quantum systems are made from outside the system; and see chapter 24 at page 141.

we'll see the new ontology may need.[a] While many wandered, lost in math, Markopoulou was blazing her unique trail to that holy grail of physics, the theory of everything.[60]

Smolin said she also was . . .

> . . . responsible for changing my views on several important aspects of quantum gravity over the last several years, which got me out of space and back into spacetime.[61]

If so, this gives us a rare glimpse of dead-ending at work in physics; neither he nor she could spot the spacetime problem,[b] what Smolin's later co-author, philosopher Roberto Mangabeira Unger called, in their work on natural philosophy, "the spatialization of time."[62]

Canned by physicist and director Neil Turok at Perimeter Institute for Theoretical Physics reportedly for being too original, Markopoulou quit physics and went into business;[63] she was worth ten Turoks; what a waste! (Also, she mentored Sundance Bilson-Thompson—of which more to come.[c])

Nobel Prize–winning physicist Roger (now Sir Roger) Penrose wrote the book on physics and reality.[64] He frolics at the fringes of philosophy. He led the charge on the impossibility of the low *entropy* (that is, the high degree of order) of the early universe arising by chance.[65] It made a deep impression as I came to see this must be an integral—even central—aspect of a real worldview. Too bad it set him to imagining a pre-universe—collapsing under its own gravity—begetting our universe with a big bounce that came after a big bounce that . . . [66] We will find a simpler way to solve his problem.[d]

Carlo Rovelli is a physicist who works on quantum gravity and time. He writes philosophic books that tackle some of physics' crazy stuff.[67] Trying to make some sense of quantum theory, he espoused an ontology he called relational[68] (in truth it did not meet the mark[e]). He put his money on quantum time.[69] He has an insightful chapter on space quanta; he gets it that they are not *in* space—they *are* the

---
a See chapter 43 at page 243; and see note 555.
b See chapter 83 at page 420.
c She was his designated mentor at Perimeter Institute; a mismatch, he would later say.
d See chapter 23 at page 137.
e See also chapter 5 at page 43.

space; but he does not tell us what *they* are.[a] Having set time on a quantum pedestal, he still seems stuck in spacetime. Nonetheless his views in recent books come closer to reality than most and so will be much quoted here.[b] And, like Smolin, he is notable for his sustained efforts to harvest fundamental truths.

Smolin—a rare physics PhD who did study philosophy—switched to loop quantum gravity some twenty years ago from *string theory* (the up-and-coming version of quantum gravity[c]), later writing a critique of it in which he (I say rightly) said,

> To continue the progress of science, we have to again confront deep questions about space and time, quantum theory, and cosmology.[70]

A dogged reacher for reality, Smolin recently came right out and said we need a new ontology.[71] In 2015, he co-wrote with a philosopher a book that circled the same issue,[72] getting the "there's one universe" concept right and the "time must be real" concept not so much. His seminal role includes preaching a real theory must be purely relational, so it must *explain* space and time (and therefore cannot, as almost all physics does, *assume* them). He worked with Markopoulou. He buys the we-are-stuck-inside view of the universe.[73] He sees that time and space can't both be real, yet still goes on about "atoms of spacetime".[74] He promotes a principled approach to fundamental physics with concepts like background independence.[75] We will draw on them.

Out of place in this company, cosmologist Neil Turok gets a mention nonetheless. He has an oar in our water with an extra universe, one he said is going the wrong way.[76] It's thin gruel; his mention here is due to his decade as director of the Perimeter Institute, a prominent hothouse for theoretical physics. It was intellectual home to Markopoulou, Smolin, and Bilson-Thompson. Of Markopoulou's departure he said,

---

a See chapter 45 at page 255.
b A word about my treatment of quotations: Where they demark conversation, they include punctuation as is conventional in the text; but where they demark other text, they include only original punctuation.
c See chapter 15 at page 95.

> She is a very fundamental thinker; she had original ideas. But at the end of the day you had to decide if those ideas are going to pan out.[77]

Unfortunately, he did the deciding.

John Wheeler was a physicist who worked both sides of the quantum-theory/relativity divide. It made him a generalist amid many specialists and from this viewpoint he saw into several silos. His worldview was wide enough to envision (a quarter century ago) that,

> Relative to the Planck length, even the miniscule entity we call an elementary particle is a vast piece of real estate.[78]

He was a leading light in Planck-scale thinking and more, we will find. (Too bad that, driven to pursue deep issues, he too dabbled in the siren spells of multiverses[a] and the *anthropic principle* that says the reason why the universe suits human life so well is it must be so because we are here to see it.[b])

Nobel Prize–winning physicist Frank Wilczek is an author whose most recent book promoted ten keys to reality. Five dealt with the ontologic issue "What There Is." They covered space and time, ingredients and laws, and matter/energy. He said,

> [S]cience teaches us what is.[79]

Unfortunately, physics, at least, doesn't. Indeed, he acknowledged we lack a crisp picture of reality, and left us with,

> The universe is a strange place, and we're all in it together.[80]

He has done better, as we'll see.

As well as a tendency to philosophic thinking, these physicists have other things in common. Like Piet Hein (and me), they want to understand reality[c] and know they don't.

In varying degree, they cling to old ontology and the related *standard theory of particle physics*[81] (fig. 1) and *standard model*

---

a And he was Everett's thesis advisor; see note 1834 and related text.
b See further, chapter 120 at page 580.
c See note 1.

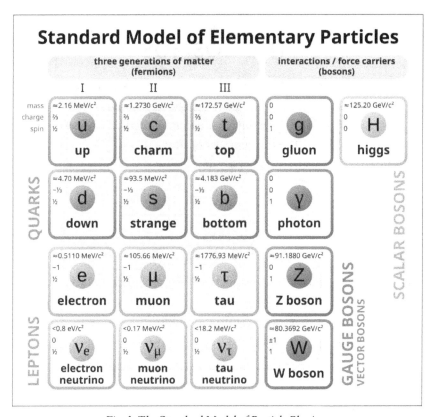

Fig. 1. The Standard Model of Particle Physics

of cosmology (fig. 63 at page 365).[82] Both models are remarkably successful. But both, we will see, are made of myths propped up by swarms of arbitrary numbers.

These leading scientists (all, save Wheeler, still with us) often say they are aware of contradictions built into the old ontology. Yet they seem unable to let go of it.

However, in the search for a world that is real, one might pick most of them as recent leaders of the pack. This book owes them more than its quotes from their works disclose.

On an entirely other plane, all seekers of the real world owe an overarching ideaistic debt to Einstein, to this day the lead trailblazer,[83] as is acknowledged in what follows by some hundreds of citations gleaned from the great corpus of his works now looming into view from Princeton and Jerusalem.[84]

2
THE COSMOS

> By the term "Universe," . . . I mean to designate *the utmost conceivable expanse of space, with all things, spiritual and material, that can be imagined to exist within the compass of that expanse.*
>
> Edgar Allan Poe (1848)[85]

Author Edgar Allan Poe's cosmogony[a] was ahead of its time in many ways but thinking of an entire universe was already a long tradition.

The concept of a cosmos, "the world or universe regarded as an orderly, harmonious system,"[86] emerged at different times in cultures with diverse modes of thought. For example, several thousand years ago in China,

> . . . ancient Chinese thinkers viewed the world as a complete and complex "organism."[87]

Centuries later, it was Greek civilization, spreading through the eastern Mediterranean, that gave birth to the long chain of thought that led to much of the science, including physics and cosmology (fig. 2), we know today.

Fig. 2. The Ptolemaic Cosmos

---

a See chapter 13 at page 76 and chapter 129 at page 617.

Anaximander was a leading Greek philosopher. He was born in 610 BCE and raised in Miletus on what's now the coast of Turkey near the Greek island Samos. Though he documented his ideas, few of his records survive.

> Anaximander conceived the idea of the cosmos.

His ideas—as we know them from accounts of others—mostly lacked what we would see as scientific method and were almost entirely wrong. Nonetheless, he can be said to be . . .

> . . . the author of the first surviving lines of Western philosophy.[88]

He owed to his mentor, Thales, the then startling idea that we can understand the world without recourse to the supernatural. He conceived the seminal idea of *the cosmos* as an open universe and he wondered about its origin.[89]

So, philosopher Charles Kahn could say of his worldview,

> Anaximander's conception of the world is . . . the prototype of the Greek view of nature as a cosmos.[90]

He thought of basic cosmologic concepts such as,

> Earth is poised aloft, supported by nothing, and remains in place because it is equidistant from all other things and thus has no disposition to fly off in any one direction.[91]

Millions, perhaps billions, of people—almost all of whom never heard of Anaximander—have since then thought about the cosmos. It entered our minds and our vocabularies in many languages.

Being concerned here with understanding, we need such language, language that enables us to conceive and contemplate and communicate ideas.[a]

Rovelli credited Anaximander with . . .

> . . . gradually founding the basis of a grammar for understanding the world, which is substantially still our own today.[92]

---

a See also chapter 72 at page 373.

Having an idea and saying so or writing it will not, alone, lead to it having lasting life. His then new view of the cosmos needed to survive and thrive in the right kind of society, where over time it would keep falling upon fertile ground. That we study it today tells us Miletus and his Greece supplied this need.

Thus, if we want a key person who set us to searching for the seminal idea of the cosmos, Anaximander is the one.[93]

And the cosmos—as an orderly and harmonious system, what it is, and how it works, and where it came from—is what we are setting out, like him, to understand.

3
THE OLD ONTOLOGY

> It is possible to repeatedly correct our worldview
> and to discover new aspects of reality that are
> hidden to the common view.
>
> Carlo Rovelli (2014)[94]

Plato may have invented both the concept and the word—the seminal idea of the idea and the word *ιδέα*.[95] It derives from the Greek verb *idein* (*ἰδεῖν*), *to see*. In this work we delve into a broader—and, today, more usual—kind of idea than Plato's (which philosophers now call a *universal*, meaning a property held in common, like the shape of circles[96]).

The idea that a real world exists—independently of our observing it—is even older than the word, maybe as old as *Homo sapiens* and older species.[97]

Each in our own fashion we try to discern what is real. Our personal quests for understanding begin early. When eyes first open, fast-forming minds start asking, wordlessly, "What *is* this?" Our infant brains are on their way to making models of the world, our private versions of reality.[98]

Each time we use our eyes or with our other senses confirm what they see, and each time we anticipate events and find they happen, we strengthen or expand the reach of our own personal idea of reality.

As our mental models grow, we think and speak in terms of them. We don't confine our conversations to bare facts or observations, we embellish. Some embellishments become as real to us as what we see.

In this way, we both partake of and contribute to a wider worldview. Tens of generations gave rise to our languages and sciences, to our philosophies and to civilizations that embody in some degree a communal understanding of reality.[99]

But how much of that understanding is true? How well does it distill the universe?

As Wilczek said,

> In our rush to make sense of things, as infants, we learn to misunderstand the world.... There's a lot to unlearn, as well as a lot to learn, on the voyage to deep understanding.[100]

So, we should not be surprised that our *idea* of reality may not match what *is*. It has always been a work in progress. Here may be the place to say to general readers, who come upon the language of a new idea, *Be of good cheer*, your path will be simpler than that of the experts who have so much to unlearn.

After millennia of building a collective idea and imparting current versions to then current generations, we are all working with some mix of reality and fiction. Looking back, we may find reason to believe the trend has been to more reality. It's harder to conclude there is less fiction.[a]

We need not go so far as philosopher René Descartes's famed elegy to skepticism of the senses, which to this day infects the body of philosophy with a well-justified unease:

> I shall then suppose ... some evil genius ... has employed his whole energies in deceiving me; I shall consider that the heavens, the earth, colours, figures, sound, and all other external things are nought but the illusions and dreams of which this genius has availed himself in order to lay traps for my credulity.[101]

Philosopher Hannah Arendt said Descartes's doubt is haunted by the nightmare that,

---

[a] *Fiction* is (in this usage) a loaded term. Please note it has no derogatory purpose. As will appear, our fictions can be and often are extremely useful, even essential. My point in using this term for our *free inventions* (Einstein's term; see note 130) is we should be clear about the distinction from reality and about which is which.

> [I]f neither the senses nor common sense nor reason can be trusted, then it may well be that all that we take for reality is only a dream.[102]

We have no reason to conclude we are all dreaming. What we seek should be achievable, a worldview that makes consistent sense of humankind's long-aggregated sense-experience despite its frailties. Once we have a consistent picture in our sights, its explanatory power may persuade us—or on the other hand may not.

Consistency is far from what we find in today's senior science's story of the world. For example, Smolin wrote that the major theories of physics lead to . . .

> . . . paradoxes, fallacies and dilemmas that plague the literature.[103]

And Nobel Prize–winning cosmologist Jim Peebles, speaking of the standard model of cosmology, said,

> How can you trust a story that keeps changing, and on the face of it seems implausible?[104]

So I do not mean to suggest that we—or even physicists—share a single ontology, one common concept of reality. Indeed, most of us steadfastly assert the virtues of our varied versions.

But almost all, even those who embrace wild disinformation, get through life using a more or less widely accepted set of interlocking concepts. In past ages, these were often hokum. For example, for hundreds of years the establishment in Europe pursued the idea of witchcraft; in consequence it murdered half a million "proven" witches.[105] That the idea was an imaginary social construct occurred to few and was proclaimed by almost no one.

These days, the central concepts underlying daily life the world over rest on what we like to think is a less fictional foundation. Although few consider it, much of that foundation rests—more or less directly—on precepts of physics.[106]

Physics' precepts in themselves comprise a worldview. Though most of us ignore them, they powerfully influence our daily lives. Whether we will or not, the canonical worldview of physics has personal consequences. Yet, these days it seems to be, to say the least, on shaky ground.

In 2013, science writer Jim Baggott sounded the alarm:

> There is as yet *no* observational or experimental evidence for many of the concepts of contemporary theoretical physics. . . . For some of the wilder speculations of the theories there can by definition *never* be any such evidence.[107]

In 2023, astrophysicist Adam Frank and physicist Marcelo Gleiser said,

> Physicists and astronomers are starting to get the sense that something may be really wrong. It's not just that some of us believe we might have to rethink the standard model of cosmology; we might also have to change the way we think about some of the most basic features of our universe—a conceptual revolution that would have implications far beyond the world of science.[108]

And later,

> We write this book with a sense of urgency because we believe our collective future and human project of civilization are at stake. . . . We believe we need nothing less than a new kind of scientific worldview.[109]

On this we are agreed.[a] But they did not go on to suggest one.

And here's the rub: Physics leads the world of science and lies closest to its philosophical foundations; but its worldview is not grounded in reality. As noted above, for the last hundred years physics has been studiously unconcerned with what is and what isn't real, even denying there *is* a reality that's independent of observers of experiments.

Putting this another way, Smolin said,

> To the extent that quantum mechanics is the correct description of nature, we are forced to give up realism.[110]

---

a See chapter 136 at page 658.

Quantum mechanics wrought a revolutionary change in the worldview of physics.[a] Physicist Alain Aspect said,

> The development of quantum mechanics in the beginning of the twentieth century was a unique intellectual adventure, which obliged scientists and philosophers to change radically the concepts they used to describe the world.[111]

The reader interested in how . . .

> . . . science is rooted in conversations . . .[112]

. . . and how they led to the quantum-mechanics revolution, could consult historian of physics Mara Beller's book.[113]

Having himself set the quantum-mechanics train of thought in motion, Einstein famously declined to take the ride:

> I still believe in the possibility of a model of reality—that is, a theory—which represents things themselves and not only the probability of their occurrence.[114]

Yet, for Einstein (who, though earlier aligned with positivism,[115] was, in his latter days, more of a realist[116]), the central concept of the old ontology was space that is continuous, and is thus made of infinitely many infinitesimally tiny points. This was (and still is) only an idea. It is not supportable by observation and—though he applied it widely—Einstein did not go out of his way to prop it up:

> I shall not go into detail concerning those properties of the space of reference which lead to our conceiving points as elements of space, and space as a continuum.[117]

As we'll see, there are reasons to conclude the points-in-a-continuum idea is wrong.[b] Early in his career, Einstein himself discerned this.[c] Yet to the end of his days he was unable to articulate his overriding aspiration, a consistent picture of reality (or, as he put it, "a closed system of thought").[118] As Holton said,

---

a The success of quantum physics in describing things in terms of probabilities is largely responsible; see the preface and the introduction.
b This part of the story starts in chapter 12 at page 74.
c This was his emerging view at least as early as 1917; see note 1268.

> [T]he very extent and depth of the advances Einstein himself helped to launch . . . eventually made it impossible for the physical phenomena to be all gathered in one grand relativistic Weltbild [worldview] of the sort he longed for.[119]

Einstein's reservations were radical and little known. The view of space as a continuum came to be deeply embedded, not only in physics, but in our shared worldview. As mathematician Louis Crane said,

> Unfortunately the classical continuum is thousands of years old and is very deeply rooted in our education.[120]

Some physicists, too, believe in their theoretic constructs, such as spacetime, as if they were reality. To them, even reduced to a formula for a probability, a particle—let's say a photon—is real: It has its name; it does its tricks; it has its particular properties; it has its standard-model place; and most of all it has its math. What more reality would one want?

At this, I cringe. When physicists *believe* their doctrines, physics takes on aspects of religion. But maybe it is not quite belief, but rather is a set of *ontological commitments*.

Quine long dominated modern study of ontology. His credo specified,

> We are ontologically committed to all and only those entities that must exist in order for the theories or statements we hold to be true to be true.[121]

And these commitments can inhere in a theory:

> The ontological commitments of a theory are, roughly, what the theory says exists; a theory is ontologically committed to electrons, for example, if the truth of the theory requires that there be electrons.[122]

Either way, beliefs or commitments, it's not easy for a new view of the world to set aside the self-affirming body of received wisdom.

For example, mathematics, as the science that waters the roots of physics, has its own ontologic issues.[a] Its widely accepted ontological

---

a We will explore some of them in part 6.

commitments include zero and infinity, whose reality has always been in question. And it has a fraught reliance on set theory, for, as philosopher of mathematics Penelope Maddy said,

> [S]et theory is the ultimate court of appeal on questions of what mathematical things there are, that is to say, on what philosophers call the "ontology" of mathematics.[123]

Our search for what is real will require us to rethink "what mathematical things there are"—or aren't—and to do this without appealing to set theory. Finding the things there are will turn out to be relatively easy;[a] setting them into their proper place may take more work.

For physicists, common ontological commitments include continuous space, the vacuum, conserved energy, and points of zero size. They may not readily be able to conceive a worldview without these concepts. Yet in due course it will appear none of these four is real.[b]

The potency—and peril—of these commitments lies in usefulness. Another example is the concept of an elementary particle, like the electron. In the nineteen hundreds this became an ontological commitment so unshakeable it borders on belief. In reaction, Nobel Prize–winning physicist Niels Bohr was led to countenance—and preach—obscure nonsense that lingers to this day and that for some takes on the character of doctrine.[124]

Perhaps the most troubled of physics' ontological commitments is the quantum for which Planck is famous. Quanta of energy form the foundation for quantum theory. We will see that, on one hand, Planck's quantum was *not* a quantum of energy;[c] and, on the other, physics has deep difficulty with the question of what energy *is*.[d] And how there can be discontinuous quanta in a continuous universe is a conundrum that deeply disturbed Einstein. Alluding to this in 1923, he went so far as to say,

> The theory of relativity was only a sort of respite which I gave myself during my struggles with the quanta.[125]

---

a See chapters 104 to 111 (beginning at page 515).
b These are strong claims; there will be strong evidence.
c See chapters 13 and 14 at pages 76 and 87.
d See chapter 86 at page 435.

By contrast, also in 1923, mathematician and physicist John von Neumann wrote with aplomb,

> [A]ll elementary processes, i.e., all occurrences of an atomic or molecular order of magnitude, obey the "discontinuous" laws of quanta.[126]

This expressed succinctly a central premise of the worldview that took root in physics a hundred years ago. It obscures key aspects of reality we know: Those "'discontinuous laws'" apply to vastly larger magnitudes than atoms;[a] the real discontinuities are far smaller than the atoms;[b] and what it *is* that is really discontinuous is unclear, but it clearly is *not* energy.[c]

Meanwhile, physical cosmologist J. Colin Hill recently said about our understanding of the universe on a far larger scale,

> The situation right now seems like a big mess. I don't know what to make of it.[127]

However, for most philosopher-physicists it is the incompatibility of our two big theories—relativity and quantum theory—that defines the ontological problem. For example, in 2001, Smolin said,

> After all, atoms do fall, so the relationship between gravity and the quantum is not a problem for nature. If it is a problem for us it must be because somewhere in our thinking there is at least one, and possibly several, wrong assumptions. At the very least, these assumptions involve our concept of space and time.[128]

This, then, outlines[d] the nature of *the old ontology*. This lies behind what Smolin (with Unger) assessed to be plagued by "paradoxes, fallacies and dilemmas."[e] Yet, because it seems (and often is) so esoteric, we tend not to realize how inextricably it is now caught up in our daily lives.

---

a See chapter 45 at page 255.
b See chapter 14 at page 87.
c See chapter 45 at page 255.
d In admittedly fragmentary fashion. Of late, many leading thinkers have published popular books and articles more systematically deploring fundamental failings of our science.
e Note 103.

Simply put, the elements of physics underlying that ontology are, almost entirely, fictions.[a] In his 1933 lecture "On the Method of Theoretical Physics," Einstein spoke bluntly of . . .

> . . . the purely fictitious character of the fundamentals of scientific theory.[129]

And, don't get me wrong, in very many ways the fictions (or, as Einstein and his friend, fellow physicist, sometime co-author and biographer, Leopold Infeld also called them, the "free inventions"[130]) of the old ontology *are* useful. That's why they are so widely and so well accepted.

Nonetheless they hold a hidden hazard: They block progress on the road to reality.

Or, as Unger, seeking to escape the limitations of what he called "the Riemannian-Einsteinian ontology," said, in more optimistic vein,

> A persistent feature of the history of science is the association between empirical discoveries and ontological programs.[131]

All this, then, is why we need—why our lives would be better with—a new ontology, a worldview that at least tries to be real. It is also why, as (in the epigram) Rovelli said is possible, we are now setting out to "correct our worldview and to discover new aspects of reality that are hidden to the common view."

---

[a] Peebles was refreshingly specific about this, speaking consistently of theories that "were useful approximations to the way reality operates" and "approximations to objective reality." See note 104 at his pages 27 and 44.

4
A NEW ONTOLOGY

> The real change that's around the corner is in the way we think about space and time. *We haven't come to grips with what Einstein taught us.* But that's coming. And that will make the world around us stranger than any of us can imagine.
>
> David Gross (2008)[132]

Our conventional worldview is largely founded on a hundred years of physics that disdained reality. Unsurprisingly, it's mostly fiction. It is filled with contradictions. Let's set out to build a worldview that at least takes aim at being real.

In 2008, Nobel Prize–winning physicist David Gross foresaw (above) a strange new worldview that he said would "change . . . the way we think about space and time." What he anticipated is what we are after here: We need to *"come to grips with what Einstein taught us"* and to discover that strange new view of the world.

| We seek a real worldview.

Others, both before and after Gross, have tried to find a way to do it. They didn't. Step by step, we shall.

So (to reiterate) a fundamental starting point is,

> [S]omewhere in our thinking there is at least one, and possibly several, wrong assumptions. At the very least, these assumptions involve our concept of space and time. . . .[133]

A real ontology will need to find and ditch those wrong assumptions. And it will need instead to find consistent answers for at least these two unanswered questions:[a]

---

a This will turn out to be easier than it may appear.

- What is space?
- What is time?

We must ask other questions that are almost as profound. For example, physicist, cosmologist, and mathematician John Barrow asked,

> How, when, and why did the Universe come into being? Such ultimate questions have been out of fashion for centuries.[134]

As he went on to say, some scientists have recently been "asking such questions in all seriousness." So will we, and we will find consistent answers.

To be clear about the expansive scope of this undertaking, let's contrast it with the new ontology physicists David Bohm and Basil Hiley worked on for two decades.[135] Though the resulting book was broadly titled *The Undivided Universe*, it focused closely on a new understanding of atom-scale quantum theory.[a]

By contrast, our aim is not understanding quantum theory, nor even understanding all of physics from quantum theory to particle physics to cosmology to relativity; it is understanding all reality over the entire history of the universe (including, incidentally, those topics too).

While this may seem a far more daunting challenge, we will find its all-embracing scope is an advantage. Solving all the contradictions will turn out to be a lot easier than solving only one or two.[b]

That's not to say it will be easy. The universe is vast. Its origins seem lost in mists of time. Its workings, we already know, are weird and hidden from our view. This creates a fundamental problem. As philosopher A. J. (Freddie) Ayer asked,

> How could it be validly determined, by reason alone, that the world is so very different from what it appears to us to be?[136]

---

a They conceive the whole universe as the essential quantum system. But *universe* does not rate a mention in their index.
b See chapter 34 at page 186.

Thus, though reason be our guide, we will need clear thinking about its hazards as we proceed. And we will come back to his question to check how we're doing.[a]

Another need is evident: We must screen all ideas physics offers. The need to do this is not itself a new idea. Physics is a human creation with no clear relation to reality. In his 1936 paper "Physics and reality," Einstein wrote,

> Physics constitutes a logical system of thought which is in a state of evolution, whose basis cannot be distilled, as it were, from experience by an inductive method, but can only be arrived at by free invention.[137]

So, getting underway, the third *seminal idea*[b] we need is that of a *real ontology*, a clear-eyed worldview that dispenses with physics' freely invented useful fictions while harvesting its real insights.[c]

Integral with that worldview, we need language we can share and use to think and speak of it.[d] Psychologist and Nobel-winning neural-networks pioneer Geoffrey Hinton said,

> [W]e have this thing called language, which we use for modelling the world.[138]

And let's be aware new language requires us to transcend bounds of the accepted worldview without getting lost in fictions of our own.

The word *ontology* itself imports a concept we need to pursue. While its roots lie in philosophy, computer scientist Thomas Gruber said,[e]

> An *ontology* is an explicit specification of a conceptualization.[139]

As used in this book it means that and a deep description of reality.[f] This involves, as we've seen Einstein (most successful ontologic

---

a See chapter 133 at page 639.
b After the *cosmos* and the *idea*.
c See chapter 39 at page 220.
d See also chapter 72 at page 373.
e From a systems perspective.
f For those who, perhaps justifiably, feel they have other or better definitions, I refer to the Humpty Dumpty rule: "When I use a word, it means just what I choose it to mean—neither more nor less."

operator of our era) put it, both "a simplified and intelligible picture of the world" along with a "cosmos" all its own.[a]

Speaking to students in Los Angeles in 1932, he said, in contrast with the narrowly predictive view of physics' purpose,

> A stronger, but also more obscure drive lies behind the tireless exertions tied to such achievements: one wants to comprehend being, reality. . . . At the basis of all such attempts lies the belief that being is completely harmonious in its structure.[140]

That is (he thought), a real ontology should be *completely harmonious in its structure*. Which is to say, its tale should be internally consistent.

Of course, ontologies may turn out *not* to describe reality. Indeed, they all *have* turned out that way so far, but one can try. And though each try is bound to be an incomplete description, it should, as ontologist Nicola Guarino said, be based on a coherent backbone:

> The backbone of an ontology consists of a . . . hierarchy of concepts.[141]

The new ontology will set out a coherent backbone. The upper entries in its "hierarchy of concepts"—its seminal ideas—are set out in this part.

A new ontology requires what philosopher and historian of science Ian Hacking called "a style of reasoning":[142]

> A style of reasoning not only brings things into view, but it also can bring them into being. It opens up a space of possibility in which new things can appear, as well as from which they can disappear.[143]

Introducing such a style will be my objective.[b] We shall see if it suits.

---

a See note 28.
b Here its elements include historical and personal settings, open engagement of the reader, extensive cross-referencing (in footnotes and the subject index), expanding and clarifying essential language, preferring meaning (keyed as *meaning* in the index) over stricter definition, a writing style, and a holistic view.

Greek philosopher Aristotle invented the idea of ontology as the study of existence[144] (though the word *ontology* was not in use until the sixteen hundreds). It settled into the realm of philosophy, and then of its unfavored offspring, *metaphysics*. It was often seen as opposed to *epistemology* (the study of knowledge[145]) and *its* handmaiden, modern physics.

In the early nineteen hundreds, Einstein, with his deeply philosophic bent, set to transforming physics,[146] which had unintended consequences for its links to philosophy. Indeed, physics soon turned its back on metaphysics. More specifically, it turned away from ontological pursuits and took on a narrow epistemologic cast.[147] Soon it was no longer concerned with reality; it was about what one could say about what one could see. This became physics' posture by the nineteen thirties; this is its prevailing posture to this day.[a]

In the same era, philosopher Martin Heidegger did not buy into the then new quantum theory's need to separate what we observe from ourselves as observers. Rather, he sought a holistic view of space and time and all their contents including people. It's all one place. Philosopher Simon Critchley summarized,

> The basic and very simple idea . . . is that the human being is first and foremost not an isolated subject, cut off from a realm of objects that it wishes to know about.[148]

In 1924, Heidegger, doyen of the positivist Berlin Circle, asked,

> What is time?[149]

He later said,

> [T]he central range of problems of all ontology is rooted in the phenomenon of time, correctly viewed and correctly explained.[150]

As noted above, *what time is* is central among many problems the new ontology must explain. We will see that physics offers a mix of rare illumination and rampant confusion about it.

Indeed, for much of physics the larger question, *What is?*, has no meaning, or at least no answer. The question's present tense implies a

---

[a] With maybe more emphasis on *say* and less on *see*.

what-is here (in space) and now (in time) that is of a piece with *what-is* elsewhere (in other space, at the same time). But, to the contrary, the theory of relativity says between here and anywhere that isn't here there is an insuperable gap in time, a slice of past we don't yet know, a present we cannot yet see.[a] All of the *what-ises* elsewhere that share this moment with our *what-is* here are undefinable.[b]

Relativity, in other words, has troubles with time that lead it to proscribe far-flung *simultaneity*.[151] It seems to say *what is* consists exclusively of *what is here*. Rovelli (summarizing it) said,

> Our "present" does not extend throughout the universe.
> It is like a bubble around us.[152]

Here he made a mistake Smolin warned about,[c] an almost universally held wrong assumption, that leads to an almost universally held wrong conclusion.

Oddly, *cosmology*—the study of the structure and development of the whole universe—whose standard model rests in large part on the theory of general relativity, does not suffer from this problem to the same degree. In part, this is because it finds, in all directions, a dull glow, cooled coals, as it were, from the world's hot beginning.[d] This cosmologic perspective supports a concept of a time that is the same everywhere.[e]

Quantum theory, too, has no problem with events both here and over there that are simultaneous; indeed it needs them. A slew of views of the wave function[f] that lies at its heart all have it collapsing

---

a That we cannot see the distant present is also an observed fact. Relativity's treatment of the distant present smacks of the positivist perspective: If we cannot sense it, it does not exist.
b Conceptually, one could consort here with a fiction known as the *block universe*. Better not unleash the monster. We will see it is seductive nonsense; see chapter 56 at page 301.
c See note 133, above.
d This is the cosmic microwave background; see chapter 17 at page 105.
e The observation that the radiation has the same temperature in all directions gives access to a special frame of reference that is not moving in an absolute sense. In turn this gives us the key to a kind of absolute time some physics rejects and other physics uses; see chapters 52 and 56 at pages 284 and 301.
f See chapter 116 at page 568.

instantly throughout the universe as soon as someone somewhere checks it out.[a]

Of this, physicist and "prophet" of the second quantum revolution,[153] John Bell, sarcastically asked,

> Was the world wave function waiting to jump for thousands of millions of years until a single celled living creature appeared? Or did it have to wait a little longer for some more highly qualified measurer—with a PhD?[154]

But then, if quantum physics could be said to have an ontology, it's surely *not* about reality. Most of its adherents say there *is* no reality until someone makes an observation.[155] There may be no easy way to fix this:

> To improve on the quantum theory, [Einstein] thought, would require starting afresh with quite different fundamental concepts.[156]

We will come to agree with him. Yet physics is now firmly founded on a slowly evolving ontology that is not much given to starting afresh. Anyone who, like I did, learned physics more than fifty years ago will tell you that its worldview and the language that it uses to describe it have both changed. Yet physics culture has a way of sticking in its rut: It throws out babes (such as, we'll see, the ether);[b] it hangs on to the bathwater (such as spacetime; it's a problematic concept we will find[c]); and it tolerates inconsistency while letting mathematics lead it by the nose.[157]

I agree with Smolin that we have been making wrong assumptions. I agree these include wrong concepts of *space* and *time*.[d] We'll find better concepts of both were long since wrongly set aside. Those

---

a Quantum theory calculates the wave function of a quantum system (usually something small, like a particle or atom). The wave function specifies all that can be known about that system—what it is and where it is and what it does—in precise terms that lead to probabilities. The calculations give the wave function's value everywhere; they show exactly how it changes over time. There is no consensus on how to understand the wave function.
b See chapter 10 at page 63.
c See chapter 83 at page 420.
d There is a surprisingly wide consensus to this effect; there is no consensus on the right concepts.

better concepts turn out to be simpler (though they may at first seem strange) and they fit together to make sense.[a]

To ease the reader's first steps down the road, this part sets up two kinds of signposts.

First, this chapter will outline six key concepts underlying the new view. They are not aimed at persuading; their aim is to begin building common language, easing readers into meanings they can use for understanding real ideas. To make best use of this "tip list," the reader is invited to take the quirks of each concept on faith until they get connected—as they will—to humankind's long search for what is real.

Second, the chapters in this part will highlight seminal ideas from great minds who wrestled through the centuries to build our understanding of the world. These thinkers sought a real ontology, one that could explain all they saw.

Over time, as science shone its light on aspects of reality, ontologic concepts edged closer to true understanding. Yet each new reach fell short because each thinker lacked key pieces of the picture. Now, at last, we may have what we need.

Looking ahead, in the next part we will see that, like a jigsaw puzzle, these disparate but key ideas come consistently together (notionally illustrated in fig. 3). As an integral whole, they give us a single picture that makes sense of many things that long have seemed surpassing strange.

Thus, the new ontology arises by combining a modest number of illustrious ideas—some ancient, some more modern—all but one of them well known at least within its circle.

Meanwhile, here is a preview of six key ideas:

Fig. 3. Jigsaw consistency (reflections on the moving statue of Kafka)

---

a See part 2.

- **Space** is something; it is not empty; it is a massive entity, made of *space quanta*, tiny bits of space. These quanta are not *in* space; they *are* space. They are linked together.
- The **space quantum** is the quintessential **atom**; it can't be subdivided; there is nothing smaller.
- **Space quanta** have a much-studied structure that itself has six dimensions. Each has a fixed volume.[a] It is extremely small.
- In the **beginning**, almost fourteen billion years ago, the universe was only one space quantum. It replicated stepwise in successive **iterations** [1, 2, 4, 8, 16, 32, . . .].
- **Matter** is made of **twists** in **links** between space quanta.
- The universe has only three dimensions. But its stepwise **iterations** give it **sequence**, which leads to a universal kind of **time**.

Each of these ideas will get grounded firmly in what is to come. Many may see them as new, though most of them are old. We'll take a guided tour through each in turn. Their deep roots in the wisdom of the ages and in recent physics will heave into view as this tale of a real ontology unfolds.

To be clear about expectations, I aim to propose and justify such an ontology. And I aim to steer clear of new physics. Real physics will be, one imagines, at least as extensive as all the existing unreal physics. Physicists, whose first instinct is to look for experimental testing need to first develop the physics; theory, not ontology, is what experiments test.[b]

As to the seminal idea of a real ontology, we may owe this to Plato, who thought he had found it (he was wrong).[158]

---

a Known as the Planck volume; see chapter 13 at page 76.
b Nonetheless, broad forecasts will emerge that, if fulfilled, will be explicable within the new ontology while adding to the list of problems of the old. See, for example, chapter 129 at page 617.

5
DISTANT RELATIONS

All the world's a stage . . .

William Shakespeare (1623)[159]

Of late, a few physicists have been pursuing physics without space and time. They hope their math will tell them what space is (and maybe time will show up too). You might say they try to feed no aspect of reality into their math because they want reality to arise out of it. Reality, they think, should be the product, not the predicate, of physics.

Shakespeare, who called his theater The Globe (fig. 4), would have understood how much those physicists are giving up: Their stage; indeed, their world. Or, more precisely, as much of it as they can manage to get by without.

Fig. 4. The Globe

Shakespeare poked fun at our perceptions of reality in *As You Like It.*[a] His tale has no tie to physics. But the subatomic particles that form the foundation of today's physics are past masters at concealing real identities. So, physicists might finish his line with . . . *and all the particles are players.*

Our story will take us to places far smaller than those particles, far tinier indeed than anything a physicist can see. To help to visualize this hidden world, this invisible reality, you will be invited to imagine you are philosopher Ralph Waldo Emerson's "transparent eyeball"; it is nothing and it sees all.[160] With its aid we will from time to time visit those players and their stage where all the action of our world is really happening.

---

a At some points in the action, he has a boy playing a girl disguised as a boy impersonating a girl.

Orthodox particle-physicists set particles to play upon a stage of space and time. Those few space-and-time-less hopefuls are looking to their math to show them what that stage is made of, and whence and when and how it came into existence along with the particles.

This is a seminal idea known as *background independence*. It says if you want to understand the world, first thing to do is ditch the stage, then—if you can figure what comes next and most of all if you are lucky—the real stage may take shape before your eyes.

Stage-free ambitions do not have much history. But one way to view Einstein's work on relativity could be as an attempt to formulate a background-independent physics of motion. He came close.[161]

> How can physics stage its play .... without a stage?

It has been almost twenty years since Smolin said a real theory must begin with no stage and rather should generate both stage and particles.[162] He long pursued an understanding of what background independence is and how to accomplish it, for example saying in 2006,

> We need a theory about *what makes up space*, a background-independent theory.[163]

And further,

> The main unifying idea is simple to state: . . . *Start with something that* is purely quantum mechanical and *has, instead of space, some kind of purely quantum structure.*[164]

There still is no such theory. But we will follow closely the italicized advice.[a]

Background independence goes hand in hand with the *relational approach*. Roughly, it says build with nothing but relationships and see what you get.[165]

Various authors propose variations on the theme of what *relational* might mean.[166] In 2006, Smolin stated it this way,

---

a It leads to another question: *What kind* of purely quantum structure should we start with? The answer—space made of space quanta—will arise from many sources. See chapters 6 to 10, following this one.

> The most basic statement of the relational view is that ... the fundamental properties of the elementary entities consist entirely in relationships between those elementary entities.[167]

And, in 2008, having long argued that *any* background is bad for fundamental physics, he offered the accurate but unfortunate observation that ...

> ... for physicists relationalism is a strategy. As we shall see, theories may be partly relational, i.e., they can have varying amounts of background structure.[168]

It calls to mind that saying: You can't be half pregnant.
More than a decade later, he said,

> [A]ll properties that refer to location in space or time should be relational.[169]

I prefer his first (i.e., 2006) and strictest way to understand the principle, which also we will closely follow. However, this leads to two practical—but also fundamental—problems.

First, what exactly *are* the elementary entities; how do their relationships arise? As Einstein perceived early in his career, these are foundational issues:

> It seems to me ... that a physical theory can only be satisfactory when it builds up its structures from elementary foundations.[170]

Second, space and time are the air and water of physics' existence. It's not that easy to set them aside and then start over. In 2003, Rovelli said,

> Everybody says they want background independence, and then when they see it they are scared to death by how strange it is.[171]

And, some years later,

> To let the background spacetime go is perhaps as difficult as letting go the immovable background Earth.[172]

And how does one even *think* about relationships without a space to hold them and a time in which they change? As cognitive psychologist Donald Hoffman said,

> If I challenge myself to imagine something—anything—outside of space or beyond time, I'm stymied. I may as well try to imagine a new color I've never seen before. Nothing happens.[173]

So, *all relational* and *background-free* sound simple. But they seem impossible to do. Philosophers Nick Huggett and Christian Wüthrich . . .

> . . . define a theory to be empirically incoherent in case the truth of the theory undermines our empirical justification for believing it to be true. Thus, goes the worry, if a theory rejects the fundamental existence of spacetime, it is threatened with empirical incoherence. . . . The only escape would be if spacetime were in some way derived or . . . "emergent" from the theory. But the problem is that without fundamental spacetime, it is very hard to see how familiar space and time and the attendant notion of locality could emerge in some way. . . .[174]

Welcome to the complex tangle we are making of our space and time. We will untangle it. We will find a simple story in its place. We will find that what we see, "familiar space and time," does indeed "emerge in some way."

When I first read Smolin's early writings on these two more or less equivalent ideas—relationalism and absence of background—I could see them as embraced into a single seminal framework that was strategic. This means making it work will be essential to succeed.

So, we need to give it full effect. We will do this too.[a]

---

[a] We will check our progress in chapter 79 at page 402.

6
THE A-TOM

> They had conceived of a kind of elementary substance of which everything was made.
>
> Carlo Rovelli (2017)[175]

Rovelli was referring to Greek philosopher Democritus and to his teacher, Leucippus. Around 400 BCE they had a seminal idea. They conceived that all matter is made of tiny bits that can't be cut into tinier bits.

It's like, take a piece of feta; cut it in half; cut one half in half. Et cetera. Quite soon you'll have a piece that can't be cut. No point in sharpening your knife. You cannot divide it.[a]

> The Greek atom was the smallest thing possible.

We call this concept *the atom*.[b] Rovelli summarized,

> The only possibility, Democritus concludes, is that any piece of matter is made up of a *finite* number of discrete pieces that are indivisible, each one having *finite* size: the atoms.[176]

For over two thousand years the atom was a conversation piece and not much more. Nobel Prize–winning physicist Werner Heisenberg summarized how the concept was seen:

> The world was reduced to atoms with empty space between them.[177]

The conversation became serious in 1905 when Einstein showed some sort of atoms do exist.[178] But it was already clear *those* atoms[179] did not qualify as *a-toms*. They could be cut.

---

a Atomos (ἄτομος) is Greek for *uncut* or *indivisible*.
b It was a simple idea; as we see in what follows, atomic and nuclear and particle physics have complicated it over the past hundred and fifty years.

Ironically, the field that undercut the import of the name came to be called atomic physics. In 1897, Joseph (J. J.) Thomson made the first cut, carving off *electrons*, for which he received a physics Nobel Prize. In 1911, Ernest Rutherford showed that, aside from electrons, the atom had a nucleus (he already had a Nobel Prize). In 1935, Sir James Chadwick got a Nobel Prize for cutting nuclei into neutrons and protons.[a]

The rush to cut things slowed, but that was not to be the end of it.[180] Even the notion that there *was* an end began to crumble. What's the meaning of a *no-cut* thing that keeps on getting cut? And so it was that atom smashers (aka particle accelerators) changed our language, robbing us of what our word *atom* means and bidding fair to stultify its seminal idea.

If we can imagine an uncuttable atom, we can contrarywise imagine cutting supposed atoms without end. Of such succession, satirist Jonathan Swift wrote in 1733,

> So, nat'ralists observe, a flea
> Hath smaller fleas that on him prey,
> And these have smaller still to bite 'em
> And so proceed ad infinitum.[181]

If a true atom does exist, it must be very tiny, smaller by far than its modern namesake. Its existence would beget enormous consequence for all our understanding of the world.

Is there a real end, a true Greek a-tom? And, a modern version of the question, could there be one truly elementary particle that makes up all the subatomic pieces of all atoms? And, perhaps most vexing, if we were to find the true atom, how could we be sure it can't be cut?

This is not a vexing problem for the modern atom-cutting industry, which is ever ready to build ever bigger and more energetic and expensive atom cutters.[b]

---

[a] This was the first step to the next level down, nuclear physics.
[b] See note 1347.

The result, stitched together with brilliant theory over the latter part of the last century, is *the standard theory of particle physics.*[a] It is an arcane body of quantum physics that gives rise to sixteen kinds of subatomic particles, all taken to be elementary. Particle detectors see the signatures of all sixteen.

Summarizing widespread perceptions of the theory's standing, a 2017 article by science writer Bernie Hobbs was headed,

> The Standard Model of particle physics is brilliant and completely flawed.[182]

Working at the particle-physics frontier, theoretical physicist Jonathan Hackett noted the history of ever-more-powerful accelerators smashing supposedly elementary particles:

> [I]t has been demonstrated repeatedly that the differences between supposedly fundamental particles are, in fact, merely consequences of the composite structure of underlying reality. . . . The difficulty is that, as such a process does not have an end, we can continue to suppose that below the currently understood structure is another set of more fundamental particles.[183]

That is, even if we *are* at the far end of the line, how can we be sure? This might seem to be the kind of question for which there could never be an answer.

If this were so, it would be frustrating. But expensive efforts to address the question could continue unabated. Costs of the current lead accelerator, the Large Hadron Collider—aka the LHC—that straddles the border between France and Switzerland, are in the ten-billion-dollar range. It aimed to explore a whole new group of particles[b] but found nothing beyond the standard sixteen.[c] It smashed none of them, so they seem elementary.[184] Probing for the next set of sub-particles would need a new accelerator that achieves

---

a See fig. 1 on page 21. It is also (and, these days, maybe more widely) known as the *standard model of particle physics*, a term I quote Bilson-Thompson using. Thus I use both, with apologies for the potential confusion.
b Supersymmetric particles; see chapters 8 and 80 at pages 53 and 411.
c The Higgs boson is another story; see chapter 92 at page 457.

higher energies.[a] It would be even more expensive to build and operate.[185]

So, it would be doubly good if we could find the final *a-tom*, and could understand it so well we would know that it *is* final.

Here, we may find a way to do just that without a new accelerator.

Meanwhile, the *a-tom* may be elusive; but it surely is a seminal idea.

---

[a] See chapter 80 at page 411.

7
THE SINGULAR INGREDIENT

> It is almost as if the stuff of which all stuff is made were reducible in the end to some simple and unique kind of substance.
>
> Teilhard de Chardin (1955)[186]

If there *is* a final atom, might there be only one kind of them?

Poet Laureate Robert Southey said little boys were made of three kinds of stuff, "snips and snails and puppy-dog tails . . . ." His children's rhyme, "What All the World Is Made Of",[187] seems to lampoon our long search for the basic ingredients that make our universe.

> Maybe everything is made of only one kind of thing.

Others, such as scientist and philosopher Teilhard de Chardin (above), pursued a seminal idea: Everything (including Southey's three ingredients) might be made of a single, literally universal kind of thing.

From sparse records we can discern that, in the sixth century BCE, Anaximander conceived a cosmos[a] that derived from a vague substance called *apeiron*.[188] This was, he thought, the source of all that is.

His substance would soon sink into obscurity but the concept had legs (as newshounds say). It inspired a lasting story and, until today, a fruitless quest: What sort of thing could such a universal substance be? It led to two and a half millennia of speculation.

Thus, in the seventeen hundreds, Western philosophy was exploring the old concept under a new rubric, *monism*. One might say monads merged the atom with apeiron. Philosopher Gottfried Leibniz, whose immaterial monads made up his matter,[b] wrote that,

---

a See chapter 4 at page 34.
b We will see this was remarkably insightful.

> A simple substance is that which has no parts. The composite is the assembly of simple substances, or monads. Monas is a Greek word meaning unity, or that which is one.[189]

And,

> The monads are the true atoms of nature; in a word, they are the elements of things.[190]

Some two hundred years later, the atom had moved from chemistry to physics, which essayed to ruin its no-cut reputation. The idea of the monad lost its luster. The nineteen hundreds saw doubt grow that we would ever find a final atom.

Yet, in his 1958 book about the quantum revolution, Heisenberg could muse,

> [A]ll different elementary particles could be reduced to some universal substance.[191]

He had no real idea what that substance could be. Indeed, he seemed to see no advance on Anaximander.

Fifty years later, Wilczek asserted (with only a shade more specificity) that,

> The primary ingredient of physical reality, from which all else is made, fills space and time. Every fragment, each space-time element, has the same basic properties as every other fragment.[192]

The vagueness of the language of this proposition tells a tale. The loose marriage of these two questions, *Is everything packaged in uncuttable atoms?* and *Is everything made of one thing?* had stayed in play for two and a half thousand years with neither finding any definitive answer.

But, in the first years of the third millennium, obscured from public view, the last needed pieces of the picture were quietly coming into view to answer both questions with *Yes* and to understand the answer to *What is the thing everything is made of?*[a]

That there is such a thing is a seminal idea. It is widely seen. Its name is spoken in the common parlance. Even so, it may astound you.

---

a Here, *everything* includes ordinary matter and energy, dark matter and dark energy; see chapters 17, 68, and 67 (at pages 105, 351, and 348, respectively).

8
THE MULTIPLICITY AFFLICTION

Not yet discouraged by the discoveries
of numerous particle species
that were yet to come, [in 1954] scientists
were still searching for
simple, elegant and universal principles in physics.

Gerard 't Hooft (1997)[193]

The idea of that single thing—of that one ingredient of everything—had staying power. But its simple singularity kept falling into messy multiplicity. Even those pursuing fewer kinds of things kept coming up with more.

Science kept finding more kinds of atoms.

For a while the Greeks held that the world is made of four elements, earth, air, fire, and water, each in turn imagined to be made of its own kind of atom.

Then, for more than two thousand years atoms were mostly seen as fiction. That is, only a few thought atoms really existed. In 1704, philosopher, mathematician, and physicist Isaac Newton was one exception, speculating atoms of multiple kinds adrift in void made up the world.[194]

Emerging from alchemy in the late seventeen hundreds,[195] chemistry brought a degree of order to the scene. But it kept conceiving of more kinds of atoms. By the nineteen hundreds, it claimed ninety-two distinct kinds in nature. Then physics got the message that matter's made of atoms and made twenty-six more kinds and counting (see fig. 5).[196]

Early atomic physics said that all these atoms are made of just two kinds of things, tiny negatively charged electrons circling far larger positively charged nuclei.[a] But it soon chopped those nuclei into positively charged protons and uncharged neutrons (the two being collectively called *hadrons*).[197]

---

a See chapter 6 at page 47.

So, for a short while, one could wonder: Is matter made of only three things?

Then the simple three-piece-atom picture slowly fell apart six ways from Sunday; once again, the number of ingredients began to rise.

Physicist John Moffat wrote,

> By the mid-1950s, the number of newly discovered particles at accelerators was so large that physicists began referring to them as the *particle zoo*.[198]

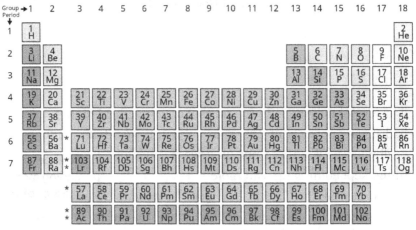

Fig. 5. The Periodic Table of 118 kinds of atoms

As Nobel Prize–winning physicist Gerard 't Hooft intimated (in the epigram) it was for many a discouraging development. Those two hadrons turned out to be made of two kinds of charged particles called *quarks*.[199] Atom smashers soon found four more kinds. The electron had already become the skinny kid in an odd triplet with the equally charged tau and muon.

Then came three kinds of ghostly neutrinos.[200] And add to the list the photon plus three other particles involved in exerting various forces.

So, in the nineteen nineties, Nobel Prize–winning particle physicist Leon Lederman could still write about . . .

> . . . a problem that has confounded science since antiquity. What are the ultimate building blocks of matter?[201]

Supposedly sorting it all out, the standard theory says the world is made of sixteen kinds of particles (see fig. 1 at page 21). Physicists have "seen" all sixteen (or seventeen, if we add the Higgs boson[a]).[202] Oh, and it also says each of the sixteen *must* have a so-called supersymmetric partner—none of which has yet showed up in a particle detector—making maybe thirty-two kinds (see fig. 71 at page 411). Or so the crown jewel of particle physics would have us imagine.[b]

As explanations go, the standard theory's even messier than that. It needs some twenty-six new parameters that must have exactly the right values to make it work.[203] One can ask *Why?* about each one of them and, so far, get no answer. Physicist Robert Oerter said (with, we will discover, clear foresight[c]),

> Ideally, physicists would prefer a single entity . . . instead of the seventeen particles, and one law with one, or maybe no, parameters to be measured. . . . All the known particles would arise from this fundamental entity behaving in different ways, like different notes played on a bugle.[204]

Each of those standard-model particles is "elementary," or so it's said, meaning it is not made of smaller pieces. The only evidence for this is negative; nobody has seen any of them fly apart. But that is just the latest story; it is where we were with hadrons—which are clearly composite—just fifty years ago.

In a way it is where we still are. Physics "explains" certain hadron observations by referring to quarks. But nobody has ever found a quark outside a composite particle. (Wilczek and two others shared a Nobel Prize for showing why.[205])

And, inconveniently for the notion that the electron is elementary, the quarks turned out to bear fractional charges of exactly one-third and two-thirds of the electron's charge.[d] Nobelist Martin Perl said,

> We cannot explain why the electric charges of the known elementary particles are zero, $\pm e$, $\pm e/3$, or $\pm 2e/3$.[206]

---

a The Higgs is a maybe; see chapter 92 at page 457.
b See chapter 80 at page 411.
c See chapters 35 and 43 at pages 190 and 243.
d See further, chapter 48 at page 268.

So, how many kinds of fundamental particles? Looks like physics says a lot. It's surely not a simple picture.

But we'll soon see that physics has (and doesn't know it has) that single entity that Oerter noted physicists prefer; and can (but doesn't know it can) then understand those pesky charges.[a] Maybe matter's multiplicity is all in our minds.

---

[a] Such problems are supposed to draw physicists and grad students like wasps to a picnic. This one seems to have been buried in the standard model.

9
SIMPLY PUT

> If "the road to reality" eventually reaches its goal,
> then in my view there would have to be a profoundly
> deep underlying simplicity about that end point. I do
> not see this in any of the existing proposals.
>
> Roger Penrose (2004)[207]

Penrose wrote the book *Laws of the Universe*. After more than a thousand pages, he said (above), reality should be much simpler. He's in good company. One could view the march of science as steps that make things simpler. Einstein codified this as,

> It is the grand object of all theory to make these [fundamental concepts] as simple and as few in number as possible.[208]

Simplicity begins with one's assumptions. The simplest story would have none. But, as Einstein, after a lifetime of experience, advised a budding scientist,

> You have to assume something.[209]

It did not keep him from finding simplicity. As Holton said,

> The heart of relativity, as Einstein noted, is indeed the discovery of far greater simplicity at the foundations than had been suspected, resulting in a unification of previously separate notions.[210]

We are here setting out to sift through lots of "separate notions"—including those found in one small (but weighty) contribution, Penrose's pages—most of which most readers might well find impenetrably complex. There may be (indeed, there is) a gem within his book, but it is buried deep amid his thousand pages. If we are to find and seize upon it, we will first require to assume something about how to choose.

Here's the plan: As Penrose, among others, indicated, we will seek simplicity; at every turn, we will assume *as little as possible* and see what that will do for us.[211]

The seminal idea that simplicity is an explanatory virtue has a long pedigree.

Thus, in the fourteenth century, philosopher William of Ockham thought of many things. For example, he promoted the strange notion that it's only things that do exist that do exist. That is, things we devise as concepts don't exist. This may seem self-evident, yet it fueled long debate. Our quest for reality will bring us into confrontation with this controversy in its many forms. We will side with William.

> Our new worldview must be very simple.

As well as philosophy, Ockham wrote of politics; he was a dissident. He also did theology; he was a heretic; he got into trouble with the pope. (It was not difficult to do this in those days.)

Yet today he's best known (by his Latin name) for Occam's razor, a rule that explanations should be simple.[212] He did not invent it,[213] but his version echoed down the years.

Newton gave the concept pride of place in his *Rules for the Study of Natural Philosophy*:

> As the philosophers say: Nature does nothing in vain, and more causes are in vain when fewer suffice. For nature is simple and does not indulge in the luxury of superfluous causes.[214]

Some six centuries after Ockham, Einstein sought to sketch new ontologic bounds for physics. He had a sophisticated version of the razor:

> The aim of science is, on the one hand, a comprehension, as *complete* as possible, of the connection between the sense experiences in their totality, and, on the other hand, the accomplishment of this aim *by the use of a minimum of primary concepts and relations*.[215]

And,

> A theory is the more impressive the greater the simplicity of its premises is.[216]

And again,

> The conceptual system . . . should show as much unity and parsimony as possible.[217]

We will apply the whole of his prescription to the choices we must make.

In the early nineteen hundreds, the philosophy called *logical positivism*[218] (founded in the eighteen hundreds by philosopher Auguste Comte though its roots go back at least to philosopher, historian, and writer David Hume[219]) took over the task of policing the canonical conceptual system.[220] It claimed sensory experience is the sole arbiter of what is real. It became a formidable movement that rejected *metaphysics*.[a]

The plague of positivism became deeply embedded in the worldview of twentieth-century physics. And, from the twenties to the fifties, it heaped opprobrium on any metaphysical concern about ontology.[221] It was a move away from simplicity. Making its case, philosopher of science Michael Friedman said,

> Why in the world should nature respect our—merely subjective—preference for "simplicity"?[222]

Yet, over those same decades, logical positivism was itself dying the death of a thousand philosophic cuts. By 1967, philosopher John Passmore could write for it a scornful epitaph:

> Logical positivism, then, is dead, or as dead as a philosophical movement ever becomes.[223]

Unfortunately, no one told the physicists, who kept its twitchy corpse on life support by earning positivist PhDs while rarely studying philosophy.[224]

Quine may be credited with rescuing ontology in the mid–nineteen hundreds. He said,

---

[a] The study of fundamental principles of being, causing, substance, space, and time, all of which are our concern here.

> A curious thing about the ontological problem is its simplicity. It can be put in three Anglo-Saxon monosyllables: "What is there?"[225]

He had a simple idea to help solve the problem:

> Our acceptance of an ontology is, I think, similar in principle to our acceptance of a scientific theory, say a system of physics: we adopt, at least insofar as we are reasonable, the simplest conceptual scheme into which the disordered fragments of raw experience can be fitted and arranged.[226]

Along the way Quine showed why, like it or not, physics is imprisoned by ontology. He explained why its versions of ontology are myths. And, having given us criteria for choosing an ontology, in the end he too, like many others, threw up his hands and said,

> The question what ontology actually to adopt still stands open.[227]

Einstein was clear about the virtue of simplicity, writing to a friend in 1938,

> The logically simple does not, of course, have to be physically true; but the physically true is logically simple, that is, it has unity at the foundation.[228]

In 2006, historian of physics and, later, director of research at Observatoire de Paris, Jean Eisenstaedt said in fewer words,

> Physics must be simple.[229]

Few non-physicists see physics to be simple. But many physicists do see physics as having special claim to a kind of simplicity. For example, physicist and science writer Jim Al-Khalili said,

> The true beauty of physics, for me, is found . . . in the deep underlying principles that govern the way the world is. . . . It is a beauty that lies not in the surprising profundity of the laws of nature, but in the deceptively simple underlying explanations (where we have them) for where those laws come from.[230]

And in a book chapter titled "First, Principles"—where the comma made his point—Smolin urged the virtues of or the need for simplicity in charting a course toward a fundamental theory. He said,

> The answer to the questions that have bedeviled us for nearly a century will be simple, and expressed in terms of elegant hypotheses and principles.[231]

We will come to agree with him.

He said if we skip this simple stage and go "right to models" we can lose our way. But that is what physics has been doing for a hundred years. (And mostly still is.)

He urged the virtue of a principled approach. But then he illustrated his point by saying,

> "It is impossible to do any experiment that can determine an absolute sense of rest, or measure an absolute velocity" is a principle.[232]

He was right; that *is* a principle. It's also wrong; such experiments exist and do measure absolute velocity.[a] There is a lesson in the fact he—a leading physicist—seemingly did not know; if we are to find our way, we must avoid such ingrained errors, especially when they are couched as principles.

And let's also note, amid the intricate complexities of today's physics, the informed chorus—that is no more than exemplified above—of approbation for simplicity.

So, *Keep it simple!* (with its extended version and KISS acronym) is a key design principle for assembling a new ontology that's real.[233] Though Ockham, Quine, Penrose, and Smolin don't disclose all their grounds for seeking simplicity, I suspect their main motivation may be similar to mine: The deepest workings of the universe *are* simple, or so I anticipate. The complexities of physics and conundrums of philosophy aren't its; they're ours.

Like North America's early human inhabitants half a hundred thousand years ago, we are venturing in virgin territory. As we look

---

[a] In chapter 56 (at page 301) we'll see the observable universe itself defines a state of rest and so provides a frame of reference that cosmologists use routinely in a way that does determine absolute velocities.

back at the world's beginning, we should find a lot more clarity about its origin than we can ever find of theirs[234]—though it was more than two-hundred-thousand-fold further back in time[a] than their existential enterprise—because the universe's origin was simple and theirs was complex.

Occam's razor is a simple rule. We'll use it to assemble new ontology that, compared with the old, will be extremely simple, a worldview that, though assuming almost nothing, explains almost everything.

Meanwhile, William of Ockham gets credit for selling us the seminal idea of simplicity. It must be our guide at every turn. To succeed, we need to choose the right insights among a veritable maelstrom of ideas.[b]

Simplicity is therefore of the very essence for our quest.

---

[a] As light takes time to reach us—one year to travel a light-year—when we see further away, we are seeing further back in time.
[b] See also chapter 39 at page 220.

10
BEING ONE AND MANY

> I myself have come, by long brooding over it, to consider it the most central of all philosophic problems, central because so pregnant.
>
> William James (1906)[235]

Much of physics is concerned with motion. As also is much of philosophy.

Around 450 BCE, Greek philosopher Zeno of Elea puzzled over deep difficulties in understanding continuity and motion.[236] Perhaps the most intriguing puzzle is his paradox of the One and the Many. At first glance it can seem a trivial or contradictory idea. Yet two and a half thousand years later, it challenges the brightest minds.

It's an ancient problem: How can motion happen?

We will see it is a seminal idea. For us too it will be, as for psychologist and philosopher William James (above), . . .

. . . the most central of all philosophic problems.

Another philosopher (also logician and mathematician), Bertrand Russell, winner of a Nobel Prize for literature, called it . . .

. . . immeasurably subtle and profound.[237]

Here it is (in the Diels translation):

> If there are many things, it is necessary that they are just as many as they are, and neither more nor less than that. But if they are as many as they are, they will be limited. If there are many things, the things that are are unlimited; for there are always others between the things that are, and again others between those. And thus the things that are are unlimited.[238]

We will find that a full resolution of Zeno's paradox calls for both an understanding of the entire universe and an understanding of the final atom. Along the way we will solve Einstein's deepest open problem. If this seems to take on too much, please hang in there; it will all turn out to be much simpler than it seems.

Zeno was a pupil of Parmenides. His paradoxes were designed to defend his mentor's philosophy,[239] in which concepts of continuity, change, motion, and plurality were intertwined. Historian of philosophy Barbara Sattler notes,

> [T]he paradoxes of motion form a systematic unity in so far as there are two basic problems underlying all the different paradoxes: how can the relation between whole and part be thought in the case of continua like movement, space and time? And how can we conceive of the relation between time and space with respect to motion?[240]

And,

> Given the way that Zeno implicitly conceptualizes motion, time, and space, there are indeed severe problems for combining time and space in an account of motion.[241]

We will find real reasons behind these conceptual problems and we will find real solutions for them.

Sattler described Zeno's view of motion:

> [T]he place at which something starts its motion ceases to be its place and a new place "comes into being" as the place where this thing is now.[242]

With this seminal idea, Zeno anticipated real ontology with an astonishing precision.[a]

And, with no evident intention, he gave us a window into the nature of time and the significance of *Now*.[b]

---

a See chapter 128 at page 614.
b See chapter 56 at page 301.

In the early eighteen hundreds, Georg Hegel laid the foundations of modern philosophy. Prominent among his own foundations were Zeno and the problem, as it was then known, of unity and multiplicity:

> [This] seems to be the contradictory beginning: out of unity, the multiplicity.[243]

We will see this too is exactly right. Yet it is the universe's quintessential problem. Its solution eluded Hegel as it had Russell—not surprisingly as it depends upon a bit of quantum craziness (on which by happenstance[244] I used to work).[a]

Zeno's paradox is a subtle problem.[b] At its core lies an unstated premise: Our understanding of the world should be consistent. Or, more precisely, it should not be internally inconsistent. This too is a seminal idea.

Here is where the ancient-Greek-thought rubber meets the road to reality.

---

[a] See chapter 19 at page 118.
[b] We will find it has an elegant solution that solves many other problems; see chapter 57 at page 307.

11
THE SUBSTANCE OF SPACE

> It is indeed an exacting requirement to have to ascribe physical reality to space in general, and especially to empty space.
>
> Albert Einstein (1952)[245]

What *is* space? For physics and philosophy, this is the most fundamental question. There will be a fundamental answer for this question too.[a]

It looks like nothing whatsoever. Returned astronaut, Roberta Bondar, said it is . . .

> . . . the unimaginable black.[246]

Yet the seminal idea that space is *something* was in fashion for a while; and then it wasn't.

Now we are about to see space isn't only something; it is literally everything. This idea is the culmination of an epic mental odyssey. It must now refashion many minds.

At first the *something* was the αἰθήρ,[247] pronounced *ay-thur*, which ancient Greeks thought gods breathed in the space beyond the Earth. Then it morphed into the *æther*, with the old-English character called *ash*. And then the aether, from the German *äther* (pronounced more like *eh-ther*) after *ash* fell out of fashion. By the nineteen hundreds it became the ether (like the anesthetic) and then that fell out of favor too.

> Space is not empty. It is full of space.

In these pages it's returning with new vigor and new substance and an old name, *space*.

---

a In chapter 40 at page 222.

In the fourth century BCE, Aristotle was not first to use the Greek word; it already meant the brightness of blue sky. When he said αἰθήρ he meant something that filled all the space he thought was out there. He could not abide the thought of *void*, dismissing it as ill conceived. He described the concept:

> The void is thought to be place with nothing in it. . . . [P]eople take what exists to be body, and hold that while every body is in place, void is place in which there is no body, so that where there is no body, there must be void.[248]

Of course, he had no grasp of the vast reach of space modern astronomy reveals. But, thinking of the space he thought was there beyond the Moon, he filled it up with something bright. He said it was something.[249] It became one of five imagined substances that made the ancient world.

Wise views and fashions about what kind of thing (or nothing) space might be long flourished as did views and fashions about matter in it. Fast-forward two millennia.

In the sixteen hundreds, philosopher René Descartes wrote of space and matter. Matter requires space, he said, and space needs must have matter. He too abhorred the notion of a void. He agreed with Aristotle, that the æther is a substance. He said it fills all space (except in solid objects). He filled it in turn with vortices (fig. 6) in a vain effort to understand motion.[250]

In 1704, having mastered motion,[a] Newton gave the æther an endorsement of a kind. Immersed in experiments with light, he asked,

Fig. 6. Descartes' space vortices

---

a Newton said, "But in what follows, a fuller explanation will be given of how to determine true motions from their causes, effects, and apparent differences, and, conversely, of how to determine from motions, whether true or apparent, their causes and effects. For this was the purpose for which I composed the following treatise." See note 214 at his page 61.

> Doth not this Æthereal Medium in passing out of Water, Glass, Crystal, and other compact and dense Bodies into empty Spaces . . . refract the Rays of Light not in a point, but by bending them gradually in curve Lines?[251]

In 1770, philosopher Immanuel Kant dissented from the space-is-something concept, saying,

> Space is not something objective and real . . .; it is subjective and ideal, and originates from the mind's nature.[252]

That is—in loose translation—he called it a fiction.

Mostly treated as distinct, these two two-sides-of-one-coin questions—whether space is real and whether there's an ether in a supposed void or vacuum—played key roles in the ontology of science that became conventional in the first decades of the nineteen hundreds.[253]

Inability to imagine what *kind* of substance ether might be led to its rapid exit from the then new worldview as soon as special relativity said we could manage space and time without it. Years later, Einstein (by then having his own new view of it[a]) put his finger on the failure of imagination:

> This ether had to lead a ghostly existence alongside the rest of matter, inasmuch as it seemed to offer no resistance whatever to the motion of "ponderable" bodies.[254]

In 1906, Lorentz gave a lecture on electrons at Columbia University. Though still speaking of space and ether as distinct entities, he solved the no-resistance problem and described the combination with, we will see, extraordinary insight.[b] He said,

> [O]ne of the most important of our fundamental assumptions must be that the ether not only occupies all space between molecules, atoms or electrons, but that it pervades all these particles. We shall add the hypothesis that, though the particles may move, the ether always remains at rest. We can reconcile ourselves with this, at first sight, somewhat startling idea, by thinking of the

---

a See this chapter, below.
b His single slip was the unnecessary—indeed undefinable—notion of the ether "at rest."

particles of matter as of some local modification in the state of the ether. These modifications may of course very well travel onward while the volume-elements of the medium in which they exist remain at rest.[255]

His brilliant insight went mostly unnoticed. Even Einstein—who surely knew Lorentz's mind—was unable to fully grasp it. But then he lacked a key ingredient they both needed to complete their understanding: How could matter—as Lorentz almost proposed—be *made of space?*[a]

The fundamental philosophic problem behind all these ethers was the seeming impossibility of action at a distance: How can something affect some distant thing without touching it or sending something to it? In particular, the waves of light needed to be waves of something.

Indeed, Newton regarded his own invention of the force of gravity as deeply troubling for this self-same reason.[b] The Moon swings around the Earth almost as if tied to it. How does the Moon (or an apple for that matter) "know" the Earth is there?

All four of the philosophers—Aristotle, Descartes, Newton, and Kant—saw their various ethers as *something* that filled what would otherwise be empty space.

Each of them was trying to imagine something that's inherently beyond experience. None of them seems to have managed the imaginative leap to thinking that their αἰθήρ-æther-aether-ether-something might *be* space rather than being *in* space.[c]

Nonetheless, we may credit Aristotle for the idea space is by its nature never empty, that there's something everywhere. Though this view, with its many later variations, was often vilified—and was completely canceled after Einstein produced special relativity without it—we will see that in this concept Aristotle was exactly right.

Einstein was (and still is) widely credited with showing, by his theory of special relativity, that there *is* no ether. What he actually

---

a The answer would not come to light until the next century; and then physics would again do its best to ignore it; see chapter 35 at page 190.
b See note 570.
c See chapter 29 at page 144.

showed (and said) was that his theory, which introduced the speed of light in a vacuum as a fundamental constant,[a] *did not need* an ether:

> The introduction of a "light aether" will prove to be superfluous, inasmuch as the view to be developed here will not require a "space at absolute rest."[256]

Even his friend, Leopold Infeld, got this wrong:

> Thus, we can describe Einstein's achievement as destroying once and for all the concept of the ether.[257]

For Infeld alone this might have been no more than an embarrassing blunder, but for a hundred years or more most physicists accepted this false story, leading to serious limitations on their science.[b]

Einstein knew Lorentz's view. He gave it slight attention at the time. But his work on general relativity persuaded him an ether of some kind must exist.[258] The awkward, much neglected fact is that theory's success gave us a demonstration that the ether must be real as it has real and substantial properties, to say the least.

In 1920, Einstein said so, plainly, though in an unfortunately quiet way: He gave a lecture, "Ether and the Theory of Relativity," at the small-town University of Leiden where his friend Lorentz had worked for more than twenty years. He said,

> More careful reflection teaches us however, that the special theory of relativity does not compel us to deny ether. We may assume the existence of an ether; only we must give up ascribing a definite state of motion to it.[259]

He acknowledged that Lorentz had got it almost right:

> [T]he ether of the general theory of relativity is the outcome of the Lorentzian ether.[260]

Summing up, he said,

> According to the general theory of relativity, space without ether is unthinkable; for in such space there not only

---
a See further, chapter 50 at page 276.
b One can see a recent example in Carroll's 2022 book, note 1538, at his pages 140 and 161.

would be no propagation of light, but also no possibility of existence for standards of space and time (measuring rods and clocks), nor therefore any space-time intervals in the physical sense.[261]

Thus, he said some old ideas about the nature of the ether—the ontological commitments that underpinned interpretation of ether-seeking experiments—were wrong:[a]

According to the general relativity theory, space is created with physical qualities; there is therefore in this sense an ether. However, this ether must not be thought of as having the property characteristic of a ponderable medium, to consist of parts that can be traced through time. The grasp of motion may not be applied to it.[262]

This insight instantly negated the canonical interpretation of the renowned experiment by Nobel Prize–winning physicist Albert Michelson and colleague Edward Morley. Assuming aether must have some state of motion they sought to detect the so-called *aether wind* of Earth's passage through it. They used an apparatus that measured the speed of light in differing directions as Earth's orbit moved them around the Sun. They found no wind.[263]

In those ten terse words—*The grasp[b] of motion may not be applied to it*—Einstein explained their observation.[264] That is, the experiment did *not* show that there is no ether; it showed the ether does not have the property they (and most everybody else) assumed for it.

In 1924, he reviewed the story of the fundamental ethers as quantum theory was about to burst onto the scene and concluded,

[W]e will not be able to do without the ether in theoretical physics, i.e. a continuum which is equipped with physical properties. . . . [E]very contiguous action theory presumes . . . the existence of an "ether."[265]

---

a In other words, physicists' interpretations of the failure to detect motion through the ether—which contributed a key element of the old ontology—were based on their shared false assumptions about the nature of space.
b Canonical translations render Einstein's *Bewegungstgriff* (second-last line in fig. 7) as the *idea of motion* and are plainly wrong.

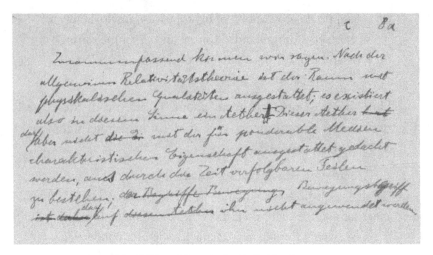

Fig. 7. Einstein's notes for his Leiden lecture

But, like Lorentz's, Einstein's lecture was too little and too late. Today, physics education widely teaches two falsities about the story: One, that the ether's dead and, two, that—with an assist from Michelson and Morley—Einstein was the assassin.

Einstein's revelatory 1920 lecture in Leiden may turn out to be his most important ontologic work. Yet almost all science writers who venture upon the subject fail to even mention it.

Emblematic of its banal burial is a comment by physicist Hanoch Gutfreund, president of the Hebrew University of Jerusalem—guardian of all that is Einsteinian—writing in a limited-edition facsimile of Einstein's handwritten manuscript on the foundations of general relativity, observing blandly,

> The general theory of relativity endows spacetime with a dynamical structure. This prompted Einstein, in later years, to modify his views on the existence of ether.[266]

To *"modify his views"*? He's speaking with masterly obscurity about the man who said,

> According to the general theory of relativity, space without ether is unthinkable.[267]

Maybe most egregious is the lack of any mention in the monumental and definitive *A History of the Theories of Aether and Electric-*

ity: The Modern Theories 1900–1926 by famed historian of science and physicist Sir Edmund Whittaker.[268]

To this day, between them, science historians and physics teachers carelessly cancel what could be the most consequential of Einstein's most extraordinary insights.

The record's clear. Between them, Lorentz and Einstein, two giants in the world of physics at the time, thinking they were rescuing the ether, came to (almost) realize it *is* space, and discerned that,

- It is something.[a]
- It pervades everything we see as solid.
- The solid things may be made of a change in ether's state.
- Those solid changes may move without moving ether.[b]

Here were four huge leaps of insight for the new ontology, all buried, waiting for the day they could be disinterred.

So, who should get the credit for the seminal idea space is something? Aristotle may have been first to imagine it. Lorentz was first to (almost) describe it. Einstein was first to (almost) nail it down. Hats off to all three.

They left us still needing to understand what kind of something space could *be*.

---

a We will soon see it is by far the most massive *something* in the universe; see chapter 29 at page 165.
b Hence the significance of Einstein's phrase *the grasp of motion*, see note b on page 71.

12
RIEMANN'S GRAINS OF SPACE

> I believe that the theory that
> space is continuous is wrong.
>
> Richard Feynman (1967)[269]

Space has been thought of—at least since the days of Aristotle—as being continuous.[270] And also, more recently—since Descartes in the mid-sixteen hundreds—it has sometimes been thought of as *discrete*, made of tiny pieces.[271]

Yet Peebles's definitive work on physical cosmology,[272] with its hundreds of differential equations revealing how utterly the standard model of cosmology depends on the assumption that space is continuous,[a] does not even mention that it might be granular. In this it was and still is true to the condition of its subject.[b]

The idea space might be granular rather than continuous has long had credible support. More than one hundred fifty years ago, mathematician Bernhard Riemann—"solitary and uncomprehended" as Einstein called him[c]—thought about both possibilities. His habilitation thesis, published after he died, was unprecedented.[273] It laid the foundation for doing geometry in space with any number of dimensions without—as was almost universal practice—assuming it to be Euclidean.[274] (Euclidean geometry is done in spaces that are flat.[d])

In a little-noted observation, Riemann said space may be granular, rather than continuous (as mainstream physics then and since chose to assume). It was a seminal idea. He said,

---

a A typical *differential equation* specifies the rate something changes in continuous time or space (or both); it may be integrated to get a continuous description of the something.
b Though there are recent signs of change.
c See note 851.
d Or, put another way, their curvature is assumed to be everywhere zero.

> [I]n a discrete manifold,[a] the ground of its metric relations is given in the notion of it, while in a continuous manifold, this ground must come from outside.[275]

He explained that, if space is discrete (or granular) one may measure things by counting granules (he even called them *quanta*[276]), thus not needing an artificial ruler (metric) brought in from outside[b] the space.[277]

Fifty years later, Einstein would build on Riemann's work to describe gravity in terms of the geometry of a continuous 4-D space that may be curved.[278]

Riemann's observation laid a foundation for Planck to stumble on a number for the granule size in 3-D space.[c] Known as the *Planck volume*, it is unimaginably tiny.

As Riemann foresaw, the math of such a space is digital. With this, Riemann took a key step toward understanding Zeno's paradox of the One and the Many and cracking the central mystery of motion. It seems that neither Riemann nor anybody else recognized this. And his seminal idea that space may be granular was almost universally ignored for more than a hundred years.

| What if space is made of tiny pieces?

---

a A manifold is a space that may differ, in properties like size or shape or number of dimensions, from the usual assumption of an infinite, flat, 3-dimensional continuous space.
b Measurement arising beyond the observed system is thus as central to general relativity as it is to quantum theory, though this is mostly ignored. See also chapter 24 (at page 141) on the significance of having to do physics inside the universe.
c See the next chapter. There appears to be no indication Planck saw the connection.

13
PLANCK'S NEW NUMBER

Physicists have long speculated that the Planck length is the ultimate atom of space.

Leonard Susskind (2008)[279]

It will be a long and winding road for us to find Susskind is saying physicists, though nearly right, are wholly wrong. There is no real Planck length.[280] But they are onto something big that's very small.

At the close of the nineteenth century, Planck was something of an outsider[a] who was inserting himself into the austere world of German physics (you can see him, second from the left in the back row, in fig. 8).

Fig. 8. Max Planck at the Solvay conference

We need to take a long look at him, the first of two physicists who found hidden windows on the unimaginably tiny world of granular space.

Like the other (whom the reader will soon meet[b]), he struggled and he mostly failed to understand—one might almost say, to come to terms with—what he had accomplished.[c] But then, so did his protégé,

---

a Even in the official photo of attendees at the prestigious, by-invitation, first Solvay Conference, Planck looks oddly as if he feels like an outsider.
b In chapter 35 at page 190.
c We will find another in chapter 17 at page 105.

Einstein, who built upon Planck's work but disagreed with him at first and, soon, with almost everybody else.

After a century, physicist and founding director of the Stanford Institute for Theoretical Physics Leonard Susskind's summary (above) could still speak only about speculation on what Planck discovered. Yet Planck did succeed in sowing a small seed of the idea that so strange a world might have some sort of real existence.

His devotion to science was, he said, the result of his realizing . . .

> . . . that the laws of human reasoning coincide with the laws governing the sequence of impressions we receive from the world about us.[281]

That is, he sought to understand the world he saw.

Planck knew precisely what career he wanted. He set out to be the leading light in the then nonexistent field of theoretical physics.[a] Drawn in by recent work on energy and on its law of conservation, he seized on the idea of entropy,[b] and became a silo sitter in what his star pupil, Nobel Prize–winning physicist Max von Laue,[282] said was . . .

> . . . a highly specialized field, in which nobody had any interest whatever.[283]

He set out to find a physical foundation for the *arrow of time*, why time goes in only one direction though physics' equations said the two directions—to and fro, as it were—share the same laws.[c]

This led him[d] to study the heat black bodies radiate over a range of wavelengths.[284] Physics had a formula for it (Wien's law, see fig. 9). It seemed to fit the data others were providing, but it had no theory behind it. Planck set out to base it on physical principles.

| Planck found a magic number that seemed too tiny to be real.

---

a He later said, "I waited in vain for an appointment to a professorship. Of course, my prospects for getting one were slight, for theoretical physics had not as yet come to be recognized as a special discipline." And further, "I was the only theorist, a physicist *sui generis*, as it were." See his autobiography, note 280.
b His 1879 doctoral thesis was on the second law of thermodynamics. In his autobiography he said, "None of the professors at the University had any understanding for its contents." See note 280.
c See chapter 59 at page 316.
d The irreversibility of heat radiation was the attraction.

He found he could do this using a concept of entropy analogous to physicist Ludwig Boltzmann's work with gases, but only by . . .

. . . restricting energy levels to a discontinuous spectrum.[285]

That is, he said energy was absorbed and emitted in discontinuous

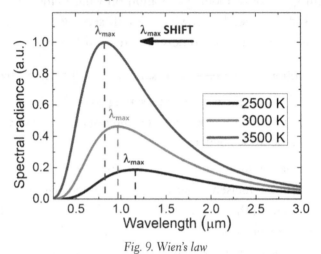

Fig. 9. Wien's law

amounts or *quanta*. Well, that's what I was taught but it's not quite right, and therein lies the heart of his story. Historian and philosopher of science Thomas Kuhn said,

> Only after studying the extended treatment of Planck's theory in [his lectures of 1905–06] was I quite able to believe that . . . his first quantum papers . . . did not posit or imply the quantum discontinuity.[286]

Some saw Planck's success as serendipity. To the contrary, in reviewing his legacy physicists Ian Duck and George Sudarshan said,

> It is clearly incorrect—but nonetheless tempting—to undervalue Planck's accomplishment as somehow a lucky trick that worked, something that fell to him by chance.[287]

Von Laue said,

> [H]e had to venture a hypothesis, the audacity of which was not clear at first, to its full extent, to anybody, not even to him.[288]

For Planck, it began as an exercise in fitting different curves to increasingly wide-ranging data. It did not work out as expected. How he wrestled with this seemingly simple problem while it morphed into a profound one holds great interest for our quest.[289] His short version of the story, in a later letter to a colleague, was,

> Briefly summarized, what I did can be described as simply an act of desperation.[290]

What was this act of desperation? Quoting Planck, science historian Helge Kragh said,

> In his seminal paper published in late 1900 . . . Planck regarded the energy "as made up of a completely determinate number of finite equal parts, and for this purpose I use the constant of nature $h = 6.55 \times 10^{-27}$ (erg sec)."[291]

Somehow—Smolin called it "this comedy"[292]—Planck had derived a new fundamental constant.[a] He calculated its value from large-scale experimental data. Having called it b (since he had another constant, a), he later redesignated it h (chosen for an odd reason[b] that reveals he did not then see—or choose to promote—its significance[293]).

In giving birth to this new "constant of nature," he authored three papers in the establishment's then leading[c] journal *Annalen der Physik*.[d]

In the first paper,[294] he proposed an equation he assumed—but was unable to derive—for the entropy of radiation that fitted the then available data. He based it on electromagnetic emission and absorption by imagined . . .

> . . . elementary oscillators which have some connection with the actual atoms of the radiating matter.[295]

---

[a] Einstein said, "Planck actually did find a derivation, the imperfections of which remained at first hidden, which latter fact was most fortunate for the development of physics." See note 215.
[b] Planck ultimately designated his new constant as h for *Hilfsgrösse*, or auxiliary quantity, which may offer some insight into what he thought of it at the time.
[c] Of this, astrophysicist Guido Fuchs said, "The heyday of the journal was between 1850 and 1920. During that time *AdP* developed into one of the leading physics journals in Europe, if not *the* leading journal."
[d] Of which he was associate editor and would soon be editor. For avoidance of confusion in the endnotes, please take notice that, at that time, its volumes restarted from 1 whenever a new editor took over, as happened in 1900.

He further said, in a statement he emphasized, of which we should take careful note:

> [The] stationary radiation state of the vacuum fulfills all the conditions of the radiation of black bodies, completely without regard to the question, whether or not the assumed electromagnetic oscillators are the actual sources of heat radiation in any particular matter.[296]

That is, in deriving the equation and his new constant, he did not know or even care about the source the heat was coming from.

Here's the rub: Even with the small energy units[a] then in use (the erg; say a small flashlight for less than a microsecond), the constant h in his equation was extremely tiny, about one hundred-trillion-trillionth of an erg-second. Such a miniscule quantum was inconceivably remote from any then contemplated physical event.

He must have understood that much. Yet he seems to have *not* seen it as a periscope into the fundamental structure of the universe.[297] This, even though he promptly[b] calculated his soon-to-be-eponymous fundamental units (at first of length,[c] mass, time, and temperature) by boldly combining h with two other fundamental constants—the gravitation constant G and the speed of light c.[d] His fundamental time and length (and related area and volume) were all absurdly tiny.[e]

Yet he *said* they were "fundamental." In ensuing years, it turned out (very slowly) he was right. And we'll see that, in the new ontology, the Planck volume[f] may be *the* most fundamental constant of the universe and its real value is one.

Yet, far from being seen as an amazing window into reality's foundations, the new units drew from Planck an offhand tone, as if they were mere curiosities tacked onto the then just-told tale of his curve-fitting success:

---

a CGS units, for centimeter, gram, second; the CGS unit of energy was the erg. Today's physics uses the International System of Units based on MKS units, for meter, kilogram, second; its unit of energy is the joule, which is ten million ergs.
b I.e., at the end of the same paper.
c Planck calculated a length. Oddly, it's the Planck volume (i.e., the length cubed) and Planck area (length squared) that turn out to be fundamental and the length itself is not.
d The Planck volume is calculated as $\sqrt{(hG/2\pi c^3)^3}$. Its value is 4.2217×10^{-105} m^3.
e The Planck length is 1.6×10^{-35} m; the Planck time is 5.4×10^{-44} s.
f The cube of the Planck length.

> On the other hand, it should not be without interest to remark that with the aid of the two constants *a* and *b* appearing in the expression for the entropy of radiation, it is possible to set up units for length, mass, time and temperature, which, independently of special bodies or substances, retain their meaning for all times and for all cultures, including extraterrestrial and non-human ones, and which can therefore be described as "natural units of measurement."[298]

"On the other hand"!, "should not be without interest"!, leading to his casually saying—in the same sentence—every non-human culture in the universe would know them as the "natural units of measurement"! Such apposition of casual diffidence and brash assertion had scarce been seen in scientific circles since Poe prefaced his cosmogony, *Eureka*.[a]

Decades later, Planck recounted he had felt he was onto something real:

> [O]n the very day when I formulated this law, I began to devote myself to the task of investing it with a true physical meaning.[299]

The second paper[300] in the series elaborated on the underlying physics in a conventional fashion and laconically concluded,

> From the measurements of Kurlbaum and of Paschen, the numerical values of *a* and *b* are:
>
> $a = 0.4818 \times 10^{-10}$ (sec. × deg.),
>
> $b = 6.885 \times 10^{-27}$ (erg × sec.).[301]

The latter was about to become known as the *Planck constant*. It would soon begin to totally transform our world.[302]

---

[a] Poe's ambivalence was a curious precursor of Planck's and Bilson-Thompson's reticence in pursuing their own insights into the quantum world. In his long prose poem—which is of interest here as a cosmogony with deep similarities to Lemaître's— Poe, a serious cosmologist in his day, asserted the universe began as a single particle that divided. In its preface he said, "What I here propound is true:—therefore it cannot die:—or if by any means it be now trodden down so that it die, it will 'rise again to the Life Everlasting.' Nevertheless it is as a Poem only that I wish this work to be judged after I am dead." See page 5 of note 85.

In the third paper,[303] he acknowledged new data showed his two previous papers must be wrong! He proposed a new rationale that led to a new formula. It fit the data well over a wider range of wavelengths; and it retained his constant *b*, now renamed *h*.

If this tale of a constant in three episodes seems more than just a bit chaotic, yes, it was.

And his new rationale for it was an astonishing departure from all the physics precepts of his time. This was his "act of desperation."

Decades later, he explained it this way:

> I imagined a system consisting of a very large number **N** of completely similar oscillators, and set out to calculate the probability that the system should have a given energy U_N. Since a probability-like quantity can only be found by counting, then it was necessary for the energy U_N to be expressible as a sum of discrete identical energy elements ε.[304]

That is, he *assumed* his imaginary oscillators held energy only in distinct packets ε that were equal to their frequency multiplied by his new "fundamental constant," *h*, which would itself become the famous elementary *quantum*[a] *of action* (a key seminal idea).[b]

Why did he assume this? It's simple: So he could just count them![c]

For six years Planck published nothing further on his action quantum. Yet his son Erwin later told of his father saying to him around the turn of the century that,

> [H]e had made the greatest discovery in physics since Newton.[305]

He tried to bring his work under the umbrella of pre-quantum physics, hoping . . .

> . . . to weld the elementary quantum of action *h* somehow into the framework of the classical theory.[306]

---

a None of his three papers used the term *quantum*.
b I.e., *not* of energy; action has dimensions of energy × time (or momentum × distance). If you don't understand this, don't worry, you are in good company for now; and its picture will slowly unfold.
c Note the relation to Riemann's suggesting if space were composed of granules, we could just count them. See chapter 12 at page 74.

However,

> The failure of every attempt to bridge this obstacle soon made it evident that the elementary quantum of action plays a fundamental part in *atomic physics*.[307]

Here we see how hard it was—with the existence of atoms barely established—for him (and for others) to conceive of even atom scale as real. Neither he nor his contemporaries could begin to imagine a far, far smaller reality.

Reflecting on all this four decades later, Planck seemed to see some fundamental depth remained unplumbed:[308]

> Now however there arose the theoretically all-important problem, to assign this remarkable constant a physical meaning. Its introduction constituted a break with the classical theory, which to many was too radical, as I had initially anticipated. . . . But the nature of the energy quantum[a] remained unclear.[309]

And, five years later,

> I now knew for a fact that the elementary quantum of action played a far more significant part in physics than I had originally been inclined to suspect.[310]

As Duck and Sudarshan said,

> [E]ven with the hindsight of 43 years, reflecting on his discovery at age 85 in wartime Berlin, Planck could barely explain his insight more deeply than at the first instant.[311]

It is unclear, too, just what Planck thought "the greatest discovery in physics since Newton" was. He knew he had invented his oscillators as a pure artifice to achieve results dictated by new data. That is, they were fictions in service to his math, guided by his own concept of simplicity;[312] he had no way to think of them as real.

To this day, understanding what the Planck constant means confronts us with deep problems.

---

a Note that, by this time, even Planck had slipped into sometimes calling it the *energy* quantum.

Let's step back to see the bigger picture: By (bad) analogy with Boltzmann's gas molecules, Planck derived an entropy from the statistics of assigning imagined equal amounts of energy to invented oscillators. Fitting data from measurements at different frequencies handed him a fundamental constant that was a ridiculously tiny, incomprehensible idea of energy times time. Combined with two fundamental constants from the world of very big things,[a] it led to a ridiculously tiny, fundamental size-scale and an even more ridiculously tiny, fundamental time-scale.

That is, it pointed straight to Planck-scale physics. But he didn't go there. For the next hundred years, neither did almost anybody else.[b] In 2003, physicist Yee Jack Ng said,

> [I]t takes a certain amount of foolhardiness to even mention Planck-scale physics.[313]

For the physics collegium, the real nature of the new constant was way out of its purview in its obscure genesis and its abstruse units as well as its tiny size. Worse, the deep implications of its scale drowned in the rising tide of atoms. Planck-scale entities were not about to become the new reality. Indeed, quantum theory would soon take reality right off the physics table.[c]

Personally, Planck was done with it.[314] His giving birth to the theory of (circumspectly unspecified) quanta earned him the 1918 Nobel Prize . . .

> . . . for his work on the establishment and development of the theory of elementary quanta.[315]

Two other physicists, less noted as players in the conventional Planck-constant story, pursued closely related work, Gustav Kirchhoff and Hendrik Lorentz.

In 1859, Kirchhoff had outlined a black-body-radiation law,[d] an unknown function based only on temperature and wavelength,

---

a They were the gravitational constant and the speed of light, both seen as large-scale properties.
b The exceptions have mainly been in the field of quantum gravity.
c See chapter 2 at page 22.
d Now known as Kirchhoff's radiation law.

showing it didn't matter *what* was emitting and absorbing radiation.[316] Already it could be seen the unknown function would need two adjustable constants, one for size and one for shape, so to speak. When Planck took up the trail a generation later, these transitioned into his two constants.

Lorentz, soon also to win the Nobel Prize for physics, perhaps the only physicist who could (and here did) out-think Einstein,[317] was on the same trail as Planck but was thinking deeper.

Unconcerned with fitting curves to data,[a] he pursued profound questions about what was going on behind the scenes. How could every substance that was tested lead to the same constant?

He said (in another passage we should note carefully),

> Without some conformity, of one kind or another, in the structure of all substances, the consequences of the second law and [Wien's] law itself cannot be understood.[318]

And he went on to take up . . .

> . . . the question what similarity in the structure of all ponderable matter must lie at the bottom of the thermodynamic theory of radiation. . . . I shall try to show that, in all probability, the likeness in question consists in the equality of the small charged particles or *electrons*, in whose motions modern theories seek the origin of the vibrations in the aether.

Probing smaller than the atom scale that preoccupied his peers, Lorentz—then the master of electron theory[319]—seized on the electron (composed as we will soon see of six Planck-scale entities)[b] as the common factor. He had no way to seek the "conformity . . . in the structure of all substances" where we are about to find it, at Planck scale, a trillion-trillion times smaller than the electron.

Looking back over this confusing scene in 2006, Wilczek summarized,

---

a He brushed aside Planck's focus on functions fitting data: "but we need not here speak of those researches."
b See chapter 35 at page 190.

Planck's proposal for a system of units based on fundamental physical constants was, when it was made, formally correct but rather thinly rooted in fundamental physics. Over the course of the 20th century, however, his proposal became compelling.[320]

So, we may say Planck started the still-emerging-today seminal idea of a Planck-scale reality that makes far larger-scale things (like electrons and atoms and heat radiation) happen. Physics was not ready for it.

14
PHYSICS MEETS THE QUANTUM

> In 1900 Planck discovered the blackbody radiation law without using light-quanta. In 1905 Einstein discovered light-quanta without using Planck's [radiation] law.
>
> Abraham Pais (1982)[321]

In the early nineteen hundreds, Planck's action quantum and its easily distorted story diffused slowly through the physics ecosystem.[a] The story followed what Einstein-collaborator and -biographer, physicist Banesh Hoffmann, called . . .

> Max Planck found a tiny place. No-one wanted to go there.

> . . . a confused and groping search for knowledge . . . illumined by flashes of insight, aided by accidents and guesses, and enlivened by coincidences such as one would expect to find only in fiction.[322]

The pivotal—but mostly neglected—question in this search was, *Quantum of what?*

The answer was entirely unclear. As Pais noted in the epigram above, Planck and his protégé, Einstein, were on different pages. Trying to make sense of Planck's result, Einstein said,

> It was as if the ground had been pulled out from under one, with no firm foundation to be seen anywhere, upon which one could have built.[323]

The action quantum's tiny size and the miniscule size of its fundamental space and time units were soon passed over; so, for example,

---

[a] Readers for whom this is new and strange terrain may find it helpful to take a quick tour of quantum history and its key people, such as the American Institute of Physics' user-friendly page, https://www.insidescience.org/second-quantum-revolution/quantum-history.

more than ninety years after their first appearance, physicist Peter Bergmann could say blandly,

> The Planck length is 1.6×10^{-33} cm, very much smaller than any known nuclear dimension.[324]

In 1905, Einstein took Planck's constant, attached it to a light particle he called a light quantum and others would come to call the photon, and ran with it. Planck himself apologized for Einstein's foolishness, for which he—Einstein that is—would soon get a Nobel Prize:

> That [Einstein] may sometimes have missed the target in his speculations, as for example in his light quantum hypothesis, should not be counted against him too much.[325]

For the physics of that era, the beating heart of the quantum was, as Nobel Prize–winning physicist Max Born summarized in 1925,

> The special character that the atom possesses is the appearance of *whole numbers*.[326]

In other words, the possibility of simple *counting*. It was, as we have seen, a possibility that Planck had built right into the theory's foundations with no sense of what if anything it really meant.[a]

At first, the physics world paid scant attention to the strange way those whole numbers had hoved into view. Kragh said,

> Very few physicists expressed any interest in the justification of Planck's formula, and during the first few years of the 20th century no one considered his results to conflict with the foundations of classical physics.[327]

Slowly an understanding emerged in some quarters that the Planck constant could have hidden implications. Meanwhile, seeping into open cracks, it had undermined and utterly overthrown the foundations of classical physics. Yet to this day, while billions of us have its byproducts in hand,[328] neither physics nor philosophy has come to grips with its deep ontologic message.

---

[a] See note 303.

By 1911, Planck began to grasp some of the (Planck-scale) significance of his seminal idea. He was invited to the first of the prestigious Solvay Conferences (fig. 8, on page 76). There he claimed,

> The hypothesis of quanta will never vanish from the world.[329]

We should bear in mind that, over the next decades, leading lights of physics focused on the hidden world of atoms and their crazy but increasingly predictable behavior for which the Planck constant seemed somehow made to measure. At the time they thought of atoms as inconceivably small[330] though they were a trillion-trillion-fold[a] larger than the scale Planck's work revealed.[331]

To illustrate how hesitantly the world of physics began to come to grips with this, sixty years after Planck's epochal paper, physicist Alden Mead, and a leading physics journal,[b] and its referees, and many other physicists, all thought the Planck scale was a novel concept. They contrived to rediscover it and did not even mention Planck.[332] Much later, Mead wrote,

> At the time, I read many referee reports on my papers and discussed the matter with every theoretical physicist who was willing to listen; nobody that I contacted recognized the connection with the Planck proposal, and few took seriously the idea of L as a possible fundamental length. The view was nearly unanimous . . . that the Planck length could never play a fundamental role in physics.[333]

So, quantum physicists set to using h in the shut-up-and-calculate[334] halls of quantum theory with studied unconcern about its fundamental meaning.[c] Philosophers of science saw the ontologic issues as whether the quantum states are real and if so what they may say about reality.[335] Historians of physics argued about what Planck had intended and about what he had done.[336]

---

a Roughly the diameter of a hydrogen atom divided by the Planck length.
b Physical Review.
c David Bohm was the leading exception; see note 135.

The governing consensus seized on atoms and on oscillators as central to the narrative, neglecting (one might almost say suppressing) Planck's explicit distance from them, with lasting damaging effect.

Thus, loose talk about it became deeply entrenched. For example, in 1951, the hard-headed Reichenbach related, doubly wrongly,

> Planck introduced the concept that all radiation . . . proceeds by whole numbers in an elementary unit of energy, which he called the *quantum*.[337]

And today, for example, the editors of *Encyclopaedia Britannica* proclaim, also doubly wrongly,

> Planck assumed that the sources of radiation are atoms in a state of oscillation.[338]

As we have seen, Planck's concept was a unit of action, not of energy; *he* did not call it the quantum; and he explicitly did *not* assume atoms were the sources of the radiation, or that anything was really in a state of oscillation.[a]

Even the brilliant physicist, philosopher and mathematician, Henri Poincaré, fell into error, saying,

> Mr. Planck's hypothesis consists in supposing that each of these resonators can acquire or lose energy only by *sudden jumps* so that the amount of energy that it possesses must always be a multiple of the same constant quantity called "quantum," and that it must be made up of an integral number of *quanta*.[339]

Planck was wise to keep his distance from such loose language and unneeded assumptions: A century later, physicists Gerhard Kramm and Nicole Mölders would derive his equations and his constant without reference to atoms or to resonators or to oscillations.[340] And as far back as 1924, in a paper that launched quantum statistics, physicist and mathematician Satyendra Nath Bose derived it based on a purely particulate view of radiation:

> [T]he combining of the light quanta hypothesis with statistical mechanics in the form adjusted by Planck to the

---

a See chapter 13 at page 76.

needs of the quantum theory does appear to be sufficient for the derivation of the law.[341]

Of this, Einstein said,[a]

> Hitherto, all derivations of Planck's formula have somewhere made use of the hypothesis of the undulatory structure of radiation. . . . The important fact that, according to the theory of Bohr, the frequency of the radiation is not determined by electrical masses that undergo periodic processes of the same frequency can only increase our doubts as to the independent reality of the undulatory field.[342]

Amid conceptual doubt and confusion, the inquisitory notion of a real quantum persisted in Einstein's mind until the end. His last words in his last published scientific work—contradicting its most basic premise—were,[b]

> One can give good reasons why reality cannot at all be represented by a continuous field. . . . This . . . must lead to an attempt to find a purely algebraic theory for the description of reality. But nobody knows how to obtain the basis of such a theory.[343]

By the nineteen sixties, there were intimations of discontent with the by then canonical atom-based quantum-picture. For example, Wheeler foresaw,

> It is quite possible that only the physics of the 10^{-33} cm region will help us to understand the physics of elementary particles.[344]

He would later coin the term *quantum foam* to describe the chaotic scene he imagined space might show us—if we could see it—at that scale (see fig. 75 at page 445).[345] This did offer us a mental image; on the other hand, it reinforced the continuous views of Planck-scale space held by almost all the few who gave it any thought.

---

a In Saunders's translation.
b This appendix was rewritten with his last postdoc, Bruria Kaufman, who may have had particular input to its mathematics. See Topper, note 535, footnotes at his pages 28 and 219.

In 1997, 't Hooft tried to move the ontologic goal posts, saying,

> Everything we now think we know about Nature will be invalid at the Planck scale.[346]

By 2004, Rovelli would say the Planck constant's offspring, quantum mechanics, itself gave the lie to its own continuous space, suggesting granular space must be a physical reality:

> [Quantum mechanics] implies that continuous space is ultimately unphysical.[347]

In 2020, Smolin subtitled his latest book *The Search for What Lies Beyond the Quantum*.[348] One might think this signaled he was hot on the true quantum trail but, to the contrary, he seemed not to grasp that Planck's quantum offered an astounding glimpse into an until then entirely inaccessible world. (We'll see another glimpse of it had come his way, right under his nose as the saying goes; somehow he missed that too.[a])

Nonetheless, in recent years some concept of reality at Planck scale has been slowly emerging.[349] (Though there are also critics.[350])

For our part, let's take a closer look at that quantum of action.

The *action* is an oddity that became prominent in physics because of its role in the principle of least (or stationary) action, which hails back to mathematician Leonhard Euler in 1744.[351] It is applied in a range of fields in classical physics and in quantum theory.[352]

In overly plain language, this principle says, when something happens, out of all possible paths it could take, it will follow the path that sums (or integrates) to the least total action. Like other elements of physics, while it became immensely useful, no one knows why it works. Physics' star explainer Feynman said,

> The miracle is that the true path is the one for which that integral is least.[353]

Like the action quantum itself, this "miracle" appears to bear a fundamental message about the real nature of energy (and so of mass; so, also, of inertia) we cannot yet decipher. Few delve into this problem; there is no notable success.[354]

---

a See chapter 43 at page 243.

But as we begin to build a new worldview we can see reasons for such difficulty: Planck's derivation of the action quantum, its embrace into quantum theory, and explorations of its meaning were all formulated in continuous space using its preferred language, the *infinitesimal calculus* (math of change[a] in steps of vanishing size). This was an inherent contradiction.

It's understandable that many (maybe most) who engage with fundamental physics resist having to—with severe difficulty—give up hard-learned math of continuous space and plunge into the unexplored new math we'll need for its granular cousin.

Notable was Hawking who said,

> Although there have been suggestions that spacetime may have a discrete structure, I see no reason to abandon the continuum theories that have been so successful.[355]

Some have better understood the need to address the nature of space. For example, in 2009 't Hooft said (in an essay titled "The fundamental nature of space and time"),

> [W]e emphasize that, in any more advanced theory for Planck length physics, the definition of what exactly [space] is, will have to require special attention.[356]

If we are on the right track, and space turns out to be granular at Planck scale, it will admit no infinitesimals. It will allow no yardsticks and support no integrals. As Riemann saw was possible, it won't be analog, it will be digital.[b] It will need brand new math.

Like Planck and those few physicists who these days give it close attention, we are left with an enigmatic quantum of action. Physics remains to this day unready to embrace it.

So, at the turn of the century, science writer Manjit Kumar summarized the status of this seminal idea:

> Max Planck stumbled across the quantum, and physicists are still struggling to come to terms with it.[357]

---

a The great majority of physics' equations are concerned with change.
b See chapters 10 and 16 at pages 63 and 103.

Recap: It is *not* a quantum of energy; it has the dimensions of energy, whatever that is,[a] times *time*, whatever *that* is.[b]

We are about to see how efforts to marry quantum theory and relativity have already given birth to a perfect way to understand the quantum but physics treats it as math fiction.

So it is that, more than one hundred twenty years after Planck came upon it, physics still has no coherent answer to the simple question that obsessed him, and Einstein right after him, *What is the quantum?*

---

[a] See chapter 86 at page 435.
[b] Chapters 28 and 52 (at pages 158 and 284 respectively) explore the real nature of time.

15
STRINGS AND THINGS

> The atoms of space are not the smallest portions of space. They are the constituents of space.
>
> Daniele Oriti (2018)[358]

Science writer George Musser quoted physicist Daniele Oriti (above) in aid of the proposition, "Building blocks of space need not be spatial."

Oriti went right on to say,

> The geometric properties of space are new, collective, approximate properties of a system made of many such atoms.

We will see that he was right in this;[a] but he did not say what those "smallest portions" or "constituents" or "atoms" are. We set out here to find a simple answer.

Musser's article asked the seminal question.[359]

> Physicists believe that, at the tiniest scales, space emerges from quanta. What might these building blocks look like?

He could find a literally figurative[b] answer on page 97. In certain circles (including his and Oriti's) it has long been in plain view.

String theory took off in the nineteen seventies, when many physicists thought it could be *the* right road to *quantum gravity*—a term that's vaguely seen as meaning the long-sought marriage of relativity to quantum theory—the holy grail for physics.[360] Some see it as a candidate for the so-called theory of everything.[361]

What *is* quantum gravity? It defies definition. In 2000, Rovelli (one of its heavy hitters) said,

---
a In Part 5.
b This is (here) not an oxymoron.

> Still, after 70 years of research in Quantum Gravity, there is no consensus, and no established theory. I think it fair to say that there isn't even a single complete and consistent candidate for a quantum theory of gravity.[362]

And, a few years later, he said,

> Quantum Gravity is therefore the study of the structure of spacetime at the Planck scale.[363]

String theory (of one kind or another), though not yet an "established theory," and being neither complete nor consistent, may now be the leading candidate.[a] The central concept that ties string theories together is they replace point particles with tiny vibrating strings of one or more dimensions.

Strings live (if they exist at all) in the unimaginably tiny Planck-scale world. To visit with a thought experiment,[b] imagine you are Emerson's transparent eyeball that, as we noted above,[c] is nothing and sees all. With it in aid, you see string theory's strings a-wiggling franticly in empty space, and what they are— electrons, photons, quarks, et cetera—depends upon the wiggles.

> A world of strings needs strange Planck-sized things.

With roots all the way back to Einstein, string theory came of age in the nineteen sixties. Its math bewitches those who live and breathe it.

However, as mathematician George Ellis said a decade ago,

> The energy scales characterizing string theory are so high that it cannot be tested by any particle collider that we can ever hope to construct. . . . If we can't ever hope to test the theory experimentally, is it really scientific?[364]

So, string theory's popularity (with concomitant funding) energizes a concept called *post-empiricism*[d]—aimed at looser ways of validating physics theories—and thereby fuels a vigorous debate.[365]

---

a See further in chapter 31 at page 176.
b See chapter 30 at page 173.
c See chapter 5 at page 43.
d Roughly, the idea that we now need to free fundamental physics from the rigors of experimental test; see further in chapter 133 at page 639.

Part 1: Some Seminal Ideas

Quantum gravity's main problem is, it's a family-to-be based on a hoped-for forced marriage but its two betrotheds have diddly-squat in common. In 2001, philosopher Jeremy Butterfield and physicist Christopher Isham said,

> The fundamentally disparate bases of the two ... theories generate major new problems when any attempt is made to combine them. ... But even to summarize these new conceptual difficulties is a complicated and controversial task ... because what the problems are ... depends in part on problematic matters in the interpretation of the ... theories.[366]

Having said this, they found a consensus on a fundamental issue that must resonate with us:

> [A] theory of quantum gravity must have something to say about the quantum nature of space and time.[367]

With this dictum, the reader will by now be unsurprised to hear, I heartily agree. On the premise such a theory is best based upon a real worldview, we are now setting out to "have something to say about the quantum nature of space and time" (and more besides).

Most fundamentally, we pluck just such a *something* from the murky world of strings and seize on what it *is* and say, *this* is the quantum character of "space"!

We seize on what that *something does* and say, *this* is the quantum character of "time"![a]

This quantum character of space (see image,[368] fig. 10) is a seminal idea born in the heartland of string theory. Yet we'll see it's also an idea that string theory does not *really* have!

String theory led physicists into a world of beautiful new math. The math showed them that, in

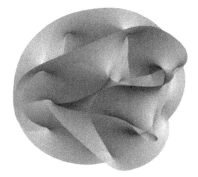

Fig. 10. Calabi-Yau manifold

---

a In this chapter we will deal with space; the quantum character of time is taken up in chapters 16 and 40 at pages 103 and 222.

order to make it work, they needed a strange entity that has six tangled dimensions. It has a funky name—the *Calabi-Yau manifold*.[369]

In our new ontology, it turns out to be the key to understanding space and time, not to mention Zeno and indeed all else. Yet it is tiny, maybe (say string physicists; surely, say I) Planck size.[370]

A manifold is just a space. It can have any number of dimensions. It could have three (like yours and mine) or twenty-two or six. Any number one might fancy is just fine.[371]

This is less off-beat than it may seem. Einstein, among others, tried to build a theory with a 5-D universe.[372] And Gross said, of string theory's 3-D + 6-D (+ 1-D, for time) space,

> If it is hard to imagine a space of four or more dimensions, [10-D] space is even weirder, but totally mathematically consistent.[373]

In the nineteen fifties mathematician Eugenio Calabi proposed this new 6-D kind of manifold. Twenty-five years later, another mathematician, Shing-Tung Yau, proved its mathematical existence. Their manifold was just what the string-theory campaign needed to get its show on the road.

Mainly for this reason, it has become perhaps the most-studied kind of space, leaving Euclid's in the dust.[374]

But in string theory it is studied, not as a reality, but as a mathematical abstraction. Much of that study is abstruse. So, for example, we know one can draw exactly 317,206,375 different curves of polynomial degree three in it.[375]

While string theory itself seems stuck in math, we have a practical objective. We'll see[a] this weird-looking manifold turns out to be just what the universe needed to *be* to get going and to finish up exactly as it is.[376]

Neither Calabi nor Yau could have guessed their manifold would be the key to understanding all that's real.

In its looks it's an unlikely candidate. Each different Calabi-Yau manifold is an oddly asymmetric critter that is impossible to picture

---

a See chapter 42 at page 236.

with precision. Depending on how one tries to depict its 6-D nature on a 2-D page it may look plug-ugly, strangely beautiful, or oddly funny.[377]

As Susskind said in 2008,

> Calabi Yau manifolds are extremely complex, with hundreds of six-dimensional donut holes and unimaginable pretzel twists.[378]

He also said,

> I have always felt that if a thing is understood well enough, it should be possible to explain it in nontechnical terms. But String Theory's need for six extra dimensions has eluded a simple explanation, even after more than thirty-five years.[379]

Fifty years have now gone by and finally there is a simple explanation, deep in the heart of the new ontology: Though we don't see the six dimensions, they may be the source of everything we do see.

String theory is not really a theory. It has a set of ontological commitments all its own, with numerous theories crowding under a shared string umbrella that has all the world made from one kind of thing—their strings. Beyond this overall idea, its strings are math abstractions. At every point in continuous space, they need those six extremely small dimensions over and above the extremely big three we see.[a]

For these string theories to hum along, the missing six *must* take the shape of a Calabi-Yau manifold.[380] Indeed, it's an idea that works wonders for the world of strings.[381] Physicist and polymath Richard Amoroso and his colleagues summarized, ten years ago,

> Essentially, Calabi-Yau manifolds are shapes that satisfy the requirement of space for the six "unseen" spatial dimensions of string theory, which are currently considered to be smaller than our currently observable lengths as they have not yet been detected.[382]

So, where are these six dimensions? Why can't we see them? Even before strings got their six, physicist and mathematician Theodor

---

a Most treat time as a dimension, so you'll find talk of ten in total; see, for example, note 373 above.

Kaluza came up with a lovely theory with just *one* extra dimension.[383] We can't see that one either.

Physicist Oskar Klein rescued that theory from early oblivion, saying its invisible dimension must have been *compactified*, an ugly word for an ugly idea.[384] It means it somehow shrank into a tiny loop and became undetectable, though no one can say how or when or why.[385]

And so, when string theorists later needed answers for more awkward questions about *their* missing dimensions, they repurposed the old ugly word to tell their tale.[a]

Keep in mind, though, a key piece of the string theory worldview: The new strings wiggle in the old continuous space. String theorists see their tiny manifold as a kind of magic math myth they can summon on demand to do their bidding wherever they are. In other words, to them, it is a convenient figment.

It's not only string theorists who found uses for invisible extra dimensions. As Susskind noted,

> Many very famous physicists—including Einstein, Wolfgang Pauli, Felix Klein, Steven Weinberg, Murray Gell-Mann, and Stephen Hawking (none of them string theorists)—have seriously contemplated the idea that space has more than three dimensions. They were obviously not hallucinating, so there must be some way of hiding the existence of the extra dimensions.[386]

Indeed, we are about to find them "hiding" in plain sight: Compact they may be when they're at home but, like June, they are bustin' out all over.[b] From the beginning, it was quite impossible to keep them in.[c] (Every word of this will come to have deep meaning.)

So, one might ask, if string theory's manifold is everywhere in space, why is string theory not *the* theory? To the contrary, it is so far from that, some say it is in trouble. Smolin—once a string-theory

---

a You know how stories go: A KISS is coming up; this ugly frog of an idea will turn into a handsome prince.

b Though it will do no good to—as Oscar Hammerstein II would have it—"Look around, look around, look around!"

c See chapter 42 at page 236.

practitioner—wrote the book on how its math gets lots of money but is, he said, not even science.[387]

Yet there seems to be something fundamental in string theory and its pivotal idea, the six said-to-be-compactified dimensions. Lots of string theorists appear to feel this in their bones. And, in 2018, physicist Sabine Hossenfelder—a critic of flawed physics who is far from a string-theory cheerleader—said,

> For all the controversy that surrounds string theory in the public sphere, within the physics community few doubt its use.... [T]he mathematics of string theory is deeply rooted in theories that demonstrably describe nature: quantum field theory and general relativity. So we are certain string theory has a connection to the real world.[388]

So, here's where I suggest string theory may harbor at least one seminal idea: Its manifold is real! It has a tiny *volume*; could it be the grain of Riemann's granularity?

If so, string theory hands us a further problem: Which of its manifolds (also known as threefolds) should we pursue? There are many candidates:[a]

> With the enormous number of candidate Calabi-Yau compactifications in hand, model builders are confronted with the challenge of isolating the set of constructions which might potentially replicate physics in the real world.... [W]hile there are 473,800,776 reflexive polyhedra, the number of inequivalent Calabi-Yau threefolds obtained therefrom is so far *unknown*.[389]

Looking back over the struggle for a deeper understanding of Planck's elementary quantum of action, Duck and Sudarshan said,

> Then this remote, ultimate, intricate "reality" and all its elements will require an existential—possibly only metaphysical—understanding.[390]

---

a The six dimensions of a Calabi-Yau threefold are grouped as three complex dimensions.

It is ironic: Physicists grasped the seminal idea—the Calabi-Yau manifold—that may be the basis of reality but they don't see it as existing.

Doubly ironic: We'll soon see it as a quantum of space, and space quanta as *everything* that really exists![a]

---

[a] See chapter 45 at page 255.

16
SMALL CHANGE

> I consider it as entirely possible that physics cannot
> be based upon . . . continuous structures. Then
> *nothing* will remain of my whole castle in the air
> including the theory of gravitation, but also nothing
> of the rest of contemporary physics.
>
> Albert Einstein (1954)[391]

If—as a long chain of thought from Riemann to Planck to Einstein (above) to (maybe) Heisenberg to Wheeler to Rovelli and many others suggests—the universe is granular[a] and its physics is therefore digital, how would it work? This leads straight to a seminal question: How could it change?

| Is the world jumping (very swiftly) 1 – 2 – 3?

Could its distinct elements transform in some smooth fashion? If so, they could not be truly digital. A digital digit is what it is; next it is either the same or it is a different digit. It has no in-between, no halfway house, no smooth segue from one digit to the next.

We have common examples of digital systems. From phones to supercomputers, they all solve the problem of change the same way: The whole system is in a certain digital state, with each bit being zero or one; next tick of its system clock, each bit is reset, some changed, some not. The system has no meaningful in-between state.[b]

The concept that the world might work something like this dates as least as far back as 1969, when civil engineer Konrad Zuse, who

---
a Curiously, while *we* thus appear to have no continuous space in our universe, the space of each granule (the Calabi-Yau manifold), though utterly inaccessible, must be continuous and so must each of the six (also, it would appear, inaccessible) externalized dimensions; see the following chapter. This may have profound implications for the math (to be developed) of nonlocality and entanglement; see chapter 25 at page 144.
b Here we can begin to understand *why* the Planck time is the smallest meaningful "time."

invented the first programable digital computer in 1938,[392] wrote *Rechnender Raum*,[393] exploring . . .

> . . . the idea of calculating space.[394]

His notion of how space could be calculated was far from fully worked out—after all, this was the era of the early MOSFET[a] logic switches—but he had the idea.

More ideas of this kind may be found in later works ranging from physicist Heinz Pagels's cosmic computer[395] through quantum engineer Seth Lloyd's serious speculation[396] to physicist Max Tegmark's mathematical-universe hypothesis[397] to *The Matrix* movie trilogy, which saw reality as a computer simulation. Philosopher Nick Bostrom said *it's so*,[398] and Gleiser argued *no*.[399]

But few seemed to grasp a key concept that follows from the digital idea: The whole universe must be in a certain digital state; next it is in a different state determined by its universal program. And simplicity says the time from one state to the next—as determined by a clock *inside* the universe—should be the Planck time, the smallest time that has physical meaning.[400]

This time limit is uncontroversial: By 1992, Davies and co-author, astrophysicist John Gribbin, could say,

> [I]t is now widely accepted that there is a fundamental unit of time, the "Planck time," beyond which intervals of time cannot be subdivided.[401]

Just why or how this should be so remains a physical mystery that's yet another we should solve.

What is not accepted, widely or perhaps at all, is that this "time" is digital, so Gribbin's "fundamental unit" is its *one*.[b]

This brings us to the seminal idea of digital change. The 3-D universe of space quanta must *do* a succession of synchronous spatial states, or *iterations*. It can't have an in-between state or an out-of-synch state, not anywhere, not ever.

---

a I.e., metal-oxide-semiconductor field-effect transistor switches that would later become the basis of the infant computer industry.

b Fundamentally, this is a sequence rather than a time; but inside the universe, we watch our clocks and assign the Planck time to each step in the sequence; see further in chapters 28 and 103 at pages 158 and 508.

17
BEING BEGUN

When I was one, I had just begun . . .

Alan Milne (1927)[402]

We are about to find the roots of the seminal idea the universe began as just one quantum.[a] But let's be clear and precise from the beginning: When *it* was one, it had *not* begun.

Of course, Milne's children's poem about growing up has no connection to *cosmogony*, the study of the universe's origin. Or does it? A person begins as a single cell that has a program that causes it to replicate. We can see there is some sort of similarity but let's not push it too far.

Milne's next line,

When I was two, I was nearly new . . .

> The universe had a beginning. Did it begin with only one?

. . . may be more readily applicable. The beginning of the universe was that one quantum turning into two.

Human imaginations have long roamed over the beginning of it all. That it *did* have a beginning was widely believed since time out of mind but is not self-evidently true.

In 1927, Russell said dismissively,

> There is no reason to suppose that the world had a beginning at all. The idea that things must have a beginning is really due to the poverty of our imagination.[403]

Of course, he meant *their* imaginations—those that (ironically, richer than his) *did* suppose the world had a beginning.

The assumption that it *didn't* begin was then much in vogue, a fashion dating maybe back to Kant.[404]

---

[a] Lemaître's idea (or was it Poe's?—see note 413) was the universe began with a single quantum (or atom) of some kind; this is a central theme of *Time One*, note 16.

Einstein long shared this assumption; he conceived the universe as being permanent and static.

The world would need to wait till 1964—when we first saw an image of the very early universe[a]—to know for sure all three (and many others) got this wrong.

Meanwhile physics tended to assume the universe was infinite in size and more or less unchanging. Then, in 1927, physicist Georges Lemaître, whom Peebles said was at that time "the physicist who best understood the implications of Einstein's theory of general relativity,"[405] solved its equations.[406] His solution revealed space itself to be expanding.[407]

Few knew of his paper. Fewer still took him or his solution seriously.[408] But Willem de Sitter, the astronomer, physicist, and mathematician who is mostly known for *his* solution of Einstein's general relativity equations, which bears the name *de Sitter space*, said,

> I have found the true solution, or at least a possible solution, which must be somewhere near the truth, in a paper . . . by Lemaître . . . which had escaped my notice at the time.[409]

In 1929, astronomer Edwin Hubble observed galaxies moving away from us in all directions, movements that could only be understood as space itself expanding.[410]

Mentally projecting this expansion back in time like a movie in reverse suggested that the universe had a beginning. In 1931, Lemaître (center, fig. 11)—who to this day is often called "father of the Big Bang"—wrote a widely neglected[411] paper in a leading journal.[412] Entitled "The Beginning of the World from the Point of View of Quantum Theory," it gave birth to a seminal idea. It said the world is made of many

Fig. 11. Robert Millikan, Georges Lemaître, and Albert Einstein

---

a The cosmic microwave background radiation, see below.

quanta; we see their number steadily increasing; so, looking back, there must have been fewer and fewer; thus, in the beginning there was only one![413]

His work arrived, with little fanfare, amid growing popular involvement with the nature of the universe. In 1925, astronomer William MacMillan had proposed a new ontology, in which . . .

> . . . new atoms are generated in the depths of space through the agency of radiant energy.[414]

This seems to have been the first glimmer of so-called continuous creation, the first version of what came to be called the *steady-state universe*. It made no place for a beginning. It matched what people thought they saw. Three cosmologists adopted it. In a 1949 radio interview, one of them, Fred Hoyle, disparaged the beginning proposed by Lemaître,[a] calling it the "Big Bang."[415]

Hoyle's catchy tag for Lemaître's beginning is still with us—now part of our language—though it has conflicting meanings and the mental image it evokes can be, to say the least, misleading.

In a 1950 book, Lemaître called his first quantum the primeval atom. It was far from elementary: Into it he packed all the universe's energy and matter. The universe began, he said, when his quantum-atom-world explosively disintegrated:

> The atom-world was broken into fragments, each fragment into still smaller pieces. To simplify the matter, supposing that this fragmentation occurred in [two] equal pieces, two hundred sixty generations would have been needed to reach the present pulverization of matter into our poor little atoms, almost too small to be broken again.[416]

We should understand Lemaître's clinging to an all-the-matter-in-one-quantum concept in relation to his aversion to the then leading alternative of continuous creation and the steady-state model:[417]

> What does this mean . . . creation? This word, creation, brings with it a whole philosophical or religious resonance that has nothing to do with the question.[418]

---

a See below.

Lemaître's version was an odd kind of exponential growth.[a] It was an ontologic insight that—despite its matter-packing failings—might have avoided the deep contradictions built into the not yet emerging[b] standard model of cosmology. Instead, it was (and is) unjustly dismissed as spillover from his religion.[c]

Thus began the separation of *the expansion* (a narrative in which Lemaître is widely credited as author of the reigning Big Bang theory) from *the beginning* (which is widely panned). So, today's standard-model storyline embraces something like Lemaître's concept from the first fraction of a second *after* the beginning.[419]

The fallout from this schism is that physics has avoided almost any thought of the beginning, the simplest and potentially most fertile field of all for it to till.

Einstein changed his view.[420] In 1934, he even praised Poe's book *Eureka*.[d] His theory of general relativity led to the conclusion the universe did have a beginning. As Hawking later said,

> If Einstein's General Theory of Relativity is correct, there will be a singularity, a point of infinite density and space-time curvature, where time has a beginning.[421]

Science writer Brian Clegg summarized more recent bet-hedging about the emerging standard model picture:

> Because this isn't accessible to our math, it's possible it didn't quite start at a point, but we are certainly talking something crammed with mass and energy in a tiny space, something that appears out of nothing for reasons that are no better explained than they are by Genesis.[422]

Already ailing, the universe-with-no-beginning died for most cosmologists in 1965,[423] with the discovery of the *cosmic microwave background radiation*[424]—a flash photo of the much stretched great

---

a It was an exponential growth of numbers but not of substance.
b North said, "[In 1965,] scientific cosmology was in many quarters not even acknowledged as having an intellectual existence in its own right." See note 411 at his page ii.
c In a Freudian slip revelatory of establishment preoccupations, Peebles (who clearly holds Lemaître's work in high regard) introduced a picture of him in a lecture as "a Roman Catholic Church" (swiftly corrected); note 405.
d He later changed his mind again.

escape of light when space became transparent[a] some four hundred thousand years after the beginning. It revealed that the beginning and expansion were both real events. It gave us tantalizing clues about—but no clear understanding of—the way the universe began.

Hawking was for long a notable exception to the general avoidance of the beginning.[425] Upon taking up the venerable Lucasian Chair of Mathematics in 1980 in long succession from Sir Isaac Newton,[426] he said,

> I think that the initial conditions[b] of the Universe are as suitable a subject for scientific study and theory as are the local physical laws.[427]

And, in 1994, he said,

> Cosmology cannot predict anything about the universe unless it makes some assumption about the initial conditions. . . . The only way to have a scientific theory is if the laws of physics hold everywhere, including at the beginning of the universe.[428]

And, in a 1996 lecture, he said,

> All the evidence seems to indicate, that the universe has not existed forever, but that it had a beginning, about 15 billion years ago. This is probably the most remarkable discovery of modern cosmology.[429]

It's too bad his last word on the subject[430] in 2010 was that the initial condition of the universe was what he called "nothing,"[c] an absurdity his admirers charitably took to be a spoof. For example, science writer John Horgan wrote,

> I've always thought of Stephen Hawking . . . less as a scientist than as a cosmic, comic performance artist, who loves goofing on his fellow physicists and the rest of us.[431]

---

a This happened when expansion cooled the dense ionized plasma—which blocked the light—until the cooling electrons and protons formed neutral hydrogen, a transparent gas.
b That is, the exact state of the universe when it began.
c See further, chapter 91 at page 454.

Like Einstein, Hawking often changed his mind, not least about this question of a beginning. Most recently we had cosmologist Thomas Hertog (who worked with Hawking) writing of combing through the ashes of Hawking's "final theory."[432] Philosopher and science historian Robert Crease said of this,

> It's a long and winding road to the "final theory" of the book's title. The closer we get, the more mathematical things become, and the farther removed from common sense.[433]

In the end we'll see Hawking was right the first time; there was a beginning.

But physicists (even Hawking) didn't want to know about the single quantum. All this may seem fanciful today but at the time there was a great debate in which the physics of reality was shunted into an allegedly religious siding.[434] In the world of physics there is no canard that's more effective than *religion* to abruptly terminate a conversation.

Thus the question of exactly how the beginning did begin gets almost no attention. For example, in his book *Until the End of Time*, Greene said,

> [W]e begin at the beginning.[435]

And then didn't. He turned straight to Lemaître and relativity equations.

What Lemaître didn't know—and couldn't possibly imagine—is that the real primeval atom was far smaller and far simpler than the one he had in mind. Looking at it—if he or we could have looked with Emerson's transparent eyeball—who would have thought such a seemingly inconsequential entity could bring forth the whole universe around us? Seeing it—if it could have been seen—who would have thought it to be the prototype for (looking far ahead) the "local *beables*" Bell hypothesized at the roots of reality.[436]

I am reminded of Arendt (who, for me, is to *doing* as Sartre is to *being*,[a] to birth as he is to death) saying,

---

a The two reflect the universe's (and our) existential motif of being/doing; see chapters 56, 57, 134, and 135 (at pages 301, 307, 647, and 655, respectively).

> This character of startling unexpectedness is inherent in
> all beginnings and in all origins.[437]

We will see what I might call the real beginning's mode of action being reminiscent of what Lemaître long ago prescribed.[438] So Lemaître gets credit for this seminal idea: The universe began as a single quantum.

But he was hopelessly confused about A *quantum of what?*

18
EXPANDING REALITY

> Let me be terse: *Our universe*
> *Grows daily more diluted!*
>
> Barbara and George Gamow (1940)[439]

Meanwhile, as is its wont, our universe is busy getting bigger. Though this is yet another mind-bending and seminal idea, it is now mainstream cosmology: Space is expanding. By 1940, polymath and cosmologist George Gamow and his spouse, editor Barbara Gamow, were putting it to verse.

Even more mind-bending than that it is expanding is the question, *Why?*

The short answer is, nobody knows. It seems few care. Reflecting on cosmology, broadcaster and "latter-day Carl Sagan"[440] Marcus Chown posed seven questions about the universe expanding but did not see fit to ask why.[441]

General relativity says space expands today because it was expanding yesterday. This truism is baked into the standard model of cosmology.[442]

We will seek a more meaningful reason (and find the truism is not true[a]).

In the early days of modern cosmology, the reason for expansion was not even a question. Einstein grew up (as did his contemporaries) in a universe that was unchanging. When his own equations seemed to say that it could be expanding, he refused to heed them. But soon observations said that it was so. Seen in the rearview mirror, the neglected reason is buried behind the Big Bang. Seen in the present it's a source of physical confusion.

---

a See chapter 69 at page 357.

It's not everything that's getting bigger. Objects like stars and gravitationally bound groups of them, such as the Solar System and the galaxies, are unaffected. But space is growing, so groups of galaxies are moving away from other groups from which they are gravitationally free.

This growth of space between them is a seminal idea. It is usually referred to as the *cosmological expansion*.[a]

Unless it's in orbit—that is, in thrall to another mass—whatever's in a region stays there for no better reason than there's nothing moving it away. That is, it stays with the space it's in and vice versa.[b] It's the cosmic version of *Dance with the one that brung ya*. Muller said it more elegantly:

> The matter, in fixed space coordinates,[c] went along for the ride—not moving, but getting farther and farther apart.[443]

Astrophysicists Tamara Davis and Charles Lineweaver set out to sort out some of its problems:

> The concept of the expansion of the universe is so fundamental to our understanding of cosmology and the misconceptions so abundant that it is important to clarify these issues.[444]

It's not an easy concept. Some serious scientists can't get their heads around it.[445] Those who assume space is nothing—no more than a void or vacuum—may rhetorically ask, *How can nothing expand?*[d] The implied answer is correct: It makes no sense to say that nothing is expanding. But, as we've begun to see, assuming space is nothing is wrong thinking.[e]

> Q: How can nothing expand?
> A: By being something.

That there are serious physicists who still think space is nothing is an object lesson: Physics education needs updating. It should be clear

---

a We will see that what is really going on is not well described by this terminology.
b At the scale of intergalactic space this needs a more nuanced analysis than has so far been attempted.
c This of course begs the question of what fixed space coordinates might be.
d See, e.g., note 1009.
e See chapter 10 at page 63.

by now: Space is a thing. Soon we will see it is by far the biggest and most massive thing there is.[a]

The evidence from many measurements is by now overwhelming that space *is* expanding. Pangalactic distances stretch 7 percent so that pangalactic volumes increase 23 percent per billion years.[446]

If space had been expanding steadily at that rate since soon after the beginning about 13.8 billion years ago, it would now be some seventeen times bigger than it was back then. The standard model of cosmology shows (aside from cosmic inflation[b]) the rate of expansion slowing for billions of years and then accelerating.[447] Cosmologist Evalyn Gates described the shock of this observation:

> [C]osmologists expected gravity—which pulls mass together—to . . . slow the overall rate of the expansion. What we have found is exactly the opposite, and it has pulled the ground out from under our understanding of the cosmos.[448]

As Peebles recently said, understatedly,

> [M]any of us wonder if there is something deeper behind it.[449]

They have not yet figured what.[c] We soon will.

Over the universe's time scale this expansion is a huge effect[450] but in a lifetime the effect nearby is too small to be evident. And so, its discovery was muddled.

Early intimations that the universe might be expanding came in 1914, with astronomer Vesto Slipher's observations of a dozen nearby galaxies. He found most were moving away from us. He reported this without commenting on its significance.[451]

In 1915, Einstein published his main paper on general relativity. It set out a revolutionary concept that at first said nothing of expansion. In 1917, he launched modern cosmology by applying its equations to

---

a See chapter 67 at page 348.
b See chapter 21 at page 126.
c We will remedy this deficiency, starting in the next chapter.

the universe. He added an arbitrary constant[a] to stop it from collapsing under its own gravity. (This was the origin of what came to be known as dark energy.[b])

In 1922, physicist Aleksandr Friedmann solved the equations of general relativity on the assumption that the universe was filled with some sort of fluid. He found many solutions, showing the universe could be expanding or contracting.[452] Einstein sharply criticized this paper, saying,

> [T]he work cited contains a result containing a non-stationary world which seems suspect to me.[453]

He soon withdrew his criticism, saying,

> [I] am convinced that Mr. Friedmann's results are both correct and clarifying. They show that in addition to the static solutions to the field equations there are time varying solutions.[454]

In 1924, five years before Hubble's famous (more extensive but less accurate) measurements, astronomer Knut Lundmark made the first recorded observations of the universe expanding.[455]

In 1927, Lemaître published his solution to the general relativity equations and a new equation that would soon come to be known as Hubble's law. It would have all parts of the larger universe moving away from each other at a speed proportional to the distance between them. He used Slipher's data to estimate the (soon to be known as) Hubble constant. He published in a relatively obscure journal[456] and his paper went essentially unnoticed.

Four years later, an English translation appeared, minus the paragraphs on Hubble's law and the constant.[457] The deletion fueled "a passionate debate"[458] about priority to the discovery.[459] It turned out that Lemaître himself excluded the missing text.[460]

His original paper showed a simple but then startling picture. Historian of science John Farrell said,

---

[a] This is sometimes called an antigravity term. It became the source of much confusion, which we will sort out later; see chapter 29 at page 165.
[b] See chapter 67 at page 348.

> Lemaître showed that . . . the radius [of the universe] *would not be constant,* as was the case in both Einstein's and de Sitter's . . . solutions. Indeed, Lemaître showed that the radius would be a time-increasing function, that distances between all points in space would increase.[461]

It's as though, on a dance floor, you find your neighbors are all moving away. Maybe you have a BO problem? But then you see their neighbors too have moved away from them; indeed, the whole dance floor is bigger. Clearly, this is not your problem.

Yet, one way or another, all those sources pointing to expansion did not make much ontological impression.

In 1929, astronomer Edwin Hubble—who had access to the new Hooker telescope—published his findings on motions of forty-six galaxies. They showed the more distant galaxies were moving away faster. Using distances measured for twenty-four of them, he derived the first (very approximate) value for the expansion rate known as the *Hubble constant.*[462]

Since then, many measurements using various techniques[463] based on various assumptions have shown discrepancies, now known as *the Hubble tension.* They led to concern that the "constant" changes over time.[464] Recent James Webb Space Telescope data [465] have intensified the debate. Of the various results, astrophysicist and team leader Wendy Freedman said,

> [I]f it led to a newer, fundamental understanding that improves our knowledge of these things that really remain mysteries at the moment, it could be profound.[466]

We will come to understand this problem is a simple one; the astrophysicists are measuring a nonexistent property of the universe.[a]

Meanwhile, this seminal idea has become crystal clear: Somehow, all over the universe, something is making new space. We need to understand how this is happening.[b]

---

a See chapter 122 at page 590
b See chapter 69 at page 357.

Physicists need to think and speak about *new space* in forthright terms. A good first step might be, start by understanding space is something.[a]

We need to discern a simple and central message for our quest: What is growing is that something we call *space*.

---

[a] See chapter 10 at page 63.

19
TUNNEL VISION

> In more conventional quantum mechanical terms, we would say that the Universe is the result of a quantum mechanical tunnelling process, where it must be interpreted as having tunnelled from nothing at all.
>
> John Barrow (2007)[467]

Quantum tunneling is a weird way for something to show up without being able to get there. For us this is a seminal idea.

In the new ontology, it is a way to get an entire universe—not from nothing, as Barrow proposed (above)—but from as near to nothing as is physically possible.[a]

Lemaître counted back to one quantum in the beginning.[b] What was that first quantum? When we connect this question with what we observe the universe having more of, it suggests a simple answer: It was a quantum of space.

> Things really can go from A to B without being in between.

We will assume the universe's beginning was "the result of a quantum mechanical process," where—less onerously, one might think, than Barrow's "nothing at all" tunneling into a universe—one space quantum tunnels into two.[c] We can conceive this as the single, simple law that gave rise to our cosmos step by step; I will denote it by the shorthand, $1 \rightarrow 2$.[d]

So, let's take a closer look at *quantum tunneling*. Physicist Paul Sutter described it this way:

---
a See chapter 42 at page 236.
b See chapter 17 at page 105.
c This seems the simplest possible assumption.
d The concept of a 3-D universe that changes in such steps has long been mooted on theoretical grounds; see, e.g., G. 't Hooft, "The Fundamental Nature of Space and Time," note 354 at his page 13.

Put a particle in a box. According to classical physics (and common sense), that particle should stay in that box forever. But under quantum mechanics, that particle can simply be outside the box the next time you look.[468]

Barrow explained it in more erudite terms:

[Q]uantum tunneling processes, which are familiar to physicists and routinely observed, correspond to transitions which do not have a classical path.[a]

You may not think much of his explanation. Yet Barrow was a great explainer.[469] His problem was, for tunneling, there *is* no explanation.

Classically (that is, without quantum queerness) it just cannot happen. How it *does* happen is both weird and simple: The particle is said to *tunnel*; but there *is* no tunnel, there is just a *barrier* (such as a gap or insulator) that it cannot cross.[470] So it ceases being on the one side of the barrier and *presto!*, it is on the other.[471] Theory and measurements suggest its move is a faster-than-light phenomenon.[472] On the other hand, conventional concepts of quantum theory say it was on both sides all along and had no need to hurry.[473]

Quantum tunneling first came to light in studies of how a particle could exit from an atomic nucleus in which it's trapped, a process called *radioactive decay*.[474] Physicist Friedrich Hund brought it to the world's attention in the nineteen twenties.[475] For example, a radium nucleus can emit an alpha particle (the nucleus of a helium atom— two protons and two neutrons) through tunneling, although the particle does not even exist as such inside that nucleus. Even if it did, it could not get out, at least not classically, because the bunch of them[b] that's needed for the alpha particle can't cross the barrier that keeps protons and neutrons in the nucleus.[476]

The phenomenon seems strange to most but is the bread and butter of all those who work in tunnel-physics silos.[477] In an odd example, Hawking invoked tunneling to posit how a pair of black holes could arise in "empty" space.[478]

---

a See note 467.
b The two protons and two neutrons.

In the nineteen sixties, I was using quantum tunneling to study superconductivity[a] that is based on particle pairing.[b] It was then still a little new and nifty.[479] Even today it continues to be cutting-edge, having, for example, been proposed as a basis for dark energy,[480] and more recently been implicated in a mechanism for high-temperature superconductivity.[481]

For things we can see, or for thick barriers, the likelihood a thing will tunnel is near zero; for all practical purposes it never happens. But if a thing is small enough and the barrier is thin enough the likelihood is large enough that it will happen often.[482] Indeed, it happens in each of us billions of times a second and our bodies couldn't function if it didn't.[483] It is found as far afield as cancer treatment.[484] Simply put, it is a way for tiny somethings to do things ordinary physics and our personal ontologies say cannot happen.

The tunnel probability increases exponentially with small decreases in the thickness of the barrier.[c] In 1986, physicists Heinrich

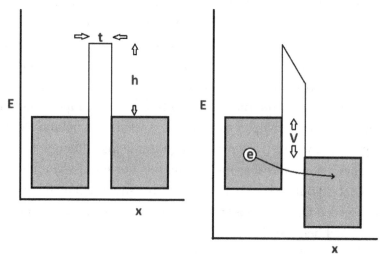

Fig. 12. An electron appears on the other side of a barrier it cannot cross

---

a A state of zero electrical resistance and complete magnetic field exclusion in some materials below a critical temperature.
b See "Theory of Superconductivity," note 1614.
c For example, the probability of an electron tunneling through a barrier increases by a factor of more than 10^{32} as barrier thickness is reduced from 5 nm to 1 nm; see note 482.

Röhrer and Gerd Binnig shared the Nobel Prize in physics for exploiting this to make the scanning tunneling microscope.[485]

In 1973, physicist Ivar Giaever shared a Nobel Prize for his work on using electron tunneling to study superconductors.[486]

When I was making tunnel junctions, with an insulating barrier of thickness t and height h between two superconducting metal electrodes (fig. 12) a one-molecule-thinner insulator could make maybe a hundred-fold more electrons tunnel. This meant if the insulator was uneven—which it always was—almost all the tunnel current came from thin spots.[487]

I point this out because we will soon see the universe may have a way to thin its "barriers," and then to thicken them. Look at the way that barrier shape changes as the energy difference V across the barrier increases. It could become very thin if V became big and so the right side dropped way down.

We won't get to understand how quantum tunneling works until we understand the quantum world. But for our purposes here, it is enough for us to understand that, despite seeming crazy, quantum tunneling of many kinds *does* happen.

Soon we will see how Hund's strange way for tiny things to do what they have no way to do gives us a key to how our world was able to begin and how it works today. With the seminal idea of quantum tunneling in aid, we will see how that simple law of nature, $1 \rightarrow 2$, could give rise to the entire universe and all the laws of physics.[a]

---

a See chapters 40, 53, and 97 (at pages 222, 291, and 478, respectively).

20
LEARNING CURVE

> Unfortunately, we don't have the slightest idea about the overall curvature of our universe on a large scale.
>
> Richard Feynman et al. (1965)[488]

In 1915 Einstein's theory of general relativity revealed space to be something (rather than a void) and showed mass causes it to be curved.[a] The whole universe is curved and local masses create local curvatures.[b] On the scale of the universe's overall curvature, local curves are tiny dimples on its giant orange.

	3-D space can be curved.

Even for a star, it's not a big effect. For example, philosopher of science John Norton calculated that,

> [A]ccording to general relativity, for each mile that we come closer to the sun, the circle [at Earth's orbit] does not lose 2π miles in circumference; it loses only $(0.99999999) \times 2\pi$ miles.[489]

Though Einstein's theory showed the whole space of the universe was curved in principle it did not specify the curvature. As of 1965, Feynman's statement was true: We did not know how curved the entire universe's space is or even if its curvature is positive or negative or zero.[490]

Today we do know,[491] thanks to Planck-satellite data, the curvature is positive, meaning 3-D space is, as Einstein thought,[c] closed somewhat like the 2-D surface of a sphere. The curvature is very small, but it's non-zero,[d] a result that is now causing consternation.[492]

---

a See chapters 10 and 12 at pages 63 and 74.
b In principle, each mass creates curvature throughout all space; in practice the effect decreases with the square of distance and so, far off, becomes very weak.
c See chapter 22 at page 133.
d With high probability.

Now we know this curvature exists, maybe we should try to understand it. Some find it hard to think of space as curved. In part this is because we are accustomed to think of it as empty; as in, how can nothing be curved?

We also think of curving as a property of 1-D lines or 2-D surfaces, not 3-D spaces. But it is a property of manifolds with any number of dimensions.

And we think of such curvatures as seen in an additional dimension. We are accustomed to see a curved 1-D line drawn on a 2-D surface or a curved 2-D panel on a 3-D car.

Thus, the conventional way to envision the *curvature of space* is to ditch one of its dimensions. Imagine yourself as a 2-D person living *in* the 2-D surface of the Earth. You see left and right around you in that surface irrespective of its curving. You have no access to the third dimension; for all you know it does not exist.

Yet, imprisoned in your 2-D world, you could determine that it is curved and you could measure its curvature. One way would be to measure triangles. The angles of a tiny triangle will add to 180°. If your world were a flat space the same would be true for all triangles, no matter how big (fig. 13).

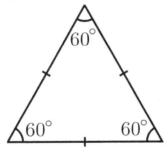

Fig. 13. Triangles of any size in flat spaces have angles that sum to 180°

But, as you measure larger triangles in your curved world, you start to find their angles adding to a bit more than 180°. This tells you that the 2-D world you live in is curved and its curvature is positive. The same approach works in our 3-D world. But the measurements must be very precise with very big triangles to detect a small deviation from perfect flatness.

The next thing to know is that the surface of a smaller sphere has larger curvature. Inside the curved 2-D world we could measure its curvature with our triangle technique. For example, an equal-sided triangle that includes an eighth of the entire world will have angles that total 270° (See fig. 14). The radius of curvature of the 2-D space can then be calculated: It is twice the length of this triangle's side divided by π (or 3.14159 . . .).

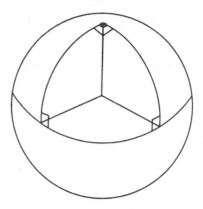

Fig. 14. A triangle in a curved space with angles that sum to 270°

Likewise in our 3-D universe we can in principle measure the curvature of its space. But we find the angles of our largest triangles add to 180° within our best precision. Thus, if our space is curved, its curvature is small. And it's difficult to draw a universe-scale triangle, so physics found another way.[493]

The latest measurements confound those physicists who thought the curvature would prove to be exactly zero. They had settled into thinking of the universe as flat, like the space assumed by Euclid, and the space used by special relativity, and the space predicted by cosmic inflation theories.[a] They all proved to be wrong: Our universe's space *is* curved. This is a seminal idea for the new ontology.

Superimposed upon that overall curvature, space is curved locally by local masses—galaxies, stars, planets, moons, and mountains, even ice sheets[494]—as described by the equations of general relativity.

---

a See chapter 21 at page 126.

These days, the greatest curvature is to be found in a *black hole*. The existence of black holes, predicted in 1784 by astronomer John Michell,[495] has been confirmed by many observations.[496]

Relativity's black holes are singularities in which huge masses are compressed by the extreme curvature of their own gravity so they collapse into a point of zero size.[497] If space is continuous, its curvature is infinite at that central point.

But, if space is granular, its curvature in a black hole is always finite though extremely large. If, as our new view asserts, space is made of Planck-sized quanta and the matter in it is just twists in links between them,[a] we can ask how big the center of an average black hole might be if its collapse was stopped when every link was twisted.

Let's follow two neutron stars as they collide to make an average black hole. (This space-quaking event has been observed.[498]) Let's give each of the neutron stars—so-called because their gravity has crushed most of their protons and electrons together to form neutrons—a typical mass of three times our Sun's.[499]

For simplicity assume the neutrons all fall into the black hole. That's more than seven billion-trillion-trillion-trillion-trillion (7×10^{57}) neutrons crammed together as tightly as each one's eighteen twisted links allow. This works out to fill a sphere about 7×10^{-16} m across, or less than one-half the diameter of a single neutron.

All its links are twisted so it *cannot* collapse further, to a point. As it can't get to zero size its gravity cannot be infinite. But with the mass of six Suns packed into that tiny size, the curvature of the adjacent space must rival that of the earliest moments of the universe.[b]

Space curvature—large local and small universal—will turn out to have profound consequences.[c]

---

a See chapter 41 at page 228.
b See chapter 42 at page 236.
c See chapters 22 and 68 at pages 126 and 351.

21
SAVING INFLATION

Understanding inflation requires breakthroughs in quantum physics and quantum gravity.

High Energy Physics Advisory Panel (2004)[500]

Lemaître deduced the entire universe began with exponential 1-2-4-8-16-... fracture of an initial "quantum."[a] In different form—with replication rather than a fracture—and with a clearer kind of quantum—space—it's another seminal idea. We can find it—hardly hidden—in more recent physics.

Recall the observation that the universe is still expanding now, though more sedately, tells us *something* has been making new space all along.[b]

Do we need two ways to make these two new-space things happen? Or could that selfsame *something* that's at work today have made new space far faster when the universe had just begun?

An early burst of exponential growth is an idea that could solve several seemingly intractable problems. So it has a name, cosmic (or cosmological) *inflation*.[501]

At first inflation seemed to be a great idea. But not for long. Math remade it into a collection of concepts and they turned out to be insupportable.

That doesn't mean that its proponents changed their minds (although a few did; see below). Maybe more would have if they knew what writer John Updike said of it,

| Inflation has mad math. We need its idea.

> The whole idea of inflationary expansion seems sort of put forward on a smile and a shoeshine. Granted it

---
a See chapter 17 at page 105.
b See chapter 18 at page 112.

solves a number of cosmological problems that were embarrassing.[502]

And here is Sutter's take on it:

> The universe goes whoosh, and inflation does its thing. But then, the inflaton[a] (somehow) changes and reduces its vacuum energy, shutting off the inflationary epoch. I'm being kind of vague here, because the physics are, well, kind of vague.[503]

And Wilczek said,

> [A] good model of inflation might enable us to test the basic idea more rigorously. . . . As yet there is no such model. There's a big opportunity for discovery here.[504]

The new ontology says Wilczek's right about the opportunity. To get its benefit, the first thing we should do is separate the seminal idea of exponential expansion from its problematic math and all the bells and whistles that went with it.

The math had the idea jumping through some arbitrary hoops. First a delay, an extremely tiny fraction of a second. Then the universe grew exponentially, far faster than the speed of light. It did this for another very tiny fraction of a second, then, for some unknown reason, its exponential growth came to an end. Fig. 15 shows one of many versions of this notion (note the logarithmic scales for the universe's size a and time t).[505]

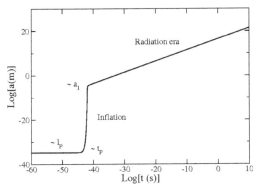

Fig. 15. Cosmological inflation vastly expanding the universe in an instant

---

a The *inflaton* is an assumed field that is supposed to have driven cosmological inflation.

With masterly understatement, Wilczek said,

> ... it would be good to have more specific ideas about how it happened.[506]

And physicist Douglas Scott said,

> [O]ne of the problems with assessing the merits of inflation is that there isn't a good alternative.[507]

Well, maybe there *is* a good alternative. Let's take a closer look at the original idea and some "more specific ideas about how it"—or something very like it—might have "happened."

When physicist Alan Guth invented cosmological inflation in 1979, he was trying to solve a problem.[508] (The problem he had in mind does not matter here; it's kind of technical and may not really *be* a problem.[509]) His idea was the very, very, very early universe had a brief burst of exponential growth.

Then he realized this same idea might solve two huge problems.[510]

His field being physics, there was no way his idea would get anywhere without it being wrapped in math. The math-wrapping turned a simple idea into an outlandish crowd of cosmologic fancies. By 2006, Gribbin could write,

> There is now a bewildering variety of different versions of inflation to choose from.[511]

And Peebles said more recently (and more reservedly) it was . . .

> . . . supported by little in the way of predictions that were not already being discussed for other reasons.[512]

What all those versions of inflation have in common is a vision of rapid expansion in, not the first, but—so to speak—the second instant of the universe. The time frame of these "instants" is insanely short: The second instant must begin, *not* at the beginning, but about one-quadrillion-quadrillion-quadrillionth of a second later. And it must stop about one-hundred-million-quadrillion-quadrillionth of a second after that!

While it is inflating, the universe grows faster than the speed of light. Vastly faster! Up to a gazillion times faster. (Yes, I realize, nobody knows what a gazillion is. But it is a *big* number.) This crazy stuff is driven by peculiarities we must assume of the already problematic vacuum energy.[a]

Why did it start? Why did it stop? Nobody knows. Physicists Andrei Linde and Paul Steinhardt invented new "laws of physics" to support its math. There was and is no evidence for those new laws.

So much—you might think—for that idea. Physics will swiftly ditch it. But no; Sutter's five-year-old appraisal is still good today:

> In the decades since Guth's initial, tentative proposal, the concept of inflation has remained frustratingly mysterious, but it still stands as our leading theory of what went down when our universe was young and exotic.[513]

There are good reasons why so many physicists embrace inflation. It seemingly can solve two hitherto unfathomable problems of the standard model. Let's take a peek.

One is known as the *horizon problem*. Look through a telescope that senses microwaves. From almost fourteen billion *light-years*[b] away, you'll see long-wavelength light, very bright but very cold. It is what's left of the *Big Flash*,[c] emitted a few hundred thousand years after the Big Bang; it is now stretched due to space expanding as it traveled to your telescope.

The stretching cooled it from its then more than 3,000 degrees. Its temperature is now 2.7255 degrees above absolute zero. (Yes, it is that precise!)

Now look in the opposite direction. You see another part of the universe, one that has had no connection with the other since the universe began—not even by exchanging radiation. Quantum fluctuations in the first instants when the universe was tiny should have created large temperature differences we should see today. Yet, everywhere you look, you see the same long-wavelength light with the same 2.7255-degree temperature in locations separated by billions of light-years.[514]

---

a See chapter 89 at page 444.
b A light-year is the distance light travels in a year, nearly six trillion miles or ten trillion kilometers.
c More usually called the cosmic microwave background radiation or CMB; see chapter 17 at page 105.

Your body has an active temperature-control system but can't come close to matching that kind of control across an inch of tissue. What could create the universe's strange degree of sameness? Inflation—so it's said—must have stretched out the early variations.[a]

The second serious problem is called the *flatness problem*. In the standard model's equations, the universe's underlying geometry is described by a number that reflects the universe's density. It is less than one if space has a negative curvature and more than one if it has a positive curvature.[515] The curvature could be either way; the number could have any value, large or small; the standard model doesn't say.[b] If the number is exactly one this would mean the world's geometry is "flat" (as Euclid assumed without even thinking); but the standard model gives no reason why it would be.

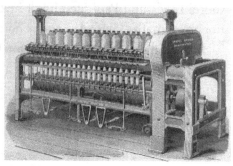

Fig. 16. Ring spinner evens yarn by stretching it

We will see the idea of this number is based on a fallacy,[c] but we are spending more than a billion dollars to find out if it is exactly one.[516] It's as if those wedded to witch stories searched for broomstick traces in the sky.

From the nineteen sixties on, measurements began delivering an unsettling message: That number seems to be strangely close to one. In 2000, cosmologist Amber Miller measured the curvature as "close to flat."[517] By 2013, WMAP satellite data showed the number to be one within less than half a percent.[518]

This began to look like more than just an odd coincidence. By 1994, Barrow would mildly say,

> Cosmologists regard the fact we are so close to this critical divide as a peculiar property of the universe which requires an explanation.[519]

---

a Like, for example, stretching pizza dough in 2-D, or evening out the thickness of wool slubbing on a ring-spinner in 1-D (fig. 16).
b See chapter 20 at page 122.
c See chapter 67 at page 348.

The coincidence is way worse than it looks. This number has been likened to a cosmological *knife edge* on which the universe is balanced in between collapsing under its own gravity or expanding endlessly because its gravity is not quite strong enough. The standard model tells us that, no matter what the number is, it soon falls off the knife edge one way or the other.

If it is within 0.4 percent of balance now, how good must that balance have been nearly fourteen billion years ago? The equations tell us it must have been balanced to a precision better than one part in a trillion-trillion-trillion-trillion-trillion. As physicist Edward Wollack said, this is "an unbelievable coincidence."[520]

Again, cosmological inflation fans say that (if their math could make it work) it would create an extremely close-to-flat universe, so that problem could go away.

Thus inflation (or some version of it) seems to neatly solve the standard model's two worst problems.[521]

But this brings us to another problem; the theory doesn't match the observations. It's so bad, one of its authors threw it under the bus: In 2011, Steinhardt wrote the article "The Inflation Debate: Is the theory at the heart of modern cosmology deeply flawed?"[522] His answer, *Yes.* And with two colleagues, in 2013, he wrote more of the same.[523] His main point was that recent observations do not support *any* of the many versions of inflation.

If we can be honest about the problems that beset cosmic inflation, no matter how handy it may seem to be, maybe we should think about a new approach: Though the standard model's equations are useful like other approximations (such as Newton's), maybe like them they don't have much to say about reality. Indeed, given that the model still has no theory or even loose description for 95 percent of the universe's contents—there is little reason why they should.[a]

Let's see if we can keep the exponential baby while we drain the murky mathematics water out.

---

a This is the total of the mysterious dark energy and dark matter; see chapters 67 and 68 (at pages 348 and 357, respectively).

The central concept—*exponential expansion* is the way that physicists describe it—is all we need to solve those pesky problems. But, if we are to solve them without inflation's hokey math, we will need a way to make it happen and a way to get it started and a way to make it stop.[a]

Almost all cosmologists think something like inflation must have happened. Its faster-than-light rate of expansion is worth saving because, on its own, it is a seminal idea.

---

[a] We will find them all in chapters 53 and 54 at pages 291 and 294.

22
EINSTEIN'S UNIVERSE

> The surface of the Earth as a whole has no boundary;
> why should it not be the same for the whole of space?
>
> Georges Lemaître (1950)[524]

The idea space is *infinite* has long held sway. But some, like Lemaître (above) and Einstein (below), thought about it carefully and arrived at the opposite conclusion.

That space is *finite* seems also to have been assumed or asserted since time out of mind.[525] To the inquiring mind it raises awkward questions, like *Where is the edge?* And *What lies outside?* These questions lead us to the seminal idea of a closed universe and, once again, it will be Einstein who will take us by the hand.

> Space is a finite sphere that has no outside.

Plato and Aristotle claimed the universe is finite but they just assumed it. Historian of philosophy David Furley said,

> For Plato and for Aristotle, there is no problem here. We have a cosmos, in their theory, and it is an organized whole with a determinate boundary.[526]

On the other hand, mathematician and philosopher Archytas reasoned that it must be infinite. In a thought experiment, he imagined himself at the universe's edge. He stuck his hand through it, establishing a further edge, from which he could repeat this trick as often as he liked (see fig. 17).[527]

Fig. 17. Outside the universe; the Flammarion engraving

It could be a perilous conviction. In the late fifteen hundreds, mathematician and philosopher Giordano Bruno thought a lot about the cosmos and said space is infinite and for his pains he was burned at the stake.[528]

The difficulty of imagining a boundary to space does give the infinite-universe idea an edge (if you'll forgive the pun). Around 1666, Newton gave it an Archytan lift, saying,

> Space extends infinitely in all directions: we can't imagine any limit anywhere without at the same time understanding that there's space beyond it.[529]

He needed space to be infinite. As his successor,[a] Hawking, said,

> Newton realized that, according to his theory of gravity the stars should attract each other, so it seemed they could not remain essentially motionless. Would they not all fall together at some point? . . . [H]e reasoned that if . . . there were an infinite number of stars, distributed more or less uniformly over infinite space, this would not happen, because there would not be any central point for them to fall to.[530]

Norton told the sorry tale of tinkering with Newton's physics—his laws of motion, his law of gravitation, and his infinite universe—all in aid of propping up the universe's math against the weight of gravity.[531]

Two centuries after Newton, Riemann included key insights in his thesis.[b] One was the idea (in less than ringing language) that our 3-D universe might be finite and yet have no edges:

> That space is an unbounded three-fold manifold is an assumption which is developed by every conception of the outer world. . . . But its infinite extent by no means follows from this.[532]

He was ahead of his time.

In 1917, Einstein wrote a paper that launched a new age of cosmology.[533] He reasoned that both assuming space is infinite and

---

a See chapter 17 at page 105.
b See also chapter 12 at page 74.

assuming space has an edge must lead to unacceptable difficulties.[534] He concluded the universe is finite and, with positive space curvature, it has no boundary.[a]

Historian of science David Topper put it this way,

> Einstein . . . deduces that the universal summation [of space and matter] bends or curves the entire universe into itself so that the resulting space is of finite extension.[535]

Cosmology calls this a *closed universe.*

In 1921, Einstein addressed the problem of the reference frame for acceleration.[b] He saw this too as pointing to a finite closed universe:

> In my opinion the general theory of relativity can only solve this problem satisfactorily if it regards the world as spatially self-enclosed. The mathematical results of the theory force one to this view.[536]

The new ontology says Riemann, Einstein, and Lemaître were right.

That is, the universe has a definite size, yet one can travel in it endlessly and never find an edge. This may seem to make no sense. It's hard to visualize. Our minds are not used to thinking of our space as being curved because its curvature extends far beyond our field of vision.

It's much easier to picture a 2-D example. The classic illustration is the surface of a sphere; for instance, the Earth. It's finite.[c] Yet we travel over it in all directions without ever coming to an end.

Or think of the Globe of Fate (fig. 18) where riding any which way never meets an edge.

Recent observations bolster Einstein's view. With high confidence, new data show a finite closed universe with a small

Fig. 18. The Globe of Fate

---

a That is, it is a hypersphere.
b In 1916, Einstein had attributed to this problem the impulse behind the general theory: Unlike velocity, acceleration requires an absolute frame of reference. Newton posited absolute space. Disliking this as arbitrary, physicist Ernst Mach had proposed instead that all the mass in the universe provided the reference frame.
c About five hundred million square kilometers.

curvature that's positive.[537] It's spherical[538] with its three dimensions looped back on themselves.[a]

Of this, physicist Paul Davies said (tongue in cheek but right in principle),

> [W]ith a powerful enough telescope, you could look right around the Einstein universe and see the back of your own head.[539]

This understanding of the shape of our universe is a seminal idea and it's a fundamental concept in the new ontology. We will soon see how it came to be closed.[b]

We can understand its finite size another way. If (as both evidence and new ontology are saying[c]) it began with finite size a finite time ago and grew with finite speed, it must be finite.

And so, the new ontology can build upon a seminal idea—a finite universe with space that has no edge and that, over vast distances, has everywhere a tiny curvature[d] analogous to that of a big sphere.[540]

And this too: Having no outside makes the universe the perfect room with no view.[e]

Oddly little noticed was that Einstein did not test his general-relativity-based universe against observation.[541] Going forward, we will find there was increasing evidence it is in almost all other respects entirely wrong.

---

a As they must therefore have been from the beginning; see chapter 42 at page 236.
b In chapter 42 at page 236.
c See chapters 17 and 42 at pages 105 and 236.
d See chapter 20 at page 122.
e With apologies to E. M. Forster; and see chapter 24 at page 141.

23
ALLES IN ORDNUNG

> There is an extraordinary degree of precision in the way that the universe started, in the Big Bang, and this presents what is undoubtedly a profound puzzle.
>
> Roger Penrose (2004)[542]

One way to express the second law of thermodynamics is the universe gets more disordered every moment, every day.[543] So, looking back, we see the early universe must have had a very high degree of order (known to physics as *low entropy*.[544]) How it could have been so highly ordered baffles physicists, including Penrose. Especially Penrose.[a]

The Big Bang portrays the universe's first moments as a roiling mess. How then could its disorder ever get so low? Seeing this too as a seminal idea, our growing new ontology can offer baffled physicists an answer. And, as I have undertaken, it will be supremely simple.

How did the universe begin so cleanly?

That the world began in some kind of state of order is an ancient concept. Bringing order from disorder is the theme of more than one creation myth.[545] But myths (even the myth we call the standard model) don't offer a real way to make the state of order physics needs to explain puzzling observations.

Surveying the scene, Lineweaver said, dryly,

> Trying to understand the low initial entropy of the universe is an important unresolved issue of cosmology.[546]

Barrow summed it up in blunter terms:

> [T]he entropy level at the beginning of the expansion of the Universe must have been staggeringly small, which

---

a See chapter 1 at page 15.

implies that the initial conditions were very special indeed.[547]

Penrose was even blunter. He called it "an absurdly low entropy."[548] He depicted the Creator poking at a box whose tiny size represented ways for the world to be the way it is. He estimated the odds of success were one in ten raised to the power 10^{123}. Else the universe would be a wasteland. If the digits in his number were atom-sized they would stretch to the edge of the visible universe and back more than a trillion-trillion-trillion times. In effect Penrose was telling us not even in your dreams.

All of which is to say the universe was highly orderly in the beginning but physics says there's no possibility for it to get to be that way. As problems go in physics-land (and these days there are many), this should get its fair share of hair-tearing.

It does. For example, Carroll said,

> The state of the early universe was *not* chosen randomly among all possible states. Everyone in the world who has thought about the problem agrees with that. What they don't agree on is *why* the early universe was so special—what is the mechanism that put it in that state?[549]

Or, as Rovelli blandly put it,

> [W]hy do the phenomena that we observe around us in the cosmos begin in a state of lower entropy in the first place?[550]

Rovelli tried to offer an "idea that provides a possible answer."[551] Kudos for a gallant try but it's implausible, to say the least. His story came to this. He posited an "extraordinarily vast universe" with *almost* uniformly high entropy that started low in our "small" part of it—*small* here meaning all we see. That is, a highly ordered chunk of universe within a bigger messy one, like an always perfectly made bed inside some stereotypical teenager's messy room.

What vexed Rovelli most was not the entropy itself, but that it always shoots time's arrow forward in defiance of equations that work equally both ways.[a] Maybe—he imagined—it works both ways in parts of the universe we cannot see. In other words, his explanation was we live in an extremely weird part of a normal universe. To be fair, he did end with,

> I'm not sure if we are dealing with a plausible story, but I do not know of any better ones. The alternative is to accept as a given of observation the fact that entropy was low at the beginning of the universe.[552]

Greene made his own contribution to what he called "the texture of reality." He said,

> The egg splatters rather than unsplatters because it is carrying forward the drive toward higher entropy that was initiated by the extraordinarily low entropy state with which the universe began.[553]

This says, in effect, the universe is an entropy machine. The new ontology agrees.[b]

Greene put his finger on the problem:

> A *splattering egg tells us something deep about the big bang*. It tells us that the big bang gave rise to an extraordinarily ordered nascent universe.[554]

How does a beginning that's supposedly a vast "explosion" make for extraordinary order? The new ontology will say, by getting going without the explosion.[c]

Rovelli was right about the alternative so we should embrace it: Accept the low initial entropy as fact! In the new ontology the early universe *began* that way. It started as a single quantum replicating. Recall that, in 1931, Lemaître said so[d] in a much neglected paper in a leading journal.[555] Of it, Farrell said,

> What's remarkable about this *Nature* letter is that—apart from discussing the idea of a temporal beginning of the

---

a See chapter 60 at page 320.
b See chapter 59 at page 316.
c It says there was a *Big Fizz*; see chapters 42 and 53 at pages 236 and 291.
d See chapter 17 at page 105.

> cosmos—it marks the first time that a physicist directly tied the notion of the origin of the cosmos to quantum processes.[556]

Unfortunately, like me, Lemaître could not quite break free from conservation of mass-energy (a mistake, as we will see[a]), and so, Farrell said, he needed . . .

> . . . all the energy of the universe packed in . . . a unique quantum.[557]

This least likely aspect of his story was soon backed by the emerging standard model of cosmology—except that it forgot the quantum.

So, Penrose gets the credit for insisting on the seminal idea that the early universe's state of order needs an explanation. (He then tried to explain it with a far-fetched cosmogony where our universe began with a Big Bounce after the Big Crunch that ended the universe before it, which began with a Big Bounce after the Big Crunch . . . et cetera.[558] To me that's reminiscent of the oft-told tale of a worldview that ends with, "It's turtles all the way down."[b] That is, he was getting close and then he blew it.)

Lemaître's claim to fame in astrophysics and cosmology was as the "father of the Big Bang," which was little more than applied general relativity. His stroke of genius, the single quantum, received scant attention.[559] Was he aware (but did not mention) that a universe consisting of a single quantum *is* a state of perfect order?[c]

---

a See chapter 85 at page 431.
b It is an infinite regress. Hawking tells a version of the turtle tale; see his *A Brief History of Time*, note 530.
c See further, chapter 59 at page 316.

24
THE INSIDE STORY

> In general, an entire spacetime ... can only be seen by an observer ... outside the universe. This is unphysical.
>
> Fotini Markopoulou (1999)[560]

For the cosmological philosopher or physicist, it's tempting to imagine one can view the finite universe from outside.[a] Of course we can't, because it *has* no outside.[b] We are stuck inside. This is a seminal idea.

Smolin, Markopoulou's colleague, brought this perspective:

> How can we, as observers who live inside the universe, construct a complete and objective description of it?[561]

As he also said,

> The problem is that in all the usual interpretations of quantum theory the observer is assumed to be outside the system. That cannot be so in cosmology.[562]

Doing real quantum theory inside the universe therefore runs into a fundamental problem. In quantum cosmology, the quantum system *is* the universe. Quantum theory *needs* an observer,[563] a classical (non-quantum) part of every experiment that *must* lie outside the quantum system.[c]

As science historian Louisa Gilder put it,

> Classical systems are paradoxically necessary to describe the quantum systems of which they are made.[564]

| Physics needs to listen up: We are inside the universe.

---

a For example, my imagined narrator in *Time One* imagines his imaginary detective gaining insights while outside the universe; note 16.
b See chapter 22 at page 133.
c The observer need not be a person; it can, for example, be a measuring device.

Bell explained it this way:

> The problem is that quantum mechanics is fundamentally about "observations." It necessarily divides the world into two parts, a part which is observed and a part which does the observing.[565]

This is no mere formality; it is essential. As Smolin pointed out,

> The information that we, as observers, have about a quantum system is coded into a construction that is called the *quantum state of the system*. . . . The quantum state is . . . a property of the boundary or interface that separates that system from the rest of the universe, including the observer who studies it.[566]

And so, as Bell pointedly asked,

> When the "system" in question is the whole world where is the "measurer" to be found?[567]

And Smolin said of this same problem,

> This . . . is the whole point. If we do not take it into account, whatever we may do is not relevant to a real theory of cosmology.[568]

Science writer Amanda Gefter, interviewing Smolin, explained,

> The universe is kind of an impossible object. It has an inside but no outside; it's a one-sided coin. This Möbius architecture presents a unique challenge for cosmologists, who find themselves in the awkward position of being stuck inside the very system they're trying to comprehend.[569]

Recognizing this as a fundamental problem that most physicists ignore, Smolin told her,

> The idea is to try to reformulate physics in terms of these views from the inside, what it looks like from inside the universe.[570]

It's not only cosmologists who are affected by our cosmic imprisonment. It has implications for our thinking about new physics. For example, in his relatively recent book with Unger, Smolin said about *causation* (the foundation of most physics),

> We also require that a theory of the whole universe be explanatorily closed. This means that chains of explanation and causation do not point back to entities outside the universe.[571]

We can now clearly see why the category "entities outside the universe" is not just empty; it is (at least for us) nonexistent.

In sum, we need to understand the limitations of studying the universe while inside it. And that it leads to a conundrum. Baez said,

> It is as if classical logic continued to apply to us, while the mysterious rules of quantum theory apply only to the physical systems we are studying. But of course this is not true: we are part of the world being studied.[572]

Our being inside is a seminal perspective so we should ponder on its meaning and attune our minds to it. It is essential if we are to understand the nature of reality.

25
A REMOTE ENTANGLEMENT

> Of the many mysteries of modern physics, few compare to "nonlocality" in quantum physics. Nonlocality means that far away objects can influence one another instantaneously (or at least much faster than the speed of light). It is as if space and time didn't exist!
>
> Marcelo Gleiser (2018)[573]

The notion that the world works locally is deeply ingrained into our perceptions. It is also ingrained into physics. And the universe refuses to cooperate.

As Gleiser said, above, this confronts physics with a conceptual mystery: What could make nonlocal influences work? His last word on this was,

> Reality is not only stranger than we suppose. It is far stranger than we *can* suppose.[574]

Newton "knew" the world *must* be local. He was aghast his gravity appeared to reach across space to the Moon:

> That one body may act upon another at a distance thro' a Vacuum, without the Mediation of any thing else, by and through which their Action and Force may be conveyed from one to another, is to me so great an Absurdity that I believe no Man who has in philosophical Matters a competent Faculty of thinking can ever fall into it.[575]

Einstein's general relativity insight—that gravity is local curvature of space and not a force—resolved that problem of apparent nonlocality.

Only recently has it been well accepted that the universe is nonetheless nonlocal as a matter of both quantum theory and observed reality.[576] The phenom-

> Spooky action happens.

enon called *quantum entanglement* is an ongoing immediate connection between widely separated particles.[577]

Physicist Rafael Sorkin saw the relation between granularity and nonlocality:

> If this reasoning [about discreteness plus Lorentz invariance] is correct, it implies that physics at the Planck scale must be radically nonlocal.[578]

The details can be messy. Science writer Jesse Emspak summarized it this way:

> Quantum entanglement is a bizarre, counterintuitive phenomenon that explains[a] how two subatomic particles can be intimately linked to each other even if separated by billions of light-years of space.[579]

Nonlocality compels a seminal idea. There must be a link between the separated particles. But saying so does not explain it: What the link could conceivably *be* has been a total mystery, till now.[b]

And please note right away how a nonlocal universe must have spatial simultaneity;[c] or, put another way, if distant events in space can be linked instantly, there must be a timeless space.[d]

Confronted with the fact that nonlocality arose out of the quantum theory he had himself originated, Einstein said it showed there was a problem with the theory. Smolin said,

> Einstein never ceased trying to find the deeper theory beyond quantum mechanics. . . . But he failed, and the simple reason was that he never understood that the central assumption behind many of his great papers—the principle that physics is local—was wrong.[580]

Science writer Amir Aczel wrote the book on it, *Entanglement: The Greatest Mystery in Physics*. His tale follows the history of entanglement's rise from quantum theory that . . .

---

a This is a loose use of "explains"; *describes* might be a better word.
b The new ontology offers a simple explanation; see this chapter, below.
c See chapter 4 at page 34.
d This conclusion does not controvert the theories of relativity; see chapter 56 at page 301.

> ... taxes our very idea of what constitutes reality. What does "reality" mean in the context of the existence of entangled entities that act in concert even while vast distances apart?[581]

The tale's trajectory runs through the epic mental struggle between Einstein and the upstart quantum enterprise. With two colleagues (Boris Podolsky and Nathan Rosen)—the three famously now known as EPR—he authored a paper, aiming to show quantum theory must be incomplete because it says events affecting one particle in an entangled pair must instantly affect the other, even at a distance.[582] Which, they thought, was obviously nonsense. (Sadly for them, turns out it's true.)

EPR's solution was the other particle already had a real property that quantum theory did not reveal. This concept came to be called *hidden variables*.[583] It seeks to explain quantum phenomena by proposing new—and maybe unobservable—components of reality.[584]

Simply put, the quantum-theory version of entanglement was, having directly interacted[a] and later being far apart, particles behave as if they continue to be directly and *instantly* connected; this is the core of what it means to be entangled.[585] Einstein ridiculed it as "spooky action at a distance."[586]

The epic struggle was about the very nature of reality. Both views spawned a sometimes confused commentary. But, as Bell explained, the nub of the matter was,

> What [Einstein] could not accept was that an intervention at one place could *influence*, immediately, affairs at another.[587]

That is, Einstein's worldview (like almost everyone's) was *local*.[588]

In 1950, physicist Chien-Shiung Wu and student Irving Shaknov published the first definite results showing entangled photons.[589]

In 1964, Bell set his oar in the troubled waters between Einstein and the quantum theory. He wrote a groundbreaking[b] paper showing a way to distinguish between the nonlocal-quantum-mechanics

---

a A typical example would be two particles (like photons) that were created in a single event.
b Pray pardon my mixing of metaphors.

and the local-EPR views of reality.[a] It was based on the statistics of repeated measurements. In conclusion, he said,

> In a theory in which parameters are added to quantum mechanics to determine the results of individual measurements . . . there must be a mechanism whereby the setting of one measuring device can influence the reading of another instrument, however remote. Moreover, the signal involved must propagate instantaneously.[590]

Short version, if you say something else is going on, it must be able to reach far away and do it instantly.[b] He impliedly thought these requirements are impossible. The new ontology will offer—as a free bonus—conceptual means to meet them both.[c]

In 1982, the local view lost the strategic battle of a long and wordy war when Aspect and his colleagues showed that "spooky action at a distance" is exactly what's observed.[591]

Entanglement has since been demonstrated many times, even with particles separated by light-years.[592] In 2024, the last possible loopholes in the nonlocal view were closed.[593]

Thus, the universe must somehow be *nonlocal*. Einstein's ontological commitment to locality suffered a fatal wound.

Physicist, philosopher, and editor of the Einstein Papers Project John Stachel summarized,

> The work of [EPR] and subsequent tests of Bell-type inequalities show that, once two quantum systems interact (in the ordinary sense of the word), they are quantum-entangled forever; they suggest that, even if quantum mechanics is modified or replaced, some such entanglement feature will almost surely characterize the new theory.[594]

Half a century after EPR wrote their potent paper, their last escape hatches were shut.[595] Aspect shared a physics Nobel Prize for this.[596]

---

a It came to be known as Bell's theorem.
b In apparent violation of the speed of light as a universal limit; see chapter 50 at page 276.
c See this chapter, below.

Meanwhile, quantum weirdness continued to baffle brains. As science writer and former *Nature* editor Philip Ball said,

> Entanglement shows us that in the quantum world separation in space does not guarantee independence, even though no measurable interaction exists between the two places. It makes a mockery of space.[597]

Some have said it's as if there is a physical connection between distant places. A few years ago, Susskind even mooted a byproduct of the old ontology, a *wormhole*,[a] as a kind of direct link to solve the problem:

> No matter how far the [entangled objects] are separated from one another, the correlations between them will behave as if there is a short connection whose length is independent of the exterior distance. In other words the correlation functions behave as if there is a wormhole connecting the systems.[598]

And physicist and *New Scientist* editor Richard Webb said,

> But as to what it all means, no one knows. Some people think we must just accept that quantum physics explains the material world in terms we find impossible to square with our experience. . . . Others think there must be some better, more intuitive theory out there that we've yet to discover.[599]

We are more likely to discover such a better, more intuitive theory if we can embrace the seminal idea of nonlocality into our worldview. The new ontology asserts the universe's fundamental structure was nonlocal from the get-go. It shows how matter may be made of tiny *quanta* so each one of them is linked directly and instantly to all others in the universe.[b] It is an inherent "entanglement feature"[c] of how it all began.

---

[a] Also known as an *Einstein-Rosen bridge*, a wormhole is math fantasy based on certain solutions to the equations of general relativity in space that is continuous.
[b] See chapter 42 at page 236.
[c] Stachel's term; see text of note 594, above.

That is, widely separated particles *appearing* to be directly and instantly connected is not absurd: They *are* directly and instantly connected.[a]

Understanding how such a connection could give rise to the quantum entanglement phenomena will call for careful study. The first step is to understand there *is* a way to make direct distant connections.

There is more to it than Bell-type experiments: For example, in a paper titled "Everything Is Entangled," physicists Roman Buniy and Stephen Hsu concluded,

> [B]ig bang cosmology implies a high degree of entanglement of particles in the universe. In fact, a typical particle is entangled with many particles far outside our horizon.[600]

All this needs new physics, which is not our purpose here, and which we can no more foresee than Planck could figure out what his quantum would do.

---

[a] See chapter 41 at page 228; however, the physics of how such a real connection works remains unknown.

26
ONE IS ALL

> People think of these things as almost separate
> ... but they're all the same story.
> Blake Desjarlais (2021)[601]

Politician Blake Desjarlais was speaking from the worldview (*tâpisinowin* in Cree) of his people.

Indeed, indigenous peoples all over the world seem to have had this view since time immemorial: They don't see themselves as *in* the world; they see themselves as *of* the world; they see the world as one.[602]

Disaggregation is, thus, an affliction of supposedly sophisticated societies. There it takes pride of place in thought and language. First step, children learn to view things (including themselves) as distinct objects *in* a world they see as just a stage for other objects.[a]

> The universe is one thing. We need one theory of it.

I too saw the world only that way.[b] It took thirty years of stories told by patient Pimicikamak and Ojibwe elders for me to start to see the way they all saw, maybe since their birth days.

Their world stands in stark contrast with the reductionist view, the modern way of understanding it all by figuring how things fit together only after taking them apart.[603]

Of this, Smolin said,

> I believe that part of the present crisis [in fundamental physics] is ... due to our having reached the limits of what we can learn solely by breaking things into their parts.[604]

---

a See also chapter 5 at page 43.
b Mostly I still do.

It's so deeply ingrained in our worldview (and especially that of physics) it doesn't rate much wider mention. In a rare illumination, business strategist and global management consulting company alum Joe Newsum said,

> One of the unknown secrets of McKinsey is how many physicists rank as the best problem solvers within McKinsey because they are experts at abstraction, disaggregation, and deduction, having gotten a Ph.D. in abstracting some element of the Universe into all of its disaggregated parts and theorizing and deducing how all of those parts interact.[605]

We will see the indigenous worldview—that it is all one—is literally true. This idea of oneness, this essentially holistic view, will prove to be key to the understanding we are seeking. That is, it's a seminal idea.

The new ontology says the holistic view is real. The concepts of parts or objects or atoms or even subatomic particles are fiction in the sense there are no such severable items.[a] This also is the subtle message from the nonlocality of quantum theory.[b]

For physics the new view calls for one theory of one universe, yet the impulse to chop up the world lies deep within its roots, embedded in its language and equations. One consequence is that its serried ranks are drawn up behind banners of two theories.

We have an extremely successful theory about the world we see: It's called relativity. And, if you don't like that one, we have another, which is even more successful: It's called quantum theory. These two bodies of theory—each a tower of strength amid the citadel of physics—are so disparate they mostly don't talk to each other.[606]

Having two theories for one world is at least one theory too many. Among many commentators, Bohm deprecated . . .

> . . . the self-contradictory attitude of accepting the independent existence of the cosmos while one was doing relativity and, at the same time, denying it while one was

---

a See chapter 130 at page 624.
b See chapter 25 at page 144.

doing the quantum theory, even though both theories were regarded as fundamental.[607]

Rovelli said,

> The paradox resides in the fact that both theories work remarkably well . . . despite the seemingly opposite assumptions on which the two theories are founded. It is clear that something still eludes us.[608]

And Smolin said,

> There is no way we can have two theories of nature covering different phenomena, as if one had nothing to do with the other.[609]

Barely discernible behind the theories lie their two immiscible ontologies.

We have one universe. We need one way to understand it. Neither of their two worldviews will ever be up to the task. Neither explains what is real. But then, neither really tries.

It's difficult for physicists who use these theories daily in their working lives to see them as the fictions that they are. Einstein understood this:

> To him who is a discoverer in [fundamental physics], the products of his imagination appear so necessary and natural that he regards them, and would like to have them regarded by others, not as creations of thought but as given realities.[610]

The truth is, in their relation to reality, both of our theories and their ontologies are plainly wrong.

They are far too successful to be set aside. So, what we do (or experts do for us) is choose, in any circumstance, which one to use.

Einstein was ever ready to discard a good theory in favor of a better one. In his Nobel Prize lecture, way ahead of his time he criticized the contradictions of a two-theory situation, saying,

> The mind striving after unification of the theory cannot be satisfied that two fields should exist which, by their nature, are quite independent.[611]

He did not see his two initiatives as fruitful starting points:

> [T]he theory of relativity and the quantum theory . . . seem little adapted to fusion into one unified theory.[612]

Famously, he failed to solve the problem. He came (like others) to conceive it in terms of the language of fields. If he had somehow succeeded, he would have run into another problem: He looked for reality; but fields (we will find) are imaginary.[a]

Smolin too has long pursued that single theory, known as quantum gravity, that bridges between relativity and quantum theory. He saw its more holistic perspective:

> [T]he problem of quantum gravity is an aspect of a much older problem, that of how to construct a physical theory which could be a theory of an entire universe.[613]

Yet Einstein's two great successes are still holding Smolin—and other adventurous physicists—hostage. Susskind said,

> I think the reason [quantum gravity] never worked very well is because it started with a picture of two different things . . . and put them together.[614]

The underlying structure of the new ontology sets out to reify the unity of all reality. We need one way to understand one universe, a single self-consistent explanation.[615] It's difficult to designate a single source but this sure is a seminal idea.

---

[a] See chapter 84 at page 427.

27
NO NOT NOW

The "present of the universe" is meaningless.
Carlo Rovelli (2018)[616]

Rovelli was restating a conclusion from the theory of relativity. It would be fine confined within the framework of the theory.

But we are in search of the reality behind the light-speed-limited observations of relativity.[a] And we will see the "present of the universe" is more than meaningful; it *is* the universe.[b] There is nothing else. In this sense, *Now* may be the ultimate in seminal ideas.[c]

> Now is real.
> Not now is not.

You may be pleased to find you can rely upon your sense of *Now*. You can hang on to it and at the same time be at peace with modern physics. Or perhaps you did not realize that physics has for many years been saying you were wrong.[617]

Rovelli paraphrased on good authority; Einstein himself dissed *Now*:

> "Now" loses for the spatially extended world its objective meaning.[618]

That is, *Now* may be okay in your head but don't imagine it works at your toes. Simultaneity, he said, cannot survive in an extended space.

Physics got this problem because it got stuck on t, the t we see in its equations. This variable's foibles create mental indigestion, which arose from special relativity. With his 1905 paper about moving rods and clocks, Einstein threw the world of physics into a commotion: A

---

[a] Physics' acceptance of relativity's proscription of *Now* is symptomatic of its ontologic roots in positivist philosophy.
[b] In chapter 56 at page 301.
[c] But I awarded that accolade to Bilson-Thompson's tweedle; see chapter 35 at page 190.

rod becomes shorter and a clock runs slower when they are moving relative to their observer. And an observer's time is not the same as anybody else's.

Einstein didn't quite say—as Gilder said he said—*time is what we measure with a clock*.[619] But—and this is fundamental to our understanding—she got the *gist* of what he said exactly right.[620]

So, common time (that is, Einstein's time and physics' time) is *not* a property of the universe; it is a property of *clocks* and their clock-watchers.[a]

But physicists grew used to treating time as a dimension. See, for example, where Rovelli—leading modern acolyte of time—shows how he's wedded to this misconception:

> [W]e can think of spacetime as being as rigid as a table. This table has dimensions: the one that we call space, and the one along which entropy grows, called time.[621]

Understanding time begins with realizing clocks don't measure any time dimension. They just measure motion. And motion by its nature is local—it is what it is where it is.

Seizing on common time as a dimension works for physics and its formulas but leads it to assert conceptual absurdities. Perhaps the worst is the so-called *block universe*.[622] It looks odd but it is a mainstream ontological commitment in the world of physics that goes back to Einstein. It regards all of time—the past, present, and future—as extant and puts it all on an equal footing, like a coexistent block of spacetime.

Supposedly we are somehow condemned to move inexorably through this block and suffer the illusion that our *now* has special meaning.[623]

This take on time is wrong, a consequence of viewing what clocks say as if it were an actual dimension. It's an abuse of special relativity that takes a concept we devise (clock time) and regards it as a property of the whole universe.

---

a So, like quantum mechanics, relativity requires an observer outside the system that is being observed.

The past is not existent; each iteration of the 3-D space that *is* the universe is gone the instant it is replaced by the next; it was not recorded; it can't be retrieved. The future also is not existent; at each instant, the next iteration is an open question till the universe constructs it.[a]

So, we see the concept physics says is *not* real is the one that *is* real, *Now*.[b] And vice versa, everything except *Now*—all the rest of time—is *not* real.[c] And we will see how this stream of new *Nows* forms the foundation for both thought[d] and action.[e]

For philosophy, the struggle to grasp ever-errant *Now* squeezed between the lost past and the undefined future baffled those who could not follow Zeno's mental leap.[f]

In modern times, Arendt, seeking in Kafka's parable of a protagonist who battles past and future the setting where thought and action should be situated, made a good start as she said,

> [T]ime is not a continuum, a flow of uninterrupted succession; it is broken in the middle, at the point where "he" stands.[624]

Special relativity tends to get a bad rap out of all this. Its touted take on time was tainted by its built-in aspiration to be a dimension. The 1905 paper was unusual. Infeld later said,

> [A] great part of this article can be followed without advanced technical knowledge. But its full understanding requires a maturity of mind and taste that is more rare and precious than pedantic knowledge, for Einstein's paper deals with the most basic problems; it analyzes the meaning of concepts that might seem too simple to be scrutinized.[625]

And Holton said,

---

a See chapters 43 and 74 at pages 243 and 381.
b See further, chapter 56 at page 301.
c See also chapter 83 at page 420.
d See chapter 132 at page 632.
e See chapter 57 at page 307.
f See note 241.

> This paper, now among the most renowned in the history of science, actually did not deal with any of the problems of greatest concern to the physicists of the time; there is not a word about the theory of matter, the nature of the electron, or the properties of the ether. Even in the space of a few paragraphs one can perceive the novelty of his mind, and can see that the true subject matter is the unification, the simplification, the rationalization of the physical world.[626]

Physicists still struggle with its deepest issues, especially when the seeming-simple theory meets granular space at Planck scale.[627] But its conceptual undoing lies in treating time as a dimension, one that is continuous to boot.

There are several well-known special-relativity "paradoxes," like the twin who travels fast and on return is younger than her brother (fig. 53 at page 287). They become easier to understand when we realize special relativity says nothing about time as a dimension of the universe; it speaks of a clock-watcher watching moving clocks.[628]

The new ontology helps us to see a clearer picture: Relativity is right—to an extremely good approximation[a]—about clock motion (the interplay between motion *in* clocks that "tells time" and motion *of* clocks that affects what the clock-watcher sees).

Those who wisely seek their understanding in views of the world around them rather than in views of passing clocks on rockets know that *Now* is real. What they may not know is *Now* is truly timeless;[b] it is simply 3-D space.[629] The real world is the literal version of *Everything Everywhere All at Once*.[630]

This is why, far from being meaningless, *Now* is such a seminal idea. Next, we will see how physics doesn't get it, but Smolin, we will also see, comes oh so close!

---

a See part 5.
b That is, *Now* is not a time of zero length; it has no useful attribute of time, it simply is.

28
TWO-TIMING

> Time, in the sense of the continual becoming of the
> present moment, is fundamental to nature.
>
> Lee Smolin (2019)[631]

These days, physics is increasingly obsessed with time. As in, *What is it?*

For example, Smolin (above) saw time as fundamental. But, even though he gave us a description, we must ask what time he had in mind. Implied in such questions is the search for something much more fundamental than clock-watchers' views of motions in their clocks.[a]

> There is a kind of time that knows no clocks.

There are many answers out there, none of which seem to succeed in making sense.[632] The new ontology sets out to understand time in a way that is consistent with all other aspects of the universe.

Like so many other fundamental questions, this one rests on old foundations. Time (and timekeeping) was long a department of astronomy, tracking time with Sun and stars. Terrestrial clocks appeared maybe three thousand years ago.[633]

Time moved into a new zone in 1687, when Newton, needing t in his equations, took a mental leap to a new seminal idea. In his epoch-marking *Philosophiæ Natu-*

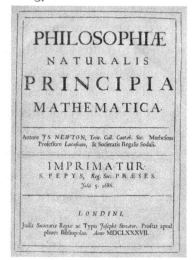

Fig. 19. Newton's *Philosophiæ Naturalis Principia Mathematica*

---

[a] Anything that changes in a systematic way can be seen to be a clock, including molecules in people.

ralis Principia Mathematica (fig. 19)[a] (in its first drafts titled *De Motu* or *On Motion*[634]), he said there are two kinds of time:

> Absolute, true, and mathematical time, in and of itself and of its own nature, without reference to anything external, flows uniformly and by another name is called duration; relative, apparent, and common time is some sensible and external (whether accurate or unequable) measure of duration by the means of motion, which is commonly used instead of true time; such as an hour, a day, a month, a year.[635]

Set forth in such plain terms,[b] this oft-quoted paragraph has been grotesquely undermined by misapplied special relativity. Newton's wisdom is now widely misconstrued.

As we weave together, thread by thread, our seminal ideas, we will see why he was ahead of his time (no pun intended) with no less than ten fundamental insights:

- That there *is* a fundamental distinction between two kinds of time;
- That *common time* (or clock time) is measured by motion;
- That there are sundry ways to measure common time;
- That *"common time"* is relative;
- That common time is a measure of true time;
- That *"true time"* is absolute;
- That true time has no measuring device;
- That true time flows;
- That true time is imperceptible;
- That true time is uniform through space.

Except—and this exception holds high consequence—true time is *not*, as we conceive of times, truly a time. The reason is, it is a property of the whole universe that has itself no manifest presence *inside* the

---

a Mathematical Principles of Natural Philosophy.
b This is Motte's 1729 translation from the original New Latin. The modern translation by Cohen and Whitman (note 213) does not differ materially in this passage.

universe where we construct our physics and must make our measurements.[a] So we will need a workaround to tap into true time.

And note he lent his weight to the assumption true time is continuous ("flows uniformly"), which will turn out to be wrong.

But now, already, thanks to Newton, our search for a new ontology achieves a sharper focus with his absolutely uniform (let's keep the name) *true time*.

Meanwhile, most (but not all) physics forcefully but wrongly insists there is no such thing.[b] For example, students are taught that the well-known experiment of the counter-traveling atomic clocks[636] demonstrates, and the success of special relativity reveals, . . .

> . . . that time doesn't work the way Newton believed it did when he wrote that "Absolute, true, and mathematical time, of itself, and from its own nature flows equably without regard to anything external."[637]

This merely reconflates the two times Newton took pains to distinguish; though widely taught, it is neither accurate history nor defensible physics.

The flying-clocks experiment was concerned with common time—what a clock says. It did not speak to Newton's absolute and true time; nor did Einstein's theory.[c]

These days, time is a beleaguered notion in the world of physics.[638] In his recent writing Smolin has assumed the mantle of its champion.[639] In the epigram above, he was writing about *causal sets*, the theory Markopoulou sought to bring to physicists' attention.[d]

It is a simple concept: A network of elements[640] (fig. 20), maybe called *events*, gives rise to—i.e., elementwise causes—new elements, and so on. Its simplicity makes it a shoo-in to depict machinery of our

---

a See chapter 24 at page 141.
b The pivotal issue is special relativity's supposed rejection of an absolute frame of reference. Cosmology shows there *is* such a frame: Satellite measurements of the cosmic microwave background provide increasingly precise access to it (see chapter 56 at page 301).
c A reminder: Einstein defined time in special relativity as what a clock says, see note 245 at his page 25; so he showed *common* time is relative.
d See chapters 1 and 43 at pages 15 and 243.

ontology.[a] Sorkin launched the concept[b] in 2007 as a world made of a network of relationships, from which space and time were to emerge.[641] Its seminal role has no doubt been inhibited by the wide acceptance that simultaneity became, with special relativity, taboo.[c] That is, feet firmly set in clock-time quicksand, we cannot speak of the events at other places as occurring all at the same time.

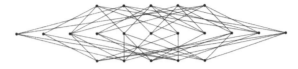

Fig. 20. Three sequential sets; lines (rising) show causal relations

The new ontology bespeaks a specific network of specific relationships—the links between space quanta. Thus, the quanta and their links offer a framework for *our* version of the universe's causal set (fig. 21); the *beings* of quanta and links are our network's elements or its events, from which a granular space emerges. And their *doings* from one iteration to the next track their causal relations with inherent simultaneity, out of which absolute time (of a kind) arises.[d]

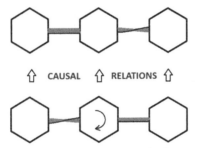

Fig. 21. A twist moves in a (1-D) causal set

Smolin liked what he called the "radical simplicity" of Sorkin's concept but, noting its failure to yield "a spacetime, with three dimensions of space,"[e] he said,

---

a See chapter 43 at page 243.
b His concept was wrong-footed from the get-go: Far from letting space and time emerge, Sorkin "quantized" spacetime; and it lacked spatial synchrony.
c See chapter 27 at page 154.
d See chapter 52 at page 284.
e The "space" is what we put "in at the beginning" (Smolin's term, see note 1949 at chapter 128), on the simple basis that it really was put in at the beginning; see chapter 17 at page 105.

> This makes it seem as if there is more to a spacetime than a rough description of a network of causal relations.[642]

That is, following a different (though related) path,[a] he arrived at almost the same place as we have with the new ontology: Its "more" are the linked quanta of space and its iterations. Don't expect space to show up when you omit its foundational element.

Further, Smolin wrote of time in the epigram, but clearly did not mean common or clock time. He wrote something similar on that same page (my italics),

> [W]e argue that time, *in the sense of the present moment and its passage,* is fundamental.[643]

And on the next page,

> Once an event has happened it is in the past, and that fact cannot be erased by a future event, even if that future event reverses the effect of the original event. This thought led us to view the passage of time as a process by which new events are steadily created from present events.[644]

Here, he was writing about his approach (with Unger) to achieving a breakthrough in fundamental physics.

His concept of time sounds more than a bit like Newton's true time.

And in a 2015 paper with a colleague proposing a new class of models, he said (my emphasis),

> There is a process continually acting in the present bringing into existence the *next* moment.[645]

With this kind of language he was leaning close to our ontology, which readers will soon see sets out the origin and elements of such a process.[b] His missing this seems to be due to ontological commitment to the continuity of time. To conceive of "the next moment," one needs a granular process (or a sequence), as opposed to "a process *continually* acting" or continuous time.[c]

---

a Note, especially, his emphasis on "radical simplicity."
b In chapters 56 and 57 at pages 301 and 307.
c Zeno could have told him that. See chapter 10 at page 63 and also note 1948.

The new ontology says—on grounds of what consistency requires—that space (though, in the everyday sense of the word, emergent) is real and time (in its everyday sense) is not. Smolin too concluded—on entirely different grounds—that only one of the two, space and time, can be real. He (as he had with Unger[646]) put his money on time turning out to be the one.[647]

Sorry, Lee, choosing *time is real* looks like a losing bet.[648] The kind of time that *is* fundamental (true time) has no natural place *inside* the universe and it is not continuous. Or, to put it another way, that kind of time is dictated by the sequential replacement of the entire universe; so, there is nowhere whence we can observe it. Or, put a third way, it lacks direct access to clocks.[a]

It's *not* time (and this is not exactly what Smolin had in mind) but his notion of "a process by which new events are steadily created from present events," when tracked (which he did have in mind) as a causal set, corresponds quite nicely to the universe's iterations in the new ontology.[b] The difference is, in the new ontology the universe *is* a causal set.

Again, Smolin was getting close. At one point, he even said the magic words,

> [T]he universe is a causal set.[649]

We will get to that causal set.[c] To do it, we will need to keep in mind two distinct kinds of time.

One kind is common time, the time of Einstein's paper on special relativity,[d] time told by a clock.[e]

The other kind is the much-maligned true time or, rather, our version of it. It has no clock other than the entire universe and we are inside that clock; indeed, we are inseparably *of* that clock.[f] That causal set will need the concept of this clock.

---

[a] We can measure the speed of photons and use it to convert their sequential motion in Planck lengths to clock time at rest in the absolute frame of the universe; see chapter 56 at page 301.
[b] See chapter 57 at page 307.
[c] See chapter 43 at page 236.
[d] See note 256.
[e] Maybe we are amenable to such a seemingly artificial definition of something we regard as fundamental because we are, ourselves, clocks of a kind; see the twin paradox, chapter 52 at page 284.
[f] See chapters 24 and 52 at pages 141 and 284.

Our true time is digital and absolute. It is the relentless iteration of 3-D space in universe-wide lockstep.[a]

We attribute time-like quality to these iterations by reference to light-speed motions (like the photon's). The attribution works only because—as our ontology *assumes*—their twist states move along to the next link each iteration.

In this way, we can count true time perfectly in principle if not in practice. We can attribute Planck-lengths of photon motion to Planck-time iterations. As the number of iterations is the same everywhere, this gives us a "time" that is, as Newton said, "absolute, true and mathematical." And it is *not* what clocks measure (though, again, we may use a clock to follow it with photon-motion calculations).

True time also leads us to anticipate a central premise of the new ontology[b] in order to restate the underlying principle of special relativity: *The speed of light is a digital property of space so it is not affected by motion in space.*[650]

Thus, by understanding what space is, we also understand what time is or, rather, the two times are. Essentially they are Newton's two times. When one considers the gestalt of ontological uncertainties and physical errors in which he was immersed, this is—to say the least—extraordinary.[651]

Please note, one key thing that he did *not* say of either time is that it is a dimension of the universe.

---

[a] Suffice here to say, this is how the universe came into existence, continued to this day; see chapter 41 at page 228.
[b] See chapter 50 at page 276 for how this works at Planck scale.

29
EINSTEIN'S BEST BLUNDER

> Einstein remarked to me many years ago that the cosmic repulsion idea was the biggest blunder he had made in his entire life.
>
> George Gamow (1956)[652]

Certainly, it seemed to Einstein that, on second thought, it was a blunder: He had tossed an extra constant into his tidy general relativity equation for the universe.[653] It turned on him. It blew his tidiness away. And when he changed his mind, it was like proverbial toothpaste; he could never get it back into the tube.

> Einstein's tame constant got away from him.

Why did he invent it? Partly just because he could; he knew he had room to add a constant without (he thought) doing any damage. But mostly because he assumed the universe's size was stable and he was concerned, without some kind of antigravity, his version of it could collapse.[a]

When observations showed it to be *expanding*, he changed his mind. He was unabashed; he always sought to learn and so he often changed his mind.[654] Indeed, he saw the humor of it (fig. 22), joking to a friend,

> Einstein has it easy. Every year he retracts what he wrote in the preceding year.[655]

Cosmology now says he may have had it right the first time; more recent data show his constant being needed (see below). This kerfuffle complicates an already confused matter. To make things worse,

---
a See chapter 18 at page 112. One can think of the constant as a number, whose size may be adjusted, that represents a kind of "antigravity" force of unknown origin that balances the universe's gravity.

Fig. 22. Einstein's humor – statue in Ulm

that confused matter has half a dozen different names amid further confusion about whether they are different things.

We'll find they are only one. And once we sort out what it is, we'll see it is another seminal idea.

It began life as the *void*, conceived as place where there was nothing.[a] Then *vacuum*, from Latin for *empty*, by assumption lacking substance (which quantum theory would later fill with fleeting particles without which it was still thought of as a void).[b] Then came *æther* and its siblings *aether* and the *ether*,[c] given substance only because it was thought the waves of light must be waves *in* something.[d] Einstein was wrongly taken to have killed all three off in 1905.[e]

But, as outlined above, in 1917 Einstein—flexing his new theory's muscles with a whole new approach to cosmology—thought up another ether with his constant.[f] He gave it a Greek letter, lambda, λ, calling it (in translation) "a universal constant."[656] It came to be

---

a From its origins in ancient Greek philosophy this has been a paradoxically substantial subject.
b See chapter 91 at page 454.
c See chapter 10 at page 63.
d The ethers tended to be thought of as *in* space, rather than *being* space.
e See chapter 11 at page 66.
f Even without the constant, general relativity established that space had ether-like properties.

Part 1: Some Seminal Ideas   167

known as the *cosmological constant*;[a] these days it gets capital lambda, Λ, along with awkward questions like, *What is it?*[657] In recent decades, some say it is the energy density of the vacuum.[658]

Its creator, Einstein, disliked it intensely, later writing to Lemaître,

> I cannot help to feel it strongly and I am unable to believe that such an ugly thing should be realized in nature.[659]

As expansion of the universe came to be accepted, Einstein tried to kill his ugly constant. By 1931, in a lecture at Oxford University (see fig. 23),

> Einstein is arguing that the cosmological constant, λ, which is a constant energy density that permeates all of space, should be ignored in this lecture, so he writes the field equation with it crossed out.[660]

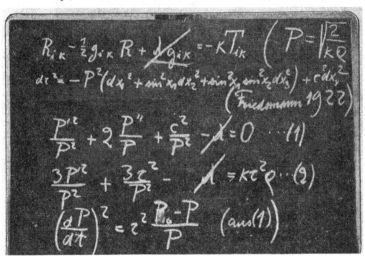

Fig. 23. Blackboard from Einstein's lecture

But cosmology kept it alive.

Confusion and controversy dog it as it endures to this day.[661] What is its meaning? Is it really needed? Is it constant? Is it positive or negative? Or, widely thought to be exactly zero, is it so?[662] If it is not, then why is it absurdly small (with various offbeat interpretations)?[663]

---

a See also chapter 67 at page 348; the constant was an arbitrary number, in units of reciprocal area, and was given whatever value was needed to try to keep the universe a constant size.

Science writer Rob Lea summarized,

> The cosmological constant has been a thorn in the side of physicists for decades.[664]

Studying billions of years' history of space expansion is not as difficult as it may sound. First, one needs *standard candles*, pioneered by astronomer Henrietta Leavitt,[665] cosmic sources of light whose brightness we know since they arise from "standard" events, such as *supernovas* (fig. 24).[a] These occur when certain stars run out of fuel and collapse under their own gravity.[666] With such candles metaphorically in hand, astronomers could look far into space and track expansion of the universe through time.

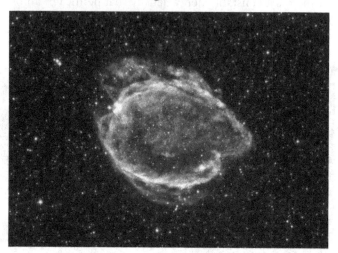

Fig. 24. *The 5000-year-old remnant of a supernova*

In 1998, astrophysicists measured Λ to be non-zero, very small and positive.[667] This told us the universe is finite and closed and, in 2011, this work won Adam Riess and two other physicists a Nobel Prize.[668]

These days, enigmatic concepts known as dark energy[b] and quintessence[669] are seen to be closely related to Λ.

In the new ontology, these seemingly disparate entities are all—if charitably viewed—the same thing in the end (though not a one of

---

a If I show a physicist a night scene where all the lights are 100-watt light bulbs, they'd say how far we are from each of them.
b See chapter 67 at page 348.

them has until now been well described). Their confusions thrived in context of old-ontological commitments, like continuous space; no absolute frame; spacetime; cosmological inflation; the vacuum; and empty space.

Despite all the confusions, cosmology these days can't do without it. As Peebles noted recently,

> We must learn to live with lambda.[670]

The new ontology may help. Thanks to its motley origins, lambda seems almost meaningless. Let's use its simplest name; let's call it *space*.

To sharpen up its image, let's acknowledge it has mass.[a] Indeed, its total mass has now been measured to be some two-thirds that of the entire universe.[671]

But, to wrap up this sketch of problems of the constant, that mass measurement depends upon the standard model of cosmology. As Einstein neatly put it, as a general proposition,

> It is the theory which decides what we can observe.[672]

In this way, the theory can have serious practical consequences. I'll illustrate from personal experience how this can work. In the nineteen seventies I worked with colleagues at Atomic Energy of Canada Ltd.; we analyzed the physics, chemistry, and biology of how mammalian cells are killed by radiation.[673] Among the cognoscenti there was then a standard way to do this, by fitting a two-parameter equation to the cell-survival data.

That equation was a good fit to bad data and a bad fit to good data. Its form had no meaningful connection to a cell-death process. As well, it implied cells can absorb a threshold dose of radiation before harm begins. This is a property the cells *do not* possess. And *that*'s a fantasy with real-world consequences.

We worked with a theory based on what radiation physically did to DNA.[674] Its equation was a good fit to good data.[675] It made sense of what was going on. We could see the DNA effects.[676] Yet, forty years (and many papers) later, the old equation is still to be found in texts.[677]

---

a Though we don't view it as matter.

To me, a good theory is one that helps to understand the world, not just fit equations to data. It allows one to extract all relevant information from a data set and no more.[a] Such a theory is, in some sense, real. Using a cell-death theory that reflects real properties can lead to more effective cancer treatments that need fewer visits and have lower costs.[678] (Radiotherapy is realizing this at last.[679])

So, with Einstein's insight about the key role of theory in mind, let's recall the current theory's related "unbelievable coincidence"[b] that Λ seems to have *exactly* the value needed to keep the universe's expansion under its antigravity "pressure" balanced on the knife edge against its contraction under gravity, to a precision better than one part in a trillion-trillion-trillion-trillion-trillion.[c] This was the picture as seen through the lens of the standard model of cosmology and its underlying theory, general relativity.[680] And it is a property the universe just does not have.

It's that same problem of curve-fitting an equation that does not reflect reality. The roots of this problem go back to conceptualizations of space in cosmology's early days. It was widely treated as a fluid; the closest anyone would come to specifying its nature and properties hinged on whether it was compressible. For example, Einstein observed,

> [T]he concept of an incompressible liquid is not compatible with relativity theory as elastic waves would have to travel with infinite velocity. It would be necessary therefore, to introduce a compressible liquid whose equation of state excludes the possibility of sound signals with a speed in excess of the velocity of light.[681]

Such conceptualizations are deeply embedded in the theory and often not explicit in papers about the standard model. But they are not grounded in reality.

---

a If it extracts more, then it is fiction.
b See note 520.
c See chapter 20 at page 122.

Space—in the new ontology—is not compressible. But then neither is it incompressible in quite the sense that Einstein had in mind. Indeed, such notions are both inapt to reality and incompatible with his insight that,

> The grasp of motion may not be applied to it.[a]

Einstein's blunder derived from his embrace of the old ontology, which led to the "unbelievable coincidence" of the knife edge.

The new ontology says there is no unbelievable coincidence. There is no knife edge! What's happening is *not* a battle between the universe's gravity and some unexplained new antigravity. The expansion must be caused by new space quanta.

Useful though it is, the theory of general relativity has no word to say about new quanta of space.

Many space quanta were made long ago; and we will soon see there are "space factories" that make more space quanta in the self-same way (though not at the same rate).[b]

These are revolutionary, back-to-the-drawing-board, rethink-our-conclusions, we-need-new-theory, reanalyze-the-data insights for cosmology. In a more elegant phrase from another Nobel Prize winner for literature, Sir Winston Churchill, they are . . .

> . . . tremendous changes cutting to the very roots of thought.[682]

For example, recent cosmology reveals most of the universe's mass is missing[683]—ascribed to the so-called dark matter[c] that we don't see, thought to contribute 27 percent of the total mass,[684] and the so-called dark energy,[685] contributing a dominating 68 percent.[686] Recalling again Einstein's dictum that the theory decides what we can observe,[d] the careful calculations of dark energy and dark matter and ordinary matter and their prevalences in the universe all stand to

---
a Re this text and, in particular, translation of the word *grasp*, see note *b* and context on page 71.
b The source will be considered in detail in chapter 69 at page 357.
c See chapter 68 at page 351.
d See note 672.

be revisited—along with their meanings and significance—when we have a theory of space based on reality.

For now, we need to better understand the seminal idea of real space that lies behind that blunder. Its best fruits may be yet to be picked.[a]

---

[a] If you want to look ahead, see chapter 67 at page 348.

30
JUST IMAGINE

> Gedankenexperiment, (German: "thought experiment") term used by German-born physicist Albert Einstein to describe his unique approach of using conceptual rather than actual experiments.
>
> Sidney Perkowitz (2010)[687]

His approach was maybe not quite as unique as physicist and science writer Sidney Perkowitz believed. But it was quintessential Einstein.

The idea of the *thought experiment*[688] in natural philosophy dates back at least to Galileo[689] and his mental tests of falling bodies and of objects moving in a moving ship.[690] It is an exercise in shared imaginations—of the thought-experimenter and of those to whom the tale is to be told.[691]

Physicist Hans Christian Oersted used the German term in the early 1800s.[692] Einstein's mentor Mach gave it wide currency.[693]

In the golden age of theoretical physics,[694] thought experiments had profound impacts on science, largely in the context of the German *Weltanschauung* (worldview) of those times when physics was a mainly German enterprise.[695]

> Imagined experiments are the least expensive and can have useful results.

Today the term *Gedankenexperiment* has been embraced (with roots separated) into English.

At its best,[696] the gedanken experiment can demonstrate a scientific proposition; a pseudo-proof invoking *We don't need to do this as we can see how it goes.*[a]

---

a But we may all get it wrong.

It is a tool that is even more consequential when, as here, we seek to understand realms of reality that lie beyond all possibility of observation or experiment.[a]

The idea one can learn about reality by thinking runs far further back, to ancient Greeks, especially to Plato. Unlike Aristotle (but like Socrates), he favored introspection over observation as a source of knowledge. His example leads to a conundrum: On one hand, most of his imagined aspects of reality turned out to be rampant fiction.[697] On the other hand, his thoughts were so impactful[698] we still cite them to this day.[b]

Einstein, who seemed to do no tangible experiments and make no measurements,[699] turned the thought experiment into a useful tool. Today, much of the world's increasingly physics-based economy derives from them too.[700]

Yet all is not well in our scientific paradise. Science struggles to define its vaunted method.[701] Worse, the senior science, physics, and its language of math are both riddled with inconsistencies.[c] Worst of all, even arithmetic, upon which math depends, itself defies all efforts to make it consistent.[d]

In days gone by, philosophy would probe ideas using thought experiments that could never lead to measurements.[702] But in this way ideas could at least be tested for consistency.[703]

For example, physicists Daniela Frauchiger and Renato Renner used a thought experiment to show that quantum theory cannot consistently describe the use of quantum theory.[704]

While logical positivism gave the philosophy of science an unwarranted bad rap in the early twentieth century,[e] more recent critics, such as Hawking, claimed with some justification that,

---

[a] See chapter 133 at page 639.
[b] A Google Ngram search (see note 914) in a corpus of English-language books published in 2019 shows Plato mentioned almost three times more often than Einstein; and he shows up in the present work in seventy-four places.
[c] See, for example, chapters 36, 40, 45, and 88 (at pages 205, 222, 255, and 442, respectively).
[d] See chapters 36 and 114 at pages 205 and 559.
[e] See chapter 9 at page 57.

> Philosophy has not kept up with modern developments in science, particularly physics.[705]

With no less justice, one might also say that physics lost its way without philosophy.[a]

In our dependency, we are like offspring whose parents divorce unwisely: We want them to remarry.

A discernible message from the new ontology would have philosophy renew its focus on fundamental physics and physics give philosophy another go,[b] not least for the benefits of those far-reaching and enlightening and inexpensive thought experiments.

They are, for us, a seminal idea.

---

a See chapter 40 at page 222.
b Some do. As these pages ofttimes show, there are physicists who find guidance in philosophy and there are philosophers who usefully consider fundamental physics. They are too few.

31
BUNS IN THE PHYSICS OVEN

> It would be madness and inconsistency to suppose that things which have never yet been performed can be performed without employing some hitherto untried means.
>
> Francis Bacon (1620)[706]

Early in the seventeenth century, philosopher Francis Bacon was writing about understanding the world.[a] Many "untried means" have come, been tried, and gone since then. Among the latest that are being tried are a collection of approaches sharing that generic label, *quantum gravity*.

Quantum gravity was once the new kid on the block. Far from it now; and yet in 2019 philosopher Steven Weinstein could say,

> Though quantum gravity has been the subject of investigation by physicists for almost a century, philosophers have only just begun to investigate its philosophical implications.[707]

Let's see if we can do a little something about that.

These days three bold programs—that flirt with visions of reality by plowing new ground in fundamental physics—occupy much of the quantum-gravity conflict-zone. They are string theory, loop quantum gravity, and causal sets.[708]

> Strings, loops and sets reach for the gold ring.

Seminal ideas are embedded in these programs, but we must be selective in choosing what we should adopt. Their up-front *bold* too easily turns *timid*. Their math can lead them deeply into new dead-ends.

---

a This part of his *Novum Organum* was titled "On the Interpretation of Nature and the Empire of Man."

That's not to say that *timid* is entirely unsuccessful. Its successes, though, seem mostly to arise from serendipity and even then they are inhibited by remnants of the theories they are seeking to supplant.

Let's look, for example, at string theory. In key ways it is the quantum-gravity lead candidate.[a]

There is in string theory no single thing that is the *string*. And there is no single theory. Rather string theory embraces various approaches that arose from tacking more dimensions onto 3-D space.[b]

Many of them need to add six space dimensions to the three we see. So where are these new dimensions? String theory treats them all as vanishingly small—said to be "compactified" to Planck scale in the form of a Calabi-Yau manifold.[c] Nobody knows how or when or where or why that compacting took place or why it didn't happen to the other three.[709] But, with 9-D plus time to play with, its math becomes marvelous to see.

The theory is derided as untestable and so is said to not be science,[710] but too many have become much too invested to allow it to just die. String theory gets most of the money, postdocs, conferences, papers, and attention but its strings are stranded, wiggling elegantly in space[711] that by default is continuous.

Here we shall shun almost all the trappings of string theory but (as we have already said[d]) its Calabi-Yau-manifold idea is seminal, so we are setting out to see what we can make of it. Not to be too coy about it, within eighty pages we will see this answer emerge, *Literally everything*.[e]

The second leading candidate was born when Wheeler and physicist Bryce DeWitt (who both worked with both leading theories) found an odd-looking equation for the universe. What was odd about it was that, unlike almost all equations in physics, the Wheeler-DeWitt equation made no mention of time.[f]

---

a Such as number of practitioners and total grants; and see further below.
b Those 3-D spaces tend to be tangled into 4-D spacetimes; see chapter 73 at page 378.
c See chapter 15.
d In chapter 15 at page 95.
e See chapters 45, 53, and 55 (at pages 255, 291, and 297, respectively).
f See chapter 80 at page 411 and fig. 71 at page 418.

Twenty years later, Smolin and physicist Ted Jacobson found it had exact solutions. The solutions were strange too. They seemed to call for loops in space. There was a solution for each sort of loop.[712] This led to math for general relativity that's granular and almost lacking input space and time (i.e., almost background independent); as Smolin has long held, such independence is a requisite for a true theory of quantum gravity.[a]

In the result, called loop quantum gravity, space and time are built from interwoven Planck-scale loops. The loops have few pretensions to reality, though as Rovelli said twenty-five years ago, it gives an . . .

> . . . intriguing physical picture of the microstructure of quantum physical space, characterized by a polymer-like Planck scale discreteness. This discreteness . . . provides a mathematically well-defined realization of Wheeler's intuition of a spacetime "foam."[713]

For us the absence of time from its basic equation suggests our timeless space may be onto something. For others, maybe not so much:

> Every few years, someone hosts a conference devoted to the problem of time in quantum cosmology.[714]

Causal set theory, the third candidate, is a kind of model of the world.[b] Sorkin is its inventor[715] and lead proponent. Of the three, it may have the best shot at being real, so, as one might expect, it looks to be the least supported (meaning money and grad students and the like).

In a 2008 overview, physicist Fay Dowker provided a candid snapshot of the causal-set field.[716] She pointed up the tension between finding a true view and hanging on to the same old same old with too much of the latter winning out.

Right off the top, her title, "Causal sets and the deep structure of spacetime," revealed a key issue. Of spacetime?[c] Really? It's a deadly addiction for physics[d] so here we'll pretend she said *space*.[e]

---

a See chapter 5 at page 95.
b As we have seen in the preceding chapter; see also chapter 73 at page 378.
c This was a problem from its beginning; see footnote b at page 161.
d See chapter 83 at page 420.
e Keen readers could scan the review by Surya, note 870, to get their own impression of how causal set theory's ontology lacks a concept of real granular space, embraces spacetime, fails to grasp simultaneity, and gets lost in math.

Dowker set up the granular-space objective and its continuity challenge:

> The concept of atomicity also has a long history as do philosophical challenges to the antithetical notion of a physical continuum. . . . [I]t is certainly now the case that many workers believe that a fundamentally discrete structure to reality is required to reconcile spacetime with the quantum.[717]

What she said is true but the effect is to lead almost inevitably to a blunder (not hers; she was reporting the story):

> The question is, "What could the discrete state be that gives rise to this continuum approximation?" and a good first response is to try to *discretise* the continuum.[718]

This is as if to say, let's start with what we now can see is wrong and chop it up to see if we can tweak it into being right. It's timid.

Einstein, supremely successful seeker after fundamental-physics, cast doubt on this sort of strategy in a letter to a colleague:

> I do not believe that it will lead to the goal if one sets up a classical theory and then "quantizes" it.[719]

Instead, we can see space quanta forming a causal set. In the new ontology, a causal-set view of the universe is inherently discrete. It gives rise to matter with its links. It has no need for clocks and time. If one accepts simultaneity (as one should[a]), each of its iterations *is* the whole universe. The complete causal set contains the universe's entire history.[b]

A common thread to these aspiring candidates for quantum gravity is that they don't say (and may not care) what's real. So their proponents literally don't know what they are talking about. They look for math that works, meaning it can spit out relativity and quantum theory when one backs up far enough from Planck scale.

The new ontology appropriates a central element from each of these three candidates, an element that in each case becomes key to

---
[a] See chapter 28 at page 158.
[b] See chapter 43 at page 243.

the real worldview's constitution (though each had no *real* place in its mini-ontology).

To recap, these elements are string theory's 6-D Calabi-Yau *manifold*; loop quantum gravity's *timeless space*; and causal set theory's *causal set*.

Selecting these three seminal ideas and fitting them into the picture are significant steps. The big step, though, is doing what the programs don't, understanding them not as mere math abstractions but as physically *real*.

32
SUFFICIENT REASON

> We must . . . make use of the great, but not commonly used, principle that nothing takes place without a sufficient reason; in other words, that nothing occurs for which it would be impossible . . . to give a reason adequate to determine why the thing is as it is and not otherwise.
>
> Gottfried Leibniz (1714)[720]

The principle of sufficient reason is "a powerful philosophical principle";[721] and it is also a simple seminal idea. We will use it as we seek an understanding that reifies reasons for everything.

Leibniz gave it life, set it front and center, and gave it its name.[722] But he did not invent it. Indeed, its kernel can be seen—with significance for us— leaping from one line of a 50 BCE epic poem by philosopher and poet Titus Carus (sometimes known by his middle name, Lucretius):[723]

The universe puts its principle into practice.

    Nothing from nothing ever yet was born.[724]

In the sixteen hundreds, philosopher Baruch Spinoza supposed something like this principle as an *axiom* (a statement whose truth must be assumed).[725]

Smolin used the principle. He said, for example, there must be a reason why we see space having three dimensions,[a] and our theories are incomplete until we know the answer.[726]

He said sufficient reason points up philosophic principles for fundamental physics:[727]

- Background independence[b]

---
a See chapter 46 at page 263.
b See chapter 49 at page 271 and note 163.

- Space and time are relational[a]
- *Causal completeness*[b] (every event has a cause)[728]
- *Reciprocity*[c]
- Identity of indiscernibles[729]

We will draw upon them later for a "reality check."[d]

Meanwhile our reified principle of sufficient reason begins at the beginning and continues with each iteration, where each quantum and each link and each twist state has its sufficient reason—arising from its parent quanta, links, and twists and what they just did that made their progeny precisely what they are.

Bit by bit, so to speak, this guarantees causal completeness.

---

a See chapter 49 at page 271 and note 167.
b See chapter 30 at page 173 and note 864.
c I.e., if A acts on B, then B must act on A.
d See chapter 79 at page 402.

33
THE WHOLE NINE YARDS

What concerns philosophy is the universe as a whole.

Sir James Jeans (1943)[730]

The universe is the ultimate definition of what is real.

So, as Jeans implies, we must explain it *all*. We need an understanding of it that is seamlessly consistent from the exquisitely tiny Planck scale through atoms and ourselves and far reaches of the cosmos to the universe as a whole entity.

There is reason to insist on this; it is the *entire* universe that *is*.[a] Not some sample of it. This is a seminal idea.[731]

Putting this another way, de Chardin said,

> Each element of the cosmos is positively woven from all the others.... It is impossible to cut into this network, to isolate a portion without it becoming frayed and unravelled at all its edges.... Structurally, it forms a Whole.[732]

These will prove to be extraordinarily insightful words.

Almost all of science is concerned with just some fragment of the universe. Even in cosmology, most of it is concerned with subsystems, such as galaxies or that part of the universe that we can see. Yet we will find the notion we can separate these fragments or subsystems from the whole is fiction. Again, often useful, but not ever real.

> Our question is: What is it?

Unger saw the significance of this and said,

> The universe as a whole is a very different kind of system from those usually studied and modeled in physics, and its comprehension will require a new paradigm.[733]

---
a See also the cosmological fallacy in chapter 133 at page 639.

The Wheeler-DeWitt equation[a] illustrates the kind of difference a whole-universe view may make.[734] Unger said (and the new ontology concurs),

> This equation suggests that, once applied to the whole universe, quantum mechanics has no place for time.[735]

This way of thinking, though not easy, is essential. It confronts us with the core of our task and its constraints. For example, Smolin said,

> The problem of how to make a theory of a whole universe is thus the problem of how to construct a theory without making any reference to anything that exists ... outside of the system we are describing.[736]

And Smolin added a new fundamental requirement. The universe must include all it needs—like resources and rules—to organize itself, to become, and to be. He said,

> [A] theory of a whole universe, if it is to be consistent with what we know of quantum theory and relativity, must be a theory of a complex, self-organized universe.[737]

This challenge, too, we shall embrace. The moment we should focus on to begin figuring this out is the first moment, the very beginning, when the entire universe was smallest and, one surely would suppose, was simplest.[b]

Physics is not helpful with this moment. Contrary to popular belief, the Big Bang—as we (vaguely) understand it—happened with the universe already well and truly underway, maybe gum-ball or maybe light-year sized,[c] and supposedly expanding under general-relativity direction. Thus, as Guth said,

> The Big Bang model ... says nothing about what banged, why it banged or what happened before it banged.[738]

---

a See chapter 31 at page 176.
b Regrettably, both physics and philosophy tend to avoid this simplest of time zones, seemingly abandoning it to religion; see also chapter 17 at page 105.
c This seeming-large uncertainty depends on the choice of model—about which controversies rage—but, on almost all views, reflects an absurdly small difference in timing.

The revelation that the universe *is* expanding caught physicists in general and Einstein in particular off-guard.[a] It threw a cosmic wrench into the workings of the nascent standard model of cosmology.

The Big Bang model shows the universe expanding after the beginning because it projects what we see happening today back through time using Einstein's general relativity equations. In other words, it says the universe was expanding because the universe was expanding. It's a damn-fool excuse for explanation, and so is not widely advertised.

This leaves a vacancy for real explanation. It's evident the key events happened *before* the Big Bang, including the real reason why expansion began. If we could peer back to the very beginning, we could ask and maybe answer, *Why did the universe start to expand?*

One can project the expansion backwards (through very early realms the theory of general relativity can't handle) until the entire universe is just a mathematical point.

If all that mass began compressed into a point—or even a large lump—the same theory says it would be a humungous *big black hole*. From which, physics tells us, nothing, let alone an entire universe, can escape.[b]

An answer to the question, why did it begin expanding, might cast a more explanatory light on why it is expanding now.

This may illustrate why explanation must begin with the whole universe. It must explain its condition from the first instant, through its growth and development, to produce the complex universe we inhabit today. It must explain all it contains. Not only the matter and the energy we see, which make up five percent of all the mass. It surely must also explain the so-called dark matter that we don't see, which contributes twenty-seven percent of that mass, and especially the so-called dark energy,[c] which contributes sixty-eight percent.[739]

We *may* ask, *What are they?* But let's get the key question right. We *must* ask, *What is it?*

---

a See chapter 18 at page 112.
b See further, chapter 69 at page 357.
c See chapter 67 at page 348.

34
THE SINGLE SOLUTION

> Truth is a whole, and the truth of physics will be found
> to link on and to be but part of that larger truth which
> is the nature and the character of the universe.
>
> Jan Smuts (1931)[740]

Field Marshal Jan Smuts was a special mix, with politician, strategist, peacemaker, statesman and philosopher added to his military credentials. His saying truth—like the universe—is a whole leads to the seminal idea that, rather than seeking many answers to our many questions, we want one answer for all of them.

Next to last on this long list of seminal ideas is that an ontology that is real should provide a single simple solution; that is it should resolve all the old ontology's problems and settle all its contradictions in one fell swoop.

This idea is more fundamental than may at first appear. Most of the many conceptual ventures reported in the literature as aiming to address the shortfalls of our current understanding take aim at a single problem; a few, such as cosmological inflation, tilt at two or maybe three.[a] Each problem may have more than one solution. So, solving separately is good strategy for having lots of publications—for finding *the* real answer, not so much.

> What is all this? is a single question that needs a single answer.

Finding a single solution to a plethora of problems is not a new idea. There is a powerful connection between fundamental physics and detective fiction. We will treat the contradictions found in physics and philosophy like clues in a detective story. The reason,

---
a See chapter 21 at page 126.

as crime-fiction aficionados know, is the detective seeks that unique insight that makes sudden sense of *all* the crazy clues.

We should not be surprised to find that Einstein saw—and, with Infeld, wrote about—this same connection between fundamental physics and crime fiction:

> In imagination there exists the perfect mystery story. . . . Can we liken the reader of such a book to the scientists, who . . . seek solutions of the mysteries in the book of nature? The comparison is false . . . but it has a modicum of justification which may be extended and modified to make it more appropriate to the endeavor of science to solve the mystery of the universe.[741]

And they saw how science too looks for an all-inclusive answer:

> [The scientist's] task is not to explain just one case, but *all phenomena which have happened or may still happen.*[742]

Their 1938 book on how physics evolves began with nine pages, chapters 1 and 2, "THE GREAT MYSTERY STORY" and "THE FIRST CLEW."[743]

To Einstein, the *clews* were all in aid of what he elsewhere called . . .

> . . . a theory which describes *exhaustively* physical reality.[744]

Smolin saw the same connection; his *Life of the Cosmos* sports an entire chapter titled "Detective Work."

The arc of the detective story's simple: Unearth the enigmatic clues and intuit the light-bulb thought that fits them all together to make sense. This is the genre's rule. As Einstein advised, it is the rule for fundamental physics too.

Yet Einstein's great theories arose in a seemingly divided way. In detective terms, he solved part of the universe's case (the space problem) in one way while crafting an incompatible solution to another part of the same case (the quantum problem). For the rest of his life, Einstein worked to find a single solution to disparate theories.[a]

---

[a] He called it the unified field theory.

It is ironic that he never saw the path he sought in vain begins with this, *The space and the quantum are the selfsame thing!*

So, let's be aware of the conceptual connections between fundamental physics and crime fiction: Both face an embarrassment of clues or evidence; both seek a new way to see them that makes sense of the whole show. That is, both need one new view to explain *all* the clues.

Though seeking the single solution *should* be physics' rule, physics fashion scorned it all the rest of Einstein's days.[745] Beyond quantum gravity programs,[a] physics tends to frown on fundamental explorations. Its mainstream method mostly explores narrow vales in landscapes of fantastic math. The vales are often beautiful but are unlikely to set physics on the road to being real.

And they are not for us. Our method of detection will be based on prolific author[b] Georges Simenon's Maigret (by contrast with those of Arthur Conan Doyle's Holmes, Agatha Christie's Poirot, or even the original, Poe's Dupin).[c] The choice is germane to our task.

We will use Maigret's strategy. One might say his is an intuitive—in contrast with an analytic—method. Maigret—one of the best-developed characters in all English literature (ironically, almost all of it in translation from French)—gives expression to this mode of thinking in some strategy-revealing scenes:

> He was not really thinking. Only scraps of thoughts which didn't add up to a coherent whole.[746]

And,

> He was drifting. Impressions formed and dissolved. He had lost all sense of time and place.[747]

And,

> He was not following through an idea. One might say he was rather like a sponge.[748]

---

a For three examples, see chapter 31 at page 176.
b He wrote more than four hundred novels that sold more than half a billion copies.
c For more on these detectives and their strategies, see *Time One*.

They signify that Maigret was absorbing, unadulterated, all he could about the context of the crime.[749]

Behind such scenes lay his conviction that this mode of thinking—emptying his mind of preconception, opening it up to comprehension—was the way to solve the crime. One might think, at first, his method was to have *no* method. Not so. He sought a single understanding of the whole.

Dear reader, that's what we seek, too. While we search physics' nooks and philosophic crannies, don't worry if you don't yet see Maigret's "coherent whole." Form impressions; try to be "rather like a sponge."

And Simenon gets credit for this aspect of the one-solution seminal idea.

35
DESPERATELY SEEKING SUNDANCE

> What is much more likely is that the new way of seeing things will involve an imaginative leap that will astonish us.
>
> John Bell and Michael Nauenberg (1966)[750]

Until very recently, there was no way our best reasoning could build a real ontology. We all were living through, as Reichenbach said, ...

> ... a passing stage when philosophic problems are raised at a time which does not possess the logical means to solve them.[751]

Or, as Updike put it,

> [T]he mystery of being is a permanent mystery, at least given the present state of the human brain.[752]

More than fifty years after Bell and Nauenberg (above) contemplated a "new way of seeing things," we are moving out of that long-lasting "passing stage" of Reichenbach's. With human brains in overdrive, we are starting to see things in a "new way." And one of them did give us "the logical means to solve" a raft of philosophic problems. As Bell and Nauenberg expected, it did "involve an imaginative leap." It did astonish me. And I expect it will astonish you.

It is time to introduce the unsung scientist who gave us that last missing insight, the key seminal idea. He is one of at least two[a] who stumbled on a window into the long-hidden Planck-scale world.[b]

---
a The other being Max Planck; see chapter 13 at page 76.
b Bell could be regarded as a third. His famous theorem is inherently statistical; but with experimental results it showed the linkages at Planck scale must somehow be nonlocal; see chapter 25 at page 144. And Einstein could be a fourth, with his special relativity's light-speed law and insight into the equivalence of mass and energy, the reasons for which we will find at Planck scale; see chapter 43 and fig. 40 at page 251.

His insight was for me the crucial missing piece of what was shaping up as a consistent picture of the universe, from the smallest to the largest scale from the very beginning. I was stitching it together from selected seminal ideas.[a]

That missing piece answered a problem that's been hovering for decades without finding a definite form. Understanding reality seems to call for a *new kind* of answer to that ancient question, *What is everything made of?*—one that simultaneously answers, *How could it all come to be?* Understanding calls for something simpler and way more explanatory than those sixteen labels for "particles"[b] and their highfalutin mathematics.[c]

> A new discovery skips town on a bike.

In a 1980 lecture called "What Is It All About?" Quine spoke of this in terms of changing ontological commitments:

> This change in ontology, the abandonment of physical objects in favor of pure space-time, proves to be more than a contrived example. The elementary particles have been wavering alarmingly as physics progresses.[753]

So, in the spring of 2010, I was building an ontology that set out to be real. I had pieces of the picture—the whole cosmos; keep it simple; space that's made of granules; Planck-size-tiny; Calabi-Yau manifolds with six dimensions; only one in the beginning; all relational; sequential time. One could see how a consistent understanding might emerge.

But I couldn't make it work. I was stuck in much the same place as so many others; I could not explain the origin of matter.

I not only needed a new answer to that ancient question, I needed an answer that would fit consistently into the new ontology's incipient consistent picture. Fat chance.

Around that time, Susskind was saying,

> Most of us particle physicists believe that if we could examine particles down to some incredibly small size-

---
a I.e., more or less as outlined in this part.
b See Fig. 1 at page 21.
c See chapters 84 and 88 at pages 427 and 442.

scale, we would begin to see the hidden machinery that makes them tick.[754]

I thought those physicists were likely right but I could not see the hidden machinery.

And I found where, in 1997, Rovelli had pointed in what proved to be the right direction,[a] but I failed to find a path to follow:

> The very distinction between spacetime and matter is likely to be ill-founded. . . . I think it fair to say that today we *do not have* a consistent picture of the physical world at all.[755]

It was an awkward kind of problem: What do you think when you cannot quite imagine what it is you want to think about? Or, as Feynman put it,

> A new idea is extremely difficult to think of. It takes a fantastic imagination.[756]

Einstein ran up against this problem early on and said so. The first sign we see shows up in a letter to a student in 1917 (soon after his continuum-based seminal paper on general relativity):

> The question seems to me to be how one can formulate statements about a discontinuum without resorting to a continuum (space-time); the latter would have to be banished from the theory as an extra construction that is not justified by the essence of the problem and that corresponds to nothing "real."[757]

Two decades later he wrote to another colleague that,

> The . . . possibility leads in my opinion to a renunciation of the time-space continuum.[758]

And, later again, to Bohm,

> My opinion is that . . . one has to find a possibility to avoid the continuum (together with space and time) altogether. But I have not the slightest idea what kind of elementary concepts could be used in such a theory.[759]

---

[a] It was the right direction if one thinks of *space* instead of *spacetime*; see also chapter 83 at page 420.

For me, the final link in humankind's long quest to understand loomed into view by chance. In 2005, physicist Sundance Bilson-Thompson came up with the "kind of elementary concepts" Einstein was unable to imagine.

It was the ultimate seminal idea.[760] He said each of the "elementary" particles of the standard model—and so all the universe's matter—could be made from six simple half-twists.[a] He wrote it up in a paper and lodged it on Cornell University's arXiv preprint site.[761]

Here was the "fantastic imagination" Feynman said was needed. And from later personal experience I can attest the author is a super speaker about physics (fig. 25). A terrific salesman for his own groundbreaking concept, not so much.[b]

Fig. 25. Screenshot of Bilson-Thompson speaking about physics

I mean, check out his catchy title, "A topological model of composite preons."

Consider how he sells you why it matters:

The significance of this model is its extreme economy.[762]

A "model"? . . . "Composite preons"? . . . "Extreme economy"? All true. But they give the world no hint his "model" founds a revolutionary new worldview built with the most profound simplicity.

---

a To be technical, it pairs half-twists in-line and braids the pairs in threes to make them all except the photon, which is the null-braid of three pairs of no-twists.
b Let me at once confess I was no better.

And his twist is not even a *thing* in the old mold of particles; it's a *relationship*.[a] Just a you-turn-that-way-relative-to-me perspective; it's perfectly relational. You'd think Smolin would have been ecstatic.

And it turns out Smolin *was* ecstatic. Years later, digging in the desert of his diatribe on failings of string theory, I came upon him saying in 2006,

> Last spring, I happened to see a preprint by a young Australian particle physicist called Sundance O. Bilson-Thompson. . . . As soon as I read the paper, I knew this was the missing idea. . . . [Markopoulou and I] invited Bilson-Thompson to collaborate with us.[763]

And collaborate they did.[764]

Here are a few fragments of what they collaborated on. (Bear with me, please.)

Bilson-Thompson discovered the tweedle, which we will come to see to be the fundamental element of all reality. And, too, we will soon see him setting out to disappear it.

Bilson-Thompson's tweedles—dum and dee—are half-twists in a *ribbon*.[b] Tweedledum is right-handed and Tweedledee is left-handed, so they are mirror images (fig. 26). Of course, you wonder what the ribbon is; he's not about to tell you. And he said little more about his tweedles. (I'll tell you what I think, later on.)

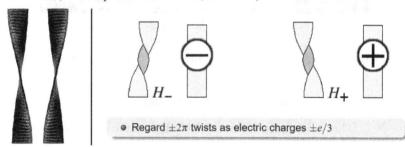

Fig. 26. Tweedles: dee and dum

Fig. 27. Making Tweedles *disappear*

• Regard $\pm 2\pi$ twists as electric charges $\pm e/3$

---

a Having composed his particles with something as real and simple as twists, Bilson-Thompson takes a swift detour away from reality: "The twist of tweedles may be viewed as a spin in some abstract space, akin to isospin." See note 762, his note 14.

b Just *what* is twisted is a question on which we will have more to say. He off-handedly calls them *ribbons*, without seeking to explain.

Part 1: Some Seminal Ideas   195

He equates twist with electric charge. The size of each tweedle's charge is one-sixth of the charge on the electron or $e/6$; the sign, plus or minus, depends on the twist direction.

Bilson-Thompson's particles are all the simple braids of three helons.[a] He depicted these braids to show their two twists as single charges, so his tweedles have vanished in these diagrams (see fig. 28). This is a bit confusing. But each of the three flat ribbons in subsequent diagrams is a helon so it means two tweedles.[b] The top and bottom bars are . . . well, let's leave them for later too.

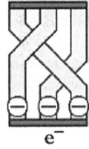

Fig. 28. Bilson-Thompson's electron made of twists (shown as charges)

For example, fig. 28[765] is his diagram for the electron. What you see is a braid of three ribbons. Each minus sign signifies two left-handed *tweedles*[c] — two half-*twists* in each ribbon through an angle of $-\pi$ (or $-180°$). That is, each of his ribbons has one full counterclockwise twist (like the left-side of fig. 29[766]). Thus, his minus signs in fig. 28 each stand for a charge of twice $-e/6$, or $-e/3$.[d] And they disappear the tweedles from our view.

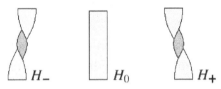

Fig. 29. The negative, neutral and positive helons

He does not take his diagrams as seriously as I do, telling me recently they . . .

> . . . are a sort of "first attempt" at representing the idea, and not a final version of what I think spacetime is actually doing.[767]

---

a And so it is referred to as a helon model.
b Except in the case of the photon.
c He makes a play on Lewis Carroll's twins, Tweedledum and Tweedledee. Bilson-Thompson's *dum* and *dee* have opposite half-twists, and so (like them) are mirror-image twins. Such wordplay is not new in particle physics (check out *quarks*).
d Here, and throughout, I use e as the *magnitude* of the electron's (and positron's and proton's) charge and show each charge's sign explicitly.

His in-line paired tweedles (or *helons*) come in three possible combinations, two left (H_-); or two right (H_+); (which each amounts to a full twist); and left+right or right+left each of which has no net twist (H_0). So helons (fig. 29) have charges $-e/3$, no charge, or $+e/3$.[a]

If his "model" were to depict only the electron it would be intriguing, maybe nothing more. But it isn't only the electron; it is all the standard model's particles physics has found in all its years of atom smashing. That is, his braided helons work for everything that makes up everything physicists have ever seen.

So, Sundance and his "model" moved to the Perimeter Institute for Theoretical Physics in Waterloo in Canada (aka PI). It was the died-and-gone-to-heaven destination for a physicist in his line of work. And he went to work with Smolin and Markopoulou—the died-and-gone-to-heaven collaborators for a physicist with his research interests.

And so too, Smolin was soon writing of how Sundance's model . . .

> . . . raises many questions, and answering them is now my primary goal.[768]

Thus, my goal seemed aligned with Smolin's. But subsequent events suggest those "many questions" may have somehow led to too few answers.

The Perimeter Institute had only recently come into existence, thanks to entrepreneur Mike Lazaridis, who saw that (as he later said),

> We need a new discovery.[769]

He kicked in a hundred million or so dollars of his pocket money in a bid to make a discovery happen. And it happened, an amazing new discovery! Maybe the most amazing ever.

Then it went away.

It may not be beyond recall. It's key to understanding where we are. Some may wonder how we can be sure that it is true. Of course, we can't.

But it is the best shot we've ever had. We *can* see how dozens of disparate seminal ideas fit together into a consistent explanation in which tweedle-particles are fundamental. We *can* see a consistent

---

a And we will see he also gets a version with no charge with simply no twists.

picture that explains things that for centuries were mired in intellectual applesauce.[a]

Bilson-Thompson's central concept is the sixteen standard model particles can all be made of simply braided[b] pairs of twists. Six twists per particle (except the photon, whose helons are all simply twistless). Whatever he is twisting must have two dimensions, but he speaks only of ribbons (or more recently of *links*).

If it is so, *none* of the sixteen particles that, as Smolin said politely, "the standard model deems elementary"[770] are in fact elementary. *They are all composite!* This is a huge—ten on the physics Richter scale—convulsion in the standard-model's worldview. A revolution in some sixty years of fundamental physics.[771] It upends our view of literally everything.

Bilson-Thompson's work may someday be seen as one of the most profound advances in science ever, bringing fundamental physics into contact with reality. Meanwhile, as can happen, it is almost universally unknown. It is going nowhere, for reasons I mostly do not know. It feels like that last line from poet T. S. Eliot,[c]

> Not with a bang but a whimper.[772]

Let's turn back the clock and take a closer look at what I do know.

To recap, in the spring of 2010, I was working on the cosmic mystery of what went on *before* the Big Bang. I was wrapping it loosely into a detective story, with a view to what a reviewer called (he meant it kindly) . . .

> . . . spoon feeding advanced physics to the everyday person.[773]

I had pinned down how *space* came into existence; I could see my way to sort through the obscurities of time; I could see almost all the pieces; I could put them all together. But I could not explain

---
a Hence, if someone were to ask me how I know it's so, I'd say, *I don't; but what's your explanation?* In other words, let's see where we can get with what we've got.
b His paper deals mainly with the bosons and the first-generation fermions, from which all stable matter in the universe is made, and on which I will focus here. His model extends to the unstable second- and third-generation fermions, which are made in particle accelerators and quickly decay. Their braids are more complex.
c He too won a Nobel Prize.

where the universe's matter came from. Clearly, *not* from a point of infinite density and zero size as the standard model of cosmology so haplessly prescribes. Neither I nor anyone I knew of thought that really happened. So, where *did* it come from? If it had no answer, my book would tell the story of an empty universe.[774]

I happened on the primal paper through a momentary error. While I was browsing the Perimeter Institute website, there was smiling postdoc Sundance Bilson-Thompson, from *Adelaide* (this is what caught my eye; it's where I was born). With a link to a paper on Cornell's arXiv preprint site—something about preons. Not sure why I clicked instead of backing out; and then I started reading. In moments it hit me; this is where the matter comes from! This smiling guy from Adelaide had figured out the answer (but with not a word about the way the universe began).[775]

He had solved Lorentz's and Einstein's problem of how matter could be made of space.[a] (He didn't say that either.)

And, though this too he didn't mention (and to this day might not concur), his concept vaguely fit in with a kind of woven substructure in Planck-scale space with roots in a 1992 paper—of which Rovelli and Smolin were co-authors:

> The basic idea is to weave the classical metric out of quantum loops by spacing them so that (on an average) precisely one line crosses every surface element whose area . . . is one Planck unit.[776]

My euphoria soon vanished as I had a whole new problem: This smiling guy was set to be lead character in my detective story. I could not write that without talking to him. I picked up the phone.

"Perimeter Institute." Bright and helpful-sounding for a Friday afternoon.

"Can I speak with Dr. Sundance Bilson-Thompson?"

"Oh," she said, "he just left."

"Will he be in on Monday?"

"No. He left."

---

[a] See chapter 11 at note 255.

"When do you expect him?"

"You don't understand. He left."

Belatedly I got the message. "Uh . . . do you know where he went?"

"He didn't say."

Getting desperate, "Does anybody know?"

"You could ask Jonathan."

Jonathan, when he picked up, was Dr. Hackett. He and Sundance were co-workers in the PI postdoc world. He too tried to help.

"I think he's on his way to Adelaide."

"Did he say when he would get there?"

"I'm not sure. . . . But he left on a bicycle."

So began weeks of search for Sundance and of total writer's block. Someone at the Special Research Centre for the Subatomic Structure of Matter at Adelaide University said they thought he might be coming there in June. But, no, no one knew where he was.

One day I bumped into a friend[a] and shared my tale of woe. Next day she called.

"Did you know he has a Facebook page?"

"What's that?"

"Well, he has three friends." She spelled them out.

Back to Google, which in those days was still building both its algorithm and its corpus. Two swings and two misses. The third was a left-field hit. It led to a blog about their travels.[777] And there was Sundance smiling at the Gulf of Mexico with his bike (fig. 30). And companion, Yana.

Fig. 30. Sundance and the bicycle

Latest post, only hours old, had them chatting with a server, Sarah, at The Point coffee and smoothie bar in San Diego. More Google. Fingers punched the phone.

"Hallo."

"I'm looking for Sarah."

"This is Sarah."

---

a Susan Guindon, formerly my executive assistant.

And suddenly it dawned on me that I was stalking Sundance. Did he not want to be found? I mumbled something harmless, mentioning his name.

"Oh," she said—same tone, same cadence, sounding like the same voice that said I had missed him as he left the PI weeks ago—". . . he just left."

I was speechless.

Into my silence, she added, "But they left some stuff at my place; maybe he'll be back."

"He doesn't know me but I'd like to talk to him about his work. Could you ask him to call me collect?"

And he did, sounding harassed. US Immigration had fined them for entering from Mexico by boat with the wrong visa, or some such. Then the boat people who offered them a ride to the Marquesas Islands—finding he and Yana were bike people, and it soon being cyclone season—now felt *boat* people might make a better crew.[a]

Conscious of the slender link, I asked where they would go.

"We might try San Francisco. Better chance to hitch a ride."

Days later I was on a plane. Then, settling in my room, laptop on the desk, I still could not write a word. In the morning, Sundance called. He and Yana were down in the lobby.

His first question was to be expected. "How did you find us?"

"Long story. Let's find the coffee shop. . . . And, by the way, congratulations."

"For what?" he asked as we walked.

"Your Nobel Prize."

"For what?"

"For discovering the final atom," I said. "Guaranteed Nobel just waiting for experimental confirmation."

"I thought I had. But I can't prove it."

"Neither can I. But I can explain it."

We talked over the next two days.

In the years that followed, we had occasional long calls. It was never clear to me discovering the final atom was something he wanted.

---

[a] From San Diego to the Marquesas is three thousand miles of open ocean.

In 2019, we spent some days in Winnipeg. He had asked for help. I tried. It went nowhere; he would lay no further claim to his discovery.

Chatting in the garden, I said, "I see your ontology as being more important than your physics."[a] In the moment he did seem to agree. Yet on the other hand it wasn't clear he saw his "model" as reflecting something real. Maybe he makes the same mistake the string theory people make with their manifold, viewing it only as math instead of ontologic message.

More than a decade after our first meeting, I am writing this because his work is worth explaining. It's the story of how making space goes hand in hand with making matter. It really is his story but he won't stand up and say so.

The closest he has come to owning his own seminal idea (and to the new ontology) was the last line of his last paper for more than a decade, in 2012:

> This brings the prospect of emergent, interacting matter from discrete spacetime a step closer.[778]

Almost right. And it was not just one step, it's more like a giant leap. What went wrong?

Over the years he told me much, but not the whole picture. He was traumatized by his experience; that much was clear. I read a 1986 interview where Hawking, asked about how science changes worldviews, said,

> [M]any scientists feel if they start talking about such questions, they'll be regarded as cranks rather than as serious scientists.[779]

Sundance never said as much to me but Hawking's observation has resonance with many conversations. And I should note that Hawking, prompted, went right on to say,

> No, not including me.

This was a conjunction made in heaven I failed to make happen. I'm not good at failing. Do Sundance and I between us (often not

---

[a] Winnipeg, 15 August 2019.

agreeing) have all this exactly right? It seems unlikely. Should more have come out of our chance intersection? I should think so.

The bigger picture looks like a PI train wreck; Markopoulou ditched physics altogether and went into business; Hackett went to work with a bank finance group; Smolin's "primary goal" vanished like the ether; Lazaridis's discovery got on a bicycle and hitched a ride to the Antipodes.

Maybe there's a simple explanation. PI spends a lot of public money.[780] Seems there are good questions that could use a few good answers.

We may yet see. Stay tuned.

PART II

GETTING REAL

In the end we are driven to search for what we hope will turn out to be the correct ontology of the world. After all, it is the desire to understand what reality is like that burns deepest in the soul of any true physicist.

Lucien Hardy (2011)[781]

36
ROAD CLOSED

> Man tries to make for himself . . . a simplified and intelligible picture of the world; he then tries . . . to substitute this cosmos of his for the world of experience, and thus to overcome it.
>
> Albert Einstein (1918)[782]

In this part, we set out to see how our simple approach to a real ontology can construct a simple way to solve a raft of problems. Successful unreal physics blocks our road. Seeing what stands in the way is our first step to find a way around it.

Given my remarks on physics as fiction, the reader may wonder why I often turn to fundamental physics in the quest for what is real. Partly it's because physics offers an almost endless stream of fundamental contradictions, i.e., clues! Another reason is it *ought* to be our best window into reality: It's no coincidence that it's mostly a few physicists these days who, like those few Greeks in ancient times, wrestle with the fundamental questions of our world. As we have already seen, key insights are concealed within its murky depths.

> Progress is blocked by seeming success.

Or, as Barrow put it,

> The question of why the Universe is as it is, is inextricably linked to that of why fundamental physics is the way it is.[783]

So, without delving too deeply, we should try to understand "why fundamental physics is the way it is."

Recall that Einstein warned us of the road-closed problem:

> Concepts that have proven useful in ordering things easily achieve such authority over us that we forget their earthly origins and accept them as unalterable givens. . . .

> The path of scientific progress is often made impassable for a long time by such errors.[a]

Did he foresee his own concepts would turn into "unalterable givens"?

In March of 1905, he wrote the paper that kicked off the theory of the world we call quantum theory.[784]

Three months later, he published another paper that kicked off another theory of the world. It was soon called relativity.[785]

He came to like the second better, while most everyone, including almost all the physicists, feasted on the first. Quantum theory was the reason we were given for his Nobel Prize.[786]

Both theories were soon stunning success stories. They went from one triumph to another for more than a hundred years. Between them they provide predictions for any experiment one can devise.[b]

The reader does not need to understand them. Rather we need to understand why we are stuck with them and so is physics; and why, singly or together, they don't offer us a way to understand the world.[c]

The message from this situation should be clear: Both theories are amazingly useful but both their mutually inconsistent worldviews are profoundly wrong.

Yet these two views continue to define the central goal of fundamental physics: To mix and match or somehow find some common ground between the two. Even its forced-marriage name, *quantum gravity*, speaks to their ongoing dominance. Despite their built-in contradictions and their failures at explaining, it is hard to think one's way past their success.

In this stymied situation, once we see the roadblocks what's the way to progress?

Should we maybe keep one foot on the seeming-solid ground of one theory while fumbling for some new foothold?[d] Should we try to

---

a See note 23.

b Note however, this is much more true of quantum theory than it is of relativity; see, for example, Peebles' dissection of general relativity as largely a social construction; note 60 at his pages 65 to 79.

c Once they left Einstein's hands, understanding reality was not even an objective.

d Much tried, so far this has failed.

find some in-between terrain?[a] Or should we set theory aside while we pursue a new worldview, hoping it may explain why relativity and quantum theory are what Peebles called . . .

. . . useful approximations to the way reality operates.[787]

Another way to put it is, does one keep following math in hopes it will throw up a real new ontology; or does one fashion an ontology that's real and then develop its new math?

One might say the former of these strategies is *timid*; the latter is *bold*.

In practice, physics (and even philosophy) tends to *timid*.

This book bets on *bold*.

---

a This has been the dominant quantum-gravity strategy for decades. That Wikipedia lists two dozen approaches speaks for itself.

37
ON NOT BEING LOST

> It is mathematics, more than anything else, that is responsible for the obscurity that surrounds the creative processes of theoretical physics.
>
> Lee Smolin (1997)[788]

As we embark on our voyage into a new view of ourselves and our world, a word of caution about essential messages from math we will encounter on the way: We will aim to decode them, but we must take care to not get lost.

It's not that we'll indulge in math—we will steer clear of it.[a] But we will set out to learn some lessons that arise from math, so let's be wary about physics' dirty secret: For more than a hundred years it *has* been lost in math.

Few insiders say so. Speaking truth to power is no better career move in physics than in any other trade. Smolin did it,[789] having switched fields, and survived. Physicist Sabine Hossenfelder did it,[790] changed careers, and said, before her book *Lost in Math* was published,

| Math is a good servant and a poor master.

> What the book really is about is how to abuse mathematics while pretending to do science. . . . This isn't a nice book and sadly it's foreseeable most of my colleagues will hate it. By writing it, I waived my hopes of ever getting tenure.[791]

Then even Hossenfelder headed for a halfway house:

---

[a] Aside from checking quantities the reader can skip, we'll get into a bit of simple arithmetic, of the $2 \times 2 = 4$ or $1 + 1 = 2$ kind; see chapter 40 at page 222 and chapter 106 at page 528.

> Some of my colleagues indeed believe that the math of our theories . . . is real. Personally, I prefer to merely say it describes reality. . . . How math connects to reality is a mystery that plagued philosophers long before there were scientists, and we aren't any wiser today. But luckily we can use the math without solving the mystery.[792]

Actually, when we use the math we tend to hide the mystery and bury the reality it almost invariably fails to truly describe. (But then, that was not its job.)

In my experience, the observation in the epigram (above) is accurate. This creates a problem that affects our quest: Math *is* an obstacle, but physics eats and drinks and breathes it. So, in distilling concepts from the world of physics we should have regard to roles math—with inbuilt assumptions—plays in revealing or concealing what is real.

The new ontology offers some insight: It shows there is real math,[793] and there's much more that is math fiction.[a] We are not there yet, but our story will explain the difference.[b] And the math fiction is where physicists get lost.

Real math is a small hall of simple pleasures.[c] Unreal math by contrast is an endless panorama in which pioneering physicists may stray all unawares, seeking pristine splendors, lured by siren songs, driven by a limiting ambition—that of finding lasting beauty in fresh fiction. We may hope real math might blaze a better trail.[d]

A real ontology must grapple with the until now neglected distinction between these two kinds of math.

It might seem simple sense for physicists—especially those who work at or near the fundamental-physics frontier—to make their own strategic choices. Yet convenience—of one kind or another—tends to tempt them into letting their math lead the way and make their most important choices for them.

---

a See part 6.
b See chapter 110 at page 543.
c But note, real math is relatively unexplored.
d I reserve the term *real math* for math that meets the test of actual existence in the universe. So, it does not here mean the math of so-called real numbers, which we will see does not deserve the name because it does not meet that test.

Einstein may have been the last great philosopher-physicist. He first sought understanding, *then* transcribed it into mathematics. Yet even Einstein, in his latter days, espoused a leading role for math in searching for new fundamental insights, saying . . .

> . . . the creative principle resides in mathematics.[794]

In the century that followed his prodigious decade (1905–15), physics feasted on his legacy while discarding his earlier approach. Today's physics searches for successful mathematics, then sometimes tries—with scant success—to explicate it as a vision of reality.

The context for these ills of the creative process is a culture within physics of disdain for philosophy, the science[795] whose task it is or ought to be to steer the other sciences away from such dead ends.[796] So many leading physicists have dissed philosophy it would be invidious to single out a few.

For they are clearly wrong. As Rovelli (not the only such voice in the wilderness) said,

> Contrary to claims about the irrelevance of philosophy for science, philosophy has always had, and still has, far more influence on physics than commonly assumed.[797]

The strategic issue here for the philosophy of physics is the halls of imaginary mathematics are vast beyond human comprehension. Getting lost in them is no mere risk; once one sets foot there with no guide it becomes all but certain.

Beyond Rovelli's call for "clearer philosophical reflection" on our method,[798] we need new vision focused on reality, vision that winnows anew the best harvests of four centuries of physics without getting lost in all its almost always unreal math.[799]

Some physicists with philosophic inclinations are coming close to that new vision. Rovelli is a lead example. He sees the purpose of scientific research as . . .

> . . . to understand how the world functions. To construct and develop an image of the world, a conceptual structure to enable us to think about it.[800]

In other words, he seeks a real ontology. But in the end, he too can get lost in the math and stuck in ontologic habit.

For hundreds of years, math has been the making of physics. But math is only language.[a] Its key virtues for expounding physics are brevity, definition, and consistency. Each has its limitations.

Math's brevity tends to conceal—even from expert users—assumptions inherent in what it is saying. For example, thousands of papers use the math of general relativity to expound on space. All of them assume space is continuous. As Riemann noted, to apply such math in continuous space one needs a *metric*, some kind of scale from outside the space[b] that enables one to make measurements within it.[c] Continuous space has no natural metric so physicists must invent one. Their studies necessarily neglect the space and endlessly explore their metric.

Even worse, most merge space and time into a mongrel concept, spacetime. As we will see, though certainly convenient—as it brings even better brevity—it is not real; there's no such thing as spacetime in this universe.[d] Yet many cleave to it and thereby cloud their thinking.

Math's brevity, in other words, may prove expensive.

Its virtue of definition, too, has limits that tend to be hidden. In the end, math's definitions are expressed in words.[e] Those words bear messages or harbor ambiguities that some math users may forget or fail to apprehend.

As to math's virtue of consistency, it is mostly an illusion. The math used in physics may appear consistent but it's not.[f] Even the simple stuff: Russell and Whitehead's epic effort to construct arithmetic on a consistent logical foundation failed.[801]

Then, in 1931, in what some see as the greatest achievement in the field of logic, mathematician and philosopher Kurt Gödel proved all systems of arithmetic embrace propositions they can neither prove nor disprove.[802] Another way to put it is, even simple arithmetic is

---

a Effectively, many languages.
b Yet everything is inextricably inside it; see chapter 24 at page 141.
c For more on metrics, see chapter 96 at page 475.
d See chapter 73 at page 378.
e See also, chapter 72 at page 373.
f See part 6.

condemned by logic to be either incomplete or inconsistent.[803] Or, said Baez,

> [A]ny system of arithmetic that can prove itself consistent must be inconsistent.[804]

A wider view only gets worse. Barrow said,

> All our surest statements about the nature of the world are mathematical statements, yet we do not know what mathematics "is": we know neither why it works nor where it works; if it fails or how it fails.[805]

Gödel's work cut to the very taproot of mathematics.[a] Fifty years on, inconsistencies in math and physics are widely accepted as necessary evils. For example, philosopher of physics Mathias Frisch said,

> Even without the certain prospect of a "correct" theory waiting in the wings, very good and interesting physics can be done with an inconsistent theory.[806]

For my part, I won't have a bar of it. The universe *exists* so it must be consistent; so too must be *its* math and *its* physics.

This leaves us facing a strategic question, *What is the ultimate creative source?* The human mind can devise axioms that give rise to a body of math like the Euclidean geometry we learn in school.[807] It may be interesting, indeed useful, but its entire enterprise is fiction; notwithstanding neat appearances, the world is not that way.

An alternative approach understands the universe itself to be the creative source. As the new ontology explains in fine detail, the universe proactively makes everything. So, we should seek the axioms the universe provides.

This will of course constrain the math, but at least the result may be real. In this book, we'll follow this path.

In sum, the search for understanding needs to keep math in its place since using math to blaze the trail is choosing to get lost.

---

[a] See also chapter 114 at page 559.

38
SEARCH FOR SENSE

> The existence of a single structure that unifies such a broad range of physical and mathematical ideas, and many others as well, is unexpected and remarkable.
>
> Joseph Polchinski (2004)[808]

Physicist Joseph Polchinski was rhapsodizing on the beauty of duality in physics.[a] There is indeed much beauty, as well as usefulness, that may be found there. But while the beauty has its boosters, it has often failed to be a good guide for new fundamental physics.[809]

As we are seeing illustrated in these pages, there's a growing crisis in the philosophy of science that targets physics specially as it demands, *What basis can we find for validating science that inhabits realms we can't ever observe?* Such are the realms that physics is (and we are, here) exploring, a whole vast universe and a vanishingly tiny quantum of space; both evermore outside our view.

> We are exploring places where eyes, ears, and noses cannot help us.

Empiricism—the precept that anything that cannot be observed by sensory experience is not science[810]—gained an enduring hold on physics[811] through the ascendancy of logical positivism a hundred or more years ago.[b]

It had many promoters; Ayer was a leading advocate amid the fashionable thirties set.[c] He said,

> [O]verall it is the mark of the empiricist that he looks to sense perception, if not as the sole legitimate source of

---

a We will find a true duality in chapter 42 at page 236.
b See chapter 9 at page 57.
c To be clear, I quote Ayer due to his leading role as English-language proselytizer, not as original source of these ideas.

> any true belief about the "external" world, then at least as a final court of appeal which any acceptable theory must satisfy.[812]

In his foundational work he said,

> To attempt to make use of causal laws in order to infer from the occurrence of observed events to the existence of things that are outside the scope of any possible observation is not merely to put forward hypotheses for which there could not be any valid evidence; it is to extend the use of the concept of causality beyond the field of its significant application.[813]

That is, if we can't see it, don't pretend to know it. He allowed it's not essential we are now able to see, but insisted we must at least foresee a way for that to become possible. This kind of concession might (or might not) help dark energy and dark matter to evade the empiricist hatchet,[814] but it can't help forever-lost-to-us spans of the universe or forever-invisible Planck-size elements of space.

Thus, for example, in his definitive tome—which his publisher claimed "founded logical positivism—and modern British philosophy"—Ayer cast aside the possibility of any ultimate solution to Zeno's paradox of the One and the Many:[a]

> It is admitted both by monists, who maintain that reality is one substance, and by pluralists, who maintain that reality is many, that it is impossible to imagine any empirical situation which would be relevant to the solution of their dispute.[815]

While this philosophical movement now seems far from the front of human affairs, it profoundly affected physics in the first half of the nineteen hundreds and invaded the wider world so far that, in his 1951 novel, *The End of the Affair,* author Graham Greene could fling off (confident his readers would all follow the allusion),

> Ayer, Russell—they were the fashion today.[816]

---

[a] See chapter 10 at page 63.

It's now out of fashion, but, as philosopher of physics Mario Bunge said, the old philosophy still flourishes within the halls of physics.[a]

Back in Ayer's heyday no one could imagine any test for anything some fourteen billion years ago, let alone a test for something so small our finest probe cannot detect decillions of them. Yet today a hot item in cosmology—though in truth it's desperately cold[b]—is a false-color picture of the almost-fourteen-billion-year-old cosmos.[c] And that small something is the central player in string theory,[d] which many physicists bet their careers could come to be the theory of everything (and would give their eyeteeth to find such a test).[e]

Both these developments began in Ayer's day. Too bad he died before they hove into some sort of view, but he did leave us the empiricist response, which took a shot at any possible Planck-scale ontology:

> The metaphysical doctrine which is upheld by rationalists, and rejected by empiricists, is that there exists a supra-sensible world which is the object of a purely intellectual intuition and is alone wholly real. We have already dealt with this doctrine . . . and seen that it is not even false but senseless. For no empirical observation could have the slightest tendency to establish any conclusion concerning the properties, or even the existence, of a supra-sensible world. And therefore we are entitled to deny the possibility of such a world.[817]

Later in life, even Ayer came to see such certainties as less than clear:

> I do not think that the distinction between theory and fact is altogether sharp. What we count as a fact is to some extent a function of our theories and even more a function of the system of concepts which are embodied in our language.[818]

---

a See note 811.
b The background radiation has a characteristic temperature less than three degrees Kelvin.
c See note 424.
d See chapter 14 at page 87.
e See chapter 15 at page 95.

Recent doctrinal developments have weakened the grip of empiricism but have not replaced it with a clear new standard for deciding what is real.[819]

The resulting intellectual paralysis—the banning, as it were, of travel beyond both frontiers—is leaving lasting imprints on the human side of fundamental physics, so we hear calls from frustrated doctors of philosophy to abandon their purported field of expertise.

The Chinese word for crisis (fig. 31) has two characters. One, *wei*, can translate as "danger." It's not quite simply true that (as is often said) the other, *ji*, means "opportunity";[820] but the crisis in physics[821] is indeed pregnant with new opportunity.

Fig. 31. The Chinese word *weiji*

My case for opportunity is that the crisis opens new ways to set science on a firm foundation. Logical positivism's empiricist principle misapprehends the nature of the human venture. It is simplistic; we are subtle. Its errors are twofold; on one hand its logic overvalues the perceptive power of the "sensory experience," and on the other it undervalues the conceptive power of our minds.[a] These are matters best not addressed in absolute terms.

To bring us "nearer to finding a solution"—in Ayer's phrase quoted below[b]—to many intractable problems, both ancient and modern, we need first to put logical positivism in its place by understanding the realities of its vaunted sense perception.

The empiricist principle says only our senses—vision, hearing, touch, taste, and smell—keep us grounded in reality. But the shortcomings of our serially superseded physics suggest that these senses are not making a great job of it.[c] And, powerful though the latter four

---

a In its defense, these concepts were not well understood when positivism was in its prime.
b See note 1531.
c This is not to deny the roles our senses do play; see chapter 2 at page 22.

senses are at times, our vision bears the bigger burden.[a] What does it deliver? A tiny fraction of the photon spectrum.[822]

The harsh fact is, we are prisoners of pitifully limited sensory input.

That our visual "sensory experience" has at best a slim connection with physical reality was brought home to me in my teens. In 1959, scientist Edwin Land's work shattered the ruling remnant of the Newtonian trichromatic theory of color vision:

> Whereas in color-mixing theory the wave-lengths of the stimuli and the energy content at each wave-length are significant in determining the sense of color, our experiments show that in images neither the wave-length of the stimulus nor the energy at each wave-length determines the color. This departure from what we expect on the basis of colorimetry is not a small effect but is complete.[823]

I was intrigued. If two colors could be seen as full color and (in other setups) two points of view could be seen as 3-D, could just two colors, each also coding one of two points of view, be seen as full color *and* 3-D? I did simple experiments whereby observers viewed through two filters a projection, through the same two filters, of two black-and-white-film images of a scene shot with 3-inch separation through the same two filters. They reported a full-color scene (that they had never seen) even when the two filters in each pair were two narrow-band yellows! That is, they observed reds and blues though only yellow photons had reached the films and only yellow photons were reaching their eyes. And they saw it in 3-D!

These simple experiments gave me a healthy respect for the brain's ability to supplement our sensory perceptions with whatever's needed to create consistent sense from partial information—without

---

[a] This was long understood: On page one of part one of book one of his *Metaphysics* (note 1982), Aristotle said (continuing the first quote in this book), "All men by nature desire to know. An indication of this is the delight we take in our senses; for even apart from their usefulness they are loved for themselves; and above all others the sense of sight." Information theorists would say he was right; the substance behind the saying A *picture is worth a thousand words* is that the information-carrying capacity (or bandwidth) of vision is far higher than that of other senses.

our being aware it is doing this.[824] They left me skeptical about that sensory-experience criterion for what is real.

And today maybe we need a broader basis for accepting or rejecting "the reality of the physical world."[a]

There is an extensive literature on what happens when we learn. My precis of it is, from birth[825] we grow and prune an intricately interconnected set of neurons[826] that receive sensory inputs and are in constant electrical interaction on a millisecond timescale.[b] The neuron is an amazingly sensitive and subtle analog to digital converter.[827]

In this way the brain sets up our internal model of the universe. The model may be further conditioned by experience of others, learning from accumulated knowledge.[c]

But our internal model is *not* only a collection of experience and readings. (Indeed, we recall few of them, and those often imprecisely.) Rather, it is our attempt to integrate experience—or, one might say, to understand it—as a whole.[828]

Infants demonstrate the power of this model-building process. We watch them learn language without being taught, merely by being exposed to speech amid other sounds including nonsense noises babbled at them. If their sound-stream bears several languages, they learn them all without confusion.[829]

Individually, the reality that we experience is the wholeness of our world, the result of a lifelong search for understanding. Collectively, it is a multi-generationally-sustained gestalt.[d] (One aspect of the latter is what we call science.)

If this sketch is at all perceptive, we can better understand both the successes and the failings of our science. We can see why the short-term, situational, sensory-experience criterion has always been ill conceived for, and is now increasingly estranged from, physics practice.

---

a Ayer's phrase; note 813.
b See further, chapter 134 at page 647.
c It is also degraded when false or misleading information is accepted.
d In using this now-English word, I do *not* mean to invoke the gestalt theory of learning or gestalt psychology, either of which might seem to have some relevance.

As we venture into Planck-scale realms, we need to face the fact that we will never see something so small with any remotely conceivable technology. Nor will we ever see most of our universe. If we are to embrace either into our worldview, we must reach them with our reason, taking such care as we may to ensure we are not merely fantasizing. This means—as our eyes appear to do—leaning on consistency of the whole picture.

As Polchinski said some years ago,

> There is a danger of defining science too rigidly, so that one might decree that any discussion of the physics at 10^{-33} cm is unscientific because it is beyond reach of direct observation.[830]

A more practical (and, dare I suggest, a more mature) definition might weigh new science by its contribution to a seamless understanding of our single universe, what one could call *explanatory power*. This we might reasonably accept as—if not clearly real—at least a move toward the right direction: It might make consistent sense.[a]

Today's physics has little to say that meets this test.

Meanwhile, let's not belittle useful fiction (even if it's based on sensory experience), a category that we'll see embraces almost all the long unfolding story-upon-story of physics built like Rome upon the layered ruins of past glories.

Let's be clear, though, it is fiction.

---

a That is, we may seek a single understanding of the world across the full range from Planck scale to the entire universe.

39
CHERRY PICKING

> Some people think we must just accept that quantum physics explains the material world in terms we find impossible to square with our experience. . . . Others think there must be some better, more intuitive theory out there that we've yet to discover.
>
> Richard Webb (2020)[831]

Science journalist Richard Webb was executive editor of *New Scientist*. Though relatively few, his "others" seem to be more numerous—or, should I say, less rare?—of late. But they may need more than just "some better, more intuitive theory." More real understanding might enable better theory to emerge.

The new ontology brings together a few dozen extant ideas. They are a mere sliver of what's sometimes called the wisdom of the ages. These few were chosen for a simple reason; interwoven they make sense. They solve a host of problems. They offer us an opportunity to understand the universe in every way at every time on every scale.

> We build a backbone with sixteen seminal ideas.

This new view hangs together where the old view falls apart. It makes sense of a range of scientific observations with their supporting equations that fit data more or less well but—as Webb (and others) note—do not make sense.

This new view may seem strange at first; so, dear reader, kindly brace yourself for some new ways to think. The new view will confirm, as the cover of Rovelli's recent book proclaims, *Reality Is Not What It Seems*.

Unlike that book (and others of its ilk), this one will say—indeed, will spell out in finest detail—what reality *is*.

At this point we can compile an ontological backbone[a] from the seminal ideas we chose with a view to being consistent and real. We are seeking something, and,

- Ockham: It is simple.
- Zeno: It is both one thing and many, and it needs to work in jerks.
- Planck: It is a quantum that's far smaller than an atom.
- Anaximander: It leads to a cosmos.
- Riemann (and Einstein): Its space is granular.
- Lemaître: It begins with a single quantum.
- Aristotle and Lorentz and Einstein: Its space is something.
- Smolin: It is made of relationships.
- Calabi and Yau: It has six extra dimensions with a tiny volume.
- Hund: It tunnels to do what cannot be done.
- Hubble: It makes space expand.
- Guth: It grows exponentially for a brief moment.
- Riemann/Einstein: It is finite and endless.
- Einstein/Smolin: It leads to a single theory.
- Bilson-Thompson: Its mass-energy is braided twisted ribbons.
- Simenon: It explains all the evidence.

This is no doubt a demanding prescription. But let's take all these ideas seriously. Let's treat them as fundamental. Let's commingle them constructively and see where we may finish up.

By their nature, many of these ideas work at scales—tiny or vast—outside experience. What we conclude must happen at those scales—where all the action is—may seem surpassing strange.

Strange is okay. But it must make consistent sense.

---

a See chapter 4 at page 34.

40
THE POWER OF TWO

> Anyone who believes exponential growth can go on forever in a finite world is either a madman or an economist.
>
> Kenneth Boulding (c. 1973)[832]

Kenneth Boulding was unusual; he was an economist *and* a philosopher. His point here was based on the seemingly sound assumption that exponential growth "in a finite world" always comes up against some limit due to finite resources. It's simple math, easy to model.

But it ain't necessarily so;[a] and to Boulding's list of true believers we should add *cosmologist*.

The new ontology's exponential growth consumed no resources. Indeed, it created exponential new resources, so it had no way to run into a finite-resource limit. Therein lie two tales; one is about the universe's amazingly fast initial growth; the other (which we will get to later[b]) is about how that amazingly fast growth came to an end notwithstanding that it had no limit on its resources.

| Exponential power grows a universe.

Unlike cosmic inflation's manic tale that starts up at some second instant,[c] the new ontology's universe grew exponentially[d] from its beginning.

Here we set out to dissect in detail what it did and how it did it. And why, starting out with almost nothing, it did not run out of gas.

We begin, as Lemaître did, with just a single quantum.[833] It was *not*—though he (backed by the then emerging standard model) said

---
a With apologies to composer George Gershwin.
b See chapter 54 at page 294.
c See chapter 21 at page 126.
d This is what the standard model terms *exponential expansion*.

Part II: Getting Real   223

it was—a quantum somehow packed with all the universe's mass and energy.[a]

Rather, it was a massless quantum, a granule of Riemann's granular space.[b] It was, as Planck would later have it, extremely small.[c] It was, as string theory would later want, a Calabi-Yau manifold.[d] What *was* packed in it was its six tangled dimensions.[e]

But keep in mind what string theories *don't* call for, *real* manifolds.[f] String theorists regard their manifolds as mathematical abstractions nestled in continuous space; for example,

> This means that each point of space-time has associated with it a 6- or 7-dimensional knot of tangled geometry which is consistent with the parameters of a Calabi-Yau space.[834]

Even Baez, my pick for one who might lead a breakthrough to the mathematics of real physics,[g] can be seen to wander at haphazard amid quantum-gravity math of cute concepts like tangles let loose in a zoo of unreal spaces.[835] His conclusion tellingly opens with,

> The problem, of course, is that we have little idea what the physical observables are in a . . . formulation of quantum gravity.[836]

Step one in the new ontology's account of the origin of the universe is a replication of a real Calabi-Yau space: One space quantum became two![h]

Looking ahead, we will see this is the *only* law of nature, on which all laws of physics depend. For this reason, I've expressed it in that shorthand, $1 \rightarrow 2$.

How did that first space quantum do this? We don't know. We don't understand the machinery of even large-scale (relative to

---

a See chapter 17 at page 105.
b See chapter 12 at page 74.
c See chapter 13 at page 76.
d See chapters 15 and 31 at pages 103 and 176.
e See chapter 41 at page 228.
f For example, one site starts its commentary with "Calabi-Yau spaces have no effect on reality." See note 834.
g See chapter 1 at page 15.
h See chapter 19 at page 118.

Planck-scale) quantum tunneling. But my work more than fifty years ago gave me an inclination to think of it replicating by this kind of quantum queerness.[a]

This concept carries deep significance in our search for what's real; the entity that *does* the tunneling is the ultimate quantum that Planck's work revealed and that was then allowed to remain obscure.[b]

Let's be clear, this is an ontological assumption. It is justified by its extreme simplicity and its plenary (as we will see) explanatory power. No other assumption comes even close while meeting our simplicity criterion.[c] No other assumption (so far) leads to a consistent physical and philosophic picture of the world.

Admittedly, there is no firm foundation for it. It's one thing for an electron to tunnel from a state of being in one place to a state of being in another without being able to be in between; it's a whole other thing for one space quantum to tunnel to a state of being two space quanta. The mental leap from the one to the other is not exactly borne out by experience. But then we are speaking of a scale that is as much smaller than an electron as the electron is smaller than the Solar System. Experience—a poor guide at electron scale—seems likely to be no guide at the Planck scale.

What we can say is, if our assumption of a single-quantum beginning is valid, something like a tunneling transition *must* have happened. Otherwise, the universe would still be stuck at quantum one.

And, for the same reason, more such replications must have followed.

We need to be clear-eyed about assumptions; they are every ontology's weak spots, yet there is no way we can do without them.

However, let's keep a sense of proportion. For example, compare the new ontology's frugal diet with the standard model's appetite. Physicist Douglas Scott noted it assumes,

> Understanding the Cosmos is possible for human beings.
>
> Physics is the same everywhere and at all times.

---

[a] See chapter 17 at page 105.
[b] See chapters 13 and 14 at pages 76 and 87.
[c] See chapter 9 at page 57.

General relativity is the correct theory of gravity on cosmological scales.

The Universe is approximately statistically homogeneous and isotropic.

The Universe is spatially flat on large scales.

The dark energy behaves like a cosmological constant, with w = −1.

The dark matter is collisionless and cold for the purposes of cosmology.

There are three species of nearly massless neutrinos.

There are no additional light particles contributing to the background.

Density perturbations are adiabatic in nature.

The initial conditions were Gaussian.

The running of the primordial power spectrum is negligible.

The contribution of gravitational waves is negligible.

Topological defects were unimportant for structure formation.

The physics of recombination is fully understood.

One parameter is sufficient to describe the effects of reionisation.[837]

True, the standard model gives quantitative predictions we can check against observations. But this needs more assumptions (six adjustable parameters).[838]

Let's pick up again tracking the universe's *steps* and its new quanta and their links with no new assumptions.

So far, step one; two quanta; two links (each quantum has two but shares them with the other quantum).

At the next step, the two space quanta replicated. Step two; four quanta, eight links.

Step three; eight quanta; twenty-four links.

Step four; sixteen quanta; forty-eight links. And so on, with always three times as many links as quanta.

Keeping track of links becomes increasingly difficult, but two things are clear; each new quantum always has three ribbons running through it, making three new links; and past step three there are no more unused pairs of dimensions.[a]

So, leaping ahead, we arrive at step ten; 1,024 quanta; 3,072 links.

And then, step twenty; 1,048,576 quanta; 3,145,728 links.[b] Though these numbers start to look large, the step-twenty proto-space was not even submicroscopic. But it was growing exponentially. The universe became big very quickly.

How quickly? Recall we assumed that—as measured by motion inside the universe—each step corresponds to one Planck time,[c] or less than one ten-million-trillion-trillion-trillionth of a second.[d] So by step twenty, less than one million-trillion-trillion-trillionth of a second after it began, the universe though still incredibly tiny was growing very fast.

Indeed, it was already growing faster than the speed of light! This may sound far-fetched. But it is almost identical with the central premise of cosmological inflation that is now widely viewed as an integral aspect of the standard model.[e] We know no reason why this could not happen. And, as Guth proposed, it solves some seemingly unsolvable problems.[f]

To distinguish this doubling phase of the universe's growth let's call it the *exponential epoch*.

If we contemplate the universe after some two hundred sixty steps—as Lemaître did[839]—it was about the size of an oxygen atom.[g] And it was growing at more than a trillion-trillion times the speed of light.

The power of exponential growth is sometimes illustrated by an old story of a Persian king who was given an exquisite chessboard. He asked what he could offer in return. The humble reply, one grain of

---

a They are neither available nor necessary.
b This assumes that each new space quantum somehow links to three neighbors; see chapter 41 at page 228.
c See chapter 16 at page 103.
d More precisely, 5.39×10^{-44} s.
e See chapter 21 at page 126.
f See chapter 21 at page 126 and note 510.
g In Lemaître's model it was far larger.

rice on the first square, two on the next, then four and eight, and so on to square sixty-four. Accepting this seemingly over-modest proposition, he soon found it would consume his empire's rice supply for something like a million years.[840]

In practice, as Boulding said, exponential growth always ends as it runs out of something needed to keep going (in the story, rice). Except for the growth in our exponential epoch, because it did not consume anything.

That exponential growth has five key differences from cosmological inflation theories:

- It has a simple physical mechanism;
- It doesn't hinge on arbitrary math;
- It begins at the beginning instead of after a fine-tuned and unexplained delay;
- It has a natural end;[a]
- Growth is (we'll see) still working the same way (but not at the same speed) today.[b]

If the exponential epoch lasted for four hundred eighty steps—some two dozen million-trillion-trillion-trillionths of a second by our clocks—the universe was by then bigger than our Solar System and growing at more than a trillion-trillion-trillion-trillion times the speed of light.[c]

However, after several hundred steps, replicating every quantum at every step ceased to be universal,[d] so the exponential growth eased, became pedestrian, then slowly ground almost to a halt.[e]

By that point it had already created a huge universe of space packed with random and unstable matter—in a veritable instant—from the application of that simple law of nature, $1 \to 2$.

---

a See chapter 54 at page 294.
b See chapter 68 at page 351.
c Again, this calculation assumes each step takes one Planck time, or rather this is how it seems to us: It is the time our clocks—inside the universe—would assign based on the Planck length deduced from black-body-radiation measurements and the observed speed of light. For more on this, see chapter 50 at page 276.
d For a reason explained in chapter 54 at page 294.
e Asymptotically as a new source of new space fell into place; see chapter 68 at page 351.

41
THINK LINKS

> It seems inescapable that Einstein's theory is only an effective theory, adequate at long distances, but to be replaced by a more fundamental theory at the Planck scale of 10^{-33} cm.
>
> David Gross (2005)[841]

Almost twenty years ago, Gross was writing of Einstein's legacy. He gave voice to a growing consensus that attempts to quantize relativity must fail at Planck scale. But the "more fundamental theory" was elusive, maybe due to an essential insight that was missing: It is to be a theory of *what*? Many might provide a ready answer, *of space*. Which leads directly to another Planck-scale question, *What is that?*

It is an inherently hard problem. We are locked *outside* the Planck-scale world as effectively as we are locked *inside* the universe.[a]

It gives rise to an odd inverted similarity: We can conceive of two realities of the entire universe; in the beginning it was very tiny and today it's very vast; the first we cannot get into and the second we cannot get out of.

| **Links are the universe's relationships.**

It is as if we are fenced off from our ultimate extremities that author and futurist Arthur C. Clarke anticipated (and exaggerated) as . . .

> . . . the unsuspected universes of the infinitely small and the infinitely great.[842]

Our isolation seems as absolute as if his "infinitely" had been accurate. Yet we may make these now suspected "universes" tell their tales.

---
a See chapter 24 at page 141.

Like Planck's quantum, Bilson-Thompson's *"ribbons"* offer us rare keys to both. As I pored over his preprint, they jumped right off the page.[a] They struck me then and strike me still as the ultimate seminal idea. *Ribbon* seems an easy-going word for such an elemental thing. We will pursue his idea and see where it goes.

My spotting his preprint was just a stroke of luck. I was exploring the first instants of the universe, when one Calabi-Yau manifold became two and then the two turned into four. Armed with toothpicks and some bits of Styrofoam I was trying to follow how the four turned into eight, a proto-space that had no matter.

I was seeing an amazing prospect in this simple version of Lemaître's one-quantum beginning. But that it held *no matter* was a seemingly impenetrable problem. Bilson-Thompson's model's braided ribbons blew all that away.[843]

So, let's crisp up their vagueness: First, to have twist states, those ribbons must have two dimensions.[b]

Now think for a moment of two manifolds as two soap bubbles abutting *in* space. The way they interrelate is they make a 2-D window between them (fig. 32).

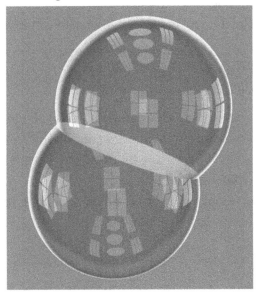

Fig. 32. Two bubbles make a window

a See chapter 35 at page 190.
b A twist in a 1-D entity is meaningless. But any 2-D entity can have a twist.

Now think instead, if you can, of the two manifolds being bits *of* space. Where they interrelate (one is tempted to say touch, but that has no meaning here) there also might be some sort of a 2-D *window*. Or you could also think of it as a 2-D *link*. And a 2-D link is the kind of thing one might (if one were being vague) just call a ribbon.

Note carefully, that while we *can* conceive of a relationship between two space quanta, and though we *can't* conceive of some fine resolution that enables us to say exactly what it is, we may think of it as either a window or a link, and this is a linguistic distinction without a difference. Either way, what we have to work with is it is a 2-D manifold-to-manifold relationship.

We can also see it as a second kind of quantum. It seems simple to assign it the Planck area. We begin to see *why* there can be no area that is smaller: At this scale we are confined to counting; we can't measure things that size as there is nothing we can measure with.[a]

The concept of discrete Planck-scale volumes and areas is not new but has never been specific. For example, in 1995, Rovelli and Smolin co-wrote a paper[844] (based on Penrose's *spin networks*[845] in continuous space) quantizing general relativity,[b] finding "discreteness of areas and volume."[c]

In 2009, Smolin said (writing of discrete quantum geometry),

> It can be asked whether the volume or area operators are physical observables, so that their discreteness is a physical prediction. The answer is yes.[846]

The new ontology much more simply gives rise to not mere discreteness but two kinds of physical quanta—volume and area—that are truly elementary. Each has a fundamental size that is the smallest possible in its dimensionality.

One is the manifold; it has the Planck volume that, en masse, gives rise to three external dimensions.[d]

---

a That is, we have no metric beyond counting; see also chapters 12 and 96 at pages 74 and 475.
b That is, the approach Einstein expressly disapproved, see note 719.
c They noted: "In the absence of Planck-scale measurements, it is of course hard to imagine how predictions for the discreteness of these observables could be tested."
d See chapter 46 at page 263.

The other is the link between manifolds; it has the Planck area, with two dimensions. There are three links for each manifold.[a]

Links have twist states that are a third kind of quanta. They can be half clockwise, no twist, or half counterclockwise.

The new ontology says that's it; that's all the kinds of inherent quanta that there are.[b]

While in the manifold's property of volume we see the first stirrings of what we call spatial extension, note we do not have a property of length. This leaves a problem of defining distance or location at Planck scale.[c]

We could of course define a length as the volume divided by the area, which is equal to the Planck length. But this is just a calculation we perform, *not* a real entity in the universe. That is, in the new ontology, there is no entity that has just one dimension.

Bilson-Thompson did not seem to conceive he was working with real space. He was not concerned about dimensions. He just needed something he called *ribbons* he could twist. He didn't mention this meant they need two dimensions. He said simply,

> It is convenient to represent the most fundamental objects in this model by twists through ±π in a ribbon.[847]

He wasn't being cagey about what they are. He didn't know; he likely didn't care. His field was *topology*, the study of properties that don't change when one deforms shapes. For example, to a topologist, a coffee cup is the same object as a donut (fig. 33). And you may recall the Möbius strip (fig. 34) that, thanks to a half-twist in its 2-D ribbon oddly has only one surface and one edge;[d] it is a topologic novelty whose half-twisted properties exist regardless of size, shape, substance, thickness, kinks, stretch, et cetera.[848] Topology is not much into details.

Bilson-Thompson wrote another paper on his model with Markopoulou and Smolin as co-authors.[849] *They* called the ribbons *links*.

---

a Each one shares six links with its neighbors, so the ratio is six half-links or three links per manifold.
b For more on inherent quanta, see chapter 45 at page 255.
c See further, chapter 62 at page 326.
d If you doubt this, check it out; it's easy to make one from a strip of paper.

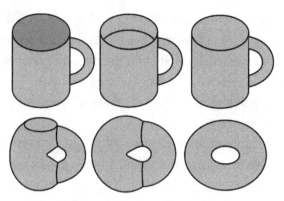

Fig. 33. *In topology your coffee cup is a donut*

Fig. 34. *A Möbius strip*

They were working on the "dynamics," as physicists say when they want things to move. Regrettably, they had their ribbons in an arbitrary space.

They were thinking topology, not ontology. Otherwise they might have had a seminal idea: Those 6-D Planck-sized Calabi-Yau manifolds are naturals for forming their ribbons and links.

If they had seen this, they surely would have also seen each manifold pins the three ribbons that, being pairs of its dimensions, run through it; and how the other ribbon ends are differently pinned— remotely (to the universe)—as each pair of dimensions forms a loop or ring.[a]

---

[a] See chapter 41 at page 228.

Bilson-Thompson's "shorthand" diagrams can be a bit confusing. So here (fig. 35) is a pair of equivalent diagrams. Both show a neutrino;[a] in the right-hand kind of diagram, Bilson-Thompson shows his twists as charges (for example, fig. 36).

He shows each particle as three strips; each strip means a "ribbon." Each ribbon (or helon[b]) either has a full twist (two tweedles, which he shows as a charge—its magnitude is $2e/6$ or $e/3$—plus or minus depending on twist direction) or no net twist and so no charge. You can think of the three ribbons as tied together (at the top and bottom; again, see the two versions of a Bilson-Thompson neutrino in fig. 35). Their crossover patterns are what he calls braids.

Two particles can merge. Here (fig. 36) his model shows what happens when an up quark collides with a W^- particle.

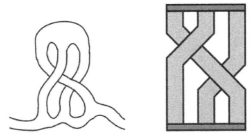

Fig. 35. Two Bilson-Thompson diagrams of the same neutrino

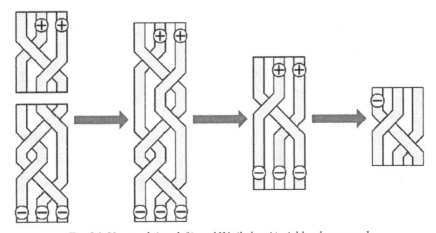

Fig. 36. Up quark (top left) and W- (below it) yield a down quark

---

a Note it is one of three kinds.
b See chapter 35 at page 190.

You can see how to just mash his two left diagrams together. Delete the bars between them. Cancel opposite twists/charges. Sort out loose kinks in braids. Squish the result to standard height and Bingo! You get a down quark. Which is what a very expensive experiment would show you does really happen.

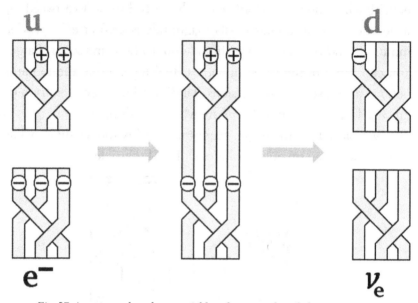

Fig. 37. *An up quark + electron yields a down quark and electron neutrino*

Each "ribbon" is made of two tweedles strung together; I call his tweedles twisted links.

All of the particles of matter and energy and all their real-life interactions play out with these links.[a]

This was the light bulb that lit up when I first saw his paper, "Matter made of space!"

Let's recheck the picture we are painting. Given our premise that, as Lemaître reasoned, the universe began with just one quantum,[b] it was a quantum of space and must have replicated—otherwise neither space nor we would be here.

---

a See fig. 37 for another example; see also chapter 43 at page 243.
b See chapter 17 at page 105.

The rest simply followed. Space *is* (as was long ago suggested[a]) granular; space quanta (string theory's C-Y manifolds) *are* the granules.[850]

We shall see how tracking paired dimensions, as that single quantum replicated, must create the ultimate relationship, the 2-D link.[b] In the chaos of space-creation links often started their existence twisted.[c] As proto-space emerged after a few tens of steps, could there already have been paired links that in threes would look like Bilson-Thompson's braided helons?[d]

These are the ingredients he used to make the standard model's particles, and have them do the things that physics sees them do in particle accelerators. Those particles could build the atoms that comprise the matter in our world. The selfsame thing that made the universe's space expand could populate it with the stars and planets.

Thus, the links between space quanta lie at the heart of the new ontology. In the one fell swoop that we were seeking,[e] they offer a way to build a consistent understanding of all we see—and for much more we don't see but conclude must be.

Tracking the links of space quanta gets confusing past the first few steps. It's easy to get lost in space; especially convulsing space that has no signposts.[f]

For this task we need topologists.

---

a See chapter 14 at page 87.
b See further, chapter 42 at page 236.
c It seems impossible for replication not to create twists.
d As Bilson-Thompson noted, braids and twists are topologically interchangeable.
e See chapter 34 at page 186.
f See chapter 62 at page 326.

42
GETTING GOING

> Only the genius of Riemann, solitary and uncomprehended, had already won its way by the middle of the last century to a new conception of space, in which space was deprived of its rigidity, and in which its power to take part in physical events was recognized as possible.
>
> Albert Einstein (1934)[851]

Einstein worked with an essential consequence of Riemann's thesis: Space was not merely a stage; it was an actor. He set out to write its script. But he could not see the last piece of the universal puzzle: Space is the *only* actor.

In this chapter we will take a closer look at what this actor *does*. Space quanta and their link relationships provide both stage and play.

Space's act began at the beginning. In his latest book, Smolin said simply,

> Space is emergent.[852]

In our new ontology, we agree in spades. What he called space emerged out of the beginning of the universe. Indeed, its emerging *was* the beginning.

Making space begins with the relationship between the first two manifolds: Whether one thinks of it in terms of ribbons linking them or windows where they kiss, it's a relationship that's based on area, which is to say it requires two dimensions. This concept is formative: It will turn out to lead to everything, so let's take a break to mind our *D*s.

A dimension is a hard thing to begin or end, maybe impossible within a finite world. By its nature, a dimension should just keep on going. This is why they come as loops.[a] Think about it. What would the end of a dimension look like and how would it work?

Becker sounds as though he's speaking of some such creative enterprise when he says,

> Natalie Paquette spends her time thinking about how to grow an extra dimension.[853]

He's speaking of physicist Natalie Paquette working with string-theory math. But she's not using it to make a *new* dimension. She's using it to expand or shrink a dimension that her theory says is there. And it is a loop. She shrinks it down to Planck size, and finds, counter-intuitively, attempts to shrink it further give rise to a larger loop.[b] (Too bad she pulls off this neat trick in space that is continuous.[c])

We too will try to keep track of dimensions. In our beginning, with two manifolds, where did those two dimensions—now external to the manifolds—come from? All that existed was two manifolds where once was one.

Occam is our guide.[d] How can we keep it simple? The only simple answer is, not new dimensions coming out of nowhere, but two of the six loops that were "inside" the first manifold got "sucked outside," to coin a grisly phrase.

| There is only one law: $1 \rightarrow 2$.

So where does this 2-D ribbon go? If you can do this in your head, you're doing better than I did. Turns out there is only one workable answer. The same two dimensions link quantum A to quantum B and run right back again, linking quantum B to quantum A, forming as it were a larger ring or loop (fig. 38). So, they must now be the *same* two in both replicas.

---

a Very large loops, as in our three dimensions—see chapter 22 at page 133; or very small loops, as in those of a Calabi-Yau manifold—see chapter 15 at page 95.
b Becker calls it *new*, but, in his account, it is merely changed.
c In which, Becker reported, she said, "If you carefully keep track of the fundamental building blocks of the theory, you can naturally find sometimes that . . . you might grow a new spatial dimension."
d See chapter 9 at page 57.

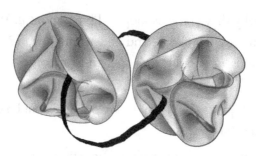

Fig. 38. Linked manifolds – the universe at step 1, with the One Ring ribbon

The replication did not create two new dimensions; it grew two that already existed. It's as if we stretched a loop that they already formed inside the original manifold.

Growing an existing dimension *is* for us a new (though not an unprecedented[854]) assumption. But it is the simplest possible assumption; and we will see it leads to an ontology that assumes little and explains a lot. Just what William would have wanted.

In what follows, keep in mind these are two of the six tiny dimensions that originally were integral to (one is tempted to think *internal* to) the first space quantum. In other words, two dimensions ran through one quantum and to and through the other and returned. It is this reentrant pair of "external" dimensions we can think of as two links.

Though not yet built out into what we would see as three dimensions, this is how Einstein's closed cosmos came into existence. It was closed from its beginning and had neither cause nor means to ever become open. It was all there was; it is still all there is.

It's so fundamental, it is worth repeating. Here is the source of Einstein's universe,[a] a closed world with no edge and no outside.[855]

And since it was so small, the curvature of proto-space was inconceivably enormous. This was the grip of gravity as it could never be again.

Let's follow the next replication steps—the kind of thing that, in their context, Bilson-Thompson and co-authors call the "evolution moves"[856] and I call universal *iterations* or *steps* for short.[857]

---

a See chapter 22 at page 133.

Part II: Getting Real   239

This is where those six dimensions—that were integral to the first quantum—begin moving out in pairs and taking over.[a]

At *step two*, both of step one's two quanta replicated. Again, that is the simplest-possible no-new-parameters assumption. So, the step-two universe comprised four manifolds (fig. 39).

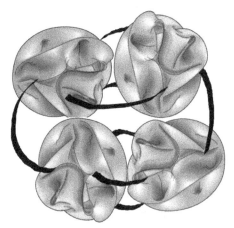

Fig. 39. Linked manifolds at step 2

Not being a topologist, I took a while to "see" how that tight loop of two dimensions passing through both step-one manifolds must run right through both offspring of each parent. In other words, it stretched again so that it ran once through each of the four manifolds. Any other outcome needs some new and far from simple supposition. To keep track (as best we may) let's call this the *One Ring*.[858]

We can then see each of the four step-two manifolds shared four links. And then envision two new ribbon-rings had come into existence. Each of the *Two Rings* ran—just as the One Ring did after step one—through the two offspring of one of the two parents. Each Two Ring took up and grew two of the four formerly unemployed internal dimensions of its parent.

As more steps proceeded, the first pair of what once were curled-up dimensions—ripped into outer space, as it were, by replications—must at each step run through each replicate quantum. So, looking ahead,

---

a Yet another irony: Far from being somehow "compactified" (see chapter 15 at page 95), the six "extra" dimensions were expanded!

the One Ring became a vast intricately meandering 2-D loop that traverses each space quantum in the universe exactly once.

The reader may well imagine that, when these universal attributes first loomed into inner vision, they brought with them a *this-is-wild-speculation* feeling. But that razor is a steady guide. Once one buys the simple premises—Lemaître's single quantum and string theory's manifold—a deviation from the course the new ontology is following demands more (and far from simple) assumptions.

And so, on to *step three*. Its universe comprised eight quanta. This step made four new ribbon-rings. Each of the *Four Rings* ran through both the two offspring of one of the four parent quanta. Each Four Ring grew the two unused internal dimensions of its parent quantum.

If your mind is up to such mental gymnastics, maybe you can see how the eight quanta were beginning to form proto-space. That is, space was starting to emerge. Each manifold had six 2-D links to neighbors. They comprised three pairs of links, three ribbons running through each manifold. These links used all six of each manifold's dimensions.

And, in your mind's eye, if it's still seeing straight, maybe you can catch how at each iteration of the infant universe each replication must have made new *twisted* links and new *three-braids*. Looking ahead, we are about to see how making new space also made the universe's matter.[a]

That is, we are on the brink of understanding everything may be made out of just one thing.[b]

Step four had sixteen quanta and this scene is complex beyond my ability to follow. There must be just seven Rings[c] as there were no free dimensions to make more.

We can now see how this space conforms to Smolin's background-free prescription . . .

---

a Courtesy of Bilson-Thompson; see the next chapter.
b Thus we see the realization of the long aspiration of Anaximander, Leibniz, Heisenberg, and the like; see chapter 7 at page 51.
c This may give rise to cracks about Tolkien's dwarves but their number is a fundamental property of the universe.

... some kind of purely quantum structure.[a]

It looms into our mental view with curious connections; from step four on, the One Ring traverses every manifold, each of the Two Rings traverses half of them, and each of the Four Rings traverses a quarter, so every manifold in the universe is traversed by three Rings, one of each order (that is, a One, a Two, and a Four).

If all this seems fanciful, I sympathize, but, again, once one accepts the simplest origin—a single manifold—at every juncture all alternatives collapse beneath their complications.

Serious study will be needed to explore potential evolutions of linkage topologies at step four and beyond. It may be a rich source of new physics.

By step ten, space was made of 1,024 quanta with three times that number of links in a tangle with no sort of symmetry. Already its growth was heading into the zone where its relatively widely separated quanta would—if clocks were ticking and there were a way to measure distance—be moving apart faster than the speed of light.[b]

Looking back from this vantage point, a still-miniscule proto-space universe less than ten zepto-zeptoseconds after its birth, we can understand precisely what was needed to create space: A manifold with six dimensions and a tiny volume that would replicate itself in incremental steps.

And that—says our ontology—is what we got. No Big Bang; just a Big Fizz, at first very fast, soon to be relatively very slow.[c]

Meantime, we already have the measure of our iterating 3-D universe: It *is*; it *does*. We can call them *Now* and *Next*.[d] Stability and change. Each depends on the other.

And thus, in its beginning, we find an intelligible source, maybe even an explanation, for quantum theory's seemingly inexplicable battle between position and motion.[e]

---

a See chapter 5 at note 164.
b See chapter 40 at page 222.
c See chapter 53 at page 291.
d Each gets a chapter (56 and 57 at pages 301 and 307).
e Also known as the Heisenberg uncertainty principle; see chapter 128 at page 614.

Here lies the true heart of our story: *Now* and *Next*, the universe's *be* and *do*, are the deep roots—and the real faces—of what we know as space and time.

43
BEING CAUSAL, BEING RANDOM

> [T]he causal set itself is simply a very large collection of causal relations.
>
> Fotini Markopoulou (1998)[859]

My short version of this chapter is that Bilson-Thompson led me to Markopoulou. Markopoulou led me to the seminal idea of the causal set. It is, as she said, all about what causes what—on some fundamental scale.

On the face of it, reality is all about causation. The past caused the present. The present is causing the future. But what does *causing* really mean?

| Is this the universe's way of working?

One may answer that it means the connection between a cause and an effect, each being some kind of event, whereby the cause necessarily gives rise to the effect. In other words, knowing the cause is enough to define the effect. Sometimes called strict or deterministic causation, this view was prevalent before 1900.

Or one may answer that it means the connection between a cause and an effect, whereby the cause influences the likelihood of the effect. This is often called probabilistic causation. This view became prevalent with the rise of quantum theory a hundred years or so ago.

Over-simply, the difference may be seen in Scrooge's question to the ghost of Christmas Yet to Come in Charles Dickens's *A Christmas Carol*:

> Are these the shadows of the things that Will be, or are they shadows of the things that May be only? [860]

Behind this seeming clarity, the language of causation can be varied and confusing. Thus, one finds von Neumann speaking of the one as "purely causal" and the other as "non-causal" or "not causal."[861]

These concepts speak to the realities of supposedly simple subatomic events where each effect may seem to have a simple cause and vice versa. If one wishes to address causation on a wider stage, things soon get messy. There are many moving parts. An effect may be the consequence of many causes. A single event may have many consequences. How may physics model this? How should we think of it?

And looming large behind these terms and their ideas, what *is* the connection between cause and event? How does it really work? There is an extensive literature. It suffers from not knowing where the action is.

Back in 1739 consideration of causation took a new direction when Hume critiqued the conventional understanding, saying our experience does not provide the reason for causation.[862] He didn't put it this way, but for him causation was no more than an idea, a fine fiction, as we can't observe it. Which is to say, again, we don't know where the action is.

Fast-forward to 2015; Unger staked a new terrain; he said we must conceive of causation as real:

> Causal connections . . . form a real feature of nature.[863]

This is a radical assertion; it conveys a Humean challenge for our understanding: *What are nature's causal connections and how do they work?*

Life seems rife with *coincidences*—meaning events seeming to be causally but really being only seemingly related.

One such began (I imagine) the day I stumbled upon Bilson-Thompson on a web page when a passing glance flagged that he came from Adelaide, where I chanced to have been born.[a] He led me to Markopoulou. Her name led me to her paper about physics needing to be done inside the universe. How could one resist?

Her paper led me to the strange new mathematics (called *causal-set theory*) she was showing physicists because (he confirmed later) it seemed to be the kind of math that's needed to get fundamental physics going. Which in turn led me—Bilson-Thompson's twists in

---

a See chapter 35 at page 190.

hand—to realize the universe *is, simpliciter,* a partially ordered[a] finite causal set.[b]

This insight leads—plainly, it seems to me, and perhaps someone will prove it—to the principle of causal completeness, that, as Smolin put it,

> Everything that happens in the universe has a cause, which is one or more prior events.[864]

Looking back, was this a chain of causation? Or of chance? That partially ordered finite causal set will soon show it was both.

We have not yet reached the curious coincidence. A staffer at my publisher was tasked to offer copies of the hardcover *Time One* (which told part of this tale) to some reviewers. Dwight Murphey was reviews editor for a refereed journal in Washington, DC; he said, "Yes, please." A while later it had not arrived. She sent another. He got both. So, when he (a former law professor) was unsure of my physics, he sent one to a space physicist named David Miller. In the end, both wrote reviews.[865]

The curious coincidence still lies ahead. Later, my publisher put me in touch with Miller. He works with Carmel Research Center, analyzing data from Voyager space probes with way-past-best-before-date systems that phone home about the space out at the Solar System's edge. We chat about his work and mine from time to time; he comes up with some great ideas.[c]

One day he said, "I am not really a physicist. My PhD was math."

"What kind of math?"

"An esoteric kind called partially ordered causal sets."

There may be a handful of such theses on the planet. *That's* a coincidence.

As author Kurt Vonnegut said, tongue in cheek,

> One would soon go mad if one took such coincidences too seriously. One might be led to suspect that there were

---

a It is *partially* ordered as the reach of causation at each iteration goes only to the next-adjacent link.

b This in turn became a barely mentioned premise of *Time One*; note 16.

c For example, see chapter 66 at page 343.

all sorts of things going on in the Universe which he or she did not thoroughly understand.[866]

That math that Markopoulou wanted physicists to see deals with the causal *sequence* of "events."[a] It is concerned with what comes before what but has no time put into it.[867] Rather, something like time should come *out* of it.[868]

This too was not entirely new. In his posthumously published work on time, Reichenbach pursued an idea "first conceived by Leibniz," setting out to show . . .

. . . that time order is *reducible* to causal order.[869]

For us, it's a seminal idea that's just waiting for the insight that simultaneous causal *beings* of space at the level of its quanta and their links *are* the real events. Each iteration of the 3-D universe is a finite set of quanta and their links. They cause the next. This is a causal set. Its sequence gives rise to a kind of time.

(For the keen reader, there's a review of causal sets in quantum gravity by physicist Sumati Surya.[870])

Causation appeared to collide with chance when the Copenhagen gang walked their new quantum-theory pup in Einstein's local park. It was an unequal confrontation.[871] In his history of quantum mechanics, historian of science James Cushing said,

> The Copenhagen group had the talent, organization and drive to carry the day in establishing the hegemony of its view.[872]

Einstein lost every argument along the way but was unbending, famously writing to Born,

> The theory delivers much but it hardly brings us closer to the Old One's secret. In any event, I am convinced that He is not playing dice.[873]

Causation might seem simple: Each cause has its effects; each effect has its causes. Well, not so fast. Causation is the subject of complex analysis and disputation.[874] There are even varied views about what Einstein's views about it were.[875]

---

a But it does not, in general, have spatial simultaneity.

But the essential problem is we see a world of causes and effects that does look causal, if by that we mean we can (sometimes) successfully predict events. Meanwhile, quantum physics shows that seeming-causal world is made of atom-scale components whose events it predicts with statistical precision based entirely upon chance.

How can we reconcile largescale causality with the quantum uncertainty? One way to view this question turns it into, *How can we reconcile general relativity and quantum theory?* That is, how can we formulate the mainstream concept of quantum gravity? Many strive; some do progress;[876] none are, so far, succeeding.[877]

Another way to come at causation might be to leave aside the rival theories and ask a basic question, *What is really going on where things are really going on?*[a]

This is our aim with the new ontology, using Occam's razor to suss out a comprehensive answer based on quanta of space that make a universe that *is* a causal set.

When I read that Bilson-Thompson paper, his ribbons leapt off the page and became links. Many insights began falling into place. So now *I'm* saying, he was twisting links between space quanta. He did not say this and I do not suppose he will.[b]

What is going on, then, is at each iteration the universe consists of the set of linked space quanta. *Next*, each space quantum continues to be a single quantum with its links, or else by chance it replicates, thus adding a new quantum and new links;[c] either way it's in a brand new universe.

At each step, each twist state stays in its link or switches to a next-neighbor in one of two directions on its ribbon. What each twist does is random[d] with probabilities that need new physics.[e]

---

a See also chapter 119 at page 577.
b Bilson-Thompson and his co-authors in another paper did call the ribbons "links" but did not say they were links between space quanta; see note 849 in chapter 41.
c And at each step, the replication of each quantum is random with probability that depends on the local curvature of space. Most places, these days, this won't happen, for a reason we will soon explore; see chapter 54 at page 294.
d By *random*, I mean determined with a probability, as with a notional fair coin toss; but therein may lie much of the new physics.
e But note the untwisted twist states of photons may move in one direction with probability 1.

So, every bit of it is based on chance. But the relation of each new universe to its predecessor is strictly causal. In other words, a way to understand causation, and indeed all of reality, is simply *See it as a flaky causal set.*

44
UNPACKING THE SUNDANCE STORY

> If nodes represent volume, and links represent area, what information is encoded by braiding/twisting?
>
> Sundance Bilson-Thompson (2007)[878]

The mystery of the Perimeter Institute's deep-sixing the keys to the universe came to be a keen preoccupation. I went looking for what happened there and then and after that.

What all left town with the bicycle?

To set the larger scene as Bilson-Thompson's work came to Smolin's attention, readers might just quickly scan the latter's 2005 review paper.[879] He was reviewing progress in loop quantum gravity research, giving us a picture of physics that is completely lost in math.

I found Bilson-Thompson's notes for a lecture he gave at a conference in June of 2007,[880] at Centro di Ciencias Matemáticas (CCM) in Mexico.[881] I thought they might offer clues to his views as of that date.

Tellingly, it seemed to me, the notes were almost all in terms of helons. Tweedles were consigned to parenthetic mention:

> N.B.: Can regard helons as composite = pairs of $\pm\pi$ twists (Tweedles)

... and did not show up again until a (still parenthetic) recap at the end, where he did mention the single-ingredient[a] point, but as if it were of only passing interest:

> The helon model has only a single type of fundamental object (tweedles).

---

a See chapter 7 at page 51.

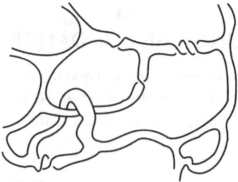

Fig. 40. Bilson-Thompson's twisted-ribbon worldview

His "body language" here may be revealing: The tweedle—his most fundamental innovation—seemed to be on its way out in ignominious fashion.[a]

The CCM-lecture notes contained a diagram (fig. 40 [882]) showing a fragment of a twisted-ribbon worldview—twists and braids of ribbons intersecting at threefold *nodes*, arrayed in continuous space.[b]

A key characteristic of his model is each node is the intersection of three ribbons or (here *he* says) *links* and (as in the epigram) . . .

. . . nodes represent volume, and links represent area.[883]

For his model, these appear as new assumptions.[884] (For the new ontology, they result from the original simplicity.[c])

Another profound aspect of his insight gets no more than passing mention. It explains the machinery behind that fraught equation, $E = mc^2$. How can we understand mass and energy to be the same thing?

It's breathtakingly easy. The twists and braids of any pair of oppositely charged, mirror-image particles straighten their braids and cancel their twists to form the photon, the massless particle of energy.

Bilson-Thompson buries this revelation in a diffident caption to his depiction (see fig. 41; insert your own opposite charges) of a particle colliding with its antiparticle:[d]

---

a I found no mention of the tweedle in his subsequent work product, which is built on three kinds of the long-standing concept of "preon."
b His threefold nodes are the equivalent of manifolds.
c See chapter 41 at page 228.
d Each particle has an antiparticle; when they collide, they self-destruct; see also chapter 71 at page 367.

Part II: Getting Real 251

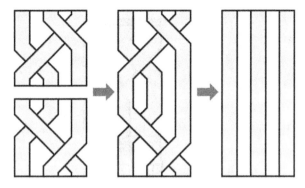

Fig. 41. Bilson-Thompson's demonstration of mass being the same as energy

Taking the braid product (joining top-to-bottom) looks a lot like particle-antiparticle annihilation.[885]

Why not promote this explanation? Why not show other examples of the power of his simple concept? Why not, for instance, answer a simple but deep question like, *How does a neutron turn into a proton by "ejecting" an electron?*[886]

It's called beta decay. A neutron consists of one up quark and two down quarks (udd); a proton is two up quarks and one down quark (uud).[887] Beta decay happens when one of the neutron's down quarks spontaneously turns into an up quark, ejecting an electron and an antineutrino in the process (fig. 42). How does *that* happen?

Fig. 42. Beta decay of a neutron

A bit of cut-and-pasting (from his paper's Figure 1) gives us the ingredients. Fig. 43 shows Bilson-Thompson's down quark next to a stack of its three beta-decay products, end to end as he says. (It doesn't matter in what order one stacks them; it works out the same.) Just cast your eye from the top down, following each ribbon, seeing what that ribbon's *net* charge is, where that ribbon ends, and what it braids

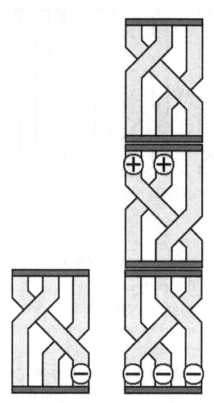

Fig. 43. Down quark decays into antineutrino, up quark and electron

over or under. You can see for yourself. The stack of decay products, reassembled as it were, turns back into the down quark.

With Bilson-Thompson's twist/braid figures we could cut and paste our way through all the particle decays and interactions known to physics.

One can imagine that, if it were more fully expounded, the extreme economy of his model[a] could take on a more obvious significance.

His CCM-lecture notes do show some things his model explains:

- The existence of the quarks and *leptons*[888]
- Why quarks and charged leptons show 1:2:3 electric-charge ratios
- Why quarks and gluons have color charge[889]
- Why (only) colored entities have fractional electric charge

---

a See note 762.

- The existence of other generations of particles
- Why mass and energy are interconvertible[890]
- How electroweak interactions work
- Why neutrinos are only left-handed
- Why electric charge is quantized (emphasized on screen)

But there could have been much more; those notes don't even mention it explains:

- How all the particles of mass and energy are composed of relationships[891]
- Why there are six quark flavors[892]
- Why there are six leptons[893]
- Why there are sixteen particles in the standard model[894]
- How various reactions and decays work
- How particles have both right- and left-handed chirality[a]
- Why neutrinos are the only particles to have only one chirality[895]
- Why the opposite charges on the proton and electron are exactly equal[896]
- How particle-antiparticle collisions create photons[897] and vice versa[898]

All in all, it stopped short of a sales pitch.

For comparison, there is (in poor video) a lecture he gave a year and a half earlier, three weeks after he arrived at the PI.[899] Smolin introduced him without the ebullience of his first take on the model.[b] In this lecture Bilson-Thompson pitched tweedles and their dums and dees, front and center.[c] He asked (well, mentioned, almost in passing) some fundamental questions, including,

> Why do electric charges of proton and electron balance perfectly?[d]

---

a Chirality is the property of a shape that cannot be made to match its mirror image.
b See chapter 35 at page 190.
c To me, he noted, "Unfortunately, referees don't share my sense of humor."
d See note 899.

The trite reason is, while the electron is regarded as a single (or elementary or fundamental) particle, the proton is known to be composite, made of three quarks.[900] Two of them have positive charge $2e/3$; the other has negative charge $-e/3$. The proton's charge is thus the result of a calculation, so to speak; $2e/3 + 2e/3 - e/3 = 3e/3 = e$.

The deep version of the question is, How can that elementary electron match this kind of fractional arithmetic with its supposed-to-be single charge $-e$? The observed "perfect balance" with its oppositely charged composite partner seems to say (indeed to shout), *That electron is not elementary!* Then we can understand the quarks' fractional charges.[a]

The elementary charge this points to is $\pm e/3$, with the electron having three negative elementary charges. Or, more fundamentally, the truly elementary charge on the tweedles (which travel in the twos that he calls helons) is $\pm e/6$, with the electron being composed of six tweedledees.[901] Were these and other fundamental implications known to his audience? He surely didn't drive home the uniqueness of his "model"; he portrayed it as a variant of other preon models.

In addition to the three papers already mentioned,[902] Bilson-Thompson was lead author of three more relevant papers in the period to 2012.[903] Taken together they look like a retreat into obscurity. Smolin seemed to become MIA. Having begun work on the tweedle with excitement about the "missing idea,"[b] somehow the excitement dissipated and the idea went away.

Far from PI,[c] physicist Niels Gresnigt remains a presence on the tweedle trail.[904]

I know too little to pass judgment. And in fairness to PI's management, whose failure to grasp the new discovery its founder sought and it long harbored in its midst might seem a scandal, one would want to know PI's side of the story. And I don't.

---

a See also chapter 48 at page 268.
b See note 763.
c At Xi'an Jiaotong-Liverpool University in Suzhou, China, which has picked up where PI left off.

45
ONE QUANTUM FOR ALL

> Might it be that . . . physicists have constructed an extremely awkward formulation of quantum mechanics that . . . obfuscates the true nature of reality? No one knows. Maybe some time in the future some clever person will see clear to a new formulation that will fully reveal the "whys" and the "whats" of quantum mechanics.
>
> Brian Greene (2000)[905]

We will see why Greene's speculation—that quantum mechanics got on the wrong track—was right. Mind you, that wrong track is preferable by far to no track.

Many think that quantum theory applies only at atomic size. It seemed a stretch when Nobel-winning physicist Anton Zeilinger and colleagues revealed quantum interference between carbon buckyballs[906] (big molecules of sixty carbon atoms about a nanometer across, for which discovery chemist Richard Smalley shared a Nobel Prize[907]). Yet when buckyballs lose their quantum coherence it is by interactions, not by virtue of their size.[908]

| Just one quantum can explain the world. |

There are far larger quantum systems. In the nineteen sixties I was not alone in having gram-sized chunks of superconducting metals,[a] and a liter of frictionlessly flowing superfluid liquid helium. Each was in a single quantum state in a cryostat I built in my lab.[909] Somehow, though, I too drank the small-quantum-system Kool-Aid.

Today there are far larger examples, such as the superconducting coils bathed in superfluid helium in the magnets of the 27-kilometer-

---

[a] Lead, tin, and aluminum; not to mention maybe a kilometer of niobium-zirconium wire.

long Large Hadron Collider, where each magnet coil and each helium bath houses a quantum system in a single quantum state.[910] Small size is not a requisite for what we regard as "quantum behavior" with all its oddities.

But super-small size is where Planck and Bilson-Thompson and a splash of string theory take us. You might say Calabi, Yau,[a] and Bilson-Thompson gave tweedles a hidden way to do their thing, long after Planck gave them a hidden place to be.[b]

Indirectly, Bilson-Thompson's paper shed a whole new light on quantum physics: Once one can see *space* as made of real quanta, one has the makings of all other kinds of so-called "quanta" because they are made from combinations of relationships between space quanta.

Once again, we need to watch out for our mangled language. In physics' hands, the word *quantum* has come to apply to an open-ended set of properties, with a tacit focus on energy:

> **quantum**, in physics, discrete natural unit, or packet, of energy, charge, angular momentum, or other physical property.[911]

Once common or garden English for *amount* (adapted from the Latin *quantus* in the sixteen hundreds[912]), the word *quantum*'s frequency in books was overtaken by physics' usage in the nineteen twenties. Meanwhile, use of the phrase *quantum of* had fallen fourfold from around 1820 to 1860 and never regained ground. But after *quantum* being introduced in the new sense of *the minimum amount that can exist* following Planck,[913] its frequency of compound use in books as *quantum+** (mostly as *quantum mechanics* and *quantum theory*) increased fourfold from 1945 to 1990.[914]

It was thus in recent times well suited by etymology to convey a sense of something fundamental and discrete while having adjustable magnitudes and being open to new attributes. Riemann was apparently alone in calling discrete granules of space *quanta*.[c]

---

a See chapter 15 at page 95.
b See chapter 13 and 14 at page 76 and 87.
c See note 276.

In an aside, one finds a fine example of its original usage by Poe in 1832. He wrote of "an undue quantum of absurdity."[915] A few years later he was first to propose the universe began as a single quantum,[a] but did not use this term in either sense, instead saying it was . . .

> . . . a particle absolutely unique, individual, undivided, and not indivisible only because He who created it, by dint of his Will, can by an infinitely less energetic exercise of the same Will, as a matter of course, divide it.[916]

No doubt Poe sounds a bit confused. But here's 't Hooft, reviewing Bohr's, Louis-Victor de Broglie's, and Schrödinger's ideas one hundred fifty years post-Poe,

> Quantum mechanics works beautifully, there is little doubt about that. However, a very peculiar question presented itself: what do these equations actually mean? What is it that they are describing?[917]

So, we need to rethink the meaning of the quantum and to this end must rethink its place in physics.

As we've seen, Planck's quantum was hard to construe as a real quantum; it was a number—a fundamental constant with odd units not commensurate with anything we see as real—that soon was showing up in almost every quantum-theory calculation.[b] Though based on large-scale observations, this number pointed not to atom scale (where physics was then starting to get busy with it) but to the far smaller Planck scale that is inaccessible to observation and even (until very recently) to thought. It is to this latter scale and to the quantum of *space* that—hand in hand with the seminal ideas of Lemaître's quantum beginning and Bilson-Thompson's tweedles—physics must now turn in order to make sense.

Understanding the space quantum is another mental challenge mostly because, as just outlined, our concepts of quanta are a century-

---

a In *Time One*, my narrator credits Poe rather than Lemaître with the idea that the universe began with a single quantum; note 16.

b See chapters 10 and 13 at pages 63 and 76. It soon became so ubiquitous it got its own special symbol, *ℏ*; see note a.

old muddle. That difficulty is doubled by the hash we're making of our thinking about space.[a]

We must also overcome our mental habit of imagining space quanta *in* space. This takes time and a degree of application. I finally got my head around it[b] but let's go with Rovelli who elegantly said in 2017,

> Quanta of space have no place to be *in*, because they are themselves that place.[918]

Grasping this and owning it are worlds apart, a bit like learning to speak a new language does not have you thinking it. So, to the reader who may get the meaning of the words Rovelli said but cannot *see* the world that way: This is a widely shared affliction; it may pass.

The next aspect of the quantum concept that we need to get a handle on is there are quanta of a kind, and then there's a whole other kind of quanta.

The first or conventional kind of quanta are those (for example, of energy) one finds in thousands of books and maybe millions of papers; they are made by *quantizing* something that is otherwise continuous in nature, impressing a constraint on it so it can take only certain values in some circumstance.

You can put on your thinking cap, select (or invent) a continuous property, and devise some math to quantize it.[919] By the latter nineteen hundreds, this had spawned a free-form sport. Late in life, in a letter (fig. 44) to Moffat,[920] who was then a student, Einstein revealed

Fig. 44. *Einstein wrote to a student about quantum theory*

---

a See chapters 2, 9, 10, 20, 83, and 89 (at pages 22, 57, 63, 122, 420, and 444, respectively).
b See *Time One*, note 16.

that, as we have already seen,[a] he was skeptical of this approach as a strategy for fundamental physics:

> I do not believe that it will lead to the goal if one sets up a classical theory and then "quantizes" it.[921]

Both in principle and in practice a conventional quantum can have any value. For example, the energies of photons vary continuously as they rise or fall through gravity.[b] They vary continuously as we move to or from them. They decrease continuously with expansion of space they traverse.

In other words, energy is inherently continuous; the "quantized" amount of energy a photon is born with is set by constraints on the electron that loses energy to emit it.

Einstein was acutely aware of the clash between the then new quantum and the old math world. The conceptual disconnect between discrete quanta and continuous space troubled him:

> [P]erhaps the success of the Heisenberg method points to . . . the elimination of continuous functions from physics. Then, however, we must also give up, on principle, the space-time continuum. It is conceivable that human ingenuity will some day find methods which will make it possible to proceed along such a path. At the present time, however, such a program looks like an attempt to breathe in empty space.[922]

He saw this issue as foundational:

> I consider it as entirely possible that physics cannot be based upon . . . continuous structures.[923]

The other kind of quanta is *inherent* and it offers us a way to "breathe in empty space." We do not create them. We cannot destroy them. They come in only a single size. The quanta of the new ontology are of this kind. In this they are like the fixed quanta of electric charge.[c] This is no coincidence; one of them (the tweedle's twist) *is*

---

a See chapter 31 at page 176.
b So the second is defined in terms of photons at sea level; see chapter 82 and note 1359.
c See chapter 43 at page 243.

the quantum of electric charge. In contrast with conventional quanta, there is no way an inherent quantum's magnitude can change.[924]

Let's take a case that seems to sit between the two. Rovelli described his *grains of space* as having volumes quantized with a *spectrum of values*. He depicted them with sundry continguous shapes (fig. 45)—maybe to fill the space he says they are not *in*?[925]

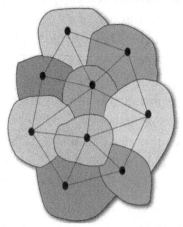

Fig. 45. Rovelli's "grains of space"

His depiction suggests continuously variable volumes. But his text makes it clear the spectrum of volume values (attributed to Nobelist Paul Dirac) is discrete, with some smallest-possible volume. He had another figure showing a discrete "spectrum" of tetrahedrons (for reasons he does not make clear) with the smallest blurred (see fig. 46); his caption reads, "The spectrum of the volume: the volumes of a regular tetrahedron that are possible in nature are limited in number. The smallest, at the bottom, is the smallest volume in existence.")[926]

It's a confusing explanation (maybe partly owing to translation). But clearly his discrete volumes arose from quantization of a continuous space; that is, they are conventional quanta. In this they are at one with all the conventional quanta quantum physics has proliferated (to our immense profit) for a hundred years, except charge.

An inherent quantum differs fundamentally from this. In its concept, in its nature, in its role, in its size, in its very being, it is unique and singular. Every instance of it has the same size; they are identical in all respects. The inherent quantum is the only kind that's real.

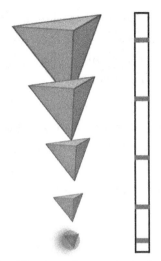

Fig. 46. Rovelli's scheme for quantized volumes of space

Recall that the standard model's quantum field theory has sixteen elementary particles. The new ontology says they are composite at Planck scale, each comprising six tweedles paired up as three braided helons.[a] Being made of inherent quanta, these particles themselves may seem in this sense to be "quanta."[b]

But we no longer can subscribe to sixteen particles with wildly varied properties as elementary components of matter and energy. We have just one, the space quantum (plus its relationships, the link, and the tweedle). It is truly elementary. And, in support of theory saying there is nothing smaller, its detectable expression is so exquisitely simple—a twist state between indivisible space quanta—it is hard to see how we could decompose them.

In other words, the long search for the Greeks' indivisible atom has arrived at a supremely elegant conclusion.[c]

This allows us to start making sense of the seemingly senseless but solidly successful quantum theory. It overcomes a fundamental obstacle. In 2015, Smolin said,

---

[a] See chapter 43 at page 243.
[b] Yet, being composite, they may stretch our concept of fundamental particles; see chapters 6 and 8 at pages 47 and 53, and for more full discussion see chapter 87 at page 439.
[c] Again, Bilson-Thompson said he thought so (calling it a *node*) but would not assert this claim; see chapter 35 at page 190 and note 866.

> While there are proposals to eliminate the spacetime continuum in favor of discrete sets of events . . . it seems a harder challenge to eliminate the dependence of quantum theory on the continuum.[927]

Inherent quanta meet this challenge; they eliminate the continuum and point to a whole new path for a real quantum theory.[a]

We'll find there is much more to the universe than those particles of energy and matter, stuff we see and feel. Almost all of it is mysterious dark energy and missing dark matter.

The selfsame space quanta will be up to dealing with them too.[b]

---

[a] See chapter 75 at page 386.
[b] See chapters 67 and 68 at pages 348 and 351.

46
IN THROES OF THREES

Why 3+1 dimensions? Why not more?

John Wheeler (2006)[928]

It's yet another troubled puzzle tucked behind the fundamental physics façade: Wheeler isn't asking why there is no extra time dimension; he wants to understand why space would have only three. Surely, he thought, there must be a reason for it. In this, he was not alone.[929]

> There is a reason why we have three dimensions.

As we've seen,[a] the new ontology says he is wrong about time being a dimension (that's his "+1," so we'll ditch it; fig. 47). He—co-author of the timeless Wheeler-DeWitt equation—could have seen this coming.[b]

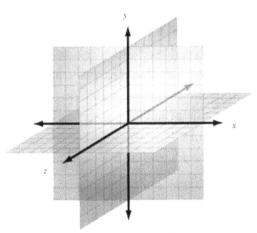

Fig. 47. *Our conventional three dimensions*

---

a See chapter 28 at page 158.
b See chapter 73 at page 378.

Smolin, too, saw our three *D*s leading to a question:

> [G]iven present knowledge, it seems that space might have more or less than three dimensions. . . . This is because all our current theories, including general relativity and quantum mechanics, would also make sense in a world with a different number of space dimensions.[930]

The new ontology shows Wheeler's thinking—that there should be a reason for the threeness of the space dimensions—was right.[a]

Wheeler's missing reason hinges on the nature of the space quantum. Let's recap how we plucked this quantum from the leading candidate for quantum gravity: string theory.[b] Its math insists there must be six extra space dimensions any place you look closely.[931] It structures them as a Calabi-Yau manifold,[932] which string theory says is Planck-sized[933] and Dowker said,

> In the case of quantum gravity we have independent evidence, from the entropy of black holes for example, that the scale of the discreteness is of the order of the Planck scale.[934]

This allows us to answer Wheeler's question. All that's needed is to see those manifolds as real.

The so-called *Calabi-Yau threefold*[c]—the 6-D manifold many string theories use—is endowed with a *volume*.[935] By its nature volume is three dimensional (though this does not mean the manifold itself has three external dimensions; the 3-D reality emerges at far larger scale).

We can relate this to how the interrelations (or links) between next-neighbor manifolds have the singular property of area that also is a real quantum.[d] Nearly two decades ago Rovelli saw that,

> Intuitively, the grains of space are separated by "quanta of area."[936]

---

a See chapters 32 and 41 at pages 181 and 228.
b See chapter 14 at page 87.
c See note 387.
d Note that no physical aspect of the Planck-scale world is a quantum of length, though when we speak of light speed as one link per step it may sound like such a quantum.

This gives rise to further triplets. Each manifold has three (6 ÷ 2) 2-D links; and, correspondingly, Bilson-Thompson's helons braid in threes.

Once there came to be many manifolds—as happened almost instantly,[a] very slightly sooner than in the old ontology's inflation story[b]—their aggregate volume came to exhibit the three-dimensionality we see.

In other words, while we can't *find* any of the three dimensions because, in that sense, they don't exist, we can *choose* them arbitrarily because we have a volume (aka a 3-D space) to work in. And we don't *see* the three dimensions; we see volume. This, then, is the explanation Wheeler sought.

Thus, while the whole of space is a real entity, the three dimensions (so understood) are emergent. Through the unimaginably violent throes of the exponential epoch, they surged into statistical existence.[c]

Newton's theory assumed space is absolutely flat. One might say he thought it had no shape. Einstein's general relativity equations are more accurate, although, in ordinary local circumstances they are only slightly more precise.

But Einstein's equations, too, approximate the shape of space. Extraordinarily precisely. Their approximations lie in the reality that, at extremely small scale, space is discontinuous. That scale is so small it's hard to see how we could ever detect the difference.

Yet the difference has deep and practical significance: Real 3-D space will make consistent sense.[d]

---

a See chapter 40 at page 222.
b See chapter 21 at page 126.
c See chapter 41 at page 228.
d The main reason for seeking a better theory grounded in reality is not the miniscule improvement in accuracy but the expectation it will lead to better understanding and new physics, see chapter 133 at page 639.

47
WHAT'S THE MATTER?

> The whole story of the world need not have been written down in the first quantum like a song on the disc of a phonograph.
>
> Georges Lemaître (1931)[937]

In 1931, Lemaître wrote that note to *Nature*[a] titled "The Beginning of the World from the Point of View of Quantum Theory." He reasoned that the universe began as just one quantum.[b] Like most everybody, he had almost no idea what a quantum *was*. He had what we may see as an extreme demand to make of his imagined version. He continued,

> The whole matter of the world must have been present at the beginning.

His first quantum was composed of matter—our whole present universe of it! He had it divide like some kind of vast radioactive decay.

Physics in his day was struggling to come to terms with his much panned 1927 revelation space should be expanding and Hubble's much lauded 1929 revelation that it *is* expanding.[c] That all of the matter was there from the beginning was the conservation mindset of the time: Matter could be neither created nor destroyed. (Much the same story is mainstream today.[938])

| Matter is a by-product of making space. |

As the reader can now see, the quantum everybody needed was a quantum of pure *space*. There was no need for matter in it. Four daunting intellectual impediments obstructed them from seeing this.[d]

---

a See note 410.
b See chapter 17 at page 105.
c See chapter 18 at page 112.
d Me too.

The first was the universal ontological commitment to continuous space, backed by the huge investment in its math.[a]

The second was mental detritus from dismissal of the ether, leaving literally nothing in its place. Though Einstein had himself in 1920 said space is something,[b] this made no mark on most minds.

The third was the difficulty of thinking about either space or matter as relational. Decades later, Smolin would insist on this but not say how to make it work.[c]

The fourth was the conundrum that, much later, sent me flying into San Francisco: If all that vast amount of matter wasn't there from the beginning, how could it arise?

Once we conceive of Bilson-Thompson's twists in action making matter out of ribbons and see ribbons as relations between quanta, the solution becomes simple. True, the details may get messy, but we don't need details now. For us it is enough to see the possibilities of matter being made along with space.[d]

When a space quantum replicates, not only is there a new manifold, but also there must be six new links to neighbors. Jamming their way into the universe's works, the links are prone to being born with tangles. That is, new space should bring new twists and braids, meat and potatoes for new particles of matter.

It is a simple concept. Making space makes matter too.

---

a See chapter 14 at page 87.
b As an unavoidable consequence of the theory of general relativity; see chapter 10 at page 63.
c See chapter 5 at page 43.
d See chapter 45 at page 255.

48
YOU'LL GET A CHARGE OUT OF THIS

> It is not known for certain why
> electric charge is quantized.
>
> Robert Foot (1992)[939]

It seems a simple question. Yet, as physicist Robert Foot said, nobody has an answer. Indeed, it seems few even ask, *Why is electric charge quantized?*

Let's add another question (also mostly both unanswered and unasked). Why is the sum of charges in the (known to be composite) proton exactly equal in size to the negative charge on the (said to be elementary) electron?[940] Bilson-Thompson asked and answered it.[a] *Sotto voce.* Nobody noticed.

| There is a reason why the electron's charge equals the proton's.

Twenty twenty-three saw the centennial of the Nobel Prize for physicist Robert Millikan's measurement of the charge on the electron.[941] Does it not seem strange more than a hundred years have passed and such a fundamental issue remains almost unexamined?[942]

In Millikan's time, it was widely *assumed* the opposite charges must be exactly equal. Indeed, if he had had his way this might have become enshrined in our language. In a footnote to his Nobel Prize acceptance speech, he said,

> When used without a prefix or qualifying adjective, the word electron may signify, if we wish, as it does in common usage, both the generic thing, the unit charge, and also the negative member of the species. . . . There is no gain in convenience in replacing positive electron, by "proton" but on the other hand a distinct loss logically, etymologically, and historically.[943]

---

a See chapter 43 at page 243.

If inflicted, this linguistic laceration might plausibly have appeared uncontroversial until 1964 when physicist Murray Gell-Mann mentally dissected the proton and other assumed to be elementary particles into quarks, which have charges -e/3 and +2e/3 (antiquarks vice versa),[944] for which work he too received a Nobel Prize when observations showed that he was right. This surely should have let the doubt-fox loose among the unit-charge chickens.

So, how are we to understand where quarks' fractions come from? Or, given nobody has ever seen an isolated quark, do such fractional charges exist on distinct particles?[945] Or, if the distinct-quark story's right, how do electrons get their charge together?

Bilson-Thompson's model offers us an easy understanding of fractional charge.

His tweedles pair (as helons) to produce net twist of plus or minus one full rotation (or none).[a] A full twist corresponds to plus or minus one-third of the electron's charge. Each tweedle's twist is quantal, so each charged helon's charge is a digital two.[b] The reason the electron's charge is exactly equal to the proton's is then simple arithmetic.

Bilson-Thompson's electron is a braid of three full anticlockwise rotations (or helons) with a total charge of three times two equals six negative tweedle charge quanta or -6e/6—a digital minus six.[c]

His proton has eighteen tweedles (or nine helons; three quarks of three each) of which eight have a total charge of eight positive charge quanta +8e/6, eight have no net charge, and two have a total charge of two negative charge quanta -2e/6. So, the proton's net total charge is +6e/6—a digital plus six (we call plus one). QED.

How charge came to *seem* quantized in units of the electron's charge e is then easy to see. That the arithmetic behind it can turn out exact follows from the tweedle's being digital. That there *is* arithmetic behind it says these particles—including the electron—must be composite. (So does the recent demonstration that a photon beam can coerce an electron into a spiral shape.[946])

---

[a] Recall each tweedle is a half rotation.
[b] Physics treats it as analog: 5.34058876×10^{-20} coulombs.
[c] This digital minus six we call minus one. Just where the twists (or charges) are at any instant, he doesn't say; but they must be somewhere within light-speed distance.

We can answer a fourth question. Amid all the chaos of twist creation, leading to a vast number of not only up and down quarks and electrons but also more exotic charged particles, why would the universe be electrically neutral to the best of our ability to detect a net charge?[947] Which is to say, how did the universe keep track?

This answer too is easy. Whatever twist contortions each ribbon goes through, it forms a Ring that's closed, so its twists must be created in matched pairs; there can never be net twist in any of the seven Rings.[a] Thus, regardless of what particles are formed, the universe cannot create net charge.

So, it's easy to understand how, over vast reaches and numberless charged particles coming into being, the universe's electric charge is balanced.

Understanding the machinery behind charge conservation[b] is a free bonus.[948]

---

a See chapter 42 at page 236.
b Aka the law of charge conservation.

49
A LIGHT VARIATION

> The fact that Nature displays populations of identical elementary particles is its most remarkable property.
>
> John Barrow (2007)[949]

Those identical subatomic particles Barrow found to be remarkable will turn out to be real.[a] They are real though each is made of twisted links that hang together in loose fashion.[b]

And while *loose* may be understatement for the far-flung[c] (Planck-scale) federation of six labile links between space quanta,[d] their digital character explains why measurements show them to be identical. It also offers opportunity to understand a little of how they behave.

| How can a neutral photon be an electromagnetic wave?

Most fundamentally, it shows how they are plainly not—as usually advertised and as we see (above) Barrow assumed—"elementary."[e]

He is far from the only physicist who seeks to understand how subatomic particles could be identical. There have been some far-out answers. Wheeler once told Feynman that the reason all electrons are identical is the universe has only one electron and it zigzags back and forth in time (positrons look like electrons going back in time).[950]

---

a See chapter 132 at page 632.
b More particularly, each comprises a scattered conclave of six real twisted links; see also chapter 132 at page 632.
c Though tightly grouped at atomic scale, they are far flung at Planck scale in the sense of Wheeler's observation about "a vast piece of real estate"; see note 78.
d See chapter 17 at page 105.
e This is not only an aspect of the new ontology; it is clear from observing fractional charges, see below.

Another way to say why those not really elementary particles, such as electrons, are identical is they are identical assemblies of identical numbers of identically twisted identical links between identical space quanta.

Our new view shows each particle—except the photon—has six tweedles that bear six electric charges. Each tweedle's charge is $e/6$, either positive or negative. They come in those pairs Bilson-Thompson calls *helons*, so, some pairs may cancel inside (as it were) the particle. Thus, the possible composite particles have charges 0, $e/3$, $2e/3$, and e, both plus and minus. All of which possible particles and charges particle physicists do "see."

So do other physicists. Wilczek recently wrote a short piece on how they have been seen in solid-state physics for forty years. He said,

> Not long ago, the outrageous idea that electrons, when injected into the right sort of material, would break into other objects seemed as far-fetched to most right-thinking physicists as the idea that the Earth moves seemed to sober natural philosophers in the time of Copernicus. Yet the Earth moves—and electrons do break apart.[951]

It will come as no surprise to readers that he went on to say,

> In the simplest case, the apparent charge is one-third that of an electron, which indicates that electrons injected into the material layer have fragmented into three equal pieces.

A benefit of new ontology is new avenues for trying to make sense of what we know, while maybe leading to new physics. For example, we can use our new understanding to take a close look at the photon (fig. 48). It is special, the simplest of all particles, a trivial or null braid made from six twistless tweedles in the Bilson-Thompson model.

It is the next thing, one might think, to plain vanilla space.

The photon may be Einstein's most significant discovery. It was the reason for his Nobel Prize.[952] He did not have it easy. His efforts to establish it ran into heavy headwinds; physicists rejected it for almost twenty years.[953]

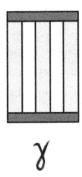

Fig. 48. The photon as three helons with no twist and trivial braid

At the time, there was no standard theory. There was no table of sixteen particles with the photon among them. Indeed, the idea light, long seen to be a wave, could *also* be a particle, strained comprehension. Einstein himself later took a softer view of it. To a friend, he wrote,

> I do not believe the light quanta have reality in the same immediate sense as the corpuscles of electricity. . . . The . . . particle-character of light will—in my opinion—be understood in a more indirect way, not as immediate physical reality.[954]

Whatever mental picture he may have had, there is no sign in his writings that he ever thought a photon might have "internal" electric charges. Yet one must wonder. How do we explain photon particles manifesting themselves as electromagnetic waves if they are in all respects charge-neutral? The concomitant of e-m waves is moving *charges*.[955]

As Einstein and Infeld themselves more broadly put it,

> How can a moving corpuscle have anything to do with a wave?[956]

However it is posed, this is a simple question that, till Bilson-Thompson brought his tweedles into view, there seemed no way to answer.

So, with our new view of the photon let's take up the challenge: How could the photon, being neutral, behave like a changing charge?

Bilson-Thompson may have offered us (albeit unwittingly[a]) an attractive possibility. It lies in his model's version of another neutral particle, the **Z⁰** boson, source of the so-called weak force. It is net neutral, but it has six charges, each of *e*/3, three positive, three negative (fig. 49). And as Bilson-Thompson noted, the photon and the **Z⁰** are related through the electro-weak theory and something called Weinberg mixing.[957] He said,

> We may . . . speculate that deforming an untwisted helon into a counter-twisted helon, or vice-versa, accounts for the Weinberg mixing between the Z⁰ and the photon.[958]

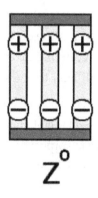

Fig. 49. *The Z⁰ as three helons with canceling charges*

In other words, his model shows a photon could transform into a **Z⁰** by "internal" twisting: Just twist the middle of each of the photon's three untwisted ribbons through a full turn.

Weinberg won his Nobel Prize for predicting the **Z⁰** particle. He said it was needed to explain how the *weak force* led to nuclear decay. In effect, his work aligned the **Z⁰** with the photon. An article summarized it this way:

> It was not just that [the Z⁰ and the photon] could be explained by similar tools. At high energies they were basically the same thing.[959]

No doubt it would take a lot of energy to change a photon into a **Z⁰**. If it had too little energy, classical (i.e., non-quantum) theory

---

[a] To be clear, nothing he published or said to me supports this speculation.

says it cannot happen. Just as it says a particle can't tunnel through a barrier.[a]

So, back to our particle/wave photon. Could the photon twist itself into a Z^0, maybe only partly, for, say, a Planck time or two—as if borrowing the necessary energy—and then untwist back again?[b] Could a higher-energy photon do it more often? What we've come to expect from quantum systems says it might be possible. And if it is possible then quantum theory says it must be so.[c]

If it were so, we could understand how our neutral photon could be cooking (as they say) with gas—or electricity!—its quantum ingredients fleetingly exposing what amounts to changing charge.

Could this then offer us an understanding of wave-particle duality?

This is pure speculation. It is also an example of the new world-view offering a way something that seemed senseless might make sense. Let's look at another.

---

a See chapter 17 at page 105.
b See chapter 100 at page 492.
c See chapter 43 at page 243.

50
TRAVELING LIGHT

> Thus we say that the speed of light is constant because it just is, and because light is not made of anything simpler.
>
> Robert Laughlin (2005)[960]

It is a fundamental mystery that's buried in an article of faith: Light speed is constant; and it sets a universal speed limit. But why? And how?

These are not the only puzzles. In vacuum its speed is 299,792,458 meters per second.[961] Why not 299,792,460? Why not 110? It will be convenient at this point to pry the answers out and bring them tentatively under our new ontologic wing.

As Nobel Prize–winning physicist Robert Laughlin said in the epigram above, the standard explanation is there *is* no explanation; "it just is." But, with the new ontology in hand, we'll see there *can be* an explanation, one that lies at the heart of quantum space and its relationship to motion.

About the same time Laughlin wrote his rambling book on reinventing physics, Bilson-Thompson wrote his too terse paper that showed a photon as six untwisted links, hooked up pairwise as three uncharged helons that are unbraided.[a] Hold up three straight fingers for the helons (or see fig. 48) and you have his photon.

Which is to say light *is*, contra Laughlin's explanation, made of something simpler.

The widely held views Laughlin adduces may be excused. How was one to know? After all, that the photon is an elementary particle

---

a Bilson-Thompson would likely take issue with my simplification that the photon's helons are not braided; rather (and no doubt correctly) he describes it as "the trivial braid"; see note 763.

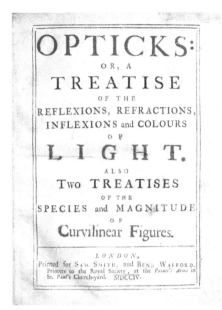

Fig. 50. Title page of Newton's magnum opus Opticks (1704)

has roots in Newton's great work *Opticks* (fig. 50)[962] and has been embraced by generations of physicists.

In 1905, Einstein rehabilitated Newton's light particles with his Nobel Prize–winning work that introduced what came to be called the photon.[963] Months later, he added a new assumption. He transformed the observation that light had a constant speed in vacuum into a fundamental law.[964] (We will see he was right. We'll also see this was a stroke of genius while flying blind.)

Thus, the old ontology was based in part on the *assumption* that— for some reason nobody understands—in a 4-D spacetime that itself is hard to understand, the speed of light in vacuum is a fundamental constant.

Of this, Planck observed,

> The velocity of light is to the Theory of Relativity as the elementary quantum of action is to the Quantum Theory.[965]

And for a hundred years or so their underlying reasons have been equally obscure.

The new ontology embraces an alternative assumption that is also fundamental: With each iteration of the 3-D universe, the non-twist states of photons move exactly one space quantum.[a]

With this in hand, we'll see each iteration of the universe registers one Planck time on our clocks inside the universe.[b]

Let's say the photon never takes a step back, that it always moves forward. The speed of the photon is the ratio of the space we "see" its non-twist states move through and the time we "see" the universe take to make them move, over many steps.[c]

At Planck scale, the speed of light in universal units is just one link per step. That is, one. Not 1.000 but $1 \div 1 = 1$.

When you work out one link per step in physics' arbitrary units (one Planck length[d] in meters per Planck time in seconds) it's exactly 299,792,458 meters per second.[e]

This brings with it a further understanding. The reason that the speed of light becomes the universal speed limit is this. It is the maximum move our seeming-smooth but Planck-scale jerky cosmos is able to make. It is the universe's penny a day; like Johnny in the seesaw rhyme, it can't work any faster.

Translated into far larger than Planck scales, this is Einstein's law of propagation of light that complements the principle of relativity:

> We shall ... introduce another postulate ... namely that light always propagates in empty space with a definite velocity [*c*] that is independent of the state of motion of the emitting body.[966]

Einstein postulated it to be a fundamental law. Now we can see why he was right: It is a fundamental property of space.

---

a They move in the direction of their propagation. How they do this leaves a lot of physics to work out.
b See chapter 52 at page 284.
c In the mind's eye the path through quantum space is kinky at Planck scale, while the path we "see" is straight. However, *straight* and *kinky* are not meaningful concepts at Planck scale. See also chapter 62 at page 326.
d This is the cube root of the Planck volume; it does not mean the links have a linear attribute of length.
e For the reason it's exact, see note 948, above.

51
MOVING PARTS

> Helons are bound into triplets by a mechanism which we represent as the tops of each strand being connected to each other.
>
> Sundance Bilson-Thompson (2008)[967]

Topology, it seems to me, may be the math that until now has cut the closest to reality. Yet, in the hands of a topologist, it can also seem ethereal.

Nonetheless, we should examine some of Bilson-Thompson's compositions of the standard-model particles, not only for what they *are*, but also for what they *do*.

> Those twisty particles must somehow move.

In his paper he said nothing of what *really* holds their braided strands together. Never mind; he's a topologist. So, he merely mentioned *mechanism* and then drew a bar across the tops of diagrams; and across the bottoms too. (See, for example, the particles in fig. 51.)

His bars seem to encapsulate the helons' twists, keeping them twisted and their braids together. But he did also let them loose upon each other. So, for example, if an electron (e^-) and its antiparticle, a positron (e^+) (fig. 51), encounter each other (see Fig. 52), the bars between them vanish, their charges cancel (that is, their opposite twists untwist each other), and their braids disentangle, becoming straight (can you do this in your head?[a]).

Thus, each becomes just three uncharged and trivially braided helons—which is to say, they *are* two photons[b] as the two particles' mass is released as pure energy. (This too is what happens in real life—maybe in *you*, while being imaged in a medical PET scanner.[968])

---

a If not, see fig. 41 at page 251.
b In practise, highly energetic gamma rays.

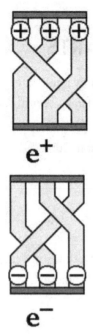

Fig. 51. Bilson-Thompson's positron and electron

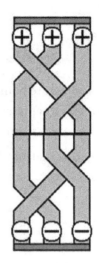

Fig. 52. Annihilating e^+ and e^- leaving two photons

Having seen how space is made, we know each quantum has three 2-D pairs—or "ribbons" as, at first, he called them—that run right through the manifold, through whatever tangle of all six dimensions is inside. Wherever else they run, those three ribbons can't be ripped out of their manifold. *This* (I say; he hasn't[a]) may be a "mechanism" (his word) that can hold his helons together.[b] It's so simple; Ockham would be happy.

Those twists are so tiny we will never see them. What—if anything—can we say of where they are or, in other words, how close to each other they may be?

Bilson-Thompson's model says no word on this. Those twists could be right next to a manifold that corresponds to his top bar. But this may not be necessary. The twists could slip by neighboring manifolds and still be on the same ribbons and have the same topological effect. And, if that's so, maybe they could wander further

---

a Bilson-Thompson did not quite say he thought this idea is wrong; but I got the message he might not embrace it.

b Until, that is, one wants to move them; see also chapter 128 at page 614 .

without risk of ever getting "lost."[a] In his last paper Bilson-Thompson contemplated that . . .

> . . . a braid is connected to a larger network ("the Universe") at only one end, which we may think of as the bottom.[969]

All this requires a rigor we can't yet achieve.

One thing that needs working out is how twists relocating along ribbons can give rise to particle motion. It's called *kinetics*, the study of how motion works. More particularly, the next key issue is the *kinematics*, how his particles *can* move (aside from what might get them going).[970]

There are two potential rules that a topologist exploring kinematics would have in mind about the Planck-scale world these links live in. One is a twist may move along a ribbon. The other is that braids and twists may interchange.[971]

The bottom line is this too needs new physics. I can't track the moving parts so I will not ask the reader to try.

In 2007, Hackett looked at this (without benefit or hindrance of the new ontology). He concluded Bilson-Thompson's particles *can* move:[b]

> Isolated substructure[s] of ribbon networks are introduced, and a theorem is proven that allows them to be relocated.[972]

And of course, the kinematics will not be the end of it. Catching up on many missing decades of real physics will require a lot of work and time and workers. The next field that will need new understanding is *dynamics, why* things move. Hackett foresaw possibilities and problems:

> The demonstration of these translations provides great promise in further developing this model into a theory that involves particle dynamics. However several key obstacles remain.[973]

---

a I will assume so.
b This is not directly applicable to the new ontology.

"The dynamics" is what Sundance said he had in mind to work on a decade ago at CSSM[a] if and when he and his bicycle arrived in Adelaide.[974] Not far below the surface of dynamics lies the problem of inertia.[b] It's a long-standing mystery all of its own.[975] It was Newton's first law in my high school physics: A body keeps its state of motion until a force makes it change.

At first glance, the new ontology may seem to only deepen the old mystery about this law. How can motion keep on going in a strictly 3-D universe where it all stops at every turn? This question reaches all the way back to the ancient Greeks.[976] Our version becomes, how can that 3-D universe somehow encode and, next iteration, implement a particle's state of motion?

A second look, though, shows the new ontology may give us new insights about why the problem has been so intractable and how we now could seek to solve it.

The Planck-scale-physics picture tells us that the question of inertia *needs* a Planck-scale answer. It's no wonder it has so long been an unsolved problem when seen only on a macroscopic scale. At that scale useful equations reflect emergent approximations—very good statistical approximations that, rather than reveal, conceal what's really going on.

Each atom of a moving body has its own momentum. Each atom would keep going just the same without its fellow travelers. Likewise, each of the subatomic particles that constitute that atom. But what may we say of the six tweedles of each particle?

Let's look at, say, a bullet shot in space: Its tweedles' moves are digital. At each iteration of the universe, each tweedle makes a one-quantum move or it doesn't. And whether it does—says our ontology—is random. And if it does, it moves at light speed. Being far slower,[c] bullet motion needs no more than a tiny bias in each tweedle's odds of vastly many light-speed random moves.

Thus the new ontology narrows the search for the source of inertia to a search for bias in Planck-scale probabilities that somehow gets

---

a Special Research Centre for the Subatomic Structure of Matter.
b See chapter 128 at page 614.
c I.e., about three millionths of the speed of light.

encoded in the 3-D structure of the universe's quanta and relationships between them—and that is conserved.

Did Sundance and his mentor and his sponsor at the PI have part of the answer? In 2007, working with a version of his model, they asked if it was able to accommodate dynamics. They said it could. They spoke about this in ambitious terms:

> Ever since the notion that geometry is dynamical was advanced in the 19th Century and, even more since that idea was realized in general relativity, there has been a dream: to unify matter with geometry and gravity by demonstrating that matter arises from singularities or topological defects in geometry. In this paper we find that this expectation may be realized in a certain class of quantum gravity theories. . . . [T]he dynamics of the quantum geometry does not destabilise the emergent particle states we have identified.[977]

More specifically, they found,

These states propagate coherently.[978]

So far, one might think, so good. How good is less than clear. For example, they set their states "in a background topological space." They leaned on work Markopoulou had done with mathematician David Kribs that hinged on broad distinctions between kinds of background independence without a coherent ontological commitment.

What could they have done with a commitment to granular space? Markopoulou's gone; so is Bilson-Thompson; Smolin seems to have moved on. We may never know.

Or maybe someone else will take a look at the dynamics in a quantum space.

52
THE COSMIC CLOCK

How . . . can cosmic time be measured,
and what would it mean to measure it?

Roberto Mangabeira Unger (2015)[979]

Behind Unger's questions lies, *What meaning can we give to time for the whole universe?* He said,

> The clock that is neither inside nor outside the universe must be the universe itself.[980]

Or, as Smolin asked, seemingly from his *insider* position[a] (but is he breaking its prime rule?),

> If there were a clock outside the universe, which notion of time within the universe would it correspond to?[981]

He answered (unhelpfully from our perspective) his own question,[b]

> Every possible notion of time.[982]

The new ontology leads us to simple answers for these questions, but it does not start with clocks.

To orient our minds, we too might mentally move ourselves *outside* the universe.[c] (But then, as Markopoulou said, our extramural observation will be unphysical.[d])

In the new ontology, outside the universe, freed from space and all relativistic foibles, one might "see" sequential iterations (if one could slow them down).

---

a See chapter 24 at page 141.
b His rationale was general relativity has undermined time's meaning: "The choice of a time coordinate in general relativity is . . . completely arbitrary."
c See also *Time One*.
d See chapter 24 at page 141. With that caution, one might see it as a thought experiment; see chapter 30 at page 173.

Thus, the universe *is* an unphysical kind of clock, or so we might conceive in such gedanken-experiment fashion. Like other clocks, it would involve observing motion, each tock corresponding at Planck scale to vast numbers of twist relocations and at grand scale to a (relatively) tiny spasm of expansion or, we might say, to the universe's tics and pulses. In this way we could "see" the universe keep a kind of absolute time that is simultaneous everywhere because it is a purely spatial object.[a] This is synchrony by being and then doing; it's too bad we will never see either of its aspects.

We can also think of the whole universe as a computer.[b] Its bits (the states of links) are stored between space quanta. Or rather, they are *trits*: They have three possible twist states, L, N, and R.[983] It runs a simple program. (I call it *Next*.[c]) The processing is both synchronous[984] and sequential.[985] That is, it has a global state that is replaced by the next global state.[986] One could also view it as massively parallel.[987] It has no central processing unit; indeed, it is the ultimate distributed-processing system, with each manifold with its links being its own processor.[988]

> The universe keeps its own kind of time.

We may imagine an outside observer—free from the strictures of space and time inside—seeing the universe go *Next*. How would such a person answer Smolin's question? Surely, they must say, it's the time of Aristotle and of Leibniz, the ordering of all events. Philosopher Ursula Coope said,

> Aristotle claims that time is not a kind of change, but that it is something dependent on change. He defines it as a kind of "number of change" with respect to the before and after.... [T]his means that time is ... a universal order within which all changes are related to each other.[989]

And, being clockless, we would see, like Newton's true time, Aristotle's has no metric.

---

a Note that, without access to clocks inside the universe, we would have no way to determine duration of each tock of the Cosmic Clock or even tell if they were equal.
b See also chapter 16 at page 103.
c See chapter 57 at page 307.

But let's not neglect the defects of our "observer" that are essential for their work: They are outside the universe;[990] and so they and their enterprise are fiction.

We have seen how the universe worked in the instants after it began, when it was easier to "follow" with few moving pieces. The universe consisted of a certain number of space quanta and that number grew. It did not grow 1–2–3–4–5. Rather, it grew 1 ; 2 ; 4 ; 8 ; 16 with no 3 or 5 and no smooth segue in between. Each Now of the universe existed; then it didn't because Next caused a new Now to be. Each Now state gave rise—even as it gave way—to its succeeding Now state with some simple rules.[a]

We'll soon see when it got big it kept on growing but did not keep doubling every step.[b] There's no reason to assume that at some arbitrary point it stopped its program. Thus, this is what we would see today if we could watch it from the outside with no light-speed law to limit vision: Kaching, kaching, kaching . . . the universe relentlessly replacing itself. No time involved; no way to say the kachings are evenly spaced. However (if we could slow their blinding speed) we could count the kachings.

It took almost fourteen billion years for someone to make a clock inside the universe so we could watch it.[c] Currently, the best clocks follow photons.[991] Watching them and what goes on we could induce (we *have* induced) existence of kachings. We can assign one link motion per kaching to be the speed of light. And from that, with a clock inside the universe, we can assign the Planck time to each iteration. This then is our fragile access to the Cosmic Clock.

Einstein rewrote the troubled tale of common time with special relativity.[992] He purported to eliminate simultaneity.[d] An event over there cannot be said, he said, to happen at the same time as an event over here.[993] We may not even say for sure one was before the other.

Observational support for his theory of watching rods and clocks is now overwhelming. But note, a clock, we say, *tells* time. The essential clockiness of Einstein's time is even buried in the usage of our

---

a See chapter 57 at page 307.
b See chapter 54 at page 294.
c I do not mean by this to exclude the possibility that we were beaten to it somewhere else.
d See chapter 28 at page 158.

language: *It's one o'clock*, one says, unmindful that *o'clock* began life as *of the clock* in the thirteen hundreds.[994] But most clocks do not *tell* the time. To find the time we need clock-watchers.

Common time is, as Einstein specified for special relativity,[995] the result of watching clocks.[a] And, yes, watched clocks appear to do just what his theory says they do. If you define time to be what a local clock reveals to you, you cannot say that something happens here at the same time as something happened over there.

You'll also see my clock going slower than yours as I move by (and vice versa). When your traveling twin returns to Earth (fig. 53), she will be younger than you are thanks to her motion.[996] Time (defined by motions of clock hands or even molecules in people[b]) is relative.

So clock time is convenient but it has conceptual problems.

Fig. 53. The Twin Paradox: When your twin returns, you are older than she is

This concept of time is a construct of our minds. It's not a property inherent in the universe. It's not real. We can find other kinds of time. Rovelli tabulated eight.[997] Of one[c] he was (as, no doubt, he had like me learned in school) especially disparaging:

> [M]any physicists, even many relativists . . . cannot free themselves from thinking of *the great big clock that ticks the universal time* in which everything happens.[998]

---

a See chapter 28 and note 620.
b See chapter 28 at page 158.
c He called it "Newtonian."

The reason he was dissing this big clock is he assumed that special relativity applies to it so it does *not* tick universal time. And yet his turn of phrase—*the great big clock that ticks the universal time in which everything happens*—is apt for our *Cosmic Clock*.[a] The which is not an ordinary clock, it is the universe. It's not a construct of our minds. It is the ultimate reality, and it is absolute.

Special relativity has nothing to say of it. The whole universe just wholly *is*,[b] though we can't see its wholeness. We must wait for light—that always bears the message of what *was*—to reach us.[c] That message travels to us through successive 3-D universes.

By contrast, as we've seen, the Cosmic Clock is by its nature inherently simultaneous throughout the universe.[d] This is not exactly news. For example, in 2008, mathematician and philosopher Peter Forrest said,

> [W]hatever objection could once have been based on Special Relativity has been undermined by the discovery of the almost isotropic expansion of the universe, which provides us with the Cosmic Clock.[999]

It is ironic that Rovelli (to whose insights we owe many debts) would so nearly get it right but then dismiss it out of hand as the epitome of error, and would later write an entire book on time that does not deem it to be worth a mention.[1000]

Even more irony may be found. In 2015, Smolin (with Unger) came out in favor of some sort of cosmic clock:

> The conclusion of our argumenta that most challenges the well-tested and well-established physical theories is that there must exist a preferred and global conception of time.[1001]

The irony is that he called relativity in aid:

---

a A farrago of nonsense has been written under this rubric but there seems to be no better name for the real thing.
b All this is good news for quantum theory, which assumes a background time with an assumed property of simultaneity. See, e.g., Smolin, "Space and Time in the Quantum Universe," note 613, at his page 244.
c This is the basis for the special relativity story.
d See chapter 28 at page 158.

This implies . . . that the relativity of simultaneity—as well confirmed as it is—give way to a preferred global time. This imperative is demanding but is supported by our ability to reformulate general relativity so as to admit such a global time.[1002]

There is yet more irony.

Some scattered few of the disciples of spacetime seek to escape its toils.[a] But, instead of casting it aside, they first embrace it, setting up their 4-D spacetimes. They then dissect the 4-D with an operation called a preferred *foliation*.[1003] This operation cuts a "spacelike" slice (or foliation) of constant time, also called a *hypersurface of the present*, through their spacetime, dropping out the time dimension to create a doubly unreal pseudo-plain-vanilla 3-D space.[1004] To cut their slice they need to call upon the Cosmic Clock.

They might as well have set their space up cleanly in the first place.

In another example, physicist Antony Valentini shows how the observed property of nonlocality gives rise to what he calls "a true slicing," only to put it back together as "space-time":

> It may now be assumed that non-locality acts instantaneously with respect to one of these foliations. . . . There is then a true slicing, and space-time is really the time evolution of the (absolute) 3-geometry.[1005]

Indeed, though saying so is doctrinally frowned on, physics would do well to do away with spacetime (which we will see is pure math fiction[b]) entirely and just work with space (with clock in hand when wanted), focused on, as Greene put it . . .

> . . . how singular and fleeting the here and now actually is.[1006]

Our clocks all measure motion of some sort (think shadow on a sundial, hands on a clockface, electrons in a digital watch, photons in

---

a See also chapter 73 at page 378.
b See chapter 80 at page 411.

atomic clocks). That is, we measure common time as Newton told us, "by the means of motion."[a]

As seen from inside,[b] the whole cosmos itself reveals no gross motion. Let's take another run at that pivotal question: How can we follow tocking of the Cosmic Clock?

Stuck inside the universe, to compute (but not observe) its iterations—in analog true-time tocks to eight-digit precision—we can divide the distance we see light move by the Planck length.[c] Thus, we can follow the Cosmic Clock's tocks everywhere.[d] It's trite to say that it takes 18,548,582 trillion-trillion-trillion tocks per second.[1007]

How do we know it's so? Because the universe explained it to Max Planck and, once he had half a handle on it, he told us.[e]

---

a See chapter 28 at page 158 and note 635.
b See chapter 24 at page 141.
c That is, to be a tad pedantic, divide by the cube root of the Planck volume.
d It has limited practical use as we have to wait to find out what is going on there.
e See note 298.

53
RUNAWAY INFLATION

*Physically expansion of the universe means
the continuous creation of space.*

Yurij Baryshev (2015)[1008]

We have known for a hundred years space is expanding. It's a concept many still find hard to understand but that it *is* expanding is now thoroughly confirmed.[a]

Indeed, conceptual confusion may be found all over town.[b] For example, when Weinberg was asked a falsely premised question,

> How is it possible for space, which is utterly empty, to expand? How can *nothing* expand?

... he came up with a false answer,

> The answer is: space does not expand.[1009]

It's not only the fact of space expanding that gets confused. We've known since the nineteen twenties that space is *not* nothing.[1010] Which is not the same as knowing what it *is*, but lots of those you'd think would know it all don't even seem to know it must be something.

As our new ontology's space is made of elementary quanta, at any literal instant there is an exact number of them that could (in principle) be counted. So space expanding means (as cosmologist Yurij Baryshev—a voice crying in the cosmologic wilderness—might have said had he been into granular thinking) space quanta are being created.[c] In quantized space, that is the only possibility because the universe's volume equals the quantum's volume times the number of quanta.[d]

---

a See chapter 17 at page 105.
b See chapter 18 at page 112.
c Regrettably, he instead went right on to say, "... together with physical vacuum."
d We have no reason to assume the quantum's volume is increasing; this would give rise to a whole new set of consistency problems.

The standard model of cosmology regards expansion as a problem that is safely tucked away in its continuous-space math. The equations show the universe is now expanding for no better reason than it *was* expanding yesterday. They offer not the slightest hint of explanation why or how expansion would begin. The mainstream version shows expansion slowing for five billion years or so due to the pull of gravity, then speeding up due to that constant Einstein added that acts like a kind of antigravity—nobody knows the reason for that either.[a]

Recall that, of this odd contrivance, Peebles said,

> This is a terribly baroque arrangement, and many of us wonder if there is something deeper behind it.[1011]

The new ontology agrees; there *is* something deeper behind it. It is terribly simple: Space expands because the universe makes more space quanta.

The same way of making them (replication by tunneling[b]) has been at work from the beginning.[c] You could say it is still the same inflation; it has just slowed down.[d]

This is a very simple story in stark contrast with the standard model's complexities of unexplained origin, mysterious expansion, dark constituents, and changing modes of growth.

Other space-expanding puzzles are:

- Why did the exponential epoch end?
- Given that expansion *did* slow down by an enormous factor, why did it not completely stop?
- Why did it speed up again after six billion years?

We will soon see how simple these three too can be.

The starting point lies in our answer to the earlier question: How are new quanta created? Let's go back to the first instants.

We saw that cosmological inflation—despite its math that makes no sense and doesn't work—does hold the kernel of a seminal idea.[e]

---

a See chapter 29 at page 165.
b See chapter 19 at page 118.
c See chapter 42 at page 236.
d See chapters 54 and 68 at pages 294 and 351.
e See chapter 21 at page 126.

We just need to ditch its space (because it is continuous) and its math (because it is superfluous).

The beginning was so small and simple, one space quantum tunneled into two. It led to exponential growth.[a] It left us, with Rovelli . . .

> . . . counting the grains of space of which the cosmos is made.[1012]

Many fail to grasp the awesome power of the exponential. It is simple math. Multiply some starting number (in this case, one) by some factor greater than one (here it is two) each period or step (here, each Planck-time iteration). The number is soon out of sight.[1013]

The universe's number of space quanta doubled each tock of the Cosmic Clock—that is, each Planck time as we figure it, each 5×10^{-44} s.

The result may seem surprising. Though the universe's initial single quantum's Planck volume is so small quadrillions of them would be far too small to be detected, after some six hundred iterations space would be as big as the visible universe is now, some four quadrillion-quadrillion (4×10^{32}) cubic light-years. Elapsed time, about thirty atto-yoctoseconds (3×10^{-41} s) *if* it continued doubling every step. This is an inconceivably short time; it is about one-hundred-million-trillionth of the time it takes light—at 300,000 km per second—to transit a hydrogen atom.

Suffice to say—and this is common ground between the new ontology and current cosmology—the universe's first burst of expansion went far faster than the speed of light.[b]

To be clear, this is based on *every* quantum replicating *every* step. Why would they almost instantly stop doing this? We are about to see a simple answer: The tiny clump of space quanta—the infant universe—was in the grip of its own extreme curvature.

And as the universe grew big this ceased to be.

The result was very fast but relatively orderly expansion that very soon was very slow. Not a Big Bang; I call it the *Big Fizz*.

---

a See chapter 40 at page 222.
b Ibidem.

54
HITTING THE UNIVERSAL BRAKES

> The goal of this notice is . . . the proof of the
> possibility of a world whose spatial curvature
> . . . depend[s] on time.
>
> Aleksandr Friedmann (1922)[1014]

The universe's exponential growth was an awesome event with no audience. As the universe grew larger, its curvature grew smaller, as Friedmann (above) famously showed.

As it grew to cosmological (say, multi-light-year) size, inconceivably huge and dense and hot masses moved apart at speeds far faster than the speed of light—speeds that (if there had been clocks to check them) doubled every three Planck times as space expanded.[a] This event dramatically demonstrated the substantiality of space.[b]

> Growing at unimaginable speed, why would the expansion slow down?

The whole exponential show lasted little more than a micro-atto-attosecond. We cannot comprehend how fast that was. Consider that mere attosecond physics is the cutting edge of today's fastest science;[1015] the emergence of the cosmological-sized universe was a factor of 10^{24} faster.

In the preceding chapter we found why those huge dense masses were flying apart so fast. Now we need to understand what almost stopped them in their tracks.

In terms of our new ontology, this is only to ask, *Why would the space quanta stop their doubling every step?*[c]

---
a See chapter 69 at page 357.
b That is, the Planck volume of a space quantum is a fundamental constant.
c They must have; else doubling would have continued.

The answer that makes simple sense is that, as we have just contemplated, the whole universe grew large. Or, to apply Hawking's language, the radius of curvature of space became "very large compared with the Planck length."[1016] Peebles said,[a]

> [T]he enormous expansion would have stretched out the mean radius of curvature ... to some enormous value.[1017]

That is to say, the universe's gravity became far less intense. Could it have ceased aiding replication? With what we know about how other kinds of tunneling proceed,[b] we can now see how this might happen.

Einstein showed curvature of space *is* gravity. We've seen how the universe is curved in on itself, or closed. When the universe had just begun, it was extremely tiny so its space was extremely curved. That is, it had extreme gravity.

If you take another look at the right side of fig. 12,[c] and imagine the shape of the barrier if we introduce a large energy difference (that extreme curvature could create), the right-hand energy level would sink out of sight, the angle at the top of the barrier would become very sharp, and so the barrier would become very thin (and the top of the barrier might be dragged down). We don't know the shape of the "barrier" for space quanta replication, but for electron tunneling barrier-thinning works for any shape and exponentially increases the probability of tunneling.[d]

So, in terms of our analogy, its tunneling "barrier" could have been at first extremely thin.[e] As space expanded it relieved the extreme curvature.[f] As it became Solar System–sized, or maybe grew to galaxy-sized, with less curvature the barrier became thick enough to drastically reduce the tunnel probability. So, space quanta failed to replicate every Planck time. The exponential epoch ended.

---

a He was speaking of the cosmological inflation equivalent; see chapter 53 at page 291.
b See chapter 19 at page 118.
c On page 121.
d We know nothing of tunneling at Planck scale but in principle the probability may approach one.
e This is at best only an analogy; "shape" and "thin" have no metric meaning at Planck scale.
f The curvature of a sphere is inversely proportional to its radius.

The Big Fizz continued.[a]

Maybe this makes it all sound gradual and in a sense it was.[b] Yet the extra time it would take to grow from Solar System size to match that of our galaxy—to grow, that is, a factor of a quadrillion in volume—while doubling at every move, is a mere fifty or so Planck times. Slowing down in such a span is the universe equivalent of stopping on a dime. Just how much the universe had to grow before the doubling slowed down is almost pure speculation. (We may get more insight when we look inside a big black hole.[c]) What's clear is there is no need to delay the slowdown any later—like the time frame favored by the fans of cosmological inflation.[d]

So, why (I think I hear you ask) would space still be expanding? The reason—coming up—is this: There are still places in the universe with extreme curvature.

---

a See further, chapter 69 at page 357.
b For this reason, a quantitative theory would likely show more very early growth than the standard model does, which could solve some problems we won't bother to go into here.
c See chapter 69 at page 357.
d See chapter 21 at page 126.

55
ROOM FOR IMPROVEMENT

> In the beginning there was an explosion . . . which occurred simultaneously everywhere, filling all space. . . . "All space" in this context may mean either all of an infinite universe, or all of a finite universe which curves back on itself like the surface of a sphere. . . . [I]t matters hardly at all in the early universe whether space is finite or infinite.
>
> Steven Weinberg (1977)[1018]

The standard model's story of the universe's origin (above) is clearly fictional and just a bit ridiculous. Around the same time, Vonnegut gave us this parody,

> [T]he Universe began as an eleven-pound strawberry which exploded at seven minutes past midnight three trillion years ago.[1019]

In 1977, Weinberg was Higgins Professor of Physics at Harvard University. He was Senior Scientist at the Smithsonian Astrophysical Observatory. He had written the book *Gravitation and Cosmology*.[1020] He was soon to win the physics Nobel Prize.[a] He was an authority on both general relativity and particle physics, and had, as he modestly put it . . .

| Making the universe expand simply means making more space. |

> . . . done small bits of research in cosmology from time to time.[1021]

That year, he wrote his popular book on cosmology, *The First Three Minutes*. Subtitled *A Modern View of the Origin of the Universe*, it was a historical ontology, a tale of what once was: It described the

---

a See chapter 46 at page 263.

still-emerging standard model's story of the way the universe began. It was an accurate description of the model; and, as we will see, it was entirely wrong.

Weinberg was himself not wholly happy with it:

> The standard model . . . is not the most satisfying theory imaginable of the origin of the universe.[1022]

He found the long-discarded steady-state model "philosophically far more attractive."[a] And he said,

> Can we really be sure of the standard model? . . . I cannot deny a feeling of unreality in writing about the first three minutes as if we really know what we are talking about.[1023]

To get a sense of how unreal it really was, let's take a quick look at the excerpt in the epigram. It sets out the then standard model's concept of what got the matter moving: an explosion. Physicists know this is nonsense; for example,

> While an explosion of a man-made bomb expands through air, the Big Bang did not expand through anything. That's because there was no space to expand through.[1024]

The notion that the initial event—whatever else it may have done, it got extremely hot[b]—happened "simultaneously, everywhere, filling all space" was silly from a causation perspective if there were already (inexplicably) some space to fill: Even on the small scale of, say, the half-sticks of dynamite I (in farm-boy guise) once used to remove gum-tree roots, there is no way such events happen instantly.[1025] At most a universal explosion wavefront might propagate at light speed.

So the question whether the universe was finite or infinite does matter. If it were finite, it might at first be infinitesimal (or, as our ontology contends, tiny and singular), having thus no problem with

---

a It has some similarities with the new ontology; see chapter 17 at page 105.
b At page 146, Weinberg said, "We can make a crude estimate that the temperature of 10^{32} K was reached some 10^{-43} seconds after the beginning." There is today no strong ground to disagree.

simultaneity. If it were infinite—or even microscopic—before the bang began, a simultaneous "explosion" would require miraculous assistance from unreal math.[a]

The early (and the recent) incarnations of the standard model of cosmology offer us insight into physics' changing story of the world's earliest moments. But at no time—not even now—has it had a serious story of how it began.

The new ontology offers a clear and simple understanding. It says space grew by replication of an initial space quantum. The twists and braids of matter came into existence as a by-product of new space quanta. Thus, the matter in space *wasn't* moving *through* it in explosive fashion. It was sitting pat in space where it had come into existence. It was the space itself that grew and took its matter with it.

And it is still so: All those galaxies and gravitationally bound galaxy clusters[1026] are simply sitting in their space. While they have local motions that arise from local gravity, their large-scale motion (over vast intergalactic distances[1027]) is due to new space that moves the bound entities apart.[b] It is somehow easier to picture this if we can keep in mind that space itself has two-thirds the mass of the universe.[c]

And is this not eerily akin to the "continuous creation" of the abandoned steady-state model?[d]

Weinberg summarized that model's key assumption:

> As [the universe] expands, new matter is continually created to fill up the gaps between the galaxies.[1028]

In comparison, the new ontology says the universe expands because new space (along with matter) is continually created and enlarges gaps between gravity-bound entities like galaxies.[e]

Thus, the creation of new space is why there is more *space* between the galaxies.[f]

---

a See chapter 21 at page 121.
b See chapter 18 at page 112.
c See chapter 29 at page 165.
d See chapter 17 at page 105.
e The sizes of entities like solar systems and galaxies are fixed by gravity.
f The authors of the steady-state model knew neither the reason for new space nor the idea space itself is something.

We do not *need* creation of new matter throughout the universe (as the steady-state model posited). We *do* need creation of space; we need enough of it to explain the expansion we observe; and we need its sources to be strewn around the universe more or less uniformly at the largest scales. (New matter is a by-product.)

So, space quanta make new space. The rate of space production is extremely low in ordinary tracts[a] but very high in space with extreme gravity.

Thus, the new ontology provides a single explanation for all the vastly varying and inexplicable convulsions of the universe's growth—it's simply quantum tunneling where curvature of space is very high.

We know how much new space is made each day.[b] In an upcoming chapter we will put this to the test. Can today's universe meet the new-space-production challenge?[c]

---

a It may well be negligible, like the chance of the out-of-gas automobile tunneling through the Tehachapi Mountains in *Time One*.
b This is, in effect, given by the Hubble constant.
c In chapter 69 at page 357.

56
BEING NOW

> The "present of the universe" is meaningless.
> Carlo Rovelli (2018)[1029]

We should not be surprised that writer Jean-Paul Sartre would offer a rebuttal of Rovelli's conventional-physics wisdom (above) about *Now*. He wrote,

> The true nature of the present revealed itself: it was what exists, and all that was not present did not exist.[1030]

He also wrote the book on being, *L'être et le néant*. Philosopher Hazel Barnes's translation is titled *Being and Nothingness*.[1031] Its subtitle is also to our point: *An Essay on Phenomenological Ontology*.

> You are right; they are wrong; now is all there is.

In her translator's introduction, she said,

> Sartre is one of the very few twentieth-century philosophers to present us with a total system.[1032]

In our terms, she meant his works embody a system of ontological commitments that speak to a worldview. In that context, Sartre's notion of *Now*—or the "true nature of the present"—could teach physics a thing or two. Here was an opportunity for physics to learn (as it once was wont to do) a lesson from philosophy.

Many others have weighed in. Arendt gathered far-flung expressions of *Now* as . . .

> . . . the Present of the thinking ego, a kind of lasting "todayness" (*hodiernus*, "of this day," Augustine called God's eternity), "the standing now" (*nunc stans*) of medieval meditation, an "enduring present" (Bergson's

> *présent qui dure*), or "the gap between past and future" as we called it in explicating Kafka's time parable.[1033]

Thus, on the one hand, *Now* is our universal observation. Less elegantly than Sartre and less eruditely than Arendt's informants, *Now* is what we know we got.

On the other hand, physics—as Rovelli summarized above—says there is no now; we all are told (I kid you not) we must imagine it.

Einstein inadvertently compounded this misapprehension, saying in a now much quoted letter of condolence (fig. 54),[a]

> For us believing physicists, the division between past, present and future has only the meaning of an (albeit stubborn) illusion.[1034]

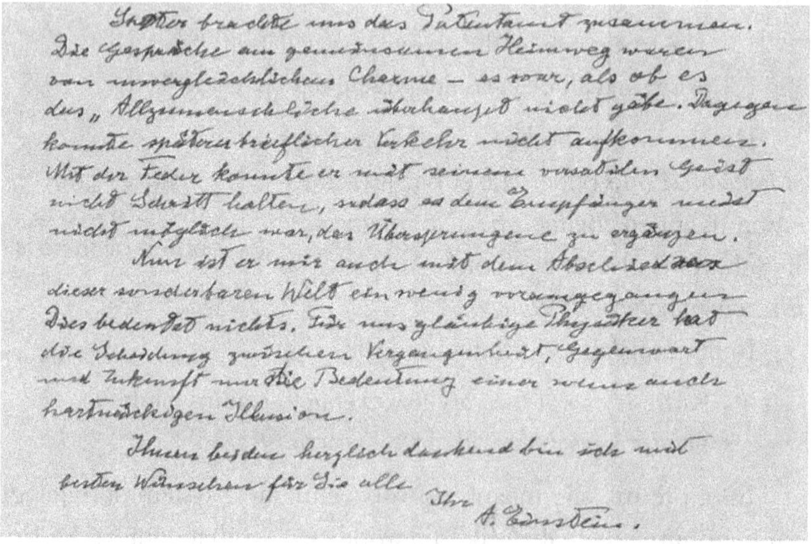

Später brachte uns das Patentamt zusammen. Die Gespräche auf dem gemeinsamen Heimweg waren von unvergleichlichem Charme – es war, als ob es das „Allzumenschliche" überhaupt nicht gäbe. Dagegen konnte später brieflicher Verkehr nicht aufkommen. Mit der Feder konnte er mit seinem versatilen Geist nicht Schritt halten, sodass es dem Empfänger meist nicht möglich war, das Übersprungene zu ergänzen.

Nun ist er mir auch mit dem Abschied von dieser sonderbaren Welt ein wenig vorausgegangen. Dies bedeutet nichts. Für uns gläubige Physiker hat die Scheidung zwischen Vergangenheit, Gegenwart und Zukunft nur die Bedeutung einer wenn auch hartnäckigen Illusion.

Ihnen beiden herzlich dankend bin ich mit besten Wünschen für Sie alle

Ihr

A. Einstein.

Fig. 54. The last part of Einstein's letter of condolence

The problem's rooted in a widespread misconception about relativity. Rovelli was a messenger for many who conceive that *relativity* provides a way to say *reality*. It's become so bad the BBC's been broadcasting misinformation like,

> Einstein's theory of relativity suggests that there is no real difference between the future and the past.[1035]

a He wrote this in a letter comforting the family of his deceased close friend Michele Besso. Though it was no doubt sincere, few serious students would read it as his considered view of the physics of time.

That program was an interview with Becker (who wrote a book entitled *What Is Real?*[1036]). The BBC's provocative headline was "Physics suggests that the future has already happened."[1037] It was shades of the block universe idea, where past and future cohabit in a 4-D spacetime "block" that exists as we wend our way through, willy-nilly.[a] Surely silly, it's a lesson in supposed reality from an equation, instead of the reverse.

Sorry, Adam; sorry, Carlo, too: That special relativity's equations don't distinguish between past and future—let alone explain our singular experience of *Now*—tells us only they do *not* describe reality. Timewise, they tell us about watching clocks.[b] Relativity is good at doing what it does; unsurprisingly it's not so good at doing what it doesn't. In other words, this is a limitation a public broadcaster might have mentioned.

The new ontology provides an understanding of our universal observation that it's always *Now*. "The present of the universe" Rovelli dissed (above) is not only real, as Sartre already told us: It is *all* that's real.

We've seen this from the beginning;[c] and noted the by far simplest assumption that, as the universe grew, it continued with its lockstep iterations.[d] But the new ontology's compelling reason for concluding that the universe is still in causal lockstep is it *must* be. The 2-D ribbons that tie the space-quantal fabric of reality together[e] would become a meaningless fragmented mess if any of the universe were able ever to get out of step.

Indeed, when we realize the nature and the role of links, we can see the universe as indivisible in a profound sense that would please Parmenides:

> Parmenides' Being does not just happen not to be divided, but cannot be divided.[1038]

And, as he also conceived, its *Now* is timeless:

> Parmenides . . . claims that Being is not in time in the

---

a See chapter 27 at page 154.
b They describe—as Einstein said—observations of measuring rods and clocks.
c See chapter 42 at page 236.
d See chapter 52 at page 284.
e See chapter 41 at page 228.

sense that there are no temporal differences—no "was" or "will be"—that would apply to Being.[1039]

The *Now* that Becker and the BBC (and, to be fair to them, also a host of physicists) want us to ditch in deference to the equations is all that ever is! The past they claim as having the same status as the future is no longer in existence, instantly discarded by a universe that keeps no copy. Each future is only an enormous panorama of might-be's until a randomly selected one of them comes into being.

In other words, each successive iteration of the universe comprises a complete, unmoving but dynamic 3-D space that "is" for its tock of the Cosmic Clock. That's it; that's all; that's over. *Next.*

It follows that the universe must be itself a special (or preferred or privileged) frame of reference. This would seem to run afoul of relativity's supposed claim that there is no preferred frame. No question, there is an apparent contradiction.

But what Einstein actually claimed (and showed) was that no *special frame* was needed for his theory. His *principle of relativity* is,

> The laws of physics are the same in all inertial [or unaccelerated] frames of reference.[1040]

In support of this, he said,

> [T]he most careful observations have never revealed such anisotropic properties in terrestrial physical space, i.e. a physical non-equivalence of different directions.[1041]

That no special frame exists was an assumption that he didn't need, that looked okay because it had *not yet* been proven wrong. What Einstein said about such observations was true when he said it.[1042]

But it is not true today. In cosmic microwave background images there is a great circle of amazingly uniform temperature, and at right angles to it there's a measurable variation from one side to the other, called the *dipole anisotropy*.[1043] It is due to the motion of the Earth relative to the universe, which is itself a special frame.[a]

---

[a] Curiously, physicist Ernst Mach conceived (said Einstein) the idea that the mass of the entire universe—then thought to mean that of all the stars—formed the special frame of reference that is required for acceleration (or inertia); see note 23 and chapter 86 at page 435.

Cosmologists correct their microwave-background images for this motion based on the simplest assumption—which happens to be Einstein's principle—that there is no preferred direction.[1044] That is, the universe is the same in all directions. This correction shows Earth moving through space at some 600 km/s toward the Leo constellation (fig. 55).[a] *With* the correction, the observed temperature in all directions of the radiation left from the Big Bang is . . .

> . . . astonishingly uniform with variations of only one part in ten thousand.[1045]

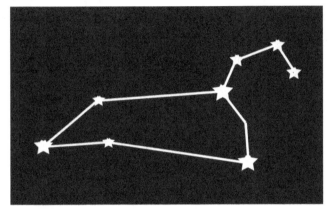

Fig. 55. Leo, where we are headed in the universe's frame of reference

In other words—as Einstein could not know but carefully allowed—the universe itself gives us an absolute frame of reference, also known as a preferred or special frame.[b] Cosmologists use it routinely.[c] They understand we are moving at some 0.2 percent of the velocity of light with respect to that special frame. This is an absolute velocity. Their measurements are increasingly sophisticated and precise.[1046]

Somewhere within the narrowing of that precision, an ideal version of this observed frame corresponds to the 3-D universe. It is

---

a Or toward the *Great Attractor*; see note 1043.
b Of course, someone may object that this frame merely gives us a velocity relative to the universe and so does not supply a special frame. However, assigning some velocity to the universe itself would violate the principle of relativity (and Ockham's razor too) as well as being daft. And too, the new ontology assigns this practical frame a new status as a surrogate for the real 3-D foliation that is in principle the universe at any instant—the very definition of absolute rest.
c It is, as outlined, an intrinsic aspect of their analysis of cosmic microwave background data.

the timeless view of space—sometimes regarded as a timeless leaf of spacetime or a foliation[a]—that is Now.

For example, philosopher of physics Tim Maudlin said,[b]

> A family of hyperplanes distinguishes one frame over all others, namely the frame in which the hyperplanes are simultaneity slices.[1047]

Thus, the whole cosmos restored simultaneity to physics after an overblown gust of relativity blew it away.

This simultaneous Now is absolute and universal. We can at last feel secure in our view—which physics has long held to be delusional—of the reality of Now.

It's worth repeating, Now is real; it's all that's real.

---

[a] See also chapters 73 and 80 at pages 378 and 411.
[b] See also chapter 52 at page 284 re slices, hypersurfaces, and foliations.

57
DOING NEXT

> We ought then to regard the present state of the universe as the effect of its anterior state and as the cause of the one which is to follow.
>
> Pierre-Simon Laplace (1814)[1048]

Contrast polymath Pierre-Simon Laplace with Smolin, who, in somewhat similar vein, said,

> [T]ime, in the sense of the continual becoming of the present moment, is fundamental to nature.[a]

Both focused on the present. But Smolin portrayed its "becoming" as "continual." While Laplace, with "the present state," "its anterior state," and "the one which is to follow," at least *seemed* to speak to a sequence of separated states.[b]

This seeming subtle difference is of the essence: To have meaningful existence, the universe must *do* as well as *be*. For thousands of years, contradictions have tormented those who asked how it can possibly accomplish both in any continuous fashion.[c]

| What is does not change smoothly; it jumps.

Thus, questions such as *What is change?* and *What is motion?* puzzled profound thinkers such as Zeno.[d] It long seemed the deeper that one dug into these questions, the more one found oneself awash in inconsistency.

---

a See note 631.
b While this sense is present in the original French text and so is not an artifact of translation, there is nothing in the rest of Laplace's essay to support a discontinuous intention.
c This chapter seeks to resolve them.
d See chapters 10 and 128, and Sattler, note 240.

The only answer to such questions that avoids these problems is the universe must *be* unchanging, and then *be* again, unchanging still but changed, in unending iterations. That is, *Now* creates the next *Now*, which is different.

In 1945, philosopher Maurice Merleau-Ponty understood what was needed to answer Zeno's questions; he just didn't seem to quite grasp that it *was* the answer, maybe because he was immersed in old ontology:

> As soon as we bring in the idea of a moving body which remains the same throughout its motion, Zeno's arguments become valid once more. It is, then, useless to object that we must not regard motion as a series of discontinuous positions successively occupied in a series of discontinuous instants, and that space and time are not made up from a collection of discrete elements.[1049]

The new ontology says precisely that. Motion *is* "a series of discontinuous positions successively occupied in a series of discontinuous instants,"[a] and space and time *are* "made up from a collection of discrete elements."

The old ontology's ideas of moving bodies calcified three hundred years ago, when Newton gave the world his laws of motion along with the math—the calculus—to follow them in continuous action. Two hundred years later, Einstein showed how Newton's answers (although almost always good approximations) were conceptually wrong; but then he didn't solve (indeed his work entrenched) the fundamental problem. Today, physics and philosophy both wrestle with the problems that weren't fixed by Einstein's better answers.

Let's start with the problem of *change*, the essential element of motion. (We'll get to *motion* later.[b])

One has only to imagine a world without change to grasp how the reality of change has deep significance. As Rovelli said,

> [T]he world is nothing but change.[1050]

So, it would be nice to know how change works. Two hundred years ago, with no knowledge of granular space or the Planck time,

---

a Or of *Nexts*.
b In chapter 127 at page 609.

Laplace in his groundbreaking work on probabilities (quoted in the epigram) gave us part of the answer: The universe causes the next universe.[a] How does this happen? And what does it tell us about time?

In 1973, Ayer set out to write of time and motion. These topics took him in turn to change; he said,

> On this view, the passage of time simply consists in the fact, *which is itself non-temporal,* that events are ordered in a series by the earlier-than relation.[1051]

It is only when needing to compare the "time" between two events with the "time" between two others at another time or in another place that—being unable to set them side by side, so to speak—the physicist requires a metric and comes up with a clock.

Here, seeking to understand the world (and not the metric) we shall set aside the clock to see what's going on.

The new ontology shows how change results from what the entire universe does. One can think of this as if the universe were running a causation app that I call *Next.*[b] It's a *do-while-there-are-quanta* loop:[c]

do {
Next();
nextnumber=nextnumber+1;
}
while (numberofquanta > 0);

With the looping of this simple program at a given Now, here tracked by "nextnumber," the entire 3-D universe executes subroutine "Next()," which replaces it with a new Now.[d]

At each step, each space quantum executes its part of "Next()" in synch; each quantum replicates[e]—or not—by tunneling with probability that depends on the shape of the barrier that depends in turn on

---
a In Laplace's hands, this resulted in a deterministic universe, an intellectual field that came to be known as *Laplace's demon.*
b See chapter 52 at page 284.
c So, it's not about to end; technically, the number of quanta is a monotonic non-decreasing function of the iteration number.
d This may seem odd; several thousand years of thought have shown the alternatives to be odder.
e See chapter 17 at page 105.

the local curvature of space;[a] each adjacent twist state relocates—or not—along its ribbon by one link.

The fundamental character of *Next* is that there are no partway places between one *Now* and the next *Now*. The universe replaces itself with a new version subject to simple rules: Each quantum replicates, or it does not; each twist state relocates, or it does not.

That is, *Next* is entirely discontinuous. The universe is digital. This (to us) strange character is of the essence if we are to understand those paradoxes about time and change and motion.[b]

I offer no new physics here. Building that will be a task of years for many physicists, and it may lead to many Nobel Prizes.

Later we will ask if it is possible to navigate—even in principle—the seemingly unsignposted landscape of Planck-scale space.[c] But let's note that, at least in principle, the links and their twist states amount to what the quantum-physics crew could label *hidden variables*. They meet Bell's prescription to a tee:

> The hidden variables should surely have some spatial significance and should evolve in time according to prescribed laws.[1052]

What we need here is the big picture from the new ontology to understand what the change we see really *is*. It is the difference between one 3-D version of the universe and its successor elaborated many times to become smooth change writ large as we observe it. That is, the accumulation of biased random Planck-scale changes more than 10^{43} times per second is what we see as change.

The primitive element of change—the single relocation at a single iteration—is digital, not analog. It is then simple to solve Zeno's paradoxes. The answers were just waiting for us to think digital.

For example, of Zeno's flying-arrow paradox,[1053] Aristotle said,

> Zeno's reasoning, however, is fallacious, when he says that if everything when it occupies an equal space is at rest, and if that which is in locomotion is always occupy-

---

a See chapters 53 and 54 at pages 291 and 294.
b See chapters 10 and 128 at pages 63 and 614.
c See chapter 63 at page 330.

ing such a space at any moment, the flying arrow is therefore motionless. This is false, for time is not composed of indivisible moments.[1054]

Now we can see Zeno was dead right and Aristotle was dead wrong. Much later, Wilczek said perceptively,

> If space and time were both discrete, then the arrow could just hop, at each tick of time, from one position to another. That simpler alternative seems more in tune with our digital age.[1055]

Simply put, thousands of years of the best thought of leading thinkers were needed for us to begin to realize—and even Wilczek did not seem to fully realize—that all the change we see as analog is digital, its bits just being—as Planck learned and yet did not himself completely realize—very small and very swift.

Thus, it's not time that's real, not even Newton's "absolute, true and mathematical time." But something very close to it *is* real, so close that one might pun and say it is the *Next*-best thing, absolute, true and mathematical succession.

58
ULTIMATE CAUSE

> We may have to get used to the idea of an absolute zero of time—a moment in the past beyond which it is in principle impossible to trace any chain of cause and effect.
>
> Steven Weinberg (1977)[1056]

Aristotle set us on the trail:

> [W]isdom is a science conversant about certain causes and first principles.[1057]

As Weinberg was suggesting, above, the "chain of cause and effect"—whatever that is[a]—presumably runs back to the beginning of the universe. Surely it must have occurred to him (though he did not say this) that the chain's origin needed to be something more consequential than "an absolute zero of time."

Smolin said,

> The single, unique universe must contain all of its causes, and there is nothing outside of it.[1058]

This sounds good but, if nothing more is said, it leads to a logical trap called an infinite regress.[1059] The sole way out of it is to begin the chain with an ultimate cause.[b]

This is how questions of causation become philosophically fraught when we get up close and dirty with a real beginning.

In his book on nothing, physicist Frank Close said of this ultimate interface,

> The paradox of creation from the void, of Being and Non-Being, has tantalized all recorded cultures.[1060]

---

a See chapter 74 at page 381.
b Of course, this leads to another question, maybe even to another book.

Close canvased the scene from a physics point of view and concluded,

> Everything may thus be a quantum fluctuation out of nothing. But if this is so, I am still confronted with the enigma of what encoded the quantum possibility into the Void.[1061]

At risk of a bad pun, I have to say that he was close, missing by only a single quantum.

In an earlier airing of the problem, logical positivist Ayer debated with philosopher Frederick Copleston on BBC radio in 1949. He said,

> Supposing you asked a question like "Where do all things come from?" . . . You are asking what event is prior to all events.[1062]

He said asking about such an event is "meaningless." Yet that is exactly what we are asking here—and we will find it may indeed have meaning.

As Close (above) concluded, the answer cannot be only "the Void" or "nothing."

A decade or so ago, science writer Jim Holt set out on an ontologic expedition seeking the ultimate cause. His book *Why Does the World Exist?: An Existential Detective Story* racks up sixty-eight index entries for *nothingness*.[1063] It cites two hundred twenty-eight sources, of which two hundred sixteen were men, seven other commoners were women, and two were queens of England, plus one dog, one cat, and one rock group. Only the men seemed to have something to say.[a]

His fun tale has a theme in common with Sartre's deeply ontologic works;[1064] it is concerned with pending nonexistence, what one may conceive as vanishing back into "the Void" from which the world, his expedition seemed to show, must have arisen. Along the way, he tied his question to the problem of ultimate cause. Yet he ended, as he began, in dalliance with nothing.

---

a What he reports about the seven female commoners fades into casual namedropping except for Simone de Beauvoir whose citations are for drinking tea and being buried beside Sartre. What does it say that Holt did not find a woman with some useful thoughts to impart about nothing?

The new ontology has no role for the void, no place—either metaphorical or literal—for nothing. It says the great chain of causation began with precisely the event Ayer claimed to be meaningless; a single Planck-sized Calabi-Yau manifold replicated. This event caused everything that was and is and will be in the universe.

There was no prior event. That first quantum simply *was*. If it was caused by an event of any sort, that event did not happen in this universe.

The first replication was, in other words, the *ultimate cause*, sometimes called brute fact.[1065] It is the final answer to all chains of curiosity.

| What's in a name? Other than it all.

Thus, the first space quantum was unique although it now has many equals. Entirely lost to view, it still has some claim on our attention. How shall we speak of it? Should it get a name?

Seemingly, we should not call *it* "the ultimate cause." If it did no more than *be*, there would be no larger universe. Clearly, it was what it *did*—its replication—that began the chain of causes.

The new ontology is just a wee bit different from all those universe-from-nothing stories. It's a universe-from-one-thing story. True, that bit was very, very wee. But it made all the difference. Lacking that wee bit there would indeed have been nothing rather than something; well, not even nothing—try thinking on that one for a while.

Its property of replication led to all the world, including (we will see) the laws of physics.[a] This then offers us an understanding of *ultimate cause*, the missing subject of so many thoughtful works, the final answer to that seeming-endless string of childlike questions, serially seeking reasons, *Why?*[b]

So, in the end (the first end of our universe that is, the one at the beginning), that string of childlike questions leaves us asking two more. One for which we can see no way to find an answer, *Why was that* there*?* (in a manner of speaking, there being then no *there* to speak of—and no *then* there for that matter). The other, more meaningful

---
a See chapter 97 at page 478.
b See the preface.

in terms of causation, *Why did it do* that? And since its successors keep on doing it, this (though we will never see it happen) may make it amenable to study.

Indeed, that wee thing, the Calabi-Yau manifold, may already be *the* most-studied geometric object, more closely considered in a few decades than triangles or spheres were in two thousand years. Mathematicians say there are many ways that its dimensions may be arranged. Thirty years ago, physicist Tristan Hübsch, the manifold's biographer, could say,

> [T]he number of known topological types of Calabi-Yau 3-folds is well in the thousands and more and more new constructions are being uncovered daily.[1066]

Our universe, with all the amazing ways it's tuned to make life possible, exists because the six dimensions of our first space quantum were just as they were, and as one might think they are in every replica.

Maybe we should see it as the *being* partner with the *doing* of that primal *Next*; maybe it's the *be-do* duo we should call ultimate cause.

59
UNIVERSAL DISORDER

> Although, as a matter of history, statistical mechanics owes its origin to investigations in thermodynamics, it seems eminently worthy of an independent development . . . on account of the elegance and simplicity of its principles.
>
> Willard Gibbs (1902)[1067]

Physicist Willard Gibbs, a fan of "elegance and simplicity" who coined the term *statistical mechanics*,[a] was more comfortable with energy than with its kissing cousin, *entropy*. Perhaps entropy did not quite partake of the same virtues; though he mentioned it a lot, he tended to say it in the same breath as "energy" or belittle it, as in "analogues of entropy."

Indeed, from its origins, the very idea of entropy seems to have been a bit loose and troubling to physics. It was, after all, first developed by an engineer, William Rankine.

Engineers seek practical solutions for practical problems. Rankine's world was mid–eighteen hundreds Scottish railroads. Their steam machines used heat to do work. By then railroad engineers knew heat as potential energy; a given amount of it could be made to do a lot of work or not so much or sometimes none at all. These days they see entropy as the availability of their heat energy to be converted into work.

Rankine studied this. At first lacking the concept of entropy[b], he sought to apply Newton's laws of motion to . . .

> . . . what is denoted by the convertibility of energy.[1068]

---

[a] The study of the statistics of many molecules.
[b] Clausius adopted the word in 1865, transliterating from the Greek ηντροπη for *transformation*.

Physics was not far behind, bearing a differential equation courtesy of Newton's calculus. Thus mathematician and physicist Rudolf Clausius defined *change* of entropy as *change* of energy divided by absolute temperature.[a] While change of this entropy has a definite value that may be zero, the entropy itself has no such point of reference.[b]

In the field of *thermodynamics* (study of the usefulness of heat), this is still the meaning of entropy.[c] But over the latter eighteen hundreds physicist and mathematician James Maxwell, with Boltzmann and Gibbs, rebuilt it on statistical foundations, defining the entropy of a gas as a constant[d] times the natural logarithm of the number of different arrangements (microstates) its molecules could have.[e] Unlike that of Clausius, this definition *does* have an absolute reference point.

> The universe knows more about disorder than we do.

Importantly for us (and for Penrose in especial[f]), a system that has only one microstate has an entropy of zero.[g]

Those who look beyond the boundaries of statistical physics will find analogous concepts, also called entropy. Thus, amateur physicist Saeed Neamati sought clarification between,

> Entropy = disorder, and systems tend to the most possible disorder
>
> Entropy = energy distribution, and systems tend to the most possible energy distribution
>
> Entropy = information needed to describe the system, and systems tend to be described in less lines

---

a The heat must be transferred to the system by a reversible process.
b In integrating the differential equation, one must add an arbitrary constant.
c It is closely related to the second law of thermodynamics; see chapter 23 at page 137.
d It is known as the Boltzmann constant k; it is also Planck's *other* constant, a; see chapter 13 at page 76.
e The reader may recall Boltzmann's approach to entropy was the starting point for Planck's journey; see chapter 13 at page 76.
f See chapters 1 and 23 at pages 15 and 137.
g That is, there is only one way to arrange its individual elements and their energies; the logarithm of one is zero; see following.

Entropy = statistical mode, and the system tends to go to a microscopic state that is one of the most abundant possible states it can possess[1069]

He received a reply from an anonymous[a] physicist—I will call him Parisi[1070]—who wrote,

Your concern about the too many definitions of entropy is well-founded. Unfortunately, there is an embarrassing confusion, even in the scientific literature on such an issue.[1071]

Parisi went on to review an even longer "partial list of different concepts, all named *entropy*":

Thermodynamic entropy

Dynamical system entropy

Statistical mechanics entropy

Information theory entropy

Algorithmic entropy

Quantum mechanics (von Neumann) entropy

Gravitational (and Black Holes) entropy[1072]

They all correspond to some concept of disorder. But they apply in various circumstances; they don't all apply to the same systems; and those that do may not give the same number.

What, then, *is* entropy? Is it a law of nature? Is it a metric? Is it a property?

As Parisi makes clear, there is no *it* in such questions. One could ask each question of each of these (and other) kinds of entropy and debate the answers. In the end one might find more embarrassing confusion.

Could the new ontology point to a single fundamental concept? Let's start at the beginning: That single space quantum, the one that

---

[a] But I think I can out him as Giorgio Parisi, who, months later, shared the physics Nobel Prize for his work on complex systems.

has no name. The one whose entropy must be so low it really bothered Penrose, among others.[a]

The most useful measure of entropy for such a physical system is Boltzmann's, his eponymous constant times the natural logarithm of the number of microstates (or possible alternative arrangements of the universe's contents). With a single quantum there is only one microstate, so the entropy is $k \times ln\ (1) = 0$; which is to say, in the beginning, the universe had no disorder.

This brings a whole new understanding to this deep and vexing problem;[b] it leads to the seminal idea that the entropy at the beginning was not only low enough to meet Penrose's "absurdly low" requirement,[c] but it was zero, a state of perfect order with nowhere to go but down.[d]

The question then is, can physics use this beginning to found a cosmological concept I will call *absolute entropy*? To do this would require an understanding of the microstates of all the quanta and their links that are the cosmos (or perhaps some part of it).

We are not there yet. But physics might get there some day.

---

[a] See chapter 23 at page 137.
[b] It should smooth deep furrows on Penrose's brow; see chapter 23 at page 137.
[c] See note 548.
[d] This is the second law of thermodynamics writ large (or, if you prefer, writ very small); see chapter 23.

60
THE DIRECTION OF TIME

> The paradox of the statistical direction [of time] remains unsolved.
>
> Hans Reichenbach (1956)[1073]

Almost all physics is symmetrical in the direction of time, with no more reason to go *to* in time than it has to go *fro*.

Clocks have, says physics, an annoying failure; their time goes only one way. And so, it seems, does ours.

An exception to time's symmetry in physics is the second law of thermodynamics, which says disorder of an isolated system must increase over time.[a] Physicist Arthur Eddington explained,

| Physics says time must go to and fro.

> Let us draw an arrow arbitrarily. If as we follow the arrow we find more and more of the random element in the state of the world, then the arrow is pointing towards the future; if the random element decreases the arrow points towards the past. That is the only distinction [as to time's direction] known to physics.[1074]

That's why physics has a problem with all those damn clocks insisting time (and all else in the world) is always going forwards in defiance of all its equations. Try as it may (and it has surely tried[b]) physics has no useful explanation.

Seen with the perspective of the new ontology, the universe offers a simple answer: The problem lies in trying to assign a real interpretation to equations that are not about reality. They were devised to give numerical predictions for statistical experiments. The universe, having no need of them, moves to a music of its own.

---

a See chapter 23 at page 137.
b See, for example, chapter 23.

Back in 1933, cosmologist and physicist Matvei Bronstein was working on the problem of the universe expanding. He was thinking of time's arrow, trying to let reason loose on the equations, giving us a graphic picture of his mental desperation:

> [T]he real universe . . . must be highly asymmetrical in ±t, and indeed it can hardly be expected that any rational human being would earnestly believe that any such things as stars absorbing, instead of emitting, energy, or soldiers rising up and marching away from the field in perfect order (but backwards) are really possible in nature.[1075]

More recently, Unger and Smolin put their heads together (while keeping their pens apart) to write an unusual book on natural philosophy. It set out to adjust the old ontology. Its twofold tendentious pitch is we should address the universe as one whole thing; and we should reassess the role of time (and so of clocks).[1076]

We've seen how Unger pondered where a clock must be to measure cosmic time, inside the universe or outside. He concluded that,

> [S]uch a clock could be neither outside nor inside the universe.[1077]

Without delving deep into this argument, we saw that Unger got it right; the Cosmic Clock, neither inside the universe nor outside, *is*, as he said, the entire universe itself![a]

With this Cosmic Clock in view if not in hand (and of course not truly in view), freed from the vagaries of earthly clocks and space and its relativistic motion, we can understand why we always see *Now* moving inexorably forward. The 3-D world that *is* keeps going *Next*. The next *Now* is the only "place" there *is* for us to be. The relentless succession of such "places" provides us (and all those clocks) our attribute of time.

That world can't do *Stop*; it can't do *Back*. What *can't do* means here is (in my loose terms) the universe has no *Stop* or *Back* app.[b]

---

a See chapter 52 at page 284.
b Compare chapter 57 at page 307.

That's not to say a stop or reverse step is inconceivable. If a reverse step should happen at some step by chance—if all the universe's just-replicated space quanta pairs were to somehow swim against the tide to merge back into single space quanta and if all links were to reverse the moves they made at the last iteration like those soldiers backing off the battlefield—there's no way we would notice. The universe's ongoing being and doing would appear to be quite unaffected. It would not even *be* a step back; it would just be an absurdly unlikely step forward.[a]

For a real reverse gear to be meaningful, there would need to be a *Back* app, a system program, that executed with a frequency comparable with that of *Next*. Observation gives no reason to encumber our ontology with such a program.

Likewise, that the world continues onwards tells us that, if there's a *Stop*, there also is a *Restart*; there is no way that between them they could ever register a real-world consequence.[b]

Seeing the world as a causal set,[c] we see it could in principle run backwards just as well as forwards. But our "be-do" universe just doesn't do that. It seems somehow anticlimactic that the long-standing *arrow of time* paradox[1078] should receive such a simple explanation.

We can see, too, it has a connection to the universe's entropy, which is sometimes said to explain the arrow. In his book about this, physicist Sean Carroll said,

> The mystery of the arrow of time comes down to this: Why were conditions in the early universe set up in a very particular way, in a configuration of low entropy that enabled all of the interesting and irreversible processes to come?[1079]

As we have seen, Lemaître told us why. The universe began with a single quantum. As space emerged, entropy increased. It (disorder) had nowhere to go but up.[d]

---

a One could see this in analog terms as a reversible process, $1 \leftrightarrows 2$, with a forward reciprocal rate constant of, say, less than one zetta-yoctosecond and a reverse of, say, more than one picosecond, which would seem to us the same; but I prefer a digital perspective that digs down to what is going on.
b See *Time One*, chapter "The Cosmic Clock."
c See chapter 43 at page 243.
d See chapters 23 and 59 at pages 137 and 316.

Part II: Getting Real   323

The Cosmic Clock tocks on.[a] Replication of space quanta is (at least effectively) a one-way street.[b]

Planck had an entropic insight.[c] Taken a little further, it could have led him to a wider understanding; the world does not run on energy; it is an entropy machine.

---

[a] See chapter 52 at page 284.
[b] See chapter 53 at page 291.
[c] Planck wrote, "While a host of outstanding physicists worked on the problem of spectral energy distribution, . . . every one of them directed his efforts solely toward exhibiting the dependence of the intensity of radiation on the temperature. On the other hand, I suspected that the fundamental connection lies in the dependence of the entropy upon energy." See note 281 at his page 38.

61
THE FLOW OF TIME

> Until we have a firm understanding of the flow of
> time . . . we will not know who we are, or what part
> we are playing in the great cosmic drama.
>
> Paul Davies (1995)[1080]

That "firm understanding" Davies sought so avidly is now in reach. It is embedded in the bones of our ontology, which we can see now mostly as a product of philosophy rather than (as he assumed) of physics.

Time flows . . . in tiny jerks.

Two decades ago, philosophic physicist Julian Barbour got lots of it right. He saw time, not as a continuous flow, but as a collection of instants, each of which is . . .

. . . like a "three-dimensional snapshot."[1081]

For my money, Smolin (who thinks time is real and almost sees it as sequential[a]) gave a more succinct account of Barbour's time than Barbour did:

> Barbour insists that the passage of time is an illusion and that reality consists of nothing but a vast pile of moments, each a configuration of the whole universe.[1082]

The reason Barbour's time deserves an honorable mention is his combination of these insights: What's flowing isn't time, it's a sequence of spaces; each is the entire universe; each is distinct; so, it's not really flowing. He even called them *Nows*. He said,

> The world is made of Nows.[1083]

The new ontology says there's a direct relationship between successive instants of the universe, or Barbour's *Nows*. Each one gives way

---

a See chapter 57 at page 307.

to (and causes very locally) the next. Another way to view this is that at each step, the universe adds new quanta and moves twists. It is a simple show. By contrast, the relationships among Barbour's "vast pile" of Nows look to me to be a kind of quantum Rube Goldberg machine (fig. 56) whose task is to fool us into misperceiving there's a flow of time.

Fig. 56. A Rube Goldberg mustache wiper

The way change happens in the universe, and the amazing rate and tiny Planck scale at which all those increments of change occur, give rise to our perception of smooth motion and conspire to have us feel we are embedded in time and it is steadily passing. We go with the flow because its smoothness is a very good illusion.

The new ontology concurs with Barbour also in another way. Time is not a property inherent in the universe. There is no time dimension—except in the sense that physics can and does invent dimensions that are often useful fictions.[1084]

Meanwhile, it is—as Barbour too would have it—always Now, not just for us but for the universe. Or, put another way, the universe *is* Now. Zen philosophy advocates living in the moment. In strict reality, there is no other way.

To understand our world, we need to see it as a 3-D space that creates change with jerks we can (in principle) enumerate. Which is to say, it's a *digital universe*.

And so, to understand our place in Davies's "cosmic drama" we must realize what we perceive is not a steady "flow of time." It is the swift stepwise successions of space.

62
THE CLOSED WORLD

> But beyond the heavens there is said to be no body, no space, no void, absolutely nothing, so that there is nowhere the heavens can go. In that case it is really astonishing if something can be held in check by nothing.
>
> Nicolaus Copernicus (1543)[1085]

It has become commonplace to view the universe as infinite and space as flat. Physicists do this too, assuming it explicitly or even inadvertently as a convenient but basic underpinning of their work.[a] Many of those who do advert to it continue to cling to their infinite-and-flat perspective even though the latest data say that this is wrong.[1086]

The new ontology concludes the universe is finite.[b] It says astronomer Nicolaus Copernicus was rightly told there's no outside. And right to think there *is* no nothing that holds space in check. The universe does not need to be held in check by anything beyond itself. It is "held in check" by being all there is.[c]

> Get your head around our finite universe.

It's one thing to assert that this is Einstein's universe; that it is finite; that it is spherical;[d] and that it has no boundary or edge.[e] It's another thing to get one's head around all this, to truly understand, to make it as familiar as, surely, reality should be.

So, let's take a longer look. Though it is ours, this universe is naturally strange to us. We think in terms of concepts formed from

---

a Working with an infinite universe simplifies some of the math of physics as there are no spatial boundary conditions.
b See chapter 29 at page 165.
c See chapter 56 at page 301.
d See note 538.
e See chapter 22 at page 133.

our experience of local bits of it. Unsurprisingly, we get the bigger picture wrong.

What do we know of it? First, there is a sphere of space around us that, at least in principle, we see. The light we see emitted by the galaxies nearest the edge of that sphere set out more than thirteen billion years ago.[a] So we see those galaxies as being thirteen billion light-years away. They show us what we might think of as a picture of the way it was some thirteen billion years ago.

But because space has been expanding all that time,[b] we know those galaxies are now more like forty-five billion light-years from us. That then seeable sphere around us (known as the *Hubble sphere*[1087]) is now some ninety billion light-years in diameter.

Thus, the most distant galaxies we see as ancient history are—as we would find if we could instantly receive their light[c]—now far beyond our seeing.[1088] That is because they and the space in which they sit are now fleeing from us faster than the speed of light, and vice versa. They are now in a "lost" zone that began growing just an instant after the beginning;[d] the light they emit can never catch us.

And so, as Smolin put it,

> So long as [it began] a finite time in the past, it will never be possible to observe most of the spatial extent of the universe.[1089]

The boundary between what we *would* see and what we *wouldn't* see—if we could see it instantly[e]—is where the space and all that's in it is now[f] moving away at light speed.[g] The galaxies we could have

---

a More technically, the sphere from which light emitted in the Big Flash is now reaching us was some 13.8 billion light-years away and is called the *particle horizon*.
b See chapter 18 at page 112.
c And if they still existed, which is unlikely.
d See chapter 21 at page 126.
e But then the speed of light would be infinite and we would see all of it.
f Note the two grounds upon which the simultaneity implicit in that "now" may be asserted: the conceptual 3-D universe of the new ontology, which is exactly simultaneous (but inaccessible to observation; chapter 52 at page 284); and the calculated 3-D universe determined from the absolute frame of the early universe (as observed; chapter 53 at page 291), which is simultaneous within the precision of the observation.
g Note that when light emitted near that boundary arrives here, its wavelength will be indefinitely long, like the low-frequency radio waves used to communicate with submarines but even longer.

seen before they got that far and fast still give off light but it will never get to us as the space it is moving in bears it away.[a] Unless, of course, expansion someday somehow slows or stops.[b]

There is more space outside that now-ninety-billion-light-year sphere. How much? We don't know for sure because we can't observe it. But we do know that there *is* space beyond it because space has been moving out of view since the beginning;[c] and we think we do know something about that.

We know space started from essentially no size. We know it expanded at a finite speed. We know it first expanded vastly faster than the speed of light so almost all of it was everywhere out of sight. We know it did that for a tiny time. We know it slowed as its growth reduced its curvature—that is, as it achieved a scale of maybe a light-day or even a light-year but much less than a billion light-years.

We know it has been expanding for almost fourteen billion years.

We know how fast it is expanding now and we're learning more about how it expanded in the past.

When we put all this together it seems to say the part we cannot see might not be vastly bigger than the part we can. (One could conjure a number but it would be no more than a guess.)

Let's explore this mental picture further for a moment. What if we could travel as fast as we like? What would we see if we went searching for the edge of space?

You are searching in a hypersphere, which is like a sphere curved in on itself, and is hard to represent in ordinary 3-D space. In fig. 57, coming to a seeming edge you find you are not at an edge at all.

Suppose you leave Earth with the latest *Star Trek* drive. You know the universe's total size. You calculate the radius of a sphere that holds exactly half its volume and you go that far in some direction. You are

---

a Given the expansion, light emitted beyond the *cosmological event horizon* (a thirty-two-light-year sphere; but see the next note re models) will never reach us; see Davis and Lineweaver, note 444.

b The reader eager to dig further into this topic is cautioned that everything they find will depend on an assumed metric as well as a model (the most common being known as the Λ-FLRW universe); the new ontology has no need to assume a metric, for reasons explained in chapter 12 at page 74.

c See chapter 40 at page 222.

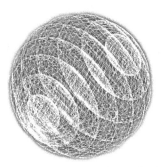

Fig. 57. A hypersphere projected on a flat 3-D space

now at the surface of a notional sphere,[a] centered on the Earth, that holds half the universe. Now take one step *outside* that sphere. You have stepped *inside* a sphere that holds the *other* half of the universe. If you travel on until you reach the center of that sphere you'll be as far from Earth as it is possible to get. Go as far again as you have come and you'll wind up (more or less) back home.

This is so, no matter which way you start out or which way you return.

If you can get your head around this journey, even make it a well-trodden trail, you may truly understand: There *is* no outside to the universe; there is not even nothing outside. All the space there is is inside.

Such is life in our closed world. Maybe we should get used to it.

---

[a] Strictly, half a hypersphere.

63
THE LOCATION ILLUSION

It is impossible to measure the position of a particle
with error less than . . . 1.6×10^{-33} cm.

Alden Mead (1959)[1090]

Alert readers may have spotted my citing Mead (above) as saying *in 1959* the Planck size sets a limit on location—while the endnote dates that same paper *in 1964*. I've already mentioned how he had a tussle with the journal (*Physical Review*) and some of its referees.[a] His paper was published more than five years after first submission. Even more remarkable, amid all that attention it made no mention of Planck![b] Those aberrations give us an insider view of silos in the world of physics. Even today, in most of them Planck's message has not yet been driven home.

> "Here" has no meaning in quantum space.

Mead's remark about what we can't *measure* at Planck scale is true. But it lacks practical significance. Our most precise position measurements, made in the LIGO instrument,[c] are five quadrillion times too crude to challenge Planck scale's strictures.

A more recent, quantum-gravity-like view offers us an alternative approach; it speaks about what we cannot make *meaningful*. This may bring us a step closer to understanding. As Bergmann put it in 1982,

> It would appear that, on a sufficiently small scale, points cannot be identified, either by their metric relationship

---
a In chapter 14 at page 87.
b Mead later said it met "years of referee trouble, eventual publication, a cold shoulder from the physics community." And, "At the time, I knew nothing of Planck's proposal." See note 333.
c See chapter 129 at page 617.

to neighboring points or by local physical characteristics. This argument suggests that in quantum gravity the concept of locality will undergo a fundamental revision.[1091]

But, forty years on, quantum gravity has yet to deliver this revision in full measure.

The new ontology sets out to point the way. At first glance, its space looks like trackless waste, a place that has no sense of place in it because there is no way a place in it can bear a label.[1092]

Or, as Einstein put it (speaking of, he thought, far larger scale),

[Space] may not be thought of as . . . consisting of parts which may be tracked through time.[a]

Imagine you are looking for a place in Fujisawa City in Japan. There are no official street names. Each district is divided into blocks or *ban*. The buildings in each *ban* are numbered, not in spatial sequence but in order of construction.[1093] You can find your number on the area guide board (fig. 58).[1094] Now imagine that it is a screen that shuffles numbers and jitters its picture faster than your eyes can follow. The plain way to describe this is there are no addresses.

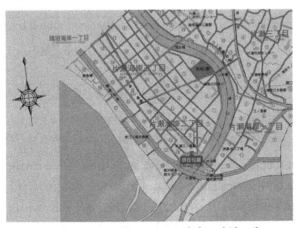

Fig. 58. *Fujisawa City area guide board (detail)*

It's something of this sort that we would see in quantum space with Emerson's transparent eyeball.[b] Being nothing, seeing all, it vaguely sees a 3-D forest of our real version of string theory's Calabi-

---
a See note 262; he referred to it as "ether" but in this context he meant space.
b See chapter 5 at page 43.

Yau manifolds. It's quite unlike string theory's picture of an empty space; its space is densely filled. And quite unlike it as well in that you see no wiggling strings.

Instead, it is the manifolds that jiggle franticly. If we should happen on a particle, there is no way to pin down where it is or, for that matter, where we are. All that our eyeball finds is here and there (and very far apart) one of its twisted links a-jiggling with its manifolds and zigzagging among them.

This is the kind of place that quantum theory told us to expect. Its uncertainty principle says something like, if we can pin down a location we don't know what's going on there; if we know what's going on we don't know where.[1095]

Quantum theory tells us, at far larger scale, *virtual particles* frenetically pop into existence and then vanish.[1096]

No one has ever observed Planck-scale space. In 2021, physicist and director of Fermilab Center for Particle Physics Craig Hogan and colleagues tried indirectly, and they reported what they found: nothing.[1097] And, as Hogan pointed out, even that result is hard to interpret, as,

> There is no rigorous theory of what we are looking for.[1098]

Writing with his colleagues of their theory, Bilson-Thompson noted,

> Locality is a tricky issue in background independent quantum theories of gravity because there is no background metric with which to measure distances or intervals.[1099]

The perceptive reader will have seen there was a further problem. He and his colleagues lacked granular space. So, they had to set their tweedles *in* some sort of space.[a] That the tweedles (with their manifolds) *are* space is not incidental here; it is the point.

Complications of the location problem could already be foreseen in the early days of quantum theory. In 1933, Einstein said,

> [I]t seems to me certain that we must give up the idea of a complete localization of the particles in a theoretical

---

a See, for example, chapter 51 at page 279.

> model.... For instance, to account for the atomic character of electricity, the field equations need only lead to the following conclusions: A region of three-dimensional space at whose boundary electrical density vanishes everywhere always contains a total electric charge whose size is represented by a whole number.[1100]

Though he had continuous space in mind,[a] he could have been reaching (without benefit of Bilson-Thompson's insight) for the new ontology. He continued,

> Not until the atomic structure has been successfully represented in such a manner would I consider the quantum riddle solved.[1101]

By their very nature, the particles of today's quantum field theory have field values that tail off into the infinite distance of continuous space.[1102] They cannot offer even the loose charge localization Einstein said would be enough.[b]

But the new ontology provides exactly that. For example, at any instant the six tweedles making an electron are between each of six definite pairs of space quanta. Each is somewhere specific, but not specifiable, within a definite region.[c]

The seeming impossibility of our ever *seeing* what Einstein was talking about—an electron at a scale far larger than Planck scale—might seem to make the prospects for fundamental physics at or near the Planck scale appear bleak.

Rovelli highlighted the magnitude of the dilemma,

> 10^{-33} cm is a dramatically short scale.... Perhaps the physics "down there" is so different from what we know, that, in the lack of experimental evidence, any attempt to guess it is hubris.[1103]

But physicists do work with subatomic particles without ever seeing one or knowing what it looks like. They intuit supposed properties to

---

[a] The context was his concern that "a field-theory... is fundamentally non-atomic in so far as it operates exclusively with continuous functions of space." See note 1414.
[b] See note 1100 above.
[c] Their random rambles are limited by light speed; see also footnote c at page 269.

make sense of things they *can* see. So, with Polchinski's words[a] in mind, a new ontology in hand, and a retuned sense of scientific method,[b] we may venture to search for new knowledge at the Planck frontier.

Let's ask what the new ontology can say about the *large*-scale structure of space based on its picture of the universe's early iterations.

First, we know that, by step three, each replicating manifold had no unlinked dimensions; that is, all six dimensions in all eight quanta were being "used" in links. At step four, there must be no loose dimension ends, as this would be physically meaningless. And, however step four happens, each space quantum can have six 2-D links. For this to work, some kind of relinking may be needed for all new quanta to become part of the universal structure.[c]

The simplest assumption is all new quanta continued to have six links to other quanta. There are two topologic structures that could possibly support this.[d] They have analogues in crystals with atomic bonds.[e]

One of these structures is called a primitive cubic lattice or *cubic honeycomb* (fig. 59). Imagine cubes stacked in layers; each layer sitting directly on the one below. Thus, there are layers at right angles to each of three axial directions. Each cube has six sides and so has six interfaces with next-neighbors. Now put a space quantum in the middle of each cube; link each one to the quanta in its four neighboring cubes to the sides and to the two above and below, so each quantum has six links.

To be clear, I am *not* suggesting that space *has* an atomic crystal structure in the sense of position of its "atoms."[f] I am suggesting that the quanta could be linked in this fashion, six links per quantum, thus constraining their relationships.

---

a See page 213 and note 830.
b See chapter 37 at page 208.
c This might be thought of as another recourse to tunneling to get from one state to another without passing through an unachievable condition in between them; see chapter 17 at page 105.
d There is also the possibility (which may be even simpler) that new links form small loops, along the lines proposed by loop quantum gravity; see chapter 31 at page 176.
e The analogy is topological, not geometrical, and does *not* imply crystalline structure.
f That would violate the principle that we can't specify anything at scales smaller than Planck scale.

Part II: Getting Real 335

Fig. 59. Cubic honeycomb lattice

Another atomic crystal structure, called *hexagonal close-packed* (fig. 60, left side), offers to lead us to a topologically different kind of space. This structure too has sheets of atoms. But, in each sheet, each atom has six neighbors. (In a crystal, each sheet is a little offset from the one beneath it, with its atoms sitting in the low spots.) Now put a space quantum in place of each atom. Link it to its six neighbors *in the sheet*, with no links between sheets.

Fig. 60. Hexagonal close-packed sheets (left)

If space is to have a topological version of this large-scale structure and remain a single entity, the sheets must fold like a vast crumpled sheet of paper.

In other words, with this kind of linkage we could understand the universe in terms of only two dimensions, though the quanta occupy three dimensions. This space could in principle be uncrumpled and laid flat. Well, not quite flat; it would be a hollow sphere far larger than the universe.

Could space—with all that's in it, subatomic particles and you and me and galaxies—comprise a crumpled sheet? This startling proposition seems to be a possible consequence of the new ontology. Is it right off the rails? Why would one give a moment's thought to the strange idea we could in principle unfold the entire universe into a pseudo-2-D sheet one quantum thick?

And what would *that* do to our notions of location?

Well, fundamental physics is already serving up something that's strangely similar. It's called the holographic principle.

64
ON A WHOLE NEW PLANE

> The idea that space might not be truly three-dimensional is rather compelling, philosophically.
>
> Edgar Shaghoulian (2022)[1104]

Physicist Edgar Shaghoulian was writing of the *holographic principle*, a relatively new frontier in fundamental physics.

One way to state it is the physics in a 3-D universe can be described by simpler physics in a 2-D universe.

Of this seemingly improbable concept, Smolin said,

[T]he holographic principle . . . may very well be the fundamental principle of quantum gravity.[1105]	What physics does in three dimensions may be done in two.

And Susskind said,

> It's not considered some wild speculation among most theoretical physicists. It's become a working, everyday tool to solve problems in physics.[1106]

Susskind was pointing out that physicists are using it to solve their problems in 2-D and then converting the solutions to 3-D. This version is often called the *strong holographic principle*.[1107]

Another version is the *weak holographic principle*. The short, sexy statement of this version is *We live in a 2-D hologram*.[1108]

The basic concept was developed by 't Hooft in his work on quantum gravity.[1109]

Susskind adapted it to string theory, and said,

> Here, then, is the conclusion that 't Hooft and I had reached: the three-dimensional world of ordinary experience—the universe filled with galaxies, stars, planets,

> houses, boulders, and people—is a *hologram*, an image of reality coded on a distant two-dimensional surface.[1110]

He called it . . .

> . . . a shocking departure from what we have been accustomed to in the past.[1111]

And . . .

> . . . a violent restructuring of the laws of physics [whose] proof requires no fancy mathematics.[1112]

Readers who digested the preceding chapter may find the "departure" just a shade less shocking. But I confess that, without forewarning of this "violent restructuring," I might not have mentioned that the universe's quanta could conceivably be linked into a single 2-D sheet.[a]

How would all this work? I do not know, I have my hands full fashioning a real ontology; so other hands can construct real physics, which bids fair to be at least as lavish as the fancied physics that we know and love.

At first, the holographic principle centered on black holes because information that had fallen into one could be conceived to be located on its 2-D spherical *event horizon*.[b]

Then, in 1998, string theorist Juan Maldacena made a startling proposal: A certain theory of spacetime that included gravity[c] was the same as another theory that had no gravity and had one fewer space dimension.[1113] (This is another example of duality.[d]) A decade later, Greene called it a "spectacular result," offering his readers,

> . . . a CliffsNotes version that doubles as a guilt-free pass to jump to the next section should, at any point, the material overwhelm your appetite for detail.[1114]

---

a See the preceding chapter.
b This is the imaginary surface from inside which the black hole's gravity prevents even light escaping.
c Note this theory uses a space (called anti–de Sitter or AdS space) that has negative curvature. Nobody has yet found a way to do the same with space that has positive curvature (as our universe evidently does; see chapters 20 and 22 at pages 122 and 133).
d See chapters 37 and 75 at pages 208 and 386.

It is not easy stuff. Here it may suffice to say that this "spectacular result" showed a way one set of strings inside a universe is equivalent to another set confined to a boundary around it. Others soon showed how to switch from one set to the other.

Notwithstanding Susskind's graphic prose (above), the thrust of physicists' opinions seems to be, not that the world *is* a hologram, but rather they can *think of* it that way.

But here please recall how string theorists see the Calabi-Yau manifold as just math they can employ, while the new ontology says we can explain much more when we see it as real. That same ontology suggests a way that Susskind's prose just might be real: Our 3-D space and all that's in it might turn out to be a topologically 2-D sheet.[a]

It is still early days. But in a recent tour de force of silo-border-crossing, physicist Philip Phillips and colleagues stitched together quantum theory, the theory of metals, superconductivity, string theory, particle physics, the curvature of spacetime around black holes, electrical resistivity, superfluidity, and gravity to show how a growing zone of mathematics tantalizes with the possibility of understanding.[1115] Its foundations rest on Maldacena's holographic duality.

Over the past twenty years, many papers have been written about the holographic principle, all cradled in a highly artificial worldview. It is mostly string physicists taking their math out for a run, without achieving any notable advance for fundamental physics. How much more might all their efforts yield if they would work with the real world?

So, yes, it does sound crazy, but the new ontology just happens to say 3-D space might really be a crumpled 2-D sheet.

---

a See chapter 62 at page 326.

65
ACROSS THE RIVER AND INTO THE TREES

> Thus galaxies with distances greater than [the velocity of light divided by the Hubble constant] are receding from us with velocities greater than the speed of light.
>
> Tamara Davis and Charles Lineweaver (2001)[1116]

Most everyone knows nothing can go faster than the speed of light. Some are more precise, saying, for example,

> Nothing having mass can be given a speed relative to a local observer that is equal to or greater than the speed of light.[1117]

That word *local* is essential here. It makes the statement right. As Davis and Lineweaver said (above), lots of space (and the matter in it) *does* go faster than light relative to us; but it is not local. It is utterly remote.

A simpler version of the rule is, nothing can go faster than the speed of light *through space*.

This may raise echoes of the there's-no-ether edict and the there's-no-special-frame-of-reference hang-up, which we inherit from bad history of special relativity. They are just misinformed.[a]

In their paper from which the epigram is drawn, Davis and Lineweaver set out to help us understand how it can be that much of the universe is leaving us faster than the speed of light.[1118]

| Entire solar systems flee from our view.

Further to the above distinction about local motion, they state (elsewhere),

> [T]here is no contradiction with special relativity when faster than light motion occurs *outside the observer's inertial frame*.[1119]

---

[a] See chapter 10 at page 63 and chapter 56 at page 301.

In the new worldview, the speed of light becomes a fundamental property of space itself. Once again, we should remind ourselves that space is something. The rule says nothing about how fast that something moves. Of course, until recently, space moving seemed to have no meaning. It still has no local meaning;[a] but on cosmologic scales space moves.

There are two situations where space, with the matter in it, may go faster than the speed of light relative to us. They are both part of the bigger ontologic picture so we should get a handle on them. But, to the dismay of galactic hitchhikers, neither offers us a way to take a ride. One is far too far away; the other also looks too hot.[b]

The first is outside the visible universe.[c] Expansion of space *is* moving this space away from us faster than the speed of light, which is why it is not visible. It's not only space that's leaving; it's the matter in that space. Numberless planets are leaving us for f-t-l limbo every minute, crossing that invisible boundary along with their stars. Farther away, billions of galaxies, each with billions of stars, are fleeing us far faster than light. This concept is often confused but is no longer controversial.[1120]

The faster-than-light matter is so far away the very question of how far away it *is* becomes confusing too. The farthest galaxy we see is nowhere near where it appears to be. The light we see it by took more than thirteen billion years to get here. While the light was on its way the universe expanded. So, as we saw,[d] now that galaxy is some forty-five billion light-years away. It's long gone from our view. The light it is emitting now will never reach us because we are moving away from it faster than the speed of light (or vice versa).

As we also saw, the part of the universe we can't see may be bigger than the part we can see; so, most of the universe may be moving faster than light relative to us. And (coming up next) so may be some of what we see.

---

a As Einstein said, in effect; see note 262.
b See chapter 66 at page 343.
c Here, I mean the edge of that part of the universe whose light will eventually reach us; see also the epigram above.
d In chapter 62 at page 326.

For my part, I keep needing to reread Davis and Lineweaver. Which leads me to a small suggestion. Don't try to explain this to your friends.

66
COOL YOUR JETS

> There appears to be an analogy here between
> the idea of matter riding along in a stream of new
> space from the black hole and the description
> in [*Time One*] of distant galaxies as not moving
> through space, as usually thought, but riding along
> as space expanded.
>
> David Miller (2013)[1121]

The most extraordinary spectacle we have found in the universe is prosaically known as an *active galactic nucleus* or AGN. It's a chaotic zone of hot infallen matter orbiting a spinning big black hole. Most spectacular are those that shoot axial jets. Science writer Daniel Clery said,

> These jets blast material up to 10 million light-years out into space—that's 100 times the diameter of the Milky Way.[1122]

Some jets seem to move faster than light. How a big black hole can shoot such jets leads to more questions than answers.[1123]

The jets could be of interest to frustrated fans of faster-than-light travel. For them, the upside is some of this f-t-l motion might be real. The downside is it will be difficult to board these jets, not least because they are too hot.[1124]

| Jets made of space might move faster than light.

As those fans know, the theory of special relativity says it's impossible for matter to exceed the speed of light. But this is qualified, . . . *relative to a local observer*; which is not where those jets are found.

And, as we've seen, general relativity says space is a thing of its own.[a] Space itself may move, indeed must move,[b] although (as Einstein put it) it's not "ponderable media."[c]

We will soon see a simple explanation for the observations showing space is still expanding. In extremely curved space at the centers of black holes, new space is made by quantum tunneling (just like it was in the beginning of the universe[d]).

The next step is an idea that space-physicist David Miller proposed (above, in a review of *Time One*). He said if—as I had proposed—black holes are space factories, the newly made space quanta might exit as axial jets of space, along with entrained matter,

> Gillespie's concept of black holes makes it reasonable to imagine that the matter in the jets from quasar black holes could to some extent be riding along in streams of new [space quanta leaving] the black hole along its axis of rotation.[1125]

Let's let that idea sit under our thinking caps for a moment.

Here is the astronomic picture, unlikely though it may appear. Astrophysicists find AGNs (the most active being also known as *quasars*) and their black holes at the centers of galaxies. Typically, each holds mass equal to millions or even billions of Suns. Gravity crushes all that matter to a tiny volume.[e] Physics figures out a black hole's mass by watching stars that whip around at stunning speeds nearby. Nothing that comes near it (inside its event horizon) can escape, not even light. Yet it somehow blasts exatons of matter every second in a jet that reaches some fifty quintillion miles away.[1126]

What is the engine that gets all those exatons of matter moving? The energies involved are stunning:

> The results are up to . . . the mass equivalent of about ten million stars like the sun! That is, one would have to

---

a See chapter 10 at page 63.
b See the preceding chapter.
c See note 262.
d See chapter 53 at page 291; in chapter 69 (at page 357) we check whether there's enough such extremely curved space in the universe to create the necessary new space.
e But not—as general relativity would have it if space were continuous—to a singular point; see chapter 93 at page 460.

(hypothetically) annihilate millions of stars and anti-stars to produce such energies.[1127]

Quite an effort for an engine that astronomer Andrei Lobanov said must be "extremely tiny."[1128]

Let's be clear about the problem. An inconceivably vast mass is heading *away* from one of the deepest gravity-wells in the universe (fig. 61). It moves at near-light-speed with violence that can shape an entire galaxy.[1129] Compared to such energies, a trillion atom bombs per second would be inconsequential. How can this be? As *Sky & Telescope*'s web editor, astrophysicist Monica Young, wrote,

> Start asking some basic questions about jets and you'll find mass confusion. How do jets form? What accelerates the flow? What is the flow?[1130]

Fig. 61. NASA's black hole and jet

Astronomers observe these jets throughout the universe. In some they see strings of huge clumps (called "cannon balls") moving at apparent speeds of up to forty times the speed of light, and . . .

> . . . a general trend of increasing apparent speed with distance down the jet.[1131]

The jets' composition is unclear.[1132] The hottest news may be some of the jet material may *not* be hot! Astrophysicist Clive Tadhunter said,

> Much of the gas in the outflows is in the form of molecular hydrogen, which is fragile in the sense that it is

destroyed at relatively low energies. I find it extraordinary that the molecular gas can survive being accelerated by jets of highly energetic particles moving at close to the speed of light.[1133]

Theories of such jets are a patchwork of astrophysical Band-Aids. Many say the engine's energy comes from the *accretion disk* of the black hole (a donut of hot matter whirling round and maybe falling in). Just how this might work is not understood. Others say magnetic fields whipped by the spinning black hole might drive jets of accretion-disk material.[1134] This kind of system is hard to define, let alone model.[1135]

In 2015, physicist Robert Antonucci reviewed fifty years of studies of AGNs. He said accretion-disk models are inconsistent with the data. Indeed, he portrayed the theory of AGNs as a mess:

> Many theory papers have already been ruled out by observations by the time they are published. Observers routinely use models to interpret their data long after the models have been falsified. . . . We aren't even close to having the correct physics.[1136]

He noted reports of apparently superluminal motion:

> The blobs appear to move perpendicular to the line of sight, often at about ten times light speed (this is called superluminal motion)! That would be very, very verboten in relativity.[1137]

Many such jets have been observed over more than thirty years.[1138] Most—maybe all—of those observations can be explained as close-to-light-speed motion toward the observer at a narrow angle.[1139] Such motion leads to underestimation of the time of travel, which affects the calculation of velocity. But angle of the jet to line of sight may not always admit this explanation.

There are other issues than the source and energy and speed and temperature and composition of the jets. For example, Antonucci asked how they stop.

> How does the bulk kinetic energy of the jet plasma get thermalized to produce the relaxed-looking giant lobes?[1140]

All these quandaries could seem much easier to explain if Miller's speculation about f-t-l jets of new *space* from *inside* those big black holes is right and if they are entraining matter.

These issues may get close study from a new NASA[a] X-ray telescope that began operating recently:

> IXPE's biggest coup might be in helping understand the mechanics of powerful jets launched by supermassive black holes in distant galaxies.[1141]

Roll on IXPE.[b]

---

[a] NASA is the US government's National Aeronautical and Space Administration.
[b] The Imaging X-ray Polarimetry Explorer.

67
DARK ENERGY IS SPACE

> Why is the dark energy so nearly zero, but not exactly zero? The dark energy puzzle is one of the greatest unsolved problems of physics today.
>
> Robert Oerter (2006)[1142]

That "so nearly zero" that troubled physicist and science writer Robert Oerter was the *density* of the dark energy. He likened it to the mass of a few electrons in "a thimbleful of empty space." Sounds like not much but there's a lot of space and so its mass totals two-thirds that of the universe.[a] It has a confused history that leads to a confusing story.

> Quantum space can solve the dark energy problem.

If we look closely at its origins in 1917, *dark energy* was an effort to make a virtue of necessity—something to push the universe apart against its inward force of gravity to explain why it had not yet collapsed. It was needed even more to explain the recently observed acceleration.[b] In this guise it was pure fiction.[c]

In 2014 Guth wrote, with disarming candor,

> What's needed is a material with a negative pressure. We are now therefore convinced that our universe must be permeated with a material with negative pressure which is causing the acceleration we're now seeing. We don't know what this material is, but we're referring to it as dark energy.[1143]

---

a See note 671.
b See chapter 18 at page 112.
c See chapter 29 at page 165; and see following.

In 2021, Wilczek wrote about what he called *Ten Keys to Reality*. He had a key devoted to dark energy. It fell under his heading "Mysteries Remain." He said,

> Dark energy could be a universal density of space itself.[1144]

He said this is the most popular idea about dark energy among researchers; other ideas being "(even more) speculative."

Those less speculative researchers—those who think dark energy reflects the density of space—have no consensus on *What is it?* Or even *Is it real?* For many, it is only a rebranded cosmic constant.

Within their theory, this makes it an absurdly tiny remnant that's supposedly left over after adding and subtracting many big components. Wilczek said simply,

> It's a big cosmic mystery.[1145]

The mystery's history is relatively brief (unless one conflates it with something that has its own history, like Einstein's 1917 cosmological constant or one of the ethers, which is what we will do here[a]). Cosmologists Michael Turner and Dragan Huterer coined the phrase *dark energy* in the late nineteen nineties.[1146]

One can find various vague descriptions, like,

> Dark Energy is a hypothetical form of energy that exerts a negative, repulsive pressure, behaving like the opposite of gravity.[1147]

One cannot, however, find an explanation for an antigravity force, or negative pressure, let alone how something with extremely low density[1148] overcomes the universe's gravity.

The new ontology offers a simple understanding, without any new assumptions (another tip of the hat to Ockham). Dark energy is simply space, just as we have already seen it. It is made of many quanta;[b] they have fixed volume; some replicate; the increased volume lifts the universe against its gravity.[c]

---

a See chapter 29 at page 165.
b Almost one hundred billion-decillion-decillion-decillion-decillion-decillion (8×10^{175}) space quanta make up the observable universe.
c And so each quantum's volume represents a tiny energy, and therefore an extremely tiny mass.

In a strict sense, dark energy is everything (though we tend to distinguish parts of it with twisty links and call that *matter*). In an upcoming chapter we will look at its ongoing origin.[a]

---

[a] In chapter 69 at page 357.

68
BLACK HOLES ARE DARK MATTER

It's a 70-year detective story. An arrest is imminent.

Michael Turner (2000)[1149]

Dark matter is a hot field that is getting lots of study (thanks to lots of money; not—I here suggest—well spent).

Recall Turner coined the term *dark energy*.[1150] But *dark matter* was the fugitive whose "imminent" arrest he could foresee a quarter century ago. Long before that, we knew matter we can see is only a small part of all the matter there must be.[1151]

> Big black holes got off to an early start.

Astronomer Fritz Zwicky said so in the nineteen thirties. So did Einstein and de Sitter.[1152]

Astronomer Vera Rubin (fig. 62) made the case for it[1153] by showing matter we see in the arms of spiral galaxies could rotate the way it does only if there is much more matter that we cannot see.[1154] She found,

> In a spiral galaxy, the ratio of dark-to-light matter is about a factor of ten.[1155]

These days there's way more evidence for its existence. Its gravity is hugely consequential, controlling galaxies.[1156] We call it *dark matter* because all we know about it is that it has gravity and it tends to hang around and we can't see it.

So, what is dark matter?

Particle physics says the matter we know about is made of particles. As Hossenfelder said,

Fig. 62. Vera Rubin found most matter is missing

Particle physicists, not so surprisingly, believe that what's creating this extra gravitational pull is really a type of particle. If they were bakers, they'd be convinced the universe was held together by cosmic croissants.[1157]

The suspect Turner favored was a weakly interacting massive particle (inevitably, WIMP). Its name reemphasizes that we don't know what it is. Nobody's ever seen a WIMP.

Many other suspects stand accused. Most are exotic, not to say fantastic, particles, each with its fans who bet careers on their pet candidate's existence. Each has been extensively (often expensively) investigated yet none has come close to conviction. Indeed, the only reason for the failure to bring in not-guilty verdicts is the perp is still at large.

So, twenty-two years after Turner, science journalist Adrian Cho could report,

> [P]hysicists stalk WIMPs by looking for recoiling nuclei in detectors deep underground, where they're shielded from radiation that can produce extraneous signal.[1158]

His headline read, "Hunt for long-sought dark matter particle nears a climax." The background to the climax was,

> [E]ver-bigger detectors . . . have come up empty so far [and] the world's biggest atom smasher . . . has yet to blast out anything that looks like a WIMP.[1159]

Hope springs eternal. For example, no one has ever seen an *axion* either and in my view it's dead in the water too; yet in 2022 a science writer could say in a leading journal,

> The hypothetical "axion" particle is gaining momentum as a candidate for explaining the ever enigmatic "Dark Matter."[1160]

In a recent review, physicist Francesca Chadha-Day and colleagues summarized key observations:

> The existence of [dark matter] is inferred from its gravitational effects, and astrophysical observations suggest that

it is "cold," i.e., it has been moving very slowly for much of the history of the universe. . . . [It] must have been present since early in the history of the universe, a year or so after the Big Bang or even earlier. . . . [W]e have very little information about its nature and properties.[1161]

Let's look through the lens of our new ontology to see what this might mean. How about *before* the Big Bang? How about a something that has indeed been "moving very slowly" ever since, a something that is difficult to make move quickly, that is almost always very cold?

How about, that is, a suspect that meets their description and that has the added virtue that we know it *does* exist, the black hole. Indeed, the same authors vaguely say,

[P]art or all of it might consist of macroscopic lumps of some invisible form of matter such as black holes.[1162]

Black holes were long since booked and then released. There didn't seem to be a way to make enough of them that would be big enough. And for a while astronomers could not find enough of them either. But these get-out-of-jail cards may shortly expire.

Black holes are being found all over space and extremely far back in time. This creates two problems for the standard model: We find evidence of big black holes before the model could have made them (see below), and we find medium black holes that the model cannot make at all.[a]

The new ontology has neither of these problems. It points to the possibility of vast numbers of *primordial black holes*, including very big ones, arising from newly minted matter in the universe's exponential epoch. In its depiction of the first micro-pico-pico-picosecond, the universe was rife with all the fixings to form big black holes.[b] How big? How many? Answering these questions will require new physics.

Primordial black holes are not news; they are widely thought to have formed one way or another. In 1966, cosmologist Yakov Zeldovich and astrophysicist Igor Novikov proposed that post–Big Bang

---
a See chapter 102 at page 499.
b See chapter 40 at page 222.

conditions might have made primordial black holes from "superdense" objects they called *cores*.[1163]

Hawking soon took up the cause. In 1970, he farsightedly[a] proposed,

> [I]t is tempting to suppose that the major part of the mass of the Universe is in the form of [black holes]. This extra density could stabilize clusters of galaxies which, otherwise, appear mostly not to be gravitationally bound.[1164]

He suggested such black holes would have been relatively small and most of them would shrink to nothing, radiating mass as heat (known as Hawking radiation) at an increasing rate, ending in "explosions."[1165] He showed a black hole's temperature should be inversely proportional to its mass, big black holes being very cold, and a mere one-hundred-million-ton black hole (i.e., a small one) being very hot.[1166]

These possibilities arose in context of the then emerging standard model of cosmology; they are still speculative. Whether—and just how—primordial black holes could have been created in the early universe depends upon precisely what was going on in the first fractions of a second.[1167] Those Hawking pictured as exploding would have come into existence by collapsing from supposed superdense "cores" or from "accretion (gravitational capture) of radiation" as the universe expanded in the Big Bang epoch.[1168]

Today's views of such black holes depend on the standard model including some version of the postulated cosmological inflation and its inflationary epoch.[b]

The new ontology sets the table for new physics and a revised standard model based on how space came to be and what it really is.[1169] It says, before the Big Bang epoch, the universe would have been extremely dense as it expanded far faster than light-speed in the exponential epoch.[c] It says the way that happened has no common ground with any of the theories of cosmological inflation that proponents seek to graft onto the standard model.

---

a This was in the early days of Rubin's work.
b See chapter 20 at page 122.
c See chapter 69 at page 357.

For independent reasons, we can now see black holes must have some source beyond what the standard model contemplates. This was seen in early results from the gravitational-wave observatory called LIGO.[a] They found colliding black holes in the sixty-to-ninety-solar-masses range—a size the standard model of cosmology cannot explain.[1170]

This observation also goes to the question of finding enough of them to be the missing dark matter: There must be many of them if they bump into each other in vast reaches of space. Where could they come from? Could they be primordial? Are they the missing dark matter?[1171]

Then there are the supermassive black holes astronomers are finding almost as far back in time as they can see, just a few hundred million years after the universe began. They are too big too soon to have been formed in ways predicted by the standard model. Astrophysicists are seeking some way to explain them.[1172] Could these black holes too have been born in the exponential epoch?[b]

As astrophysicist Nico Cappelluti and colleagues said recently,

> [Primordial black holes] . . . could also serve to account for early massive black hole seed formation and address the intriguing origin of the [supermassive black holes] with mass of the order [10 billion solar masses] powering detected luminous quasars already in place . . . when the Universe was [less than 800 million years] old.[1173]

In other words, black holes may have existed from right after the beginning and could explain what dark matter is:

> "Our study shows that without introducing new particles or new physics, we can solve mysteries of modern cosmology from the nature of dark matter itself to the origin of supermassive black holes," Nico Cappelluti said.[1174]

---

[a] The Laser Interferometer Gravitational-Wave Observatory; see chapter 129 at page 617.
[b] See chapter 53 at page 291.

Their study simply *assumes* there were vast numbers of primordial black holes including many of more than a million solar masses.[a]

Trying to find how this could happen, in the last two decades physicists delved into their potentially primordial creation as other dark matter candidates fell by the wayside.[1175] Maybe earliest was astrophysicist Karsten Jedamzik in 1997.[1176] But in the end this field of inquiry is all about what *may* have happened in the first second or so. So far, that story rests largely on failing ontological commitments and "untested models."[1177]

However, change is in the wind, both theoretical[1178] and observational.[b] Observation found a seven-solar-mass black hole only five thousand light-years away. Finding that one took a lot of work,[1179] so if we did more work we might find many more.

In sum, it looks like most of that matter we don't see because it is "dark" might just be black holes formed in the universe's first few instants. If so, it would explain a lot.

New physics could help here and cost less than WIMP-hunting expeditions—which the new ontology suggests may keep on coming up with nothing.[1180]

---

[a] As does the new ontology; but it shows how they could have come into existence.
[b] See chapter 129 at page 617.

69
A PRODUCTION CHECK

> So what is it that is speeding up the Universe? It is called dark energy and is a challenge for physics, a riddle that no one has managed to solve yet.
>
> Annika Moberg (2011)[1181]

The Royal Swedish Academy's science editor Annika Moberg was writing about the 2011 Nobel Prize for physics. Her unanswered question brings us yet another riddle with both black holes and space quanta hidden in its depths.

We've known for about a hundred years space is expanding.[a] But of course it must be slowing down. This had become a pillar of the cosmologic story. As physicist Adam Riess and Turner said in 2008, setting the scene,

> Can black holes make enough space to keep the universe expanding?

> Until recently, astronomers fully expected to see gravity slowing down the expansion of the cosmos.[1182]

Riess shared the 2011 physics Nobel Prize for discovering that the expansion rate is now *increasing*.[b] Physicist Olga Botner said,

> In a universe which is dominated by matter, one would expect gravity eventually should make the expansion slow down. Imagine then the utter astonishment when two groups of scientists headed by this year's Nobel Laureates in 1998 discovered that the expansion was not slowing down, it was actually accelerating.[1183]

The acceleration is attributed to that antigravity effect physicists call dark energy. So how can it begin, grow stronger, and take over?

---
a See chapter 18 at page 112.
b See chapter 29 at page 165.

As explanations go, saying it's dark energy is like invoking magic. For example, NASA—whose Planck satellite gave us the latest data on it—said,

> We know how much dark energy there is because we know how it affects the universe's expansion. Other than that, it is a complete mystery.[1184]

And what *do* we know about how much dark energy there is? In terms of mass-energy, it is about two-thirds of the universe.[1185] That is, two-thirds of everything somehow got lost. As author Oscar Wilde might have had Lady Bracknell say, *To lose one third may be regarded as a misfortune; to lose two looks like carelessness.*[1186]

Fortunately, the new ontology will have a word or two to say about this missing mass, whose growth at an increasing rate offers us two tests.

One test is to find enough extremely curved space in the universe today whose replicating space quanta could explain the rate of expansion we observe.[a] The other is to explain how that rate could increase.

In comparison, the standard model has no explanation for expansion beyond saying, *It is expanding because it was expanding.* Cosmologist Neta Bahcall says the expansion of space is . . .

> . . . caused by the stretching of space-time itself.[1187]

One might as well say the stretching of spacetime[b] is caused by the expansion of space itself.[c]

The new ontology offers what appears to be a complicated four-point explanation:

- There must be some special space in the universe that still has extreme curvature.

- This special space must be widely distributed (as the expansion is).

---

a We punted this issue in chapter 55 at page 297.
b We'll give the fashionable "spacetime" reference a pass here (see chapter 80 at page 411); it is almost impossible to escape this term in papers where often, as here, it just means "space."
c This is not to accuse Bahcall of tautology but to illustrate that the standard model *is* a tautology.

- There must be enough of it to generate the new space we observe.
- Its rate of space production must be growing.

But this is not complicated. Indeed, it is simple. We already know there is a special space that meets the first requirement. It is the space at or near the centers of black holes. And big black holes are widely distributed. The central issue is: *Is there enough highly curved space to meet the third requirement?*

A black hole is mass whose gravity has caused it to collapse. This is the fate of many stars that are somewhat more massive than the Sun.[1188]

Black holes are thought to come in two main size ranges: relatively small ones of three to fifty solar masses (symbol $M_\odot$), and bigger ones of fifty thousand to more than a billion $M_\odot$.[1189] The bigger ones of course have stronger gravity.

We know much more about the biggest of them. It's becoming clear that most or maybe all big galaxies have an extremely big black hole, some with many billion $M_\odot$, at their centers:

> Astronomers believe that supermassive black holes lie at the center of virtually all large galaxies.[1190]

So, the first test question reduces to, Can extreme space curvature in big black holes make enough new space to match the space the data say is being made?

We only need a rough-and-ready calculation, so we can simplify it. Let's look for space with extreme curvature that has *all* its quanta replicating every iteration. The question then becomes, Can we find enough of *that* space inside the universe's big black holes? It's like, Is that factory floor big enough to turn out all those cars?

From astronomic measurements we know how much new space we need. The Hubble constant gives the answer. Its conventional value is 67 in the usual units—km/s/Mpc—which are both antiquated and absurd.[1191] Converted to intelligible units, space is expanding by 23 percent per billion years.[a] So, pick any big volume to track. It will be 23 percent bigger in a billion years. Let's scan the entire universe

---

a See also chapter 18 at page 112.

that's visible, about 13.8 billion light-years in radius as we see it. But since that light left it's been expanding so it is much bigger now, with radius some forty-six billion light-years.[1192]

(This is where we get into some very small or big numbers. So, to avoid long strings of zeros, we'll use *scientific notation*.[1193] In this notation, 10^{-2} stands for 0.01, 2.5×10^3 stands for 2,500, et cetera.)[a]

Converting light-years to kilometers[b] and calculating volume,[c] we find the volume of the visible universe is 3.45×10^{71} cubic kilometers. So, to grow by 23 percent in a billion years it will need to grow by 3.9×10^{-63} percent each Planck time. That works out to 1.36×10^{11} or 136 billion cubic kilometers per tock of the universal clock. That looks like a staggeringly huge amount of space and a staggeringly tiny time to make it. But let's see.

Recent estimates show about two hundred billion galaxies in the visible universe,[1194] each, we assume, with its own big black hole. So, on average, we find each big black hole needs only 136/200 = 0.68 cubic kilometers—a sphere of radius a mere 550 meters—of fully active replicating space to punch out all the new space we know we need to keep our universe expanding at its present pace.

So, for what it's worth, the answer to our test question is, Yes! A modest volume of extremely curved space in each of the big black holes at the centers of the universe's galaxies could make enough space to explain expansion data.

To put this in proportion, we can take a typical black hole in a typical galaxy—our galaxy, the Milky Way. Its black hole's mass is measured as 4.3 million $M_\odot$. There is a direct relation between a black hole's mass and the radius of its event horizon, the boundary from which nothing, even light, can manage to escape the black hole's gravity: The event-horizon radius is 2.95 km per $M_\odot$.[1195] This means our galaxy's black hole has an event-horizon radius of 12.7 million km.

Its gravity is proportional to the inverse square of distance from the center. A simple calculation shows the gravity at 550 meters from

---

a Those who break out in a rash at decimal arithmetic can skip the details.
b 9.46×10^{12} km/lyr.
c $4\pi r^3/3$.

that black hole center is 6×10^{14} (or six hundred trillion) times more extreme than the gravity at its event horizon—from which light cannot escape.

This tells us the universe may have plenty of space where extreme curvature could drive a working new-space factory.

Thus we may credibly suggest the Big Fizz is continuing today.

There's one question that's left hanging from the fourth requirement of our explanation. *Why is the expansion speeding up?*

The volume of the extreme-gravity regions where space is made increases with the cube of a big black hole's mass. It has a voracious appetite, consuming stuff that comes too close.[a] Most of its diet is gas, which it eats sparingly.[1196] But stars and even other black holes are fair game.

This means each one of the universe's big black holes is growing. The number of black holes grows too, as large stars consume their fuel and become supernovas whose invisible remnants are black holes. Black holes collide and merge.[1197] And even small black holes hold volumes of extreme curvature.

Bigger space factories and more of them need no encouragement to keep on making even more new space. The Big Fizz, one might say, is slowly getting fizzier.

This offers yet another view of the extreme economy the new ontology provides: We can understand the two great cosmologic mysteries of our times—dark matter and dark energy—as two sides of one coin. It is just the universe's population of black holes—their *being* and their *doing*.

They are being what we know they are, big masses we can't see.

They are doing what we know they do, curving space, consuming energy and matter. And inside their extreme curvature they do what *it*'s been doing all along since the beginning—making space.

Thus, a single well-known phenomenon, quantum tunneling, suggests a simple way for space to do what we can see it must have done to get to be the way it is. Grow exponentially for a to-us absurdly

---

[a] But note its gravitational reach is only as long and strong as that of any object with the same mass.

tiny time,[a] then slow dramatically, and then grow more slowly for billions of years, and then accelerate. It could do all of this by replicating quanta exponentially wherever space has extreme curvature and very slowly where the curvature is small.[b]

Thus all four kinds of stuff whose composition furrows brows right now—ordinary matter, ordinary energy, dark matter, and dark energy—may all be made of just one thing, all of them made in the same way. Our name for it is *space*.

All of which suggests the standard model of cosmology may need to be rejigged.[1198]

---

a In this respect, it is similar to cosmological inflation precepts.
b In having a natural explanation, it differs from cosmological inflation models.

70
ESCAPE FROM A MONSTER BLACK HOLE

A universe so constituted would have . . . no centre.

Albert Einstein (1917)[1199]

As Einstein said, there is no center. It is like a stack of spheres (fig. 57 is an attempt to depict it[a]) whichever way you want to go.

So, there was nowhere to flee *from*; there was nowhere to go *to*. Yet it was the ultimate escape. Much of us—that is to say, the pro-to-matter that would later lead to us and all our real estate—arrived *out* of the fixings for the universe's biggest-ever big black hole. It was like a real black hole in the sense that its gravity could at first crush it. It was also the entire universe.[1200]

The reason we escaped is that black hole never had a chance to get its act together. Gravity cannot act instantly; its reach moves at the speed of light.[1201] Almost all the black hole's fixings were fleeing far faster.[b] By the time gravity caught up, matter[c] was distributed through a large volume and, large scale, its gravity was pulling in all directions.[d] (It did fall into black holes, but they were purely local.[e])

> For an instant the universe looked like a huge black hole.

Now here we are; that's how we know that it (or something like it) happened.

The new ontology gives us a way to think of this. With it in aid let's take a closer look at the dramatic action of the universe's early growth using a thought experiment.[f]

---
a At page 329.
b See chapter 40 at page 222.
c *Matter*, here, includes energy, which was much of the mass.
d Again, it is a hypersphere, with no edge and no center.
e See chapter 102 at page 499.
f Re thought experiments, see chapter 30 at page 173.

Think, if you can, of that not-even-infant universe, as its volume doubled every Planck-time. Watch it grow for 456 iterations (almost twenty-five atto-yoctoseconds). By then, its radius was more than fifty million kilometers—roughly the size of Mercury's orbit—but, unlike that planet, it was not *in* space; it *was* space; it was all of it; and it was still extremely tightly curved in on itself.[a]

Its mass had yet to settle into standard theory particles. Much of it was extremely energetic radiation, yet it was very dense and randomly uneven.

It would seem to have had all the fixings for a big black hole, the biggest black hole ever. But three key fixings for that fate were missing. First, it hadn't had time to collapse (although local patches were about to do that). Second, its growing matter was moving "apart" far faster than the speed of light along with its space.[b] And third, when gravity's effect began to noticeably spread, it was pulling in all directions.

As to the first, its growing gravity had extended its reach no more than a few pico-zeptometers. It took far longer—microseconds[c]— before gravity began to pull matter of random denser patches into local collapses that became primordial black holes. They would have had a range of sizes. They may have sucked up lots of the initial energy and matter, and so disposed of much of the entropy.[d]

We might think of these black holes, forming moments later, as vast chunks of matter in flight, with inconceivable momentum flinging them out into space. But that's not the real picture.

The new ontology says that picture is constructed in the wrong universe: Our universe already had no *out*; there was no *into space* there either. Again, as Einstein said, it had no center. There was nowhere for those black holes to go, no *flinging* direction that was different from any other. They sat in still-expanding space where they had come into existence.

---

a See chapters 22 and 62 at pages 133 and 326.
b At first there was no clear distinction between space and matter. And "out" had, as Einstein implied in the epigram, no direction.
c At one microsecond, the reach of gravity was 300 meters.
d A black hole is an almost perfect cosmic garburator; see also chapter 23 at page 137.

As the universe grew exponentially, its curvature decreased accordingly. So, after some atto-yoctoseconds, replication and expansion slowed (relatively speaking).[a]

In the first fraction of a second, most of the particles of matter annihilated with their antiparticles, making a sea of energetic photons.[b] Matter particles remained (as there was more matter than antimatter[c]). Mostly they were quarks (which quickly coalesced as hadrons) and electrons[d] (in total, perfectly charge-neutral[e]) which, much as the standard model says, cooled as space kept expanding and in less than four hundred thousand years became mostly neutral hydrogen gas.[1202]

The standard model (fig. 63) follows the clumping of that gas—due to nonuniformities of density and so of gravity—to form stars and galaxies. It uses the equations of general relativity to track how expansion kept on going, slowing for maybe five billion years, then (we now see) starting to accelerate.[f]

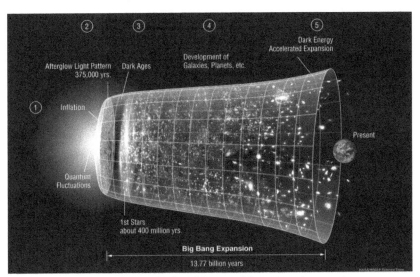

Fig. 63. The standard-model story of the universe's expansion

---

a See chapter 54 at page 294; note that this is a different trajectory from that of cosmological inflation and there was no sharp demarcation of the ending of the exponential era.
b See chapter 43 at page 243 and fig. 40.
c See the next chapter.
d Plus vast numbers of low-mass neutrinos.
e See chapter 48 at page 268.
f See chapter 67 at page 348.

We've known for some time that the first part of this picture has its problems. For example, as new telescopes see further back in time, it is becoming clear there were big galaxies long before gravity and random density variations could make them from that gas.[1203] Black holes have come to be regarded as galaxy organizers.[1204] In effect, those telescopes are finding big black holes millions of years before the earliest the model says they could have been organized.[1205]

In 2024, the new space telescope[a] had already found a large galaxy that existed sooner than the model can explain. Astronomers Stefano Carniani and Kevin Hainline said,

> [H]ow can nature make such a bright, massive, and large galaxy in less than 300 million years?[1206]

In 1971, Hawking speculated that primordial black holes could have formed in the very early universe.[b] He conceived there could be vast numbers of them but he thought they would mostly be small.[1207] However, that likely was a consequence of using wrong cosmology (the standard model).

Later, some saw larger possibilities. Astronomer and mathematician Bernard Carr and colleagues said,

> Black holes with a wide range of masses could have formed in the early Universe as a result of the great compression associated with the big bang.[1208]

The tale of our "escape" suggests it could have been accompanied by a host of primordial black holes, including many very big ones.[c] That's all we'd need to explain the twin puzzles of the big black holes and of the galaxies that came into existence so much sooner than the standard model's physics could have made them.

---

a See chapter 129 at page 617.
b See chapter 68 at page 351.
c It's not clear what would limit their size. See also chapters 94 and 102 at pages 467 and 499.

71
THE ANTIMATTER MATTER

> We live in a Universe dominated by matter, containing very little antimatter. The laws of physics, however, seem to include an almost exact symmetry between matter and antimatter.
>
> Helen Quinn (2003)[1209]

For each kind of matter particle there is an antimatter particle,[a] its mirror image with charges reversed.[b] When a particle and its antiparticle get together, they annihilate. Their mass is transformed into energy, a pair of photons.[c]

Both the standard model and the new ontology depict a sea of particles and antiparticles in the universe's first instants, and a massacre—that's emblematic of what's loosely known as the Big Bang—as most of them annihilated, leaving a hot sea of energetic photons.

The standard theory of particle physics tells the tale of matter particles and antimatter opposites. For the reason particle physicist Helen Quinn gave above, it led to the conclusion that we shouldn't be here; the universe should have very little matter as it should have (almost[d]) all annihilated. Here, too, something is amiss with standard theory physics.

> Physics says our matter should have met its antimatter match.

The model is based on some simple symmetries, like *parity*, the symmetry of mirror images.[1210] Another, *charge conjugation symmetry* pairs up each particle with its own antiparticle. This symmetry says their numbers should always be equal.[1211]

---

a The chargeless neutrino is an exception; see chapter 116 at page 568.
b Here, "charges" includes electrical charge and other kinds of charge depending on the particle.
c See chapter 43 at page 243.
d See below.

That is, according to our most successful physics, the universe should have begun with a bigger bang and then turned into a short, boring story rather than the epic we inhabit.[1212] But the new ontology would beg to differ. Let's see how this could arise.

First, let's be precise, physics does *not* say there should be a perfect particle-antiparticle match. Physicists Makoto Kobayashi and Toshihide Maskawa shared the 2008 Nobel Prize for explaining how a miniscule asymmetry gave rise to a few more particles than antiparticles.[1213] But, as mathematician Marianne Freiberger summarized,

> The symmetry violations established by Kobayashi and Maskawa are far too small to account for the large-scale triumph of matter over antimatter.[1214]

With this out of the way, the heart of the matter is none of the universe's matter was created according to laws of physics; it was created by that single law of nature, $1 \rightarrow 2$.

Physics' problem with it starts with the standard theory building on the symmetries that led it to predict the particle-antiparticle balance. The new ontology tells a different story: While it has electrical-charge balance from the get-go,[a] it does not begin with those same symmetries that cause predicted particle and antiparticle numbers to match.

Note that the standard theory did not *predict* symmetries. Their role in physics dates to work by Nobel laureate Pierre Curie.[1215] They took hold before the standard theory's origins, some sixty years ago. One can see the success of the standard theory as a badge of honor for some symmetries.

These days such symmetries are big in physics.[1216] For example, Gross said,

> [I]t is hard to imagine that much progress could have been made in deducing the laws of nature without the existence of certain symmetries.[1217]

The new ontology tells us a slightly different story. Its fundamentals have no symmetries at all, other than charge. It says searching in

---

a See chapter 48 at page 268.

symmetries may be useful but will not lead to understanding. In this it stands with Smolin, who said (for quite different reasons),

> Symmetries are properties of fixed backgrounds, and the occurrence of a symmetry in a theory is a clear sign that the theory is background dependent.... We have posited that the fundamental theory is background independent, which means there are no symmetries.[1218]

The new ontology's absence of symmetries begins with the strange asymmetric shapes of those Calabi-Yau pictures one finds online.[a] Or just look at that space quantum image on page 97. It seems to exclude all possibility of spatial symmetry.[b]

At Planck scale the twists and tangles in links between space quanta burst into existence. Sight unseen, we can say with confidence emerging space was quite an asymmetric mess (I've called it knots in Nature's knitting[c]).

The symmetries that form foundations for today's particle physics emerged as large-scale space emerged and Planck-scale structure vanished from imagined view.[d] Such symmetries are just large-scale approximations. They have no role—not even approximately—at Planck scale.[e]

When photons become electrons and positrons, they make matched pairs and therefore are in equal numbers.[1219] But, when knots in Nature's knitting gave rise to particles, some were quarks—both positive and negative—and some were positrons and electrons, for example. There was no machinery to make each quark a partner antiquark, or each electron a paired positron.

So, while charge neutrality resulted,[f] symmetries that led the standard theory to predict equal numbers of each particle and antiparticle had no application where the action was. Explaining in detail why we are here may need physics to take a closer look at those first instants.

---

a See, for example, Hanson's, note 375.
b Though no such image is itself meaningful at Planck scale.
c In *Time One*.
d See chapter 40 at page 222.
e Other than the twist (or dum/dee) mirror symmetry, which is charge symmetry; see chapter 35 at page 190.
f See chapter 48 at page 268.

Meanwhile the universe is sending a strange message with two kinds of antihelium that are unexpected and are generating even stranger explanations.[1220]

This brief essay is of course no substitute for a real theory of particle formation. We may have to wait a bit for that. Meanwhile, the new ontology offers a way[a] to understand how there could be so many unmatched matter particles left over.[1221]

---

[a] With no new assumptions.

PART III

MEN AT WORK

It is clear that some radical change in our understanding is required, but we disagree as to what that change needs to be.
Lee Smolin (2019)[1222]

72
IN A FEW WORDS

> Language can become a screen which stands
> between the thinker and reality.
>
> Arthur Koestler (1964)[1223]

In this part, we take a closer look at our new view, a world that—at Planck scale—is both strictly causal and strictly random, and check how it addresses a small sample of puzzles people[a] work on. Then, as promised, we'll check how the whole thing stacks up against some principles Smolin laid out as needed for new fundamental physics.

But first, we are here using words. Let's check out our exposure to linguistic hazards and then clarify some terms we'll need to use.

Linguist (and son of a linguist) Arthur Koestler spoke six languages and wrote prolifically. In the epigram, he was writing of "the snares of language."

In 1620, philosopher and lawyer Francis Bacon wrote,

> [M]en converse by means of language, but words are formed at the will of the generality, and there arises from a bad and unapt formation of words a wonderful obstruction to the mind.[1224]

Our language is (as Bacon said) collectively produced so no one is responsible.[1225] Journalist and writer Jane Jacobs said of this more recently,

> Speakers make a language and yet nobody, including its speakers or scholars, can predict its future vocabulary or usages.[1226]

Nonetheless, we can try to provide a supply. As we literally get real, we need *real language*. This is no new need. For example, Reichenbach excoriated the philosopher . . .

---

[a] Mostly men, still, I'm afraid.

> ... [whose] language has lacked the precision which is the scientist's compass in escaping the reefs of error.[1227]

Language shapes thought.[1228] The language we use creates both possibilities and limits for what we *can* think. For example, we will see how often almost thoughtless use of one word, *spacetime*, stultified progress in thought for several generations.[a]

Thus, terms coined and shared, or words given new and special meanings, come to have far wider consequence than the mere messages they bear in sentences that use them. In the process, words also partake of the inherent ambiguity of language,[b] at best airing ambiguities inherent in experience or subtly testing ground, at worst confounding content of communications.

Then, as little is conveyed with one-word utterances, we put words together, adding further ambiguities we then dissect as elements of grammar (such as syntax or semantics).

For the reader who might like to get a better feel for what a truly messy business it can be to think or read or speak, try a quick dip into philosopher Jaakko Hintikka's efforts to find logic in our simple statements about what we think we know or maybe we believe.[1229] He has, for example, an entire chapter on the vexations of statements about knowing what we know, adducing views from Plato, Aristotle, St. Augustine, Thomas Aquinas, Benedict Spinoza, and Arthur Schopenhauer.[1230]

| **Words affect what we can think. We need real words.** |

Physics has long sought refuge from such tribulations in its math. Yet, as astrophysicist and cognitive scientist Piet Hut said, there is no escape:

> Even in the more idealized world of mathematics, the notion of a purely logical full description of a mathematical system had to be given up for all but the simplest systems.[1231]

It is amid all this uncertainty that language shapes our thoughts about what's real.[1232] Ideally, as Rovelli said,

---

a See following, and chapters 72 and 83 at pages 373 and 420.
b See chapter 134 at page 647 and note 2059; it is, for example, the lifeblood of the cryptic crossword.

The value of a novel idea or a novel language in theoretical physics is not in the fact that old physics cannot be expressed in the new language. It is simply in the fact that it is more effective for describing reality.[1233]

But often it is not. And then using terms that express unreal concepts as if they *do* describe reality spawns and reinforces worldviews that may be far from real, weaving webs of fiction. They may also obscure tensions that could otherwise have been provoking real questions.

In the end they can exclude us altogether from reality, condemning us to watch their shadows while mentally chained (fig. 64) in Plato's cave,[1234] imagining this is (in poet Piet Hein's words) "the whole show."[a]

Fig. 64. Plato's cave

Familiar usage of fictional concepts—like spacetime or multiverses—and misleading images—like a Calabi-Yau manifold *in* space or the universe afloat in infinite *nothing*—conspires to exclude real thoughts from minds.

Here I find an ally in Unger, who (for example) valued . . .

> . . . alternative vocabularies, free from the taint of the spatialization of time.[1235]

Yet even Einstein could slip up and add to linguistic confusion. Having shown gravity is not a force, he was known to speak of it in those terms; historian of science John (J. D.) North noted,

> In speaking of "gravitational potentials" we saw that Einstein kept some of the older terminology, and . . . he does, after all, mention "force."[1236]

---

[a] See note 1.

Translation elevates potential for linguistic confusion especially with those who, as Einstein did, like to make play with words.[a] Even pros can get it wrong. For an (admittedly dramatic) case in point, in 2023 researchers at the Institute for the Study of War chided four leading Western news media for claiming a key Russian warmonger was urging Russia to stand down—pitched as a major news development. In fact, he urged the opposite. The ISW crew explained,

> Much of the nuance included in Prigozhin's speech is lost when translating Russian to English. Prigozhin has an idiosyncratic rhetorical and writing style that relies heavily on deadpan sarcasm, selective ambiguity, aphorisms, vulgarity, and ironic slang.[1237]

Amid extant linguistic burdens, this book is about understanding some new fundamental concepts. For this we need new language.[1238] Educators Neil Postman and Charles Weingartner said,

> [W]hat we call a subject is its language. . . . What is "history" other than words? Or astronomy? Or physics? If you do not know the meanings of "history words" or "astronomy words," you do not know history or astronomy.[1239]

Here is a glossary of key words. We need to grasp them well enough that thinking *with* them becomes possible.

First, these are terms for things that do exist:

- *space*, a real object that is most of the universe's mass[b]
- *space quantum*, an indivisible Planck-volume of space
- *link*, one of six connections each space quantum has with neighbors
- *ribbon*, a loop or ring of links that, braided in threes, make particles
- *tweedle*, a half-twist in a link
- *helon*, two links in tandem
- *particle*, a simple braid of three helons

---

a Hence my somewhat obsessive cares with translated text.
b Admittedly, this begs the question of what mass is; see chapter 86 at page 435.

- *step*, an iteration in the sequence of 3-D universes
- *tock*, a time-like version of a step
- *Big Fizz*, the way new space quanta made the universe grow

Then, these are terms in wide use whose limitations we should keep in mind:

- *time*, whether absolute or common, is not a dimension
- *elementary particles* are not elementary
- *emergent properties* arise from many quanta and so are inexact
- *empty space* does not exist, even as an ideal
- *laws of physics* are approximations to reality
- *spacetime* is a mongrel mix of emergent and fictional dimensions
- *vacuum* or *void* are names for space that imply a nonexistent property
- *Big Bang* is a hodgepodge of concepts, some of which are wrong

With improved language now in mind, we may find more common ground.

73
ALL IN A GOOD CAUSE

> I admire the ambition and radical purity of causal set theory. It is a completely relational description of spacetime, in which each event is defined completely in terms of its place in the network of causal relations.
>
> Lee Smolin (2019)[1240]

In 1998, Markopoulou wrote a paper on causal sets subtitled "What the universe looks like from the inside"; it was published in 2000.[1241]

Markopoulou's paper is a primer for physicists on math language she thought to be well suited to fundamental physics. She was unusual in her grasp of the significance of their (and our) imprisonment,[a] and of the intuitionistic logic[1242] (with its links to language[b]) underlying causal sets.[1243] And, we will see, she was promoting the right language, *causal set theory*.

> Causation must be built into the universe's bones.

Then, in 2009, Markopoulou—in my view one of the most deeply original thinkers in fundamental physics—still in her thirties, quit physics altogether. It's a confounding episode in my grappling with Bilson-Thompson and his "model."[c]

I came to appreciate Markopoulou's perspective on fundamental physics and her math-language proposal for it. In a world where timing is everything, she was ahead of her time. We'll see that the universe *is* (as she maybe suspected and as Smolin would later assert)[d] a causal set.[e]

---

a See chapter 24 at page 141.
b An important trail we will not follow here.
c See chapter 1 at page 15.
d See chapter 28 at page 158.
e Strictly, though he did not say so, a finite partially ordered causal set.

In other words, the kind of math she tried to plug is real.[a]

A causal set is just a bunch of discrete things that *are* and then cause things to *be*. The things may be called *events*. This, we will soon see, is a fine description of everything the universe is and does.

It's not immediately evident, but the flip side of her perspective—in line with that of her colleague, Smolin[1244]—is that real physics must explore the universe as an entire entity. That is, we should address its entirety explicitly, although we are inside it and we cannot see a lot of it.[b]

The first thing one normally needs for any exploration is the fixings for a metric. The usual culprits are measuring sticks and clocks in an imaginary continuous spacetime. At least as early as 1999, Markopoulou reached implicitly for the digital metric that is inherent in granular spaces that are well matched with causal sets, speaking of . . .

> . . . elementary Planck-scale systems that interact and evolve by rules that give rise to a discrete causal history.[1245]

Riemann might have liked this.[c]

Causal set theory's potential as a candidate for quantum gravity[d] has been the focus of a modest research program since the nineteen seventies.[1246] "Modest" is the operative word. In 2013, amid his pitch for the reality of time, Smolin was content to lump causal sets in with "some other approaches to string theory."[1247]

Six years later, within a few pages of his bold assertion that the universe *is* a causal set, Smolin still seemed oddly off-handed about it:[e]

> The causal set hypothesis is one of several competing hypotheses concerning the properties of spacetime atoms.[1248]

Continuing, he did give causal sets a shout-out:

---

a Or, strictly, can be real.
b See chapter 24 at page 141.
c See chapter 12 at page 74; see also Meyer, note 649.
d See chapter 31 at page 176.
e By "spacetime atoms," he appeared to mean, in effect, space quanta in some kind of mashup with a Planck time. Yet another wander down the dead-end alley with the *spacetime* signpost; see chapter 83 at page 420.

> Compared to the others, such as spin foam models, [causal set theory] enjoys the great advantage of its utter simplicity, in that the only properties of events are their causal relations. This greatly narrows down the possible forms that a fundamental law of spacetime atoms could take.

The new ontology narrows down those "possible forms" more tightly. It prescribes,

- They are space atoms, not spacetime atoms;
- The set's events are space quanta with their links;
- The events are set-wise simultaneous and sequential;
- A kind of absolute time emerges from the set's sequence;[a]
- The causal relations of its "atoms" are: Each iteration one stays one or becomes two; and twists in links between them stay or move across one quantum;[b]
- In the first kind of causal relations, making new space quanta is extremely rare except in special circumstances;[1249]
- The causal set is finite and closed because the universe is finite and closed;[c]
- The set is *partially ordered* because at each step the causal relations of each element affect no more than its next-neighbors.[1250] (This is equivalent to the light-cone limitation in continuous space.[1251])

As Markopoulou said, it is . . .

> . . . the set of events in a discrete [space], partially-ordered by the causal relations.[1252]

Thus, although we are of it, and so see it from inside, we can understand the entire universe as a finite, partially ordered, causal set.

---

a See chapter 52 at page 284.
b See chapter 57 at page 307.
c See chapter 22 at page 133.

74
CAUSING PROBLEMS

> Causal thinking spontaneously arises in a child at about the time when she or he realizes that by exerting forces on nearby objects, the child can make these objects move according to their will.
>
> Časlav Brukner (2014)[1253]

As quantum physicist Časlav Brukner observed, causal thinking begins early in our worldview training. Waving infant arms, we grasp that we provide those early causes. Later on, causation takes on a range of meanings for serious thinkers.[1254]

In 1888, philosopher Henri Bergson wrote of causation and the nature of time in defense of free will.[1255]

In 1912, Russell said its meanings are all obsolete and best eliminated from the lexicon.[1256]

Planck—who one might say planted his constant wrench firmly in the causal works—knew it had many shades of meaning. Near his life's end, he wrote,

> In the fight currently raging about the meaning and validity of the Law of Causality in modern physics, . . . everything depends on a clear understanding of the sense in which the word "causality" is used in the science of physics.[1257]

He favored the then best accepted definition,

> An occurrence is causally determined if it can be predicted with certainty.[1258]

His protégé, Einstein, took the same view. It led to a deterministic world with no place for free will. Thus they (and many others) saw causation as opposed to chance. Though Einstein himself later recognized,

> It is probably out of the question that any future knowledge can compel physics again to relinquish a statistical theoretical foundation in favor of a deterministic one.[1259]

More recently, a growing literature has been trying to make sense out of causation in a chancy world amid the wider debate about what causation *is*.[1260] But until now no one has framed the causal issues in terms of a primal causal mechanism like the one the new ontology presents. It combines strict causation with absolute chance. In sum, its concept of causality is based on space made of quanta, with matter made of twisted links moving randomly between them with all changes strictly caused.[a]

> Chance is a close companion to causation.

In this locally random space, the certainty of relativity—as Einstein, long its lead defender, understood it[1261]—has no place. For example, general relativity is usually seen to be deterministic.[1262]

Recall Einstein's dismissal of dice.[b] (He lost that argument.) By contrast, we roll the dice for every link at every step.[c]

In many postulated universes that are run by relativity,[d] causation lives in light-cone forests. The *light cone* (see, e.g., fig. 65) is a spacetime specialty that, in the past of any point event in spacetime, embraces all events in space and time from which a light-speed particle could have reached that point; and in the same point's future embraces all events such a particle emitted from that point at that time could reach.[1263] Thus, all causes affecting and all effects caused by an event lie within or on its light cone.[e]

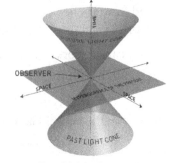

Fig. 65. The past and future light cones

In the new ontology we don't need light cones because we know what is really going on. So, we define

---

a See chapter 43 at page 243.
b See note 873.
c Ditto.
d That is, with special or general relativity, in a flat or curving continuous space.
e The light cone concept gets bent out of shape in the faster-than-light growth of the exponential era; see chapter 40 at page 222.

all causal chains from or to the event, with twist states moving locally from link to link at most one link per tock.[a] Which is to say, up close the speed of light becomes a property of space. There is thus no need for an approximate relativistic view of it.

The old ontology's version of causation runs into trouble with quantum theory too. For example, Brukner said,

> It has been repeatedly pointed out that the notion of time might be at the origin of the persistent difficulties in formulating a quantum theory of gravity. But how do we formulate quantum theory without the assumption of an underlying causal structure and background time?[1264]

The new ontology says he's asking a good question. It answers with an inherently causal structure for its space, using sequence rather than time. Once they get over that, the quantum theorists may find they can apply real quanta to reformulate their theory as a good approximation.

And so, as Planck observed,

> [T]he indeterminist discovers a statistical root in every law of physics . . . he regards them, one and all, as laws of probability.[1265]

The vast number of iterations in any time frame accessible to measurement and the vast number of space quanta in any volume accessible to observation give us our sense of large-scale causation and provide a basis for the causal laws of classical physics as well as quantum theory's. Thus, causation, as we see it, is emergent. Its strictness is well camouflaged by coin flips.

For the slightly math inclined, think of it this way. The expected number of heads for a single coin toss is 0.5 ± 0.5. The standard error of the mean of a sample of random coin tosses is inversely proportional to the square root of the number of tosses. For example, the mean of a hundred tosses will have an error tenfold lower, 0.50 ± 0.05.[b]

If you do 10^{44} tosses, the standard error of the mean is lower by a factor 10^{22}; all variation vanishes from view. That is, aggregated coin

---

[a] See chapter 52 at page 284.
[b] This is illustrative at the expense of rigor.

tosses each step of the universe will look like an exact 50 percent "law" to more decimals than we will ever see.

Thus, we can see how all the causal laws of classical and quantum physics are emergent.[a] They rest on statistical foundations. They are analog approximations to a digital reality.

To illustrate how this can explain the inexplicable, consider an electron passing through a hole to a detector screen.

In the classical view the electron is an elementary particle—with a single charge of -e and with a definite position—taking a definite path; its ballistic trajectory passes through the hole like a bullet and hits the detector screen at a predictable location.

In the quantum theory view, the electron has no definite position, it takes no definite path, and it has no objective properties; it is described by a wave function[b] that fills the universe and then "collapses" in an instant when it strikes the screen at an unpredictable location with predicted probability.

In our new view (fig. 66), the electron is six tweedles—each Planck-sized with charge -$e/6$; at Planck scale they are separated with a tendency to work a region sized to the electron's wavelength,[c] which is inverse to its speed[1266]—each meandering randomly, though not quite like what is shown in the sketch.

Fig. 66. *Six tweedles of a new-ontology electron pass through a slit S to a screen*

At any instant in this view, each tweedle could, with certain probability, be anywhere that's possible. But (as Einstein wanted *particles* to be) each tweedle *is* somewhere definite at every instant.[d]

---

a See chapter 94 at page 467.
b See chapter 116 at page 568.
c Or so I speculate.
d Notwithstanding it's a somewhere we cannot assign a metrical location; see chapter 62 at page 326.

However—as quantum ontology insists—the *electron* has no definite position or path; it's all over the map.

Until, that is, it arrives at its destination (whether this is set up as a "measurement" or not), which requires the tweedles to converge. Their arrival is thus an event whose causation physics does not understand, but seemingly they have more of a map than we can see.

Yet we have made some progress toward understanding causation by recognizing as real something—multiple undefinable and random pathways—physics takes to be imaginary and recognizing as imaginary something—the elementary particle—physics takes to be real. And so it is that, as Einstein said in 1950,

> The problem is not so much the question of causality but the question of realism.[1267]

At the level of principle, we can reconcile the seeming contradictions of the classical and quantum-theory views with a new understanding. At Planck scale, in the heart of realism where the real action is, all events are strictly causal and are also strictly random.[a]

---

[a] See chapter 73 at page 378.

75
THE TWO-SLIT MYSTERY

In reality, it contains the *only* mystery.

Richard Feynman (1961)[1268]

It's a famous mystery. Maybe, as Feynman said, it could contain the *only* mystery.[a] It certainly has helped us make a hash of how we think about reality.

Polymath Thomas Young first reported the phenomenon of interference in 1803. He illustrated (see fig. 67) how water waves coming through slits in a screen at A and B would "interfere" to give minima at C and D and E and F,[b] with maxima in between. He showed similar behavior with light.[1269]

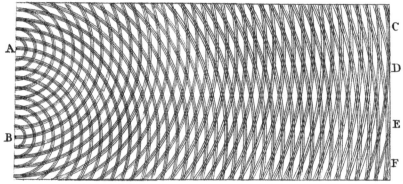

Fig. 67. *Young's diagram of water-wave interference*

Einstein got the mystery going in 1905. He said light was composed of photons[c] that behave as particles *or* waves, depending on how we observe them.[d] He stubbornly held to this view for years before even his friends embraced it. As Becker said,

---
a In the sense that it symbolizes the ultimate question: What is the true quantum?
b In his diagram, these locations are a bit out of line.
c The term *photon* was coined two decades later.
d See note 785.

For nearly two decades, almost nobody other than Einstein believed in photons.[1270]

With the success of quantum mechanics, wave-particle duality switched from a crazy observation to a fundamental principle.[1271] Not only photons, all particles have wave equivalents.[a] Whether you get to see a wave or particle depends upon not only the experiment you do but how you "watch" it.

Quantum mechanics quantifies what happens. Feynman said nobody understands it.[b]

Einstein and Infeld summed up the conundrum:

> It seems as though we must sometimes use the one theory and sometimes the other, while at times we may use either. We have two contradictory pictures of reality; separately neither of them fully explains the phenomena of light, but together they do![1272]

And Einstein said,

> [W]hat nature demands from us is not a [particle] theory or a wave theory; rather nature demands from us a synthesis of these two views which thus far has exceeded the mental powers of physicists.[1273]

Wave-particle duality is laid bare in the two-slit experiment: A stream of particles (most often photons or electrons) strikes a screen through two narrow slits (lower planel, fig. 68) A detector screen shows what reaches it, a typical wave-interference pattern (upper panel).

Ball explains,

> One fundamental aspect of quantum mechanics is that tiny particles can behave like waves,

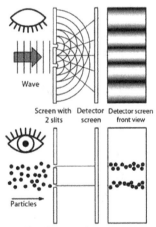

Fig. 68. The two-slit experiment without/with the path "observer"

---

a In principle any object is dual to a wave but the wave behavior is undetectable for macroscopic objects.
b See note 1600.

so that those passing through one slit "interfere" with those going through the other, their wavy ripples either boosting or canceling one another to create a characteristic pattern on a detector screen.[1274]

He continued,

> The odd thing, though, is this interference occurs even if only one particle is fired at a time.

How so? A single particle that's going through *both* slits and interfering with itself? That's what happens when the photons or electrons act as waves. And this weirdness gets worse (see the bottom panel of fig. 68):

> Measuring which slit such a particle goes through will invariably indicate it goes through only one—but then the wavelike interference (the "quantumness," if you will) vanishes.[1275]

That is, if we "watch" which slit it goes through—for example, with a path detector (the "eyeball" in the figure)—it behaves like a particle. The interference pattern vanishes; the particles go straight like bullets to the screen. When we turn off the path detector, we see the interference pattern.[1276]

Note that it's not our knowledge of the path that makes the photons or electrons behave like particles; it is the measurement detecting where they go.[1277]

This Jekyll and Hyde behavior circles round our question about what is really going on. It is known as complementarity. It is central to a plethora of physics fictions and, from early days in the quantum revolution, has led to a lot of loose talk. Bohr—the central player in construction of the quantum theory story and its language—was the chief culprit. Thus, Beller variously said,

> Bohr had an intense lifelong attachment to complementarity.

> The complementarity principle was a device of legitimation—it led to no new physical knowledge.

> Bohr's writings on complementarity are metaphorical in a strong sense.
>
> Bohr's voluminous improvisations on the theme of complementarity [were] filled with affective analogies, subjective associations, and allusions to "harmonies," expressed in "common language."
>
> Part of the Bohrian myth is that he thought very clearly and only expressed himself obscurely.
>
> Bohr's discussion of existential matters resonated strongly in the tender souls of his disciples.[1278]

His loose thoughts are embedded in what's taught today. Caveat emptor.

In the epigram, Feynman—who sold more than a million copies of his famous Caltech lectures—was speaking of the Jekyll and Hyde behavior Bohr was on about. Elsewhere he said physics is baffled by the inexplicable results of the experiment:

> The question now is, how does it really work? What machinery is actually producing this thing? Nobody knows any machinery. Nobody can give you any deeper explanation of [the two-slit experiment] than I have given; that is, a description of it.[1279]

Feynman's comment has held true for half a century, but, thanks to Bilson-Thompson, it may not be true much longer. His tweedles point us to a way to understand the mystery, not at atomic scale, where Feynman (and physics) was looking, but at Planck scale where its action is.

Let's pause to think. In retrospect, if one were disinclined to think in terms of quantum magic surely this experiment was striking evidence the photon *isn't* elementary!

Einstein and Infeld had the key in their hands in 1938, when they wrote,

> Since one particle is indivisible we cannot imagine that it passes through both the holes.[1280]

That "one particle is indivisible" was mere belief built upon the absence of evidence. As Becker said,

> Nobody's ever seen half an electron, or anything less than a whole electron in one well-defined place.[1281]

Of course, nobody has ever seen a whole electron either, but one can observe its consequences and measure its charge. And one can also observe the consequences of fractional electrons and measure their charges. Such observations of charges $-e/3$ were well established[1282] two decades before Becker's bold assertion (which nonetheless most physicists might echo).[a]

For a hundred years, physicists' inability to smash any of the growing list of standard theory particles in their accelerators fed the mindset and the language of an ontological commitment to their being elementary. That the electron *is* elementary continued to be unassailable despite the implications of quarks (in its charge-size counterpart, the proton) proving to have charges of $+2e/3$ and $-e/3$.[1283]

In the new ontology, that "one particle"—the one Einstein and Infeld (and a host of others) took to be indivisible—is not merely divisible, it is divided. For Feynman's only mystery, this surely changes everything.

Visualizing the two-slit experiment, we can address the literally inconceivable picture of that single particle somehow going through both slits so it can interfere with itself. Instead, we see a cloud of six tweedles that might pass through the two slits in any of the combinations 6+0, 5+1, 4+2, 3+3, 2+4, 1+5, or 0+6.[b] The middle five of these seven combinations support interference patterns.[c]

> There is no mystery if the particle's not elementary.

Now we switch on our path detector. It tells us which slit each particle traverses. The interference pattern vanishes. A simple

---

a What Becker would consider to be "one well-defined place" is unclear, but both quantum theory (to which he was appealing) and the new ontology say that at the scale of the electron there is no such thing; see also chapter 62 at page 326.
b Maybe a closer look at screen-hit data could shine a new light on the mystery.
c Simple statistics tell us these five combinations account for almost 99 percent of events.

explanation is at hand. The detector's interaction with the tweedles requires them to behave as a particle before they reach the detector screen (where their arrival accomplishes the same thing). The path detector causes all six tweedles to pick one slit or the other. Hence no interference.

It is ironic that Feynman was the author of the path integral method for the quantum theory that replaced classical electrodynamics.[a] It sums (or integrates) over all possible paths the electron might take.[1284] And all possible paths are exactly what the tweedles sample.

So, we can see a simple way to make sense of the mystery—the two-slit experiment's "two contradictory pictures of reality"—if tweedles have something to say about it.

---

[a] See also chapter 14 at page 87.

76
ALL THE SAME

Are there indeed such building blocks?
Why are they all of equal magnitude?

Albert Einstein (1917)[1285]

Writing to a student (Walter Dällenbach, who became an electrical engineer[1286]), Einstein was considering the pros and cons of "a discontinuum" instead of the canonical continuous space. It's clear from his letter Dällenbach had written to him raising the issue. Einstein replied,

> You have correctly grasped the drawback that the continuum brings.[1287]

Einstein went on to say,

> Yet I see difficulties of principle here as well. The electrons (as points) would be the ultimate entities in such a system (building blocks). *Are there* indeed such building blocks? Why are they all of equal magnitude? ... But this is devilishly difficult.[1288]

The devilishly difficult thing was to make sense of it. And—as he wondered then and we are left to wonder still—why *are* all electrons equal, at least to the precision (a few parts in ten billion) of our best measurements?[a]

| A simple solution to the electron identity crisis.

Our new ontology offers an answer to Einstein's question: Electrons are not "points" or even "building blocks."[b] It says his conclusion

---

[a] The current best value is $1.602176634 \times 10^{-19}$ coulombs.
[b] Nor are they the quantum fields today's physics imagines, useful though that view may be; see chapter 84 at page 427.

that electrons would be the ultimate entities was wrong. Electrons are composite, made of six ultimate building blocks.

But their being composite does not in itself answer his other question, "Why are they all of equal magnitude?" By this he must mean their mass and charge (their charge to mass ratio having by then been measured[1289]), not their size as he takes them to be points. The new ontology does not address their mass. Let's look at their charge.

Einstein's question leads directly to our new understanding of the electron as six tweedledees. The charge of each subunit is a topological half twist in a 2-D link between space quanta. Bilson-Thompson depicted his untwisted "ribbons" as lying flat on the page so one might think a half twist would perforce be exactly 180°. But it isn't easy to see it that way in the 3-D Planck-scale chaos of the new ontology.

However, there are three ways to see this leading to the electrons' identity of charge.

One way is to simply postulate (that is, a new assumption, but it's the assumption Bilson-Thompson made[a]) the tweedle's twist is digital in units of π radians or one-half turn. This, in effect, *defines* all twists to be the same. But getting there by definition won't keep Ockham happy.

Another way allows the half twist to be physically analog (as is a computer bit). Then, given the vast number of moves that take place during even the fastest measurement, what's measured will be averaged to a digital half turn within our best precision.[b]

A third way would result if, as suggested could be possible, space quanta and their tweedles form a 2-D sheet.[c]

But, having set up these scenarios, is there a real distinction here? Are we not attaching meanings to concepts that are beyond the pale at Planck scale?

---

[a] In his first paper; note 755.
[b] See chapter 74 at page 381.
[c] Chapters 62 and 64 (at pages 326 and 337, respectively) explore the possibility the linked space quanta form a 2-D sheet, which would result in digital twist.

If we can't give real meaning to sizes smaller than Planck volumes—an increasingly accepted view—how can we give real meaning to small changes in twist angles in the links between them?

Curiously it is this *lack* of meaning, inherent in trying to probe Planck scale, that lends the best substance to the meaning Bilson-Thompson assumed for his tweedles. Either they are twisted or they're not.

Twist may get to be digital by Planck-scale default.

77
BIG BUSINESS

> Both possibilities—a cosmos that stretches infinitely far, and one that is huge but finite—are compatible with all our observations.
>
> Brian Greene (2011)[1290]

We've seen how something that started very small got to be very big. But how big is that?

For a long time, the issue was—as Greene implied—whether the universe was infinite or merely very big. Observations since Greene wrote the epigram convincingly conclude it must be finite,[a] but stop short of telling us how big.

> The Universe is VERY big but we can contemplate it all.

The new-ontology version of the life story of the universe has so far come out in fits and starts. Let's size it up in overview.

The last hundred years have seen a complete transformation of our understanding of the universe. Fifteen years ago, cosmologist Turner said,

> The universe of 100 years ago was simple: eternal, unchanging, consisting of a single galaxy, containing a few million visible stars.[1291]

Today, its picture is far vaster and more complex and, although it's still evolving, there's an emerging (though still somewhat conflicted) consensus.

In 1986, Peebles originated the notion of cold dark matter (CDM), some kind of "massive, weakly interacting particles" that dominated the universe.[1292] Add lambda[b] to get the lambda cold dark

---
a See chapter 20 at page 122.
b See chapter 29 at page 165.

matter (ΛCDM) story that's the current *standard model* of cosmology. Its picture is not all that different from ours.[1293] That's no surprise. Though named for a body of theory,[1294] that picture mostly summarizes observations that reveal what happened.[a] What's new about the new ontology is not so much what happened, it is mostly why and when and how.

So, again let's check out fig. 69 (a repeat of fig. 63 for your convenience), NASA's version of that model. It depicts two of the universe's three dimensions (one vertical and one into the page) closed in a circle,[b] expanding left to right in time with a roughly billion-year grid.

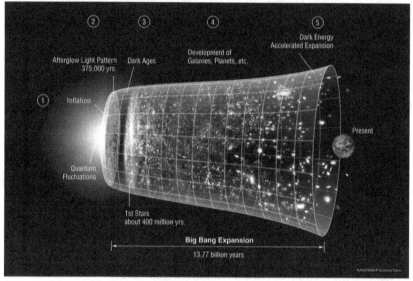

Fig. 69. *The ΛCDM-standard-model story of the universe's expansion*

It also shows the usual culprits. A beginning that's obscured from view;[c] exponential growth labeled "Inflation"[d] with its unexplained slowdown;[e] the "Afterglow" that's now the cosmic microwave background radiation;[f] then darkness for a few hundred million years until gravity could gather local density bumps (that supposedly began as early "Quantum Fluctuations") in the mostly hydrogen gas, and com-

---

a But note, the observations are interpreted through the lens of theory.
b Appropriately for a finite universe.
c See chapter 17 at page 105.
d See chapters 21 and 53 at pages 126 and 291.
e See chapter 54 at page 294.
f See chapter 17 at page 105.

pressed it so it became very hot and ignited nuclear reactions—the first stars.

Expansion kept on going but it slowed down for about five billion years.

Only twenty years ago, Peebles (who literally wrote the book on physical cosmology[a]) co-wrote with cosmologist Bharat Ratra a review that contemplated the expansion could keep on slowing, possibly forever, or could even begin shrinking and collapse.[1295] However, as we've seen, more recent data showed that, as depicted here, about six billion years ago expansion started to accelerate.[b]

Unlike most of us today, our ancestors a hundred or more generations ago saw the Moon and countless stars transit and planets gyrate in clear night skies all their lives. It made them think. We inherit the fruits of their thinking. With not much more to go on, we think of the universe as big.

Given a clear sky free of light pollution, sharp eyes can make out the Andromeda Galaxy, made up of about a trillion shining stars some two million light-years away.[1296] A privileged group—a few astronomers—see further. They explore the entire visible universe[1297] and, with modern telescopes, can see far further than the rest of us. Even fewer look close to the boundary beyond which nobody can see.[c]

It's easy to get confused about this. Light from the farthest galaxies, now reaching us, took nearly fourteen billion years to get here. So, one might envision the visible universe as a sphere almost twenty-eight billion light-years across. That's hard enough to get one's head around.[d] But we know all that time the universe has been expanding.[e] The furthest stars that we see at that distance are not *there* anymore; the expansion of the space between us and them has moved their space and them away. That almost-twenty-eight-billion-light-year sphere now[f] has a diameter of ninety-one billion light-years.[1298]

---

a See note 82.
b See chapter 68 at page 351.
c With the Webb telescope; see chapter 129 at page 617.
d With *Star Trek* warp drive almost ten—let's say one million times the speed of light—we would take forty-six thousand years to reach the "edge" of what we see.
e See chapter 17 at page 105 and chapter 69 at page 357.
f Special relativity theory cautions us to be careful about using words like *now* on a cosmic scale. But the new ontology gives *now* new instantaneously universal meaning; see chapter 56 at page 301.

To make matters more confused, the light *now* leaving galaxies that are the furthest anyone has seen will never reach us. That light is traveling through space that is itself now moving away from us at more than light speed.[a]

A simple sphere-volume calculation tells us the size of the universe astronomers can see is some four hundred million-trillion-trillion cubic light-years. If their seeing can be our believing, we should figure this big universe is real.[1299] But the new ontology already told us even more. When we set out to follow its growth we concluded it must be bigger than the visible universe, but the part we cannot see might not be vastly bigger than the part we can.[b]

There is some support for this understanding of the whole universe's size. Analyzing a model based upon various views of the universe, conventionally plausible assumptions and—more than a decade ago—then-best data, astronomer Mihran Vardanyan and colleagues estimated the missing part may be two hundred fifty times bigger than what is visible.[1300]

So, the whole universe is very big; it's bigger than what we can see but maybe not by all that much.

From my (admittedly unsystematic) reading, few physicists or astronomers or even cosmologists seem to be thinking in these terms so far. Even fewer provide better estimates.[c]

These days, cosmology is big business but it is mostly business as usual.

---

a See chapter 65 at page 340; again, the use of *now* depends on the simultaneity the new ontology supplies, see chapter 56 at page 301.
b In chapter 62 at page 326.
c Based on citations of the Vardanyan et al. paper.

78
BEING POINTLESS

Points of space have no existence in themselves.

Lee Smolin (2001)[1301]

Points play a leading role in physics. It's not too much to say they are where all its action is supposed to be. This has troubling consequences. For example, points infect the math of physics with a plague of meaningless infinities.[a]

Smolin was speaking (above) about points in the continuous space of general relativity. In 1973, Hawking and Ellis said general relativity has the universe beginning as a point of space.[1302] Simply put, like Lemaître,[b] they rewound its movie of the universe expanding.

> The universe has no points but it has a point.

They ducked the infinite-density problem of the resulting singularity (neither more nor less absurd than the infinite density of the classical point-like particle[1303]), concluding,

> The results we have obtained support the idea that the universe began a finite time ago. However the actual point of creation, the singularity, is outside the scope of the presently known laws of physics.[1304]

The Big Bang theory begins with that point.[1305] Thus, in his recent book about reality, Rovelli explains,

> Fourteen billion years ago, the universe was concentrated almost to a single, furiously hot point.[1306]

---

[a] See chapters 15 and 88 at pages 95 and 442; string theories and granular spaces solve all such problems.
[b] See chapter 17 at page 105.

I imagine he inserted *almost* because he does not believe in points. He buys into a view, based on a long line of supportive evidence, that says there are no points in this universe.[a] (We agree.[b])

A large part of string theory's promise is that it is pointless. A point is defined to have zero size and no dimensions. The smallest thing in a string-theory universe is a Planck-scale string with at least one dimension.[c]

In 1988, cosmologist Robert Brandenberger and string theorist Cumrun Vafa described how string theory could solve the universe's hot-point-beginning problem. Setting the scene, they said, with brazen understatement,

> The initial singularity of space is a major issue which has not been totally believable from a physical point of view.[1307]

They discarded the "ugly" compactification of the six extra dimensions and instead said—as does the new ontology—they were compact when the universe began. They concluded that,

> [S]trings could resolve singularities which appear in standard cosmologies.[1308]

Solving the beginning problem this way would have been a giant leap for cosmology—if anyone could make a string theory testable in the real world. There is no evidence for strings. Maybe, say some (like Smolin), there never will be.[1309]

The new ontology appears to leave no conceptual space for real strings at Planck scale where they must (if they exist) belong. But, on the other hand, it appropriates string theory's star math-fiction character, the Calabi-Yau manifold, and says it's real.[d] With one fell stroke this accomplishes string theory's objective of abolishing those points; we don't need all the rest of it.[e]

A small digression. In his 1977 book on how the universe began, Weinberg famously concluded,

---

a See, e.g., notes 844; 918; and 1012.
b See chapter 12 at page 74.
c But the theory has the string in a continuous spacetime made of points.
d See chapter 15 at page 95.
e Nonetheless, there may yet be more value in the string theory story.

> The more the universe seems comprehensible, the more it also seems pointless.[1310]

Of course, he didn't mean (as I do, above) there are no points *in* the universe. He meant there is no point *to* the universe.[1311] With the benefit of a half century of observations and with our ontology at hand, I beg to disagree.[a]

Weinberg was far from alone in this. The idea even had its own animated TV comedy series, *Rick and Morty*.[1312] Then there was author Mark Manson and his book.[1313] And author Joris-Karl Huysmans explored pointless existence in the eighteen eighties.[1314]

My thesis here is simply this linguistic twist: The same understanding of reality that points us to there being no point *in* the universe gives us reason to dispute there being no point *to* the universe. Indeed, it gives a literally-whole new meaning to the relationship between the universe and our own existence.[1315]

It says of every one of us: You are not *in* the universe; you are *of* the universe; your life is an intimate participation in its continuing creation; without you, the whole universe would now and evermore be less.[b]

What more point could one want?

---

[a] And see chapters 134 to 136 at page 647.
[b] And see chapter 136 at page 658.

79
A REALITY CHECK

> As the theory goes, when the Universe was born 13.8 billion years ago, all matter was condensed onto a single point of infinite density and extreme heat. . . . Suddenly, this point began expanding, and the Universe as we know it began.
>
> Matthew Williams (2018)[1316]

Science writer Matt Williams set out the familiar canonical story of the universe's birth. We know that story is *not* how the universe began; few (or maybe no) cosmologists would say it is.

That story says no word about how it all came to be infinitely condensed in the first place, or why it would suddenly expand. It's based on extrapolating the size of the universe back in time using the theory of general relativity, which theory has no application to a single-point universe or even to a tiny one and may not work at cosmologic scale.[1317] In Williams's defense, that is the best theory physics has and that is the short version of its story,[a] a story it's impossible to understand.

| | Let's get our story straight. What's real and what's not? |

Our new ontology tells a sometimes similar yet fundamentally different story.[1318] The biggest difference is that we can understand it. In *Time One*, it emerges from the travails of an amateur-detective narrator whom a reviewer called,

> . . . a dope-smoking beach bum computer hacker (and philosophy Ph.D. dropout) a fictional character at least as far out on the fringe of society as John D. MacDonald's

---

a Grafted into this model, almost at its front end, we have cosmological inflation, which looks intriguing until one finds it simply doesn't work; see chapter 21 at page 126.

houseboat-dwelling narrator and detective, Travis McGee.[1319]

That was where the writing took me and, I must admit, it is an unlikely setting for the search for ultimate reality. But it may be well suited to exploring physics theories that are fictional.

By contrast, this book aims for the straight story. It takes up the challenge left by Baggott in the epilogue of his *Guide to Reality*, having canvassed from the ancients to the modern theories,

> There is simply nothing we can point to, hang our hats on and say *this is real*.[1320]

We will say *this is real* (also *this is not*) based not on this theory or that but on extant seminal ideas we can fit together into one single consistent worldview. Let's review its key concepts and, as promised, check them out against some principles Smolin proposed.[a]

It all starts with one key assumption: The universe began as a single *real* Calabi-Yau manifold, based on the math-fiction concept that is central to string theory.[b] It was a quantum of space and it replicated.[c]

The entire universe we see arose from replication of space quanta. We have followed its evolution with the fewest and the simplest possible assumptions. This led inexorably to a natural marriage of two old ideas: Space is something; and it's made of quanta that are indivisible, atoms that truly can't be cut.[d] They lie at the heart of Planck's discovery:[e] *The space quantum is Planck-sized.*[f]

In the stress of at first extremely curved space, the quanta replicated by quantum tunneling,[g] their numbers at first doubling each sequential step,[h] then slowing down as space grew large enough to have much smaller curvature.[i]

---

a He based them on the principle of sufficient reason; see chapter 32 at page 181.
b See chapter 15 at page 95.
c See chapters 17 and 42 at pages 105 and 236.
d See chapters 10 and 6 at pages 63 and 47.
e See chapter 13 at page 76.
f See chapter 46 at page 263 and note 933.
g See chapter 19 at page 118.
h See chapter 42 at page 236.
i See chapter 54 at page 294.

Each space quantum has six links to other quanta;[a] like ribbons, they may have a twist this way or that.[b] All subatomic particles are made of six paired-and-triply-braided mostly-twisted links.[c]

The quanta and their links are not *in* space; they *are* space.[d]

Twist states of links move randomly, at most one space quantum per sequential step, which equates to the speed of light.[e]

The space quantum's volume gives the universe its emerging three dimensions.[f] Its sequential replicative evolution gives it the property of—not time—but sequence, from which we derive two kinds of time.[g]

With this summary in hand, let's check how well the new ontology accords with several principles for fundamental physics that Smolin enunciated,[h] saying,

> Our aim is to combine quantum physics and spacetime at the level of fundamental principles. I believe the right principles to shape this unification are the following.[1321]

He had an inapt aim;[i] but that does not invalidate his principles, which arise from long insightful struggle with how to progress in fundamental physics. The first is "background independence."[j]

Of this, Smolin said,

> A physical theory should not depend on structures which are fixed and which do not evolve dynamically in interaction with other entities.[1322]

The new ontology reifies this principle. Any physical theory arising from the new ontology *must* be devoid of fixed structures because

---

a See chapter 41 at page 228.
b See chapter 35 at page 190.
c See chapter 43 at page 243.
d See chapter 62 at page 326.
e See chapter 50 at page 276.
f See chapter 46 at page 263.
g See chapters 28 and 52 at pages 158 and 284.
h Smolin added two more principles—reciprocity and identity of indiscernibles—that we do not need to pursue here.
i See Einstein's comment, chapter 25 at page 144 and note 606; and see chapter 36 at page 205.
j See chapter 5 at page 43.

it has none. Every structure in its universe is made of interactions between space quanta that all evolve dynamically at every iteration.[a]

Smolin's second principle is "space and time are relational."

> A relational observable, or property, is one that describes a relationship between two entities. In a theory without background structures, all properties that refer to location in space or time should be relational.[1323]

The new ontology reifies this principle. Its space is purely relational and its common time is a measure of motion that is also purely relational. The second principle becomes the flip side, as it were, of the first. Any physical theory set in the new ontology's space must rest upon relationships (twist states of links between space quanta); all locations are relational.[b] Such a theory must derive its time from motions of locations.[c]

Smolin's third principle is "causal completeness."

> If a theory is complete, everything that happens in the universe has a cause, which is one or more prior events. It is never the case that the chain of causes traces back to something outside of the universe.[1324]

The new ontology reifies this principle. Any complete theory in the new ontology's universe must describe its causal set. Each event in the set has its cause or causes in immediately prior events. The first or ultimate cause—a challenge Smolin ducks—is the first step;[d] it has no prior event; and it has no causal chain tracing back to anything "outside of the universe."[e]

Thus, the new ontology not only conforms to these principles, its universe gives rise to them.

In his 2019 book, Smolin scanned the fundamental-physics scene and said he was . . .

> . . . at times deeply frustrated by our lack of definite progress on fundamental physics during this last half century.[1325]

---

a See chapter 57 at page 307.
b See chapters 41 and 62 at pages 228 and 326.
c See chapter 57 at page 307.
d See chapter 58 at page 312.
e Indeed it has no "outside"; see chapter 22 at page 133.

He talked candidly about the way the system works:

> [T]he academic world was modeled on monasteries, which were designed to perpetuate old knowledge while resisting the new.[1326]

He said he is convinced that . . .

> . . . something very new is needed.[1327]

The something very new that's needed seems to me to be quite plain to see—a new physics based on reality, a physics that connects with the world's two existential levels, Planck-scale quanta and the universe in its entirety as well as ordinary levels in the middle.

In this, as with his principles, we are on the same page as Smolin. It was an unlucky toss of a metaphoric coin that led him to bet on time as the fundamental element of reality.[a] If it were not already evident in Smolin's prior writings on this question, we might have fingered his co-author for this misstep. Unger wrote:

> Time is the fundamental aspect of reality—of all of nature—by virtue of which everything changes.[1328]

Let's be clear here on the basic points of departure between the new ontology's reality and Smolin's and Unger's suggested ontological commitments.[b] They say time is not emergent, leaving space[c] to fill that role:

> We . . . must treat time as non-emergent, global or cosmic (in the sense of preferred time), irreversible, and continuous.[1329]

The new ontology says it's space (the whole vast granular-at-Planck-scale entity, not its continuous approximation) that's real.[d] So is its sequential iteration. It says time (of any kind) derives from space's sequence and the measurement of motion. Its true time is indeed global, cosmic, and irreversible, but it is *dis*continuous. And any measure of either true or common time must be emergent.

---

a See chapter 28 at page 158.
b In the same sentence, Unger hastened to disclaim his own "proto-ontology" label.
c Or spacetime; they are not consistent.
d Its macroscopic properties are emergent.

However, there is more and deeper similarity with "recent Smolin" than may at first appear. Indeed, Unger and Smolin came *so* close, with their hypothesis that,

> [A]ll that is real is real in a moment, which is one of a succession of moments.[1330]

They lacked two concepts: "all that is real . . . in a moment" is *space*; and the "succession of moments" is *discontinuous*.

It may seem surprising that our new ontology is almost all composed of concepts that were long extant. Much of the change is due (as promised[a]) to a reappraisal of what is real and what is not, coming to see some former fictions as being real and vice versa. The cumulative effect is transformative.

Here's a summary of key conceptual/linguistic reassignments, and where relevant discussion may be found.[b] This may illustrate how, using old ingredients, we are now building a new physical worldview and why it offers fresh fields for real physics.

REALITY	FICTION	CHAPTERS
→	BIG BANG	42, 53, 69
BIG FIZZ	←	42, 53, 69
C-Y MANIFOLD	←	40, 41, 45, 58
CAUSAL SET	←	43, 31, 73
→	INFINITY	22, 110
NOW	←	52, 56
→	POINT	62, 76, 77
→	POSITION	62
QUANTUM	←	12, 13, 14, 73
RIBBON/LINK	←	41, 43
SPACE	←	10, 28
→	SPACETIME	10, 82, 83
→	TIME	56, 57, 82
→	FIELDS	84
→	VOID	62, 72, 89
→	ZERO	88, 89, 108

---

a See the introduction.
b This includes some reassignments we have not yet reached.

But let's get real; the latter's up against an institutional inertia that is tied into the base of physics—its financing. For example, in 2023, science writer Flora Graham wrote (shades of the sad Markopoulou story),

> [F]unding systems favour scientists who aim for incremental advances—research that is likely to produce publishable results—rather than "blue sky" projects with uncertain outcomes.[1331]

And the chance of "something very new"[a] faces a further challenge: It is unavoidable—and problematic—that ultimate understanding of our world must be based on entities we cannot observe.[b] If this is the price of understanding, we need to get over it.

Thirty years ago, Maddy, leading advocate for realism, said,

> [D]espite philosophical qualms about unobservable entities, we should admit they exist if our best science tells us they do.[1332]

Here lies the heart of the matter for the test of what is real. Science is no simple or straightforward story; so how shall we choose *what* our best science tells us exists?

There is no preordained answer. My argument is incremental: Aiming for reality will make for better science, not for worse.[c] As it does today, science will need to fend for itself.

So, with space quanta at one end of the scale and lost reaches of the universe at the other, we are deep into unobservable zones. In trying to embrace them in a real ontology, the practical approach for fundamental physics is, I suggest, seek three qualities—consistency, simplicity, explanatory power.

With some such attitude, philosophy of science may help physics find a path to understanding.[d]

---

a See Smolin, note 1327, above.
b Indeed, we passed this threshold some time ago; for example, cosmology is firmly based on unobservables like spacetime and dark energy, and no one has "observed" a quantum field or quark.
c See also Riemann, note 1549.
d See further, chapter 133 at page 639.

PART IV

ON THE WAY OUT

For what I call my real world is something other than Reality. It is a construction, required for certain ends and true within limits, but beyond those limits more or less precarious, negligible, and in the end invalid.

Francis Bradley (1914)[1333]

80
DESPERATELY SEEKING SUSY

> Finally, there is the news—or lack thereof—from the latest particle accelerators, which have not found any evidence for the extra particles predicted by supersymmetry.
>
> Adam Becker (2022)[1334]

Becker's laconic comment announced a particle-physics-business disaster. String theory and the standard theory (fig. 70) need a thing called *supersymmetry* (SUSY to the trade).[1335] Each particle, they say, has a supersymmetric partner. This doubles the number of "elementary" particles the universe must have. Not sixteen but thirty-two.[a]

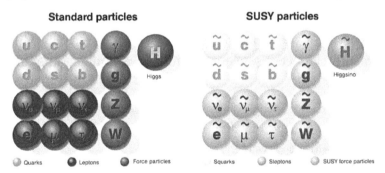

Fig. 70. *The standard model particles with their missing supersymmetric partners*

The new ontology has no place waiting for more particles: The standard theory's sixteen use up all the simply braided double-tweedle triplets. If their supersymmetric partners exist, we'd need some more complex braids (or it could be the tweedle model's wrong).

Finding no supersymmetry set off "a widespread panic."

More particles permit more explanations; with supersymmetry one could even get gravity into the theory.[1336] It

---

a Neglecting the Higgs; see chapter 92 at page 457.

would let spacetime mix with other symmetries.[1337] It could support assumptions that the standard theory needs.[1338] It could even furnish the dark matter physicists can't seem to find.[a] (And it would need somewhere between nineteen and more than a hundred new parameters that would each call for its own new explanation.[1339])

Seeking evidence of supersymmetry was a prime reason for the European Organization for Nuclear Research (aka CERN) to build the world's biggest and most complex machine, the Large Hadron Collider. CERN said,

> If the [supersymmetry] theory is correct, supersymmetric particles should appear in collisions at the LHC.[1340]

Already there was trouble brewing back in 2014. Two CERN physicists were quoted saying,

> [R]esults from the first run of the LHC have ruled out almost all the best-studied versions of supersymmetry. The negative results are beginning to produce if not a full-blown crisis in particle physics, then at least a widespread panic.[1341]

A decade ago, one of supersymmetry's creators, physicist Mikhail Shifman, told a conference it was time to read the writing on the wall. But,

> Of course, people do not give up their dreams easily. They hasten to modify [supersymmetry] in a contrived way to keep it viable.[1342]

LHC physicists have run up more than forty trillion high-energy proton collisions. They found not a single sign of supersymmetry. The panic signals peaked and went off air.

Supersymmetry has not expired—well, not officially—but it sure looks like a dead man taking an extended stroll.

---

a But see chapter 68 at page 351.

81
A PARTICULAR PROBLEM

> The whole thing's held together by entities we don't know exist at all and [that] have no real physical basis.
>
> Mike Disney (2008)[1343]

Astrophysicist Mike Disney was speaking of the standard model of cosmology. A BBC interviewer introduced him with,

> For Professor Disney, the standard model is simply an unproved theory.[1344]

This model (aka the *standard muddle*) is an indigestible pea soup that's made from fanciful ingredients in place of water, peas, and hambone:

- Fourteen parts *dark energy*, a label for the mystery of antigravity that makes the universe expand despite its gravity;
- Five parts *dark matter*, a label for the mystery of seeing much more gravity in action than ordinary matter can explain; and
- One part *ordinary matter* that we know and love, consisting of fictions we don't understand.

Disney's criticism of the soup and its ingredients is sound; the interviewer's *unproved theory* is excessive praise. After all, in this model almost all of everything is labeled *dark*, which tells us scientists do not know what it is. It's not that they have no idea; they have too many ideas, none of which seem to be panning out. And they don't really know what ordinary matter's made of either.

On the other hand, as we have seen, the new ontology explains all three ingredients—both what they are and how they got here. It says they are all versions of one single thing.

Of course, this likely is less than exactly right. But, let's be clear, it is the *one* coherent explanation of the three ingredients we've got. And it explains a lot.

This sets a standard to which physics should at least aspire. But, as we've seen, the leading candidates for that missing dark matter are new particles.[a] And the well-trodden path to find new particles is bigger atom smashes that need bigger atom smashers.[1345] Likewise, there are all those missing particles of supersymmetry.[b]

So, the standard theory industry[1346] is gearing up for more expansion, the Future Circular Collider (or FCC) with costs calculated, at predesign stage, as more than twenty billion euros (with room for expansion).[1347]

In 2020, industry whistleblower Hossenfelder said,

> [P]article colliders are currently the most expensive physics experiments in existence. . . . But while the cost of these colliders has ballooned, their relevance has declined.[1348]

The reader may be unsurprised to find that I concur.

The purpose of the LHC and now the FCC is to find a way to move beyond the standard theory. It is an expensive strategy that, so far, isn't working.

To be clear, I share the widely held view that, at some higher energy, there may be a window to be found into new physics. But, given all the higher-energy experiments the universe is laying on for us in space for free,[c] constructing a big new machine down on the ground (or under it) would be, it seems to me, misspending money.

> Making faster particles is getting too expensive.

And, as we have seen already happened with the LHC,[d] building a bigger one in hope its energies will reach to that new-physics

---

a See chapter 68 at page 351.
b See the preceding chapter.
c Cosmic particles achieve energies more than a million-fold higher than those in our best accelerators; see also chapter 129 at page 617.
d See the preceding chapter.

window could court more disappointment.[1349] For all we know, its window may again be higher than the new accelerator's ceiling.[a]

The new ontology points to far less expensive and more promising paths to new physics.[b]

Most of all, we should aim money at the fundamental-physics frontier. These days it's not down here, it's out there.

---

[a] Efforts are made to predict the requisite energies; it's not clear the predictions have a lot of value.
[b] See also chapter 129 at page 617.

82
TAKING YOUR TIME

> Only by keeping in mind that our beliefs may turn out to be wrong is it possible to free ourselves from wrong ideas, and to learn.
>
> Carlo Rovelli (2017)[1350]

Few concepts are as confused and as pervasive as the idea of time. It lies at the core of our lives. Time, we think, is all we have.

You may discern I have not yet delivered on my undertaking to answer the question *What is time?*[a] The answer is, it is any one of several ideas we invent.

At least since Newton tied his common time to then new math,[b] that idea of time has been at the heart of almost all of physics and a great deal of philosophy. One might almost say that physics morphed into the study of things changing over time. Yet physicists' efforts to grasp the nature of their time seem to lead to bafflement and even, of late, flat rejection.[1351]

> Your idea of time is more real than most physicists'.

One of my two lead living philosophic physicists, Smolin, said,

> More and more, I have the feeling that quantum theory and general relativity are both deeply wrong about the nature of time.[1352]

The other, Rovelli, said,

> The nature of time is perhaps the greatest remaining mystery.[1353]

---

[a] See the introduction and chapter 4 at pages 7 and 34.
[b] I.e., the calculus; see chapter 57 at page 307.

Part IV: On the Way Out 417

In his book *Now: The Physics of Time,* physicist Richard Muller said it is . . .

> . . . remarkable that we understand so little about the fundamentals of time—what it is and how it relates to reality.[1354]

Time once seemed to be a simple thing. Sundials may have been the first clocks some thirty-five hundred years ago.[1355] They measured the apparent motion of the Sun which, since time out of mind, everybody watched.[a]

Global navigation led to better (and more portable) devices whose regular internal motions kept good time for months to track one's longitude.[1356] (Instruments called sextants and the Sun tracked latitude.)

Today, physics' definition of time is what a clock measures.[1357] Naturally, physics needed, and now has, more precise clocks. Check any of them out and you will find they measure motion.

The second started out in life as one eighty-six-thousand-four-hundredth of a rotation of the Earth relative to the Sun's mean apparent position.[1358] Now it is 9,192,631,770 electromagnetic oscillations of light from an electron changing orbit in a cesium atom at sea level at absolute zero temperature.[1359]

Muller, seeking to break free from clocks, began his chapter titled "Time Explained" with a strategic question,

> What makes time move on?[1360]

His unsatisfying answer, amid some religious speculation, was,

> Time moves forward because our current state is so highly improbable.[1361]

The new ontology tells us the universe has no time.[b] It just *is*.[c] And *next* it *is* again. By contrast with the almost universal (no pun) assumption, it is not continuous. So, what are we to make of what we thought was time?

---

a Of course, it's really the not quite even rotation of the Earth.
b See chapter 52 at page 284.
c This is, in principle, the perfect expression of simultaneity; though, as special relativity explains, the finite speed of light denies us direct access to it; see chapter 56 at page 301.

Note that the foundation of that definition of the second is the relentless *Next*.[a] Not just because each iteration *takes* time as our clocks compare things but because *Next* is the heart of the machine that runs and regulates the light-speed motion of those photons from the cesium atom. Without the *Next*, they could not move.

One could close one's eyes and try to watch in gedanken-experiment slow motion — as if from outside — the whole universe as it jerks its sequential *Next*s. So it's the universe's *Next*s that drive all of our times along.

The notion such a vast array performs this choreography does seem fantastical and yet it is the only concept that can stitch more than two thousand years' insights into a consistent tapestry so we can understand space, time, and motion. But it's not contrary to anything that we observe; it violates no laws. It seems fantastical because it lies so far beyond all our experience.

That tapestry tells us the universe does not itself possess the property that we call time. The new ontology is not alone in this.

In 1967, DeWitt, Wheeler, and physicist Peter Bergmann explored the math of what would become early quantum gravity. They found a simple new equation for the entire universe (fig. 71). While in some sense it seemed to be the quantum-gravity equivalent of quantum theory's famous Schrödinger equation — the quantum equivalent of Newton's laws — which governs how a quantum system changes over time, strangely the new equation had no time![1362]

$$\hat{H}(x)|\psi\rangle = 0$$

Fig. 71. *The Wheeler-deWitt equation for the universe*

Smolin explained,

> Because there is no clock outside the universe, the quantum state of the whole universe cannot change in time. ... The quantum universe simply is.[1363]

---

a See chapter 57 at page 307.

Smolin and colleague Ted Jacobson and, later, Rovelli found solutions to the new (now known as Wheeler-DeWitt) equation,[1364] that led to loop quantum gravity.[a]

More than fifty years after its birth, the equation gets mixed reviews because it is not testable. But, Rovelli said,

> [It] . . . opened the world of background independent quantum gravity, a unique source of inspiration, and a powerful conceptual tool that has forced us to understand how to actually make sense of a quantum theory of space and time.[1365]

Physicist Tatyana Shestakova considered whether it is more fundamental than the Schrödinger equation.[1366] Equivocally, she concluded,

> [I]t is too early to deprive the Schrödinger equation of its fundamental status.[1367]

The new ontology would beg to differ. It is way past time.

Inside the universe, clocks are for cooking in the kitchen and for making it to meetings. It's true local clocks can measure spans of time with exquisite precision. But even for tracking time of day, they can run into conceptual problems.[1368] As a foundation for physics, clock time is a source of endless complication and confusion.

For a new physics of the universe, maybe we should track events in terms of the universe's iterations.

At any instant, an exact integer gives the current count of them.[b] And we know (though we cannot see this) it is the same instant and same integer throughout the universe.

This is not at all like Einstein's time. For understanding all the action, it may be much better.

---

a See chapter 31 at page 176.
b That we don't know the integer exactly should not present serious problems.

83
SHEDDING SPACETIME

We have a lot of hints from physics that spacetime as we understand it isn't the fundamental thing.

Natalie Paquette (2022)[1369]

Becker quoted Paquette in support of questioning if space and time are real, saying,

> At the deepest level of reality, questions like "Where?" and "When?" simply may not have answers at all.[1370]

They were both right. Yet neither seemed to see how spacetime made a conceptual mess for physics. This is not a new perspective. Its roots were in plain view a hundred years ago. For example, in 1927, Reichenbach said,

| Spacetime is a handy meaningless concept.

> The treatment of the problem of time as parallel to that of space has been detrimental.[1371]

Hume, in 1748, wrote of the deep distinction, noting . . .

> . . . the different properties of space and time. . . . [A]ny one may easily observe, that space or extension consists of a number of co-existent parts disposed in a certain order, and capable of being at once present to the sight or feeling. On the contrary, time or succession, though it consists likewise of parts, never presents to us more than one at once; nor is it possible for any two of them ever to be co-existent.[1372]

For physics, their conflation has been a convenience and a disaster. Once we understand the blend we'll see it mixes everything that's real with a quite different thing that is invented; it's a mashup that confounds clear thought.

Hermann Minkowski was a number theorist, a teacher who taught Einstein. He swiftly embraced special relativity.[1373] He conceived the notion of multiplying the time variable t by the speed of light c and using minus ct as a kind of oddball extra space dimension—thereby treating time itself as a dimension—in Einstein's equations.[a] In a dramatic presentation, he proclaimed,

> From now on, full space for itself and time for itself, sink completely into shadow and only a kind of union between the two should be allowed to preserve independence.[1374]

He disarmed critical evaluation of this devious device by promptly dying. His new entity came to be called *spacetime*. Its four dimensions came to be the basis of relativity's math. This gave the equations simpler form and seemed to lend them a new meaning. Many physicists embraced it so entirely in their ontological commitments they no longer spoke, or maybe even thought, in terms of space.

Spacetime was a doubly mongrel gadget. It glued a time-based pseudo-dimension—a concept physics with good reason fails to understand—onto an unreal (continuous) version of the three-Ds of our real space. It left physics astray in a confused light-cone forest.[b]

Physicists could soon be found trying to unglue the gadget. For example, as early as 1933, cosmologist Howard Robertson dissected a "Minkowski universe":

> But such a centered universe . . . does in fact imply a cosmic time. . . . The curved 3-spaces τ = const. may then be described in terms of 3 spatial coordinates.[1375]

In 1998, Wheeler said,

> In general relativity, for instance, it is sometimes easier to talk of the three-dimensional geometry of space evolving through time.[1376]

---

[a] These equations commonly have four variables, x_1, x_2, x_3, x_4, the latter being calculated as $-ct$.
[b] See chapter 74 at page 381.

In a more recent example, we find physicist Matías Hersch solving a key quantum-gravity equation by slicing general relativity's

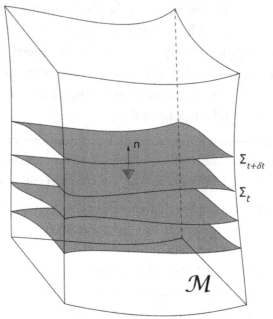

Fig. 72. *A foliation of spacetime into slices of space*

spacetime into what looks like the new ontology's 3-D closed spaces (fig. 72):[a]

> In [a] formalism of GR spacetime, a 3 + 1 dimensional Riemannian manifold . . . is foliated into a one parameter family or trajectory of space-like slices (leaves). Unless otherwise noted we will assume the spatial slices to be closed.[1378]

And in his 2022 inside story of cosmology's search for reality, Peebles has two dozen references to "space sections," saying,

> [W]e can define a three-dimensional space through four-dimensional spacetime by all the spatial positions at a chosen value of the time elapsed since expansion began.[1379]

---

a See figure 1 of note 1377.

Back in 1908, Minkowski's first try—with a different definition for time—did not work. He soon found a fix; he evidently did not grasp its implications. He defined time to mean what's measured by a clock or more precisely, *proper time*,[a] a then notable innovation.[1380]

Einstein (who also used clock time) understood the fundamental weakness this approach imported into theory:

> It stands out that the theory (except for the spacetime) introduces two kinds of physical things, namely (1) measuring rods and clocks, (2) all other things . . . This is inconsistent in a certain sense; strictly speaking, measuring rods and clocks should actually stand revealed as solutions of the basic equations . . . not as being, to a certain extent, independent of the theory.[1381]

At first, he dismissed Minkowski's 4-D pseudo-space as "*überflüssige Gelehrsamkeit*" or "useless erudition."[1382] But then, he and almost everybody else embraced it.[1383]

It soon became entrenched. As Valentini said in 2008,

> Unfortunately, after 1905, the dogma of Newtonian space-time was quickly replaced by the dogma of (local) Minkowski space-time.[1384]

Convenience it has for sure. Spacetime sets up tidy shorthand for equations.[1385] But (to harp on about the central point) it mixes something almost real and three-dimensional—continuous space—with something neither real nor a dimension—clock-based time—and so spoils our ability to think straight about either.

Indeed, spacetime's legacy is that for a century it shunted much of fundamental physics into a conceptual siding.

For example, Muller said,

> The elusive meaning of *now* has been a stumbling block in the development of physics.[1386]

In this, he was plainly right. But when he put forward his own concept of *now* he embedded it in spacetime, the biggest stumbling

---

[a] Think of proper time as time shown by a watch; in other words, by a clock that follows its observer.

block of all. The possibility that *now* is the real 3-D space aspect of his abstract 4-D spacetime—and so it is space's forced marriage to a nonexistent time dimension that creates that stumbling block—did not seem to occur to him. Millions stumbled with him down the same false trail.

The heart of the matter is that spacetime has no physical reality; but physicists tend to treat it as real.[1387] Many think and speak and write and work as if they see it so; they forget that Einstein (to remind you again) cautioned,

> Concepts that have proven useful in ordering things easily achieve such authority over us that we forget their earthly origins and accept them as unalterable givens. . . . The path of scientific progress is often made impassable for a long time by such errors.[a]

Thus, they become the victims of their own ill-gotten ontological commitment.

By corollary, relativistic physics does *not* regard 3-D space itself as real. It echoes Minkowski, who (also to remind you) said,

> From now on, full space for itself and time for itself, sink completely into shadow.[1388]

And sink into shadow space soon did. Some may have been relieved, as spacetime seemed to sink time too—the conceptual problems of time were ever a thorn in the side of those concerned with fundamental physics.

As we saw, Einstein was for a while an exception to the rule, delving as deeply into fundamental physics as he could while saying time is given by a clock.[b] Special relativity says saying what a clock (or, rather, its clock-watcher) says is not so simple if it's moving, or if it is far away.[1389]

There was the occasional naysayer such as Smolin, as far back as 1988, doubting spacetime's wisdom:

> I believe . . . time *is* different from space.[1390]

---

a See note 23.
b See note 245 and chapter 82 at page 416.

But Smolin—nudged back into line by Markopoulou, it would seem[a]—spent decades in thrall to what one could call the spacetime state of mind.

In 2009, Markopoulou wrote a powerful pitch for a then new approach to quantum gravity,[1391] saying,

> What is interesting for us is that this mathematics contains no reference to any background spacetime that the quantum systems may live in and hence it is an example of [background independence].[1392]

Her definition of that background independence was,

> [A] theory is background independent if its basic quantities and concepts do not presuppose the existence of a given background spacetime metric.[1393]

Her math assumed no spacetime at all. She went on to say,

> This picture implements the idea that spacetime geometry is a derivative concept and only applies in an approximate emergent level.[1394]

She too came so close!

And Smolin, having concluded space and time cannot both be real, sometimes seemed to walk away from spacetime. In 2019, he said of his new model,

> [T]here is no spacetime, fundamentally.[1395]

Once freed from spacetime's clutches, physics must address the problem of motion in the space of a relational universe[b] where anything that moves is made of link-state relationships between space quanta that exist but lack linear location.[c]

Some may say this linked-space-quanta space looks like another background. To this there are two answers. One, courtesy of Baez, is that background independence is an objective; let's not get hung up on it.[1396] The other is, in a theory of the whole universe, this "back-

---

a See chapter 1 at page 15 and note 61.
b See chapter 5 at page 43.
c See chapter 62 at page 326.

ground" begins as one quantum of space, with all that follows being purely relational, and that is as close to background independent as it is possible to get.

After centuries of physics' focus on it, motion is still an enduring mystery.[a] Spacetime—which in even its general-relativity form *is* in some degree a background[b]—is a key-log for this logjam. With it out of the way, one can come to grips with the real problem. A universe that is relational requires a physics that is as relational as it is.

As Euclid showed, and Riemann elaborated, one can do useful geometry in imaginary manifolds or spaces. Spacetime, too, turned out to be useful. The new ontology explains machinery behind its math.

This does not make it real.

Let's give Unger the last shot in my rant.

> Call it the Einsteinian-Riemannian ontology, marked by the project of rendering time a shadowy department of geometry: the spatialization of time.[1397]

Spacetime's crime is it corrupts real thinking.

---

a See chapter 127 at page 609.
b For example, it imports a metric.

84
FIELDS OF DREAMS

According to quantum field theory,
fields alone are real.

Gary Zukav (1979)[1398]

A physicist can make a field for almost anything you care to name except potatoes. A field in physics is conceived as something that exists throughout an infinite continuous space. It has a value at each point in space at every time.

Fields got their start in life in the early eighteen hundreds. Iron filings around a magnet seemed to say something was *there* (fig. 73). Physicist Michael Faraday developed that field concept about 1832.[1399] Others have been adding to it ever since. It's another example of a useful fanciful idea coming to be seen as real.

> The idea of a field is a very useful fiction.

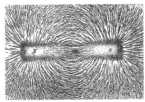

Fig. 73. "Field" revealed by iron filings

By the eighteen sixties, physicist and mathematician James Maxwell's theory seemed to say light is a wave in the "electromagnetic field."[1400]

These days fields are physics' explanation for literally everything (see popular author Gary Zukav, above). Well, for everything except the field.

In other words, a field can be a fancy way to say we don't know what is going on, but let us not let that get in our way.

Fields and their theory are difficult to describe in plain language, maybe because the fields *are* their math. In a recent book aimed at the bold layperson, particle physicist Matt Strassler gave it his best shot.[1401] And then he said,

> Part of the problem with talking about fields and what they are, is that fundamentally we don't know. It's always harder to explain something you don't really understand.[1402]

Today, quantum field theory "may be the most successful theory ever," according to mathematician Steven Strogatz, interviewing David Tong, a physicist who said it can achieve spectacular agreement with our observations, but also with disarming candor said,

> [T]he mathematics we're dealing with is not something that's on a rigorous footing. It's something where we're playing sort of fast and loose with various mathematical ideas.[1403]

In 1938, Einstein and Infeld wrote a book about how physics developed from Galileo's day. Almost half the book was about fields. They acknowledged,

> Starting as a helpful model the field became more and more real.[1404]

A hundred pages later, they said,

> There would be no place, in our new physics, for both field and matter, field being the only reality.[1405]

Fields solve existential problems for a physics that has long lost contact with reality. They may be the ultimate expression of the continuity compulsion. But the new ontology gets by without them. It's odd to think the whole concoction might have never come to be if Newton and Leibniz had not contrived the calculus.[1406]

Physics math portrays a subatomic particle as a kind of wrinkle in a quantum field. Thus, it is an inherently analog entity; it has an infinite array of values at all points in the universe. But all this instantly collapses at a single point when it's "observed."

These well-known obscurities have not impeded the hegemony of fields. For example, quantum physicist Art Hobson wrote an article entitled "There Are No Particles, There Are Only Fields."[1407]

And science writer and astrophysicist Ethan Siegel said,

[It's] more accurate to view the entire Universe as a complicated quantum field that, itself, contains all of physics.[1408]

Though many venture blithely into physics' fields while brandishing the language of reality, few seriously try to demonstrate that they exist. Their substance seems to rest on little more than lined-up iron filings. But Feynman thought fields must be real and explained why:

> The fact that the electromagnetic field can possess momentum and energy makes it very real.[1409]

However, such indicia depend upon the theory one uses to describe the situation. Thus, in the field theory of gravity, there is an energy-momentum tensor for the gravitational field.[1410] But when gravity is described by general relativity, that energy-momentum tensor vanishes. Meaning not that its value is zero, meaning that it has nowhere to be.[1411]

Einstein played a leading role in extending the field concept. In 1915, his theory seemed to give it real substance. Or rather, as he wrote in 1952, his field (which he took to be the metric field[1412]) gave substance to the theory's concept of space:

> There is no such thing as empty space, i.e., a space without a field. Space-time does not claim existence on its own, but only as a structural quality of the field.[1413]

Behind the scenes, he came to think the field and its continuous space could be the wrong road for physics. In his 1933 essay on the method of theoretical physics, he said,

> [Field] theory is fundamentally non-atomic in so far as it operates exclusively with continuous functions of space.[1414]

In 1954, he wrote to a friend,

> I consider it entirely possible that physics cannot be based upon the field concept, that is, on continuous structures. Then nothing will remain of my whole castle in the air including the theory of gravitation, but also nothing of the rest of contemporary physics.[1415]

Indeed, as we have seen, Einstein quietly favored the idea of discontinuous space for almost forty years.[1416] But he never could quite get his hands on (so to speak) its elementary ingredients. Having in his latter days come to the view that math was the essential language of physics, he was frustrated by his inability to conceive a mathematical approach to granular space, saying,

> But for this we unfortunately are still lacking the mathematical form. How much I have toiled in this direction already![1417]

As was often so, his second thoughts were closer to the mark than were his first. But he also lacked the Calabi-Yau manifold; and what may be the needed math hove tentatively into view some seven decades later.[1418]

Einstein's post-relativity career was mainly concerned with searching for a field that embraced both electromagnetism and gravity, a personal obsession that was ahead of his time.[1419] He failed.

Physics is now heavily invested in the field and in its math. This is a problem as they both evaporate when physics moves to quantum space.[a]

Fields are sweet dreams. Though they are functional, we should not fall into thinking of them as real.

---

a And see chapter 133 at page 639.

85
A CONSERVATION CONVERSATION

> The conclusion, whether we like it or not, is obvious:
> energy in the universe is not conserved.
>
> Edward Harrison (1981)[1420]

Cosmologist Edward Harrison's conclusion was obvious to him but it was also likely a minority opinion. Almost everyone knows energy *is* conserved.[a] And this is said to be a fundamental law.[1421]

The new ontology—and lots of evidence—agrees with Harrison that almost everyone is wrong. Of special interest, the energy of the whole universe is *not* conserved.

For starters, the new ontology's universe held no matter before it got going and its energy content was nonexistent. Right from the get-go, it was making space—and we know space has now achieved a mass that is some sixty-eight percent of everything and is described in terms of energy.[b] It was making lots of matter too along with ordinary energy.[c] This is how it built itself; it keeps on doing it today.[d]

> Conservation of energy is only a local law.

In other words, from a standing start the universe's energy became enormous in a heartbeat. Though it soon settled down, several Nobel Prizes confirm space is still being created.[e] As well, the new ontology says tangled links in the new space make braids that are new particles of energy and matter.[f]

---
a Here, "conserved" means "is neither gained nor lost."
b See chapter 29 at page 165.
c See chapter 47 at page 266.
d See chapter 69 at page 357.
e See chapters 18, 40, 55, and 69 (at pages 112, 222, 297, and 357, respectively).
f See chapter 47 at page 266.

The universe's energy, this picture says, is inexorably on its way up.[a]

The old ontology leads to a deep reason energy should be conserved in a continuous space. In 1918, mathematician Emmy Noether[1422] showed in a famous paper[1423] that this law results from a symmetry of the universe[b] that can be expressed as *the laws of physics are the same over time*.[1424]

This law of conservation of energy was dinned into me in high school physics, along with its twin, mass conservation, and their $E = mc^2$ equivalence with its sly message via Einstein:

> It followed from the theory of special relativity that mass and energy are both but different manifestations of the same thing—a somewhat unfamiliar conception for the average mind.[1425]

"Of the *same* thing"—that took some comprehending. But of *what* thing?[c]

So, fifty-some years on, the concept of the universe making more space[d]—and more matter with it—gave me mental indigestion. Once that was overcome, I was confronted with the laws of physics themselves—including that one—plainly changing over time. There were no such laws in the beginning; they emerged from vast numbers of space quanta *after* the universe began.[e]

It gets much worse. It turns out that conserving energy can be another messy business.

Noether's theorem derived from the local flat Minkowski space of special relativity.[1426] We now know its concept of space isn't real; but at most times and most places in the universe,[f] it's a good approximation. For this reason conservation of mass-energy is an approximate and local law. What then may we say when we push its boundaries?

---

a This neglects the question whether matter and energy in a black hole remain "in" the universe.
b Note that the existence of such symmetries and their implications for physics depend entirely on the assumption that space and time are continuous.
c That defeated me until I stumbled onto Bilson-Thompson's model.
d See chapter 67 at page 348.
e See further, chapter 97 at page 478.
f Times and places where it is not a good local approximation include before the Big Bang and near a black hole.

When push came to shove in 1915, having published three versions of general relativity that used Minkowski space and turned out not to work, Einstein ditched it and went with a requirement known as *general covariance.*[a] Thus refounded, his theory effectively abandoned conservation laws.[1427] Indeed, it excluded any mention of the energy of gravity itself, to the dismay of physicists who work with it,[1428] said Baryshev,

> Rejecting the Minkowski space inevitably leads to deep difficulties with the definition and conservation of the energy-momentum for the gravity field.[1429]

Long before 2008 when Baryshev wrote this, I had to come to terms with cosmologist Edward Harrison's blunt pronouncement that energy is not conserved.[b] Peebles arrived at the same conclusion in another standard text.[1430] Their reasoning was that, as space expanded, the wavelength of the photons in it lengthened so, with no corresponding increase in their number, their total energy decreased.[c] And—in the standard-model paradigm—no corresponding work was done (and energy increased) by space expanding.

Though slow, this decrease in the universe's energy alone would be enough to undermine the conservation principle; but that is the least of its troubles. The expansion corresponds to the creation of new space and we now know that space has lots of mass (or energy).[d] Thanks to long-standing confusion concerning the cosmic constant,[e] few seem to have wondered whether that new mass offsets or even overwhelms the loss of radiation energy due to expansion.

Then there's the increasing potential and kinetic energy of all the matter—galaxies and gas and the black holes that are dark matter[f]— that are being lifted (so to speak) in each other's gravity, and accelerating too, gaining kinetic energy. And, as Baryshev said (thinking of speedily receding galaxies),

---

a See chapter 99 at page 485.
b See note 1420.
c It may not sound like much but this is a *lot* of missing energy.
d See chapter 67 at page 348; see also note 671.
e See chapter 85 at page 431.
f See chapter 68 at page 351.

> Another puzzling consequence ... is that in exact relativistic expansion dynamics of the universe there is no relativistic effects [sic] due to the velocity of the receding galaxy.[1431]

As well as the mass-energy of new space itself, mentioned above, one must wonder to what degree newly minted space quanta's links create new knots in Nature's knitting—new particles—that may add much more mass to this mad mess of energetic uncertainty.

Finally, there is this caution from Davis:[a]

> The total energy of the universe is neither conserved nor lost—it is just undefinable.[1432]

The new ontology can bring a certain clarity to this confusing picture: On the scale of the whole universe and its whole history, new mass-energy from the creation of space and matter (forget the rest of it) makes obvious nonsense of the conservation law as any kind of fundamental principle.

Speaking of the local view in his famous physics lectures, Feynman said,

> [T]hat there is no perpetual motion at all is a general statement of the law of conservation of energy.[1433]

Having had a peek into its workings, we can see the universe is *the* perpetual motion machine par excellence.

---

[a] Based on the standard model.

86
MASTERING MASS

> Matter is not what it appears to be. . . . The mass of
> ordinary matter is the embodied energy of more basic
> building blocks, themselves lacking mass.
>
> Frank Wilczek (2008)[1434]

One physics concept we all grasp intuitively is the idea of mass. We buy stuff by the pound or kilogram. We feel its heft and so we think we know what mass is.

But we don't really know. Neither does physics. Understanding mass turns out to be a rat's nest of uncertain problems.

Mass used to sit upon a seemingly solid foundation. It was a kilogram of platinum[a] in Paris.[1435] Einstein, the world's finest physics-mischief-maker, undermined that solid kilogram by making it the same as energy. He knew this was a problem and that aspect of it didn't seem to trouble him.

> Mass is energy . . .
> whatever that is.

He never fixed it. Others tried, but we'll see their partial successes only made it messier.

On our way to the problem, let's take a closer look at those kilograms of stuff we buy, the stuff that Wilczek said (above) is "not what it appears to be."

To a physicist, matter starts with atoms and their mass consists almost entirely of the protons and the neutrons of their nuclei. (Electrons add much less than 1 percent.)

Our physicist will tell us that the proton and the neutron each consist of three quarks plus some particles called *gluons* that help keep the quarks together. The masses of the quarks are known. The gluons, like photons, have none.

---

[a] It was alloyed with 10 percent iridium.

Simple sums offer us insight into what mass isn't. The separate masses of the quarks add up to far less than the masses of the protons and the neutrons that our physicist says they comprise.

What's missing from this picture is their energies. Each quark is in frantic motion—near the speed of light—inside its *baryon* (proton or neutron).[1436]

Indeed, one can explain more than 95 percent of the masses of the baryons based on the *energies* of their constituents, using Einstein's energy equation in reverse: $m = E/c^2$.[1437]

So much for mass at atom scale. At cosmic scale, we have those Planck-satellite measurements showing dark energy (aka space) is two-thirds of the universe's mass.[a] As Wilczek put it,

> The primary ingredient of reality weighs, with a universal density.[1438]

This two-thirds of the mass is generally not included in the *m* in physics equations. When it does show up, it's in the guise of that cosmological constant that, in general relativity, acts as an antigravity force that overcomes the universe's gravity. We know that's not antigravity; that's just space. And it shows up as energy.

So, short story, mass *is* energy. Whatever that is.[b]

Now back to that definition problem. Physics runs into an even more fundamental problem trying to combine the pieces and define that *m*, which, like *t*, shows up with seeming simplicity in some equations. In the world of relativity, that mass is a mess. Science writer Steve Nadis said,

> Einstein recognized the challenges involved in quantifying mass and never fully spelled out what mass is or how it can be measured.[1439]

One problem arises from an unavoidable source of extra energy. General relativity shows how mass gives rise to gravity, by curving space. This curvature gives rise to its own energy. And that energy too is mass.[c] Which gives rise to gravity. Et cetera.[1440]

---

a See chapter 29 at page 165.
b See chapter 85 at page 431.
c It is inconsequential for the mass of an apple but may be a big deal for a black hole.

Thing is, these many energies and their equivalent masses don't keep themselves neatly sorted into their categories. So these days physics tries to define the whole ball of wax, again using general relativity. It seeks to define mass in a region of space by reference to the curvature of the region's boundary. With great difficulty, this leads to a version of what mass is (called—in case you come across it—the *quasilocal mass*).[1441]

The flip side of the mass mess is *inertia*. It's another fundamental concept that we do not understand. Feynman said,

> The law of inertia has no known origin.[1442]

In his biography of Einstein, Pais elaborated,

> The origin of inertia is and remains *the* most obscure subject in the theory of particles and fields.[1443]

All this confusion about mass and its aliases should not be surprising. After all, the impetus for the theory of general relativity arose—Einstein then said[a]—from the relationship between inertia and the mass of the universe (a principle he attributed to his "precursor,"[b] physicist Ernst Mach, who certainly knew nothing of dark energy).[1444]

In sum, general relativity itself—the theory of space and its relationship to mass—is grounded in a deep misunderstanding of what space is[c] and no understanding of what mass is.[d]

At Planck scale, the new ontology explains the "particles" with mass—quarks and electrons mostly—are made of braided twist relations between emerging space quanta. In a sense—a sense that was, for me, essential[e]—this explains the origin of massy matter in the universe and of the particles matter is made of. Yet in a wider and more fundamental sense, it fails to explain why those particles have mass.

---

a Like other aspects of the theory's foundations, this too has been the source of much confusion; see chapter 99 at page 485.
b This is Pais's term for the relationship; note 44 and see further at his p. 284 ff.
c Essentially all accounts, including Einstein's, say so; see chapter 10. The theory led Einstein to conclude that space had substance, a conclusion physics has since largely ignored; also see chapter 10.
d See further, chapter 99 at page 485.
e See chapter 17 at page 105.

It would seem mass must, like all else, be relational in nature. So far, we can't decipher how. (Though there is now much talk about the Higgs field,[1445] it just gives us yet another name for something we don't understand.[a])

Likewise, once we understand mass, it will be seen to be emergent and, so, statistical in origin. That understanding—and so too that of energy—needs must await a real physics of quantum space and a Planck-scale description of what energy is.

---

[a] See also chapters 83 and 128 at pages 420 and 614.

87
ELEMENTARY MY DEAR WEINBERG

> We will not be able to give a final answer to the question of which particles are elementary until we have a final theory of force and matter.
>
> Steven Weinberg (1997)[1446]

Weinberg was an uncommon kind of physicist in this (among other things): He questioned the assumption that the standard theory's particles are elementary. He explored the issue a quarter century ago and concluded that, when we have a final theory, ...

> ... we may find that the elementary structures of physics are not particles at all.[1447]

We don't have a final theory but we can already see that he was right about that.

It's not clear that Bilson-Thompson knew of Weinberg's view, published in a by then defunct in-house SLAC[a] mag to which Weinberg was a frequent contributor.[1448] His view that the status of the known particles was an open question was shared by few. By the same token, Bilson-Thompson's unveiling of those particles as composite met not so much rejection as indifference.

> Only space is elementary.

On the other hand, having no role for space quanta,[b] Bilson-Thompson was reluctant to claim tweedles are elementary.[c] Indeed, by 2012, he and his co-authors would explicitly back away from their unique qualities:

> [I]nstead of treating the helon model as yet another model of elementary particles, one can encode it in

---
a The Stanford Linear Accelerator Center.
b Beyond *nodes*, an equivalence he did not identify.
c The term *elementary* does not appear in his seminal paper.

LQG and SF[a] models to make a theory of both space-time and matter.[1449]

Bilson-Thompson's nodes and tweedles—our quanta and links—have no claim to be particles. Would Weinberg have seen *them* as elementary?

Oddly, the answer is uncertain. Whether they are *composite* depends on what we mean by that. *Merriam-Webster* says "made up of distinct parts or elements."[1450] It's not so unclear with space quanta themselves. Or not exactly, but each link's existence does depend upon two quanta.

The tweedle's a strange critter, having no enclosure in the particle zoo. A twist in a 2-D link 'twixt two 6-D space quanta, it has no volume of its own; it only has an area. By 2004, Rovelli saw this:

> Intuitively, the grains of space are separated by "quanta of area."[1451]

So, each standard-model particle is made of six quanta of area? Whether or not we call this composite, at least we can no longer blithely call it elementary.

In this way—even without "a final theory of force and matter"—Weinberg's question finds "a final answer." There are no elementary particles; there is and always will be no such thing.

And we can conclude with confidence there never will be some smaller or simpler sub-twists that compose the tweedle; it is, I'd argue, irreducible by virtue of its primitive and wholly relational nature.

Likewise, though it may not be exactly what Greek thinkers had in mind, the Calabi-Yau manifold is the ultimate uncuttable entity. Given the variety of compositions that have, over the centuries, masqueraded as the *atom*, maybe it's not too much of a stretch to call the space quantum the *final atom*.[b]

I like to think Democritus (fig. 74)—who, speaking of the world of difference between what senses experience and what he believed exists, said . . .

Fig. 74. Democritus

---

a Spin foam models (don't go there).
b See also chapter 58 at page 312.

[B]y convention sweet and by convention bitter, by convention hot, by convention cold, by convention color; but in reality atoms and void.[1452]

... and so got the fundamentals half right long before the rest of us—would be well pleased.

88
FUDGE IT

> The shell game we play ... is technically called "renormalization." But no matter how clever the word, it is still what I would call a dippy process!
>
> Richard Feynman (1985)[1453]

Assuming space is continuous leads physics into fertile fields that are far from reality. This leads in turn to artificial problems.

Emblematic of such problems is the physics fudge known as *renormalization*. It is long-standing and successful if you want to reach a number that may work. But it fails to reveal—indeed it totally conceals—what's really going on. And it has a credibility problem.

The fudge works something like this. You do a calculation. It is the sum of many numbers. Some of them turn out to be infinite, so the sum is infinite. To get a finite answer you throw out all the infinities. You tally up the numbers that remain.

| Throw out infinity so little is left.

In 1963, Nobel Prize–winning physicist Paul Dirac said,

> It seems to be quite impossible to put this theory on a mathematically sound basis.[1454]

And, later,

> Sensible mathematics involves neglecting a quantity when it is small—not neglecting it just because it is infinitely great and you do not want it![1455]

Feynman added to his "dippy process" comment more soberly,

> I suspect that renormalization is not mathematically legitimate.[1456]

More recently, the process has become a little more sophisticated.[1457] Doubtless it is useful. But if not the worst of the inconsistencies that plague our physics it may be the most notorious. As we see, it made famous physicists like Dirac and Feynman cringe.

The new ontology dictates switching to a granular space; this has a serious downside of needing new math. A compensating upside is *that* math—that is to say, real math—will certainly be digital with none of those infinities.[a]

For physicists this won't be easy. Getting there means ditching all they learned through many years of study. It means starting over.

Perhaps more realistically, it means recruiting students. But let's look on the bright side. Throughout history physicists have blazed new trails by picking up new math.[1458]

And for most research physicists the pot of gold at the end of their rainbow is *new physics*. Fudge-free calculations may turn out to pave their way.

---

a See chapters 77, 104, 108, and 110 (at pages 395, 515, 535, and 543, respectively).

89
A SUITABLE SPACE

> We do not know how time and space
> behave at very small scale.
>
> Carlo Rovelli (2017)[1459]

It is worth saying plainly. The Planck-scale world is real. String theory's smallest possible entity[1460]—a space quantum—is real. It exists. Its relations to its neighbors exist. And so it is the key to understanding the small-scale structure of space.

Thus, the Planck volume is real.[a] The Planck area is real.[b] They are entities that are expressed in the universe.[c]

By contrast, the Planck length is *not* real. It is a number one can find by dividing the Planck volume by the Planck area, or by taking the cube root of the former or square root of the latter.[d] It is a useful idea. But it is not something that is expressed in the universe. So, we are trying to envisage space where volume and area are meaningful while length and location are not.[e]

> It's hard to comprehend space comes in pieces.

I confess to trying at one time to make a model of the Planck-scale world, showing how linked quanta could fit together. It was an ill-starred enterprise. I was trying to construct granular space *in* continuous space with a preconception granular arrangement should reflect some sort of structure.

---

a See chapter 53 at page 291.
b See chapter 41 at page 228.
c See chapters 13 and 14 at pages 76 and 87.
d Equivalently, one may calculate it, as Planck did, as the square root of the Planck constant times the gravitational constant, divided by 2π times the velocity of light cubed.
e See chapter 62 at page 326.

Another such approach tries to model granular space's structure after the fashion of natural foams with bubbles that have equal volume. Foams already had a growing literature when, in 1887, Lord Kelvin posed what came to be known as the Kelvin problem: What is the structure of a foam with equal bubble-volumes that has the smallest total surface area?[1461] He proposed an answer.[1462] More than a hundred years later physicist Denis Weaire and student Robert Phelan came up with a better answer.[1463] No one has yet beaten that.[a]

Of course, we see bubble foams *in* quasi-continuous space. They hold no lessons for us about Planck-scale space quanta. Quantum space has no such structure. In the new ontology its structure consists only in the topological relations—that is, the links—of manifolds.

It is surely not at all like Wheeler's quantum foam[b] (which also is conceived *in* continuous space or spacetime; e.g., fig. 75). Its intrinsic energy density (or calculated cosmological constant) is absurdly higher than the value we observe, which then leads to wild theories trying to obliterate it. Physicist Steven Carlip described the problem:

> We do, in fact, observe an accelerated expansion of the Universe that could be due to a cosmological constant. But a Planck-scale cosmological constant is some 120 orders of magnitude too large, making it what has been called "the worst theoretical prediction in the history of physics."[1464]

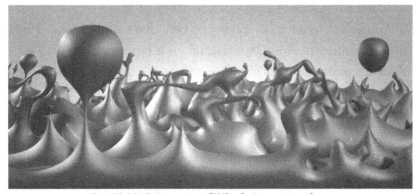

Fig. 75. *NASA's version of Wheeler's quantum foam*

---

a Nor has anyone proved it has the smallest possible area.
b See chapter 14 at page 87.

As Hossenfelder explained, it was neither one hundred twenty orders of magnitude too large nor the worst prediction.[1465] But it does confirm there is a problem with our understanding. Peebles said,

> It was and remains profoundly puzzling, a "dirty little secret" of cosmology.[1466]

In decades past, some came to think about the small-scale structure of space (or spacetime) in terms of string theory and quantum theory's uncertainty principle.[1467] Of course, that view too was set in continuous space and time.[a]

In that same time frame, Smolin espoused a "*lattice*" view, which he saw as pointing to a discrete kind of space. (He conceived this as an alternative to viewing space as a strange kind of superconductor.[1468]) He optimistically said,

> [I]f we put together the key ideas and discoveries [from quantum gravity approaches], a definite picture emerges of what the world is like on the Planck scale.[1469]

A key piece of his "definite picture" is that in some of those quantum gravity approaches space comes in Planck-sized quanta.[1470] But they do not explain how or why.

These examples illustrate Rovelli's recent we-don't-know take on small-scale space in the epigram above. Theoretical physics has long been at a loss to describe space at the Planck scale and seems still to be making little headway.

Observations at that scale are even more problematic. Physicist Giovanni Amelino-Camelia and colleagues proposed a way to observe a possible spread of pulses from very-high-energy gamma-ray bursters arriving from far across the universe to probe the Planck-scale structure of space, *if* that structure conforms to some versions of quantum gravity.[1471] Two decades later, that slim possibility remains unrealized.[1472]

Though we can't see the structure of space, a gedanken-experimental trip may offer us some conceptual insight.

---

a Compare with the Planck-scale view; see chapter 42 at page 236.

We shall once more require Emerson's eyeball.[a] Changing nothing, missing nothing, you can freeze-frame and inspect space at Planck scale. You can see space quanta do exist. You can, that is, see they are there, but cannot see where *there* is. There's no lattice; there's no reference frame. The best resolution of your eyeball lets you count quanta.[b] That is, frame frozen, you can distinguish one from none or two. You see nothing smaller as there *is* nothing smaller there to see.

The next frame is not quite the same, without your being able to compare. Twist states perform a kind of random walk, in one-link steps. You cannot specify what changed.

You expect the space quanta to be linked in the appropriate relationship from a *topological* perspective. But you cannot identify locations; it's a concept without meaning.[c] And so, paging through frames, there's no meaning to be given to the motion of the quanta. Your eyeball view is *not* the arty roiling foam of fig. 75, on page 445.

What roils, then, is the twisting (and the braiding) of the links and the corresponding flitting in and out of existence of all kinds of exotic particles, manifest at larger scale, maybe lasting just a step or two, fleeting realities that don't rise to the level of observable events. This, then, is our space at the smallest scale that has a meaning. It is teasing, seeming as if it should have some sort of structure.

Emerson's transparent eyeball shows you that it never quite deserves the word.

---

a See chapter 5 at page 43.
b Actually "seeing" at this scale would require photons of inconceivably high energy.
c See chapter 62 at page 326.

90
THAT'S NOTHING

> In any more advanced theory for Planck length physics, the definition of what exactly the vacuum state is, will have to require special attention.
>
> Gerard 't Hooft (2009)[1473]

One could rephrase 't Hooft as saying, *We should figure out what nothing really is.*

It may seem odd that many serious authors have written about *nothing*.[1474] Readers should be clear (as not all authors are) about what kind of nothing is in contemplation.

Some thinkers about nothing think it equals *empty space*. That's a concept without meaning, with space, as we have seen, being chock-full of all the Planck-scale quanta that make up the universe.[a]

We are all used to flinging the word *nothing* carelessly in all directions. Those who take more pains with it, who try to think about its actuality, conclude it leads to endless problems. The new ontology reveals a reason: Nothing is a pseudo-something—like a one-word oxymoron—that does not exist.

> It's difficult to achieve nothing and our universe can't do it.

It might seem like a simple, well-defined idea: Enclose a volume, empty it, you have a vacuum, free of matter, even nearly free of energy if the walls are kept very cold; surely it should be almost absolutely void. It seems simple but it is extremely complex; it sounds right but it is wrong. Its seeming simplicity misled generations of physicists (me included) to imagine it is real.

It's not only impossible to create this emptiness in practice (I did try). It is a nonexistent thing *in principle*. One hundred percent sci-fi.

---

[a] Which is to say, full of itself; see chapter 67 at page 348.

Sixty years ago, I was boiling high-purity lead to make an atomically clean single-crystal surface and needed a high vacuum so the surface atoms would not react with the atmosphere. This meant a very fast vacuum pump with a low ultimate pressure. So, with help from a great workshop, I built a cryogenic pump, the heart of which was a liter of liquid helium in stainless steel at four degrees above absolute zero.

At 4 K, gas molecules (except maybe helium) don't ricochet; they stick.[1475] The pump worked to spec.[a]

But even though essentially all gas molecules in that top chamber (fig. 76) were on a one-way trip to the pump chamber bottom-right, the top one held not vacuum but a thin gas because it was constantly replenished by outgassing from the inside surfaces of glass and metal.[1476] It was surely not as empty as is interstellar space. And even that's not empty.[1477]

Fig. 76. My 1960s high-vacuum system

It was good enough to serve my purposes, and bad enough to make the point that emptying a volume may be easily said but it is never done.

Yet we can imagine it. Of this, philosopher Bede Rundle cautioned,

---

[a] It reached some fifty-trillion-fold less than one atmosphere.

> Attempts to think away everything amount to envisaging a region of space which has been evacuated of its every occupant, an exercise which gives no more substance to the possibility of there being nothing than does envisaging an empty cupboard.[1478]

Nonetheless, the *concept* of that vacuum has remained lodged firmly in the minds of both philosophers and physicists since their twin avocations came to be. It has an extensive literature—much written about nothing, it might seem—that we will no more than skim to get a sense of how the concept came to its present condition.

Democritus—to whom we assign credit for those atoms—needed a good place to put them. Requoting him, the other half of his reality was nothingness:

> . . . but in reality atoms and void.[a]

This was, of course, an exercise in pure imagination. Lawyer and philosopher Pinhas Ben-Zvi observed that, over a span of some two thousand years,

> Some of the greatest minds of all times like Plato, Democritus, Aquinas and Leibniz had no trouble to "represent to themselves" a concept of nothingness which means absence of space as well.[1479]

What these great minds had in common was a vague notion that—shades of current physics and the new ontology—space is something and so *its* absence is needed for there to be nothing.

Following the work of physicist and mathematician Evangelista Torricelli in the mid–sixteen hundreds, the vacuum concept was refined and the notion it is nothingness was reinforced.[1480]

In 1783, having over the years expressed other views,[1481] Kant said there's no such nothing, although his conclusion sounded like a perfect void that (like my vacuum) suffered from practical imperfection:

> For . . . between every degree of occupied space and of totally void space, diminishing degrees can be con-

---

a Democritus, note 1452.

ceived.... Hence there is no perception that can prove an absolute absence of it.[1482]

Fast-forward and it has become a basic reference, the *vacuum case*.[a] In quantum field theories, for example, it means all components of the stress-energy tensor must be zero.[1483] But it is chock-full of quantum action.

Quantum theorists have their reasons for insisting there are some things in their nothings. This led them to spoil the empty fun. They made a prediction that led to a test.

Even a perfectly emptied vessel, quantum theory said, will not be empty. Not even close. It will be filled with virtual particles that pop from nothing into a fleeting but consequential existence and then pop back again.[1484] Physical chemist Walther Nernst pointed this out in 1916.[1485] Physicist Ahmed Almheiri recently said,

> The emptiness of the vacuum in quantum theory belies a sea of particles ... that conspire to make empty space feel empty.[1486]

The theory says the sea of particles has mass. The theory says how much. More recently we've had the means to measure it and confirm the theory's prediction.[1487]

The problem comes when we compare it with the value of the vacuum energy based on the cosmological constant, Λ.[b] The resulting clash is called the *vacuum catastrophe*.

Physicist Michael Hobson and colleagues wrote of the quantum theory prediction,

> This gives an answer about 120 orders of magnitude higher than the upper limits on [the vacuum energy] set by cosmological observations.... Nobody knows how to make sense out of this result.[1488]

---

a A notional region of space and time where the energy may be defined as zero, though the situation is more complex; for example, that energy may be equated with the not-quite-zero cosmological constant.
b See chapter 29 at page 165.

We've seen this before.[a] Surely (even though it's not quite right[b]) this says there's something to this nothing that we do not understand.

A false assumption is the source of the catastrophe. As we've seen, the standard model has the universe expanding due to finely balanced antigravity represented by that constant, Λ; it's the knife edge back again.[1489] The new ontology says this is not what is happening.[c] Rather it is driven by inexorable replication of quanta of space.

Thus there is *no* nothing in the universe—not anywhere, not even a Planck volume of it. Quite aside from wayward molecules and quantum fluctuations and imaginary points and fields, even so-called laws of physics, space (despite its empty-sounding word) is never empty.[d] Where there is no matter, it is always fully filled with that most massive object in the universe, the space itself.[e] The catastrophe results from treating space as something other than it is.

This clash of concepts has far-reaching consequences for various fields of human inquiry including logic and physics.

For example, the specter of emptiness hovers over logic. In his influential 1950 text, *Methods of Logic*, Quine explored how the validity of logical statements about a "universe" depends on the chosen universe not being empty. (Here, "universe" means merely some place chosen for consideration.)

> [T]he concept of validity has been left dependent upon the choice of universe of discourse. . . . Some schemata fail . . . when the universe is construed as empty.[1490]

The astute reader may hear echoes of the convoluted problems math and physics find with zeroes.

And those problems reach back to the real universe and Russell, who toyed with their existential significance for logic in arithmetic:

---

a In chapter 89 at page 444; see also note 1465.
b See Hossenfelder's comment in the preceding chapter and note 1465.
c See chapter 22 at page 133.
d There *is* a small industry working with spaces that are assumed to have topological holes in them.
e And where there *is* matter there too there is space.

> There does not even seem any logical necessity why there should be even one individual—why, in fact, there should be any world at all. [Footnoted:] The primitive propositions in *Principia Mathematica* are such as to allow the inference that at least one individual exists.[a] But I now view this as a defect in logical purity.[1491]

In other words, logic is unavoidably involved with existential questions like, *What is?*[b] And has deep problems with its counter-question, *What is not?*

Like Russell, Quine left these questions with a sense of logical compromise:

> It behooves us therefore to put aside the one relatively inutile case of the empty universe, so as not to cut ourselves off from laws applicable in all other cases.[1492]

How much more suitable to the requirements of logic might it be to rely (as Maddy would have us do[c]) upon a scientific ontological commitment?[d] To wit, in a universe made of space quanta, there is no basis for even an imaginary empty "universe."

---

a By "individual" he meant an item that could be counted.
b See further in part 6.
c See Maddy, note 1332.
d But then we would need our science to pursue reality.

91
SOMETHING FOR NOTHING

> The idea of getting something out of nothing may sound absurd, but absurdity is not the worst allegation made against quantum mechanics.
>
> Ahmed Almheiri (2022)[1493]

Almheiri was concerned with getting little bits of energy out of a black hole, the so-called Hawking radiation,[a] not with making an entire universe. But Hawking had long thought about that too. How could the universe begin from nothing, without even time beginning? He had a thing about it.[1494]

He was not alone. For example, there was physicist Lawrence Krauss who wrote a book, *A Universe from Nothing*, claiming,

> [W]e all, literally emerged from quantum nothingness.[1495]

Philosopher David Albert, seeking—with acerbic pen[1496]—to set some of its confusion straight, said,

> Where, for starters, are the laws of quantum mechanics themselves supposed to have come from? Krauss is more or less upfront, as it turns out, about not having a clue about that.[1497]

Ouch.

Some physicists would fall for almost any crazy story, seems to me, if it would help them think about the universe beginning without calling for an ontological commitment they must take on faith.

Yet Hawking had his own faith. He believed in nothing. His belief in it ran counter to what Sartre, long the chief curator of our personal ontologies, had cautioned,

---

[a] See chapter 68 at page 351.

[W]e must be careful never to posit nothingness as an original abyss from which being arose.[1498]

So, Hawking co-wrote (with author and physicist Leonard Mlodinow) a book that said the universe began with nothing. M-theory, they said, says it was so.

M-theory? It's not even a theory. It's a grab bag of *five* string theories that are themselves not real theories.[1499] It is math run amok. It cries out for Nobelist Wolfgang Pauli's perfect put-down:

> **Those who say it all came from nothing don't know nothing.**

It's not even wrong.[1500]

Of their something-from-nothing story, physicist and mathematician Peter Woit—who wrote the book[1501] on *that* short sentence—said,

> If you're the sort who wants to go to battle in the science/religion wars, why you would choose to take up such a dubious weapon as M-theory mystifies me.[1502]

Woit noted M-theory does not meet even one of the authors' own four requirements for a good physical model. He quoted them as saying,

> People are still trying to decipher the nature of M-theory, but that may not be possible.[1503]

He cited Hamish Johnston—online editor of *Physics World*—who said,

> There is just one tiny problem with all this—there is currently little experimental evidence to back up M-theory. In other words, a leading scientist is making a sweeping public statement on the existence of God based on his faith in an unsubstantiated theory.[1504]

He quoted with evident approval Horgan's saying that he thought of Hawking as "a cosmic comic performance artist."[a]

Was Woit too hard on Mlodinow and Hawking? With M-theory in hand, they claimed the universe created itself out of nothing:

> Spontaneous creation is the reason there is something rather than nothing, why the universe exists.[1505]

---

a I concur with Woit's take on *The Grand Design*. Some of it is so awful it seems Hawking must have been, like Banksy, just having us all on.

With multiverse in other hand, they said that this explains—if this is explanation—why this universe is perfectly contrived for us, without recourse to a creator. It's our old friend the anthropic principle, of which Horgan cheerfully observed,[a]

> The anthropic principle has always struck me as so dumb that I can't understand why anyone takes it seriously.[1506]

First prize for the most telling shot at universe-from-nothing physics goes to talk show host Larry King who, interviewing Mlodinow about the book, astutely asked,[b]

> Where did the nothing come from?[1507]

Such nothing-nonsense may seem not to warrant the attention it was getting. But physics-fun-and-games aside, this kind of glib misinformation tends to spread unnoticed like a backlot bushfire.

The final insult from Mlodinow and Hawking was their telling us we'll need five theories to describe the universe.

Five? The central aim of fundamental physics is to find *one* theory because two are too damn many.[c]

---

a I agree with this view too.
b The less than enlightening reply: "When [physicists] started to look deeper, a lot of counterintuitive and hard to understand concepts came up which can be understood and described with mathematics. One of them is this concept of nothingness. According to quantum theory, there is no such thing as nothingness. You can have nothingness in quantum theory. But from that, things will arise." See note 1507.
c See chapter 26 at page 150.

92
HIGGLEDY-PIGGLEDY

> There has been quite a history of "observed" new particles; research groups at colliders would report the discovery of a new particle, and subsequent data would eventually show that these new particles did not exist after all.
>
> John Moffat (2014)[1508]

The biggest news (the only big news?) in the particle-physics world in the last decade was the announced detection of a new particle, called the *Higgs boson*, in 2012 at the Large Hadron Collider.[a]

No question it was news. But was it a particle? I was a skeptic then. And so far I still am.

The Higgs has for fifty years been a prediction of a body of theory that, strangely, has roots in superconductivity[1509] and its quantum cousin, superfluidity.[1510]

In mid-2011, my publisher, who had a long career in marketing, was planning a year ahead for the release of *Time One*. He thought the fundamental-physics world would greet it with cries of glee (I did my best to disabuse him). He asked if I could foresee what that world might look like by mid-2012.

> Higgs if necessary but not necessarily Higgs.

I said they would discover the Higgs boson.

"How could you know that?"

"Ten billion dollars in funding up for renewal; twenty-some countries must agree; none of the supersymmetric particles their big machine was built to find. They need the Higgs."

---

[a] See fig. 78, showing the CERN data supporting the Higgs discovery.

"What would it take?"

"A few hundred out of a few quadrillion collisions, looking like a particle that promptly vanished. Messy pseudo-statistical analysis of a vast amount of data on hard drives spread around the world."

"So why do you say they'll find it?"

"They could find it or not find it; what do you think they'll do?"

"How do you know that it will be the Higgs?"

"They could call it Smith; what do you think they'll call it?'

You may see this as cynical. So, I think, did he. But he was a bit taken aback next year when the world heard about the Higgs.

I'm not the only one who's from Missouri. Nor the only one who doesn't like to rain on a parade.

For a moment let's hark back to Moffat, who had built his own particle theory. One of its virtues is it needs no renormalization.[a] Another (could it be a vice?): It has no Higgs.

As he said (in the epigram above), mistaken claims of new particles are not new. He cited some notorious examples. One was an earlier detection of the Higgs:

> [I]n 1984, the experimental group working . . . at SLAC announced the discovery of the Higgs boson.[1511]

The data in support of that discredited discovery[1512] look (as Moffat noted) "uncannily similar" to those for the revival of the Higgs.[b] (Compare figs. 77 and 78, top to top and bottom to bottom.)

It's a formidable problem with data-analysis that is inherently arcane. Moffat said,

> Because the Higgs is so short-lived, we cannot detect it "directly" at the LHC. Its presence can only be inferred by observing the lower-energy particles that it can decay into. . . . Detecting a Higgs-like signal within the large backgrounds is like the proverbial search for a needle in a haystack.[1513]

---

a See chapter 88 at page 444.
b As the error bars show, the LHC data have a higher level of significance.

Part IV: On the Way Out 459

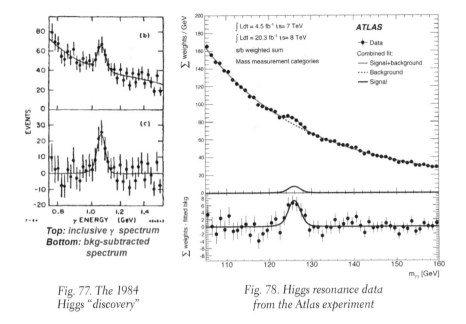

Fig. 77. The 1984 Higgs "discovery"

Fig. 78. Higgs resonance data from the Atlas experiment

The present circumstance is unprecedented. We are ten billion dollars and two Nobel Prizes[a] down the road.[1514] It will take much data and analysis, more money and great courage, to retract the second coming of the Higgs boson if ever there is need.

If, on the other hand, it is solidly confirmed as it may be by more data, Moffat's theory will stay on the shelf.[1515] And the new ontology may take a hit; it would need a new way for another fundamental but not elementary particle to form so the Bilson-Thompson model might lose some of its "extreme economy."[b]

Meanwhile we are left to echo particle physicist and cosmologist Stephon Alexander, who in 2022 said,

But the question is: What *is* the Higgs?[1516]

---

[a] The prizes were awarded for the theory, not the discovery, but of course the discovery mattered.
[b] But his model is in a sense open-ended in that more kinds of particles may be made. For example, he said, "We can represent higher generation fermions by allowing the three helons to cross in more complicated patterns." See note 762.

93
A SINGULAR DIFFERENCE

> Of all the entities I have encountered in my life in physics, none approaches the black hole in fascination. And none, I think, is a more important constituent of this universe we call home.
>
> John Wheeler (1998)[1517]

The new worldview will come to concur with Wheeler's thought on the black hole.[a] All the more important that we grasp just what it is—and (here we tackle) what it isn't.

Picture this if you can: A giant star runs out of fuel. Its internal heat can no longer hold up its weight so gravity takes over. It collapses within seconds.[1518] Much of its enormous mass falls into its core at speeds maybe a tenth the speed of light. In the crunch, atoms collapse; their protons and electrons are crushed into neutrons. Then the neutrons collapse too. So do their quarks and gluons. A vast blast of detritus is ejected by the violence. What is left is a black hole.[b]

> One may squash empty space but not space quanta.

Penrose shared the 2020 Nobel Prize for proving general relativity requires that black hole to contain a singularity[1519]—its mass packed down into a point of zero size.[1520]

The supposed existence (we don't get to see it) of that singularity is entirely a consequence of our assuming space to be continuous. But what if space is made of quanta?

The first thing that is clear is there would be no singularities, not in black holes, not anywhere, not ever. Quantum space cannot have points of zero size. This offers some relief for physicists dividing

---
a See chapter 102 at page 499.
b Core-collapse is not the only kind of supernova.

masses (finite) by their volumes (zero) to get infinite densities. Few thought that made sense.

If our giant star had been a little smaller, then its gravity might not have been enough to crush the neutrons. The result would be an extremely hot neutron star.[1521] (Some call its matter *neutronium*.[a]) Once cooled it would be hard to find.[1522]

If you could travel to a neutron star, you would be on a cold fast-spinning world (up to 40,000 rpm,[1523] at which any auto engine would rapidly self-destruct), no bigger than Los Angeles. You would not survive its gravity; your weight would be about ten billion tons; even your atoms would collapse, leaving a surface skim of (mostly) neutrons.[1524] Neutronium is strong enough to stop you and the neutron star from falling further, but only just.

We have direct evidence of neutron stars. We've even seen the unique gravity-wave signature of their demise when two collide—in one of the universe's most awesome, space-quaking events.[1525] Most of their matter collapses into a black hole. LIGO detected one of these collisions, and then led others to examine it with telescopes. The outflung debris included a few hundred Earth-masses of freshly minted gold.[1526]

Back to the black hole. Would it be different in quantum space? Its gravity is near enough the same as its continuous-space cousin's. It too crushes even neutronium. What's to stop it from collapsing to a point?

The new ontology leads to a kind of matter that is much more crush resistant than neutronium. It has no name so let us call it *ultramatter.*

Recall, at Planck scale, matter is made of half-twists in the links between space quanta. Even in neutronium, those twists are "far" apart.[b] As neutronium collapses, its twists must be forced closer. Once they get to be next-neighbors, they have nowhere else to go.[c] Simplest conclusion is the black hole's matter stops collapsing.

---

a The term is not widely used in scientific literature as it is not a well-defined material; for example, its structure and composition must vary widely from the center to the surface of a neutron star, and from the poles to the equator of a fast-spinning neutron star, whose surface at the equator may move with speed comparable with the speed of light.
b They would on average be millions of space quanta apart.
c They may well resist getting that close; more new physics needed.

This is pure speculation but, in this way, we can understand how a black hole could come to be without that singularity. In its stead would be an ultramatter core of finite density, where all the links are twisted.[a] And, by the way, this offers an elegant answer to the vexing problem of *Where does a black hole's vast entropy reside?*[1527]

We can investigate that ultramatter core using simple arithmetic.[b]

Let's pack those two colliding neutron stars into their black hole and figure out what size its matter would be with all links twisted.

A neutron star typically has one-third more mass than the Sun's 2.0×10^{30} kg.[1528] So, the mass of one neutron star is 2.7×10^{30} kg and the mass of our two colliding neutron stars is 5.4×10^{30} kg.

Let's keep it simple and assume all their mass fell into the black hole (in practice, some of it—such as that gold—got blown away by the sheer violence of the collision).[1529] Next let's figure out the number of neutrons. The mass of one neutron is some 1.7×10^{-27} kg. So, we divide the mass of the two stars by the mass of one neutron to get the number of neutrons; it comes to 3.2×10^{57}.

That's a lot of neutrons, but then it was a lot of mass and, if it were ordinary star matter, it would have been bigger than our Sun, with diameter perhaps two million kilometers. Packed by the black hole's crushing gravity, it would be a lot smaller. How small?

Each neutron is made of three quarks. Each quark is made of six twists. So, there are eighteen per neutron. We had 5.8×10^{58} twists. There can be up to three twists per space quantum.[c] Thus, we would need a sphere of at least 1.9×10^{58} space quanta to house all of the crushed neutrons.

Last question, then, is how big that sphere of ultramatter would be. The volume of each quantum (the Planck volume, equal to the cube of the Planck length[1530]) is 4.2×10^{-105} cubic meters. Multiply this by the number of space quanta we just calculated and we get 8.2×10^{-47} cubic meters.

---

a Or maybe only many of the links.
b It involves some numbers that are very small or very big, so we'll use scientific notation, as before; see chapter 69 at page 357.
c Each twist is associated with two quanta. You may say I have no way to show the twists don't double up—two twists per link, a kind of super-tweedle—or cancel out; if you want to so assume, please be my guest, but I'm not going there.

It certainly sounds small, but it's not easy to imagine. We can use the formula for the volume of a sphere[a] to find our ball of compressed neutrons is 5.4×10^{-16} meters in diameter. Compare this with the diameter of an ordinary neutron, 16.0×10^{-16} meters.

In other words, in the black hole the matter of two neutron stars could crush down to an ultramatter sphere one-third the diameter of a single neutron. But no further. It is tiny—a lot smaller than an atom—but it's *not* a singularity.

Ultramatter is difficult to envision. But it's not as difficult as infinite density. Meanwhile, those unsingular black holes may need new physics.

---

[a] $4\pi r^3/3$.

PART V

THE EMERGING PICTURE

> Problems which were raised by Plato and Aristotle in the fourth century BC are still discussed, and the work of all the intervening centuries has brought us no nearer to finding a solution of them.
> Alfred Ayer (1982)[1531]

94
IN THE BEGINNING

> The vainest dreams of those who think the world
> could have been created by some fortuitous chance,
> or by some necessity of fate, or that it existed by itself
> from eternity, consisting of its necessary laws, are
> utterly destroyed.[a]
>
> Rogerio Boscovich (1763)[1532]

Book One of this work[b] follows one aspect of the new ontology, its depiction of how the universe began. It is written as an arc of personal discovery by a fictional narrator with a view to easing readers into new words and new ways of thinking. As reviewer Robert Leverton wrote,

> This book is about spoon feeding advanced physics to the everyday person. It completely skips the math (sweet!) and talks about the strengths and weaknesses of Quantum Mechanics, String Theory, the Calabi Yau Manifold and many other interesting physics topics and even non-physics topics to paint a picture of the beginning and I mean the beginning.[1533]

The trade-off was the book's narrator's questing testing of ideas never set out shortly what the new ontology says of the beginning. Nor have I so far done so in these chapters. So, shared language in hand,[c] here it is.

---

[a] Rogerio Boscovich was a notable physicist, astronomer, mathematician, philosopher, diplomat, and poet. He was also—as would be anyone with such credentials in his day—a theologian. In his famous work on natural philosophy, he argues the world must have been created.
[b] *Time One*; note 16.
[c] See chapter 72 at page 373.

The universe began about fourteen billion years ago.[a] As polymath Rogerio Boscovich noted (above), more than two hundred years ago, extant answers had their problems. The new ontology offers an entirely different answer.

Initially there was no space,[b] there was no time, there was one Planck-sized[c] Calabi-Yau manifold, with six dimensions. It had the property of volume.[d]

So far as we can ever tell, there was not even nothing else. It was a state—physics would later say—of perfect order.[e]

The universe began when that manifold tunneled into two manifolds, or what we will call space quanta, each identical with the original. I call this first iteration of the universe *step one*.

The two quanta were linked by two of their dimensions. Each had the property of area. The "size"[f] of each (whether conceived as a window or ribbon-like link), physics would later indicate, was the Planck area.

> This is the short story of how the universe began.

The two linked quanta were in a state of extreme curvature or what our physics calls gravity.[g]

By its nature, it was finite and closed.[h]

The two quanta tunneled into being four. This was step two.

The four quanta were each linked to the others.[i] There were eight links.

The four quanta tunneled into being eight, with twenty-four links; this was step three.

One can see this as a proto-three-dimensional space. It shows why and how what physics calls *space* is emergent: It was at first crudely granular and came to seem something like continuous only after there were very many quanta.

---

a See notes 19, 1287, and 1297. Real physics may revisit the calculation.
b And so, no gravity.
c But there was no metric by which one could say what its size was.
d See chapter 15 at page 95.
e That is, no entropy; see chapter 23 at page 137.
f This is a conceptual assignment rather than a measurement.
g See chapter 20 at page 122.
h It would continue to be that way.
i See fig. 38 at page 239.

At step three, *time* was not yet meaningful. But our physics could later say (pretending to have been there with an impossibly precise clock) that step three corresponded to three Planck times.

After step three, the topology of linking became more complex, but there were always three new links per new quantum.[a]

Exponential (doubling) growth continued. By step ten, the tiny infant universe was expanding faster than the speed of light.[b]

The number of manifolds continued to double at every step until the universe grew large enough to ease the extreme curvature that gave rise to virtual certainty of tunneling.

That easing off began when the universe was maybe a light-minute or even a light-century[c] in size. By then, there had not been time for light to move even an atom-sized distance. Physics would later say the elapsed time was less than 10^{-40} second. This is the awesome power of exponential growth.[d]

Twists in links arising from chaotic replication gave rise to particles and extreme densities of energy and matter;[1534] if there had been time for gravity to take effect—it acts at the speed of light—the universe would be only a huge black hole.[e]

In the absence of symmetry, there were more particles than antiparticles;[f] those that met their match annihilated; the universe became awash in high-energy photons.

The uncanceled twists settled into standard theory particles (and maybe more besides)[g] in random clumps.[h]

In the first microseconds (by later clocks), clumps were collapsing into a horde of primordial black holes in an increasing range of sizes as expansion slowed.[i]

---

a That is, six links per quantum, each shared with a neighboring quantum.
b That speed increased by a factor two every three steps.
c I.e., twenty million to one quadrillion km—a wide latitude.
d See chapter 40 at page 222.
e See chapter 69 at page 357.
f See chapter 71 at page 367.
g See chapter 43 at page 243.
h See chapter 40 at page 222.
i See chapter 69 at page 357.

At the one-second mark the universe—far larger still—was replicating far less fast. Its twists had settled into particles. Its quarks were tightly tucking into protons and neutrons. It was still very hot, a state of matter called a plasma: Electrons were too hot for the protons to catch.

Around 370,000 years later, the plasma cooled enough for the protons and electrons to condense as neutral hydrogen.[a]

Aside from those primordial black holes, which played a role whose results we can see today,[b] the universe set about its business much as the standard model of cosmology describes.

Those black holes got down to their business right away—long before the hundred million years or more the standard model says would be needed to make them—on one hand compressing gas clouds to make stars and organize the early galaxies, on the other hand making more space and so taking on the task of keeping the expansion going.[c]

More than thirteen billion years later, Webb-telescope[d] astronomers—gazing back across that now expanded reach of space and its vista of time—would see those galaxies were there far sooner than the standard model could explain.

---

a The plasma blocked photons; the neutral hydrogen allowed photons to take flight; we see them—stretched and cooled—as the cosmic microwave background radiation; see chapter 17 at page 105.
b See below.
c See chapter 69 at page 357.
d See chapter 129 at page 617.

95
PROPERTY DEVELOPMENT

> It is surely a cultural deprivation to be unaware of the chain of events through which some mysterious genesis nearly 14 billion years ago triggered the emergence of atoms, galaxies, stars, and planets.
>
> Martin Rees (2009)[1535]

Emergence is an ordinary word for the process of coming into view or into existence. It also has the special meaning we use here: *Emergence is the arising of a new property from a multiplicity of things that lack that property.*

Readers will by now have got the drift but let's lay it squarely on the table. The universe is made of many space quanta; we have no way to detect properties of a single quantum; we can only detect very many quanta; and very many quanta can exhibit properties no single quantum has.

| The laws of physics are statistical descriptions.

We say such properties *emerge*.

In the new ontology, the properties of everything we see are all emergent.

Cosmologist and astronomer Martin Rees was speaking of emergence (above). He was not just saying atoms arose from protons, neutrons, and electrons and led to larger things, though that would be true. Nor did he mean that they arose from vast numbers of space quanta; that's true too, but he didn't know it.

He was speaking of the serial emergences that made a vast universe of atoms and led to life or, if you prefer, how astronomy gave rise to us.

Life is thus the result of emergences that span cosmology and physics and then chemistry to finish with ecology, becoming ever more complex along the way. But, let's not forget, they all are made of space.

Because an emergent property is the averaged result of sheer numbers its math is always an approximation. It can be a very good approximation when the numbers become very large. It can be such a good approximation that even our most precise measurements will seem to show it as exact.[a]

An example of emergence is the property of pressure. Think of a molecule of air (say nitrogen) confined in a box. It bops back and forth just like a 3-D version of Atari Pong,[1536] flying straight until it hits the sides, where surface molecules absorb its blows and give it kicks. There is no pressure as we understand it, only random impacts.

Now double the number of molecules. The rate of impacts doubles but the picture is the same.

If we keep doubling the number of molecules every second, in a minute or so the box will begin to experience pressure we might measure. (Soon after that it will explode.)

For an ideal gas the pressure in a fixed volume at a fixed temperature will increase in proportion to the number of molecules.[1537] But pressure is a property that has a useful meaning only when that number is extremely large, much more than millions.

That is, it is emergent.[1538]

This example speaks to the classical view of particle behavior, a view that excludes quantum considerations. There is also a quantum version of a particle in a box, well-known because we can solve its math exactly.[1539] Useful though each of these two views may be, they both conceal the real workings of our universe.

The real view of our molecule sees a thin wisp of twists, each of which jerks this way and that at light speed, with a bias in the jerks that drifts the wisp across the box; the box wall, too, looks like a mist of manic twists. The interactions of the wisp twists with the mist twists are probabilistic and complex.[1540]

---

a See chapter 74 at page 381.

It's not a useful view for almost any purpose except one. It reveals what is really going on. We then have opportunity to build on that. Maybe it will help us build physics that's totally new.

Both of physics' two main theories—relativity and quantum theory—are emergent.

From this perspective, the two sub-theories of relativity, special and general, may best be seen as statistical theories of motion and of space, respectively.[a]

Special relativity is a large-scale statistical theory of motion of objects including clocks in a local continuous Euclidean space—both motion *in* clocks that tells "time" and motion *of* clocks relative to a clock-watcher (aka relativistic observer). Its depictions of local motions of objects emerge from many random moves of many twist states in (local, and so almost flat) granular quantum space. Its light-speed limit sets the watching rules; it is a Planck-scale property of space.[b]

General relativity is a large-scale statistical theory of continuous Riemannian space (including its contents). Its depictions of motions emerge from many random moves of many twist states in curved granular quantum space.

Quantum theory is a large-scale statistical theory of particles and their motion in a local continuous Euclidean space. Its depictions of particles and their motions emerge from many random moves of many twist states in locally almost flat granular quantum space.

The difference between continuous space and granular space has numerically inconsequential but fundamentally profound implications.

For example, if space were truly continuous, all observations would be inaccurate. As Riemann said,

> [In] a continuous manifold, every determination from experience remains always inaccurate: be the probability ever so great that it is nearly exact.[1541]

---

a This is not how Einstein saw them; see chapter 99 at page 485.

b Special relativity is based on the same property at large scale; see chapter 50 at page 276. This is why special relativity—a theory based on watching moving rods and clocks—stumbled into the equivalence of mass and energy, a Planck-scale phenomenon; see chapter 43 at page 243.

By contrast,

> [In] a discrete manifold, the declarations of experience are indeed not quite certain, but still not inaccurate.[1542]

Once it is clear space itself is emergent and that particles are properties of space, the emergence of everything we see—including Rees's atoms, stars, and planets, including their large-scale properties—follows much as he would see it, with a few tweaks (like big primordial black holes) that would help him solve the biggest problems.

The history of the universe may be elegantly summarized in emergent terms: Space emerged from space quanta; all else emerged from space.

96
GET WITH THE METRIC

> It would appear that, on a sufficiently small scale, points cannot be identified either by their metric relationship to neighboring points or by local physical characteristics.
>
> Peter Bergmann (1968)[1543]

Physicist Peter Bergmann—Einstein's colleague who wrote a 1976 text on general relativity—was here writing about quantum gravity at the Planck scale. He continued, suggesting that . . .

> . . . in quantum gravity the concept of locality will undergo a fundamental revision.[1544]

We've seen reasons to conclude that he was right.[a]

The concept of locality is kissing cousin to the concept of measurement. At any scale, we say something is *there* using some measure of its separation from something that's somewhere else.

| Physics needs a ruler to study its space.

We have two requirements for the measure: It must be portable, and its size must be fixed. Such a measure is known as a metric.

Measurement in space is easy at our own scale in the world; indeed, our bodies long defined some common metrics.[1545] And even Neolithic societies could do geometry to millimetric accuracy over many meters.[1546]

Today one may lay some sort of ruler on whatever one would like to measure, just like Euclid never thinking that the space itself is curved.

A non-Euclidean space calls for a non-Euclidean metric. In 1854, Riemann set out a fundamental choice: If one finds space to be gran-

---
a See chapter 62 at page 326.

ular one may measure by counting granules, which are its natural metric; if one finds space to be continuous one must introduce an artificial metric from outside the space.[a]

As noted earlier, cosmology has long assumed the latter. Four famed cosmologists devised the artificial metric that's in general use.[b] It's a solution of the general relativity equations, and as such it is a product of imagination.[1547] It in effect defines the model of the universe that dominates cosmology. Space studies that employ it mostly fail to note they are not really studies of the space; they are just studies of the metric.

In cosmology, the metric is not merely the basis for measurement: It has physical significance; it is an essential ingredient of reality. Thus Wilczek could summarize (with—as is almost always the case—no mention of Riemann and his metric option),

> The primary ingredient of reality contains a metric field that gives space-time rigidity and causes gravity.[1548]

Until recently, *small-scale* meant the domain of atoms. There—amid confusion about underlying reasons[c]—the Planck constant ruled the roost for the last hundred years. Yet, despite all indications that the Planck volume and Planck area are fundamental,[d] they are not usually seen as having metric meaning. This reflects the choice to work in continuous space.

Riemann said the question of which space (and kind of metric) to use should be decided on the basis of facts:

> The answer to these questions can only be reached by starting from the conception of phenomena which has hitherto been justified by experience, and which Newton assumed as a foundation, and by making in this conception the successive changes required by facts which it cannot explain.[1549]

---

a See chapter 12 at page 74 and note 275.
b They were Aleksandr Friedmann, Georges Lemaître, Howard Robertson, and Arthur Walker; hence, the FLRW metric.
c See chapter 14 at page 87.
d See chapter 89 at page 444.

A half century later, Planck and the experimentalists whose data he digested did reveal facts which continuous space cannot explain.[a] We have yet to make the changes they require. In a quantum world, the threshold question is *What is the quantum?*[b] With a coherent answer that's concordant with those facts, we may find our way to a real metric.

Studies of the universe at very small or even very large scale—for which a real metric really matters—call for an ontological commitment to the answer to this question. Properties of the metric should be properties of the universe rather than being artificially defined. Only with a real metric can we explore reality rather than just our favorite fiction.

As we've seen, the heart of the metric problem becomes evident at Planck scale: Physics likes to measure distance.[c] Distance is a metric relation real space does not possess at Planck scale. Fundamental physics should count quanta of volume and area, the measures of real space.

We should heed Riemann's admonition:

> We are therefore quite at liberty to suppose that the metric relations of space in the infinitely small do not conform to the hypotheses of geometry; and we ought in fact to suppose it, if we can thereby obtain a simpler explanation of phenomena.[1550]

We can and must.

Real physics requires a real metric.

---

a See chapters 13 and 14 at pages 76 and 87.
b See chapter 45 at page 255.
c It can be even more problematic; in general relativity, the "distance" metric is a 4-D pseudo-distance in spacetime.

97
INLAWS AND OUTLAWS

> There is in fact a kind of chicken-and-egg problem
> with the universe and its laws. Which "came" first
> —the laws or the universe?
>
> Dennis Overbye (2007)[1551]

Where did the so-called laws of nature come from? Did laws of nature make the universe or did the universe make the laws? Science writer Dennis Overbye reviewed both sides. These questions underlie our understanding of what's going on so let's take a closer look at his chicken and his egg.

Physicists tend to think of physics laws as laws of nature. Most take the source of laws for granted. Implicitly they (almost always) treat them as having been there all along. The new ontology insists there hasn't always been a *there* for them to be.[a] So, what were the laws of physics at the instant that the universe began?

> **There were no laws of physics when the universe began.**

Lederman claimed the laws of physics were all in place . . . or maybe not:

> The laws of [physics[b]] must have existed before even time began in order for the beginning to happen. We say this, we believe it, but can we prove it? No.[1552]

We can see there is a reason why we cannot prove it. It is wrong.[c]

Some thinkers have proposed the laws were there when there was nothing and that indeed the universe was in some way created *by* those laws from nothing (or, rather, nothing else).[1553] Aside from our inability to give real meaning to *nothing*, the laws of which they

---

a See chapter 17 at page 105.
b He said *laws of nature*, but the context makes it clear that he meant *laws of physics*.
c See chapter 95 at page 471.

speak turn out to depend upon there being something, an internal contradiction of their proposition.[1554]

Some physicists with philosophic inclinations speak of physics' laws in terms of *place* of origin. Did they arise inside or outside the universe? *Outside* leads to bootless speculations such as lots of universes, each with different laws.[1555]

Smolin spoke of this in terms of place and time, saying (without favor),

> If the laws of physics are timeless, if they are true everywhere and for all time, any explanation of them must lie in something that is not in the universe.[1556]

The old ontology gives birth to febrile questions such as *How can laws arise inside the universe?* What kind of lawlessness would have been rampant before they arose?

Nonetheless, Dowker liked them *inside* in an intrinsic sense (another thought of hers that is spot on[a]):

> [I]t would be more satisfying if the laws themselves were, somehow, physically real; then the physical universe, meaning everything that exists, would be "self-governing" and not subject to laws imposed on it from outside.[1557]

Smolin too came to be of the *intrinsic* persuasion. Implicitly regarding the laws as emergent, he said,

> The single, unique universe must contain all its causes, and there is nothing outside of it. . . . [T]here are no immutable laws, timeless and external to the universe, which sometimes act as if from the outside to cause things to happen inside the universe. Instead, laws of nature must be fully part of the phenomena of nature.[1558]

Chemist and author Peter Atkins gave us a succinct "inside" view:

> There are no laws in a universe that does not yet exist, for laws come into existence as the behaviour they summarize emerges with the emerging universe.[1559]

---

a See chapter 19 at page 118.

All three (Dowker, Smolin, and Atkins) were saying three important things: Physics' laws came from within the universe; they arose *after* the universe began; and they arose as part of the natural order.

As we've seen, the new ontology supports all three propositions. It provides a more specific reason. It says one quantum (being) plus one law of nature (doing) made the universe. That first law we denote as $1 \to 2$. It conveys the universe's two essential elements: what it is; and what it does. You can't get much simpler than that.

It says what we call the laws of physics then *emerged* from the space quanta and the iterations of the universe as increasingly accurate descriptions of its being and doing in (so to speak) bulk.[a]

This is an archetypally Humean understanding.[b] For example, Carroll summarized,

> To the Humean, all that exists is the actual universe. . . . [T]he "laws of nature" are just helpful summaries of patterns we see. . . . Humeans think that laws simply describe the world, they don't govern it.[1560]

Here, again, language matters for our understanding. In his book on physics' building blocks, 't Hooft said,

> [T]he laws of Nature are complicated.[1561]

But he was speaking of the laws of physics. We have seen that the law of nature is amazingly simple.

Carroll's "laws of nature" (he too meant the laws of physics) describe consequences of the character of the universe's primitive elements—the quanta—and their link relationships. What we call *laws* emerged when the numbers of quanta (and, later, of composite emergent entities, like particles or molecules) grew large enough to make collective properties statistically evident.

So, we may safely say all laws of physics[c] are statistical and therefore inexact. We are accustomed to see some laws this way. For example, physicist Richard Fitzpatrick lectured that,

---

a See chapter 95 at page 471.
b That is, akin to the empirical philosophy of David Hume.
c Other than charge conservation, which is digital.

> We can easily appreciate that if we do statistics on a thermodynamic system containing 10^{24} particles then we are going to obtain results that are valid to incredible accuracy.... [W]e can forget that the results are statistical at all, and treat them as exact laws of physics.[1562]

Thus, emergence of behavior that the laws of physics describe is a real phenomenon. But the emergent laws of physics are themselves not real; they are descriptive statements we contrive for our convenience. We should be unsurprised to find we can contrive them in various versions.[1563]

They are not nature's; they are ours.

98
THE SESSILE SOURCE

> From the very beginning there has always been present the attempt to find a unifying theoretical basis for all these single sciences, consisting of a minimum of concepts and fundamental relationships, from which ... the single disciplines might be derived by logical process.
>
> Albert Einstein (1940)[1564]

With our broad answer—*space*—in hand we can dig deeper into the long-standing debate about *Where do laws of science come from?*[1565]

Einstein sought a single source for all the laws, the "logical foundation," as he put it,[1566] from which laws of science would all directly arise. He was plowing barren fields—continuous space and already emergent laws.

The new ontology offers the space quantum as the basis for this task. It says there was no physics (as we know it) in the beginning: There were no conservation laws; there was neither special relativity nor general, nor any quantum theory.[a]

> Laws of physics arise from the space quanta.

It was not merely that such laws did not exist (though that's true, they didn't). In the beginning, there was neither mass nor energy to be conserved, there was neither motion nor space to move in, there were neither subatomic particles nor atoms let alone large objects. So all the laws of physics that speak to matters of this kind had then and there no meaning whatsoever.

This alone would be enough to tell us that the laws of physics are emergent stories.

---
[a] Other than $1 \to 2$.

We can think this through a little further in light of our understanding of precisely what the laws of physics along with all else emerged *from*, namely, a single manifold and single law, *1 → 2*.[a]

There would seem to be no room to tweak this law. It has an ineffable quality. Does this mean there was only one way such a world could work out in the end? Only one set of laws that could emerge from it?

Not so. String theory's investigations of the Calabi-Yau manifold—the centerpiece of its math we have appropriated into our new ontology—show there are many ways that manifold's dimensions could have been arranged. Smolin said,

> There are literally tens of thousands of ways to curl up those six dimensions, and each one of them leads to the prediction of a different set of particles and interactions.[1567]

That is, our world could have emerged in any of so very many ways.[1568]

To be clear, Smolin was referring to work done in continuous space with string theory. By contrast, we are looking at granular space and there is no theory for it yet.[b] But, in light of the string theory result, it seems likely different physics with different laws could emerge from different ways to curl up those dimensions.

So, the simple answer, says the new ontology, to why the laws of physics are the way they are, rather than the many other ways they might as well have been, is, as the laws emerged from the sheer number of space quanta and their links and tweedles, they were shaped by the unique tangle in the manifold's dimensions.[c]

There are other, relatively messy answers to this question. Some of them strive to hide the elephant in their room: Why would laws lead to such a *suitable* universe?[d]

Thus, in 2021, seven physicists (Smolin among them) asked,

---

a See chapters 19 and 40 at pages 118 and 222.
b Here lies the next horizon for our exploration; here may be found new fundamental physics.
c One might regard the tangle as "the universe's DNA."
d See further, chapter 120 at page 580.

> [W]hy are these—and not others that seem equally consistent mathematically—the actual laws?[1569]

Disconcertingly, the authors went on to answer their own (perfectly good) question with a (to me wacky) neural-network model they said showed a universe could *learn* the laws of physics . . .

> . . . by exploring a landscape of possible laws, . . . without supervision.[1570]

The "without supervision" bit is of the essence here; some sort of supervisor is the elephant whom they hope will be (if I may mix my metaphors) missing in action.

Readers who don't feel the need to read that paper for themselves might instead check what physicist Tim Andersen said of it in a user-friendly review.[1571] Noting the "stellar academic credentials" of the authors, he went on to say,

> [I]t does not read like the scribbling of mystics nor the hasty product of an over enthusiastic Friday lunch conversation.[1572]

(That he saw the need to say this says it all.)

We should just bite the bullet. The laws of physics of this universe are plainly very special and must have arisen directly from the beginning.[a]

---

a See also chapter 120.

99
RELATIVELY UNFOUNDED

> When Einstein completed his theory, his own account of the foundations of the theory was adopted nearly universally. However, among the voices welcoming the new theory were small murmurs of dissent [that] are now some of the loudest voices of the continuing debate.
>
> John Norton (1993)[1573]

Understanding Einstein's theories of relativity has long been fraught, with their foundations sometimes seeming as obscure as those of quantum theory. With new ontologic insight of a kind that he long sought but lacked, what can we make of them?

> General relativity is approximate geometry.

Though he linked the theories with the label "relativity,"[a] the principle of relativity may be the least of what they are about.

His principle of relativity is an assumption that the laws of physics are the same in all unaccelerated coordinate systems.[1574] On its face this principle is not new physics.[1575]

Yet while special relativity began life as a theory about relatively-moving rods and clocks, it turned out to be about relations between mass and energy and local motion in a universe where (by virtue of the principle) the speed of light in space became a fundamental constant rather than merely an observed fact.[b]

That led to a whole lot of new physics! It's physics whose foundations our ontology portrays at Planck scale. That's a scale where Einstein's thinking never went.

---

a The linkage was his; but *relativity* was contrary to his preference for *invariance theory*; see Holton, note 210, at his page 6.
b See chapter 95 and note *b* at page 474.

And general relativity, which began life as extending the special theory to accelerating rods and clocks,[1576] turned out to be a theory of gravity and space.

Much more new physics! Same comment about its foundations and their scale.

If understanding this is starting to seem tricky, it's because—as many said—the meaning of the theory is mired in confusion:

> The voices of dissent proclaim that Einstein was mistaken over the fundamental ideas of his own theory and that the basic principles Einstein proposed are simply incompatible with his theory.[1577]

We could just shrug and say, like quantum theory, relativity is perfect twentieth-century physics. It works; don't ask it to make sense; shut up and calculate.[1578] Yet we may have an opportunity to understand it better.

For starters, our ontology agrees with Einstein's undercover thought that space is granular,[a] but that's not what his theories said. And it sees matter as a property of space, which, also, they did not.[b] These concepts go right to the heart of relativity but were entirely absent from its origin, development, and exposition.

For those who were the later sources of our seminal ideas, Einstein's theories of relativity loomed large. It was not just the theories themselves that set the scene, but what came before—his struggles; and what came after—his own (and others') efforts to grasp what he'd done.[1579] We still don't really understand this. The evidence suggests neither did he.[1580]

Recall it started out as a theory of motion. Newton's laws were okay if things don't move fast. But when they do, Einstein's special theory is more precise than Newton's laws, and it becomes essential near the speed of light. But more precision was not what drove him to work on it; he was pursuing questions about the nature of light itself. He soon found his simple questions had deep implications:

---

a Though it is not so in any of his theories.
b See chapter 43 at page 243.

> The special theory of relativity, which was simply a systematic development of the electrodynamics of Maxwell and Lorentz, pointed beyond itself, however.[1581]

And to further complicate the picture, it led him to that serendipitous discovery—equivalence of mass and energy.[1582]

The *special* theory was special in that it dealt only with *inertial*—that is, unaccelerated—motion. In 1907, he began to conceive his *general* theory[1583] as being general in that it included any kind of motion whatsoever. He never evolved beyond this *special/general* terminology. Yet, later, looking back, he wrote,

> 1909–1912 ... [G]rübelte ich unablässig über das Problem nach. (From 1909 to 1912 ... I brooded incessantly over the problem [of equivalence of acceleration and gravity].)[1584]

That is, in four of the years between his two key relativity papers, the focus of his general theory was shifting from acceleration to gravitation even as he came to see them as the same. This shifting objective was further complicated when that theory led him to another serendipitous discovery, namely that space is a real entity—a concept that was almost universally disdained.[a]

In the end, his general theory kept its old name but became the new theory of gravity. Thus, in 1966, physicist and mathematician John Synge said,

> Relativity? I have never been able to understand what that word means in this connection. ... [I]t is now apparent that nobody ever understood it, probably not even Einstein himself. So let it go. What is before us is Einstein's theory of gravitation.[1585]

The principle of general covariance came to lie at the heart of Einstein's struggles to formulate the theory. This principle said arbitrary changes of coordinates do not change the form of the laws of physics. Or, as he said (in a for-him-rare paragraph in emphatic italics),

---

[a] Most of the dismissive sources wrongly cited Einstein himself and his special theory; see chapter 10 at page 63.

> *The general laws of nature are to be expressed by equations which hold good for all systems of co-ordinates, that is, are co-variant with respect to any substitutions whatever (generally co-variant).*[1586]

Norton said,

> In developing general relativity . . . [Einstein's] efforts were dominated by a single theme, covariance.[1587]

And he explained,

> Einstein offered the principle of general covariance as the fundamental physical principle of his general theory of relativity and as responsible for extending the principle of relativity to accelerated motion. This view was disputed almost immediately.[1588]

Though this principle loomed large in Einstein's own view of his theory (forged as he struggled over strange new math—aided by his friend, mathematician Marcel Grossmann[1589]), its significance was challenged as early as 1917 by physicist Erich Kretschmann.[1590] Kretschmann's point (less than clearly made) was that general covariance is a mathematical formality for *any* theory and therefore contributes no physical meaning to it. Einstein tried to defend his view; but Kretschmann had it right.[1591]

As Norton summarized,

> Through [general covariance], Einstein proclaimed, the theory had extended the principle of relativity to accelerated motion. Einstein's critics responded that general covariance had nothing to do with a generalization of the principle of relativity. Worse, general covariance was physically vacuous, a purely mathematical property.[1592]

To add to the confusion, even deeper than the special theory's, the general theory's math seemed to merge space irretrievably with time.[a] Though this began as mathematical convenience, most physi-

---

a To confuse things even further than Minkowski had, there is no observable time variable in general relativity; see note 982.

cists gave it physical meaning; many tended to conflate it with reality. They were wrong.[a]

All these, together with other muddles, long succeeded in obscuring any understanding of what space really is, let alone time. Norton summarized the underlying situation:

> In November 1915, Einstein completed his general theory of relativity. Almost eight decades later, we universally acclaim his discovery as one of the most sublime acts of human speculative thought. However, the question of what Einstein discovered remains unanswered, for we have no consensus over the exact nature of the theory's foundations.[1593]

Einstein's own views about its foundations are unhelpful; they were contradictory and often changed.[1594] And, as early as 1917, he anticipated his theory's description of the world could soon be superseded:

> But I do not doubt that the day will come when that description, too, will have to yield to another one, for reasons which at present we do not yet surmise.[1595]

Today, the theory is what it is and we must find our own views of its foundations.

What we might call the real foundations of the general theory's description of space are protected from hands-on exposure by the impossibility of observing events at Planck scale, where space's goings-on go on.

Yet, from the vantage of the new ontology, we can see four fundamental reasons why, as Einstein foretold, the old description must yield to a new one:

- Space is not, as his description assumed, continuous.
- Space comprises real quanta rather than an arbitrary metric.
- Relativity's time is not a dimension of the universe.
- Matter and energy are properties of space, not distinct items in it.

---

[a] See chapter 83 at page 420.

Einstein himself privately surmised the first two of these reasons. For example, near the end of his life, he told a close friend he had made no progress on explaining "the atomistic character of nature" and said,

> My opinion is that... one has to find a possibility to avoid the continuum (together with space and time) altogether. But I have not the slightest idea what kind of elementary concepts could be used in such a theory.[1596]

How then should we now understand the meaning of his theory of general relativity? Seven observations seem to stand out:

First, it is a theory of space[a] and gravity.

Second, its general covariance is only a formality.[b]

Third, Einstein's evolving ideas about what space *is* were wrong.

Fourth, its provision for new space—the driver of the growing universe—was an arbitrary afterthought (the cosmological constant[c]) that does not reflect reality.

Fifth, it works, to a good approximation; it matches measurements within observational error in a range of situations.[d]

Sixth, that it works despite deep misconceptions of reality reveals its content is—as Einstein said from the beginning—about measurements of motion; which is to say, it is about doing geometry in an arbitrary space and not about reality at all.[e]

Finally, that it nonetheless did lead us to new fundamental insights is due to its roots tapping into the Planck-scale behavior of light.[f]

---

a *Space* here includes matter and energy made of space.
b That is, it is concerned with the *form* of the equations.
c See chapter 53 at page 291.
d In context of my comments about curve fitting at page 170, it is important to our whole story to recognize the limitations on our grasp of just how good (or bad) a match this is; see especially, Peebles, note 60 at his page 65 ff.; see also his page 162 ff.
e Note that Riemann, whose work was always at the foundations of relativity's equations, explicitly said it was about doing geometry in an arbitrary space and all but urged the virtue of a granular space; see chapter 12 at page 74. Einstein chose an assumed continuous 3-D space, awkwardly wedded to an assumed time dimension based on the observed behavior of clocks, choices maybe suited to a physics of measurements but not to understanding what is really going on.
f See chapter 95 and note b at page 474.

In overview, we can see that Einstein was, like Planck with his heat radiation, deeply into the business of distilling large-scale[a] grasp of space and time without understanding either of them, indeed, without setting out to truly understand them. Another way to put this is they were deriving large-scale consequences that arise from Planck-scale physics without bothering to turn their minds to Planck scale. This handicap makes their successes all the more astounding.

We are left trying to understand the origins of mass and inertia. Neither the theory nor the new ontology explains them.[b]

We are left, too, with another open question. In his sometimes desperate struggles with his field equations, Einstein focused on a measure of the curvature of space as the generally covariant entity.[1597] The *Ricci curvature*, as it is called, measures how a *volume changes* as it moves along a geodesic—the shortest path—in a continuous Riemannian space.

It's not clear how much Einstein's choice of Ricci curvature was influenced by chance. He had a desperate need for help in developing his theory's math and it was met in large part by a text. In the eighteen nineties, mathematician Gregorio Ricci-Curbastro developed math that later turned out to be what Einstein thought he required. Ricci-Curbastro and a student wrote the text and Einstein, together with Grossmann, seized upon it.[1598]

What if Einstein had instead focused directly on *volume*?[1599] This might have led him to the other kind of Riemannian space, a granular one made of tiny volumes, one that needs no arbitrary metric because (at least in principle) one can count granules,[c] an alternative that might have eased his path.[d]

It could have led him—leading ontologist of his time—to something like the new ontology.[e] If he had succeeded, who can say what the foundations of relativity would be seen to be today?

---

a Here, *large-scale* covers all the way from atoms to the universe.
b But the latter does tell us that the origins of both must be emergent; see chapter 86 at page 435.
c See chapter 12 at page 74.
d But maybe granular math would still have defeated him.
e On the other hand, it might well have just led him back into frustration.

100
NEW LIGHT ON THE OLD QUANTUM

> I think I can safely say that nobody understands
> quantum mechanics.
>
> Richard Feynman (1965)[1600]

By 1965, few other than Feynman would venture to say *understands* and *quantum mechanics* in the same sentence. Even he said a safe sentence.

Planck's constant[a] and Einstein's photon[b] let loose the quantum theory's dogs of war whose devastation of the very concept of reality overshadowed the next hundred years of endeavors to understand our world.

For three decades, Einstein stubbornly resisted the onslaught. By the end of his life, reality had become an increasingly obscure—even beleaguered—idea; the theory of the energy quantum was a numerically stunningly success and a conceptual disaster area.

| The real quantum needs new thinking.

If he could have lived to see the new ontology, would it have helped? He was concerned about causality:[c]

> The thing with causality torments me a lot too. . . . I would be very, very unwilling to give up causality entirely.[1601]

As we've seen, the new ontology provides for strict causality along with total randomness,[d] each with its role at every quantum at each iteration of the universe.[e] The randomness would be a shock to him; the strict causality would be a reassurance.

---

a See chapter 13 at page 76.
b See chapter 50 at page 276.
c Re causality, see chapter 43 at page 243.
d Einstein would not have regarded this as strict causality.
e One may then look for new theory to show how this leads to quantum theory's probability.

Even more, Einstein was concerned with reality. He insisted it existed. He wanted to understand it.

With probing insight, he always doubted how, in principle, any kind of elementary granularity could exist in a continuous universe. Stachel called it . . .

. . . his half-century-long struggle with the quantum.[1602]

His frustration shows through in his saying to his friend, engineer Michele Besso,

> The whole fifty years of conscious brooding have not brought me nearer to the answer to the question "What are light quanta?"[1603]

He went on to tell Besso everybody thinks they know the answer and they are all wrong.

I like to think Einstein would be thrilled by any confirmation space itself is made of quanta and so can supply the elemental granularity he lacked in trying to make sense out of light quanta. He might have seen how photons made of quantum space could behave as both waves and particles as he for years insisted in the face of overwhelming opposition, saying (with Infeld),

> We have two contradictory pictures of reality; separately neither of them fully explains the phenomena of light, but together they do.[a]

In quantum space, the contradiction disappears.[b]

So, maybe the most significant implication of the new ontology for quantum physics is its new foundation for the theory's central premise that things come in quanta. Well, not so new; Riemann explicitly contemplated quanta of space more than a hundred sixty years ago.[c] But mainstream physics totally ignored this insight even as it (led by Einstein) built that quantum "castle in the air"[d] with Riemann's alternative, continuous space as its foundation.

---

a See note 1272.
b See chapter 75 at page 386.
c He even called them *quanta*; see note 276.
d See note 1415.

Quantum theory's quanta and space's quanta of volume and area are not the same, not even close.[a] Yet they are related. A real quantum can be manifest in multiples, which can exhibit quantum theory's quantum behavior because they are *composed* of quanta.

No one has observed a single space quantum—or anything else at Planck scale for that matter—and no one is likely to do so soon. We do detect single photons, quantum theory's original quanta. Their bundles of energy are, in our terms, loose null-braided assemblies of three pairs of untwisted links or quanta of area.[b]

However, it is the particle's much-larger-scale relationships that exhibit quantum properties we can observe. It seems to follow that the photon's "quantum," for example, is a property of the whole region of space quanta in which its non-twists wander. We must find new physics to explain how their behavior in this region relates to the photon's wavelength.[c] Even for a tiny (ultra-high energy, very short wavelength) photon, this region includes a huge number of space quanta.[d] One remembers Wheeler's observation about "a vast piece of real estate."[e]

The new ontology's one-link-per-step foundation for the speed of light suggests each non-twist cannot wander further than will allow it to make landfall at its photon's destination.[f] Within this wide limit, though—as quantum theory's canonical interpretation says, and we agree—conceived as a particle, the photon has no definable location until it is made local by arrival.[g]

Some readers will be raising an objection. How can the photon's twist states "know" its destination? Short answer, we don't know. But this question leads to an intriguing speculation, one that is resonant with Einstein's best-known thought experiment:

---

a See chapter 45 at page 255.
b As noted in chapter 49, this sounds like a recipe for plain vanilla space.
c Note that, other than the one-quantum-per-step speed-of-light limit, this region has no defined boundary; but see note below.
d Ultra-high-energy gamma rays with energies of a peta-electron-volt have wavelengths of order 10^{-21} m. Two cubic wavelengths would comprise some 10^{42} space quanta.
e See note 78.
f That is, the boundary of its permitted region would comprise a kind of double-ended granular "light cone."
g See chapter 116 at page 568.

Einstein recalled how, at the age of 16, he imagined chasing after a beam of light.[1604]

What if he had chased a photon instead? Through the lens of special relativity, he would have seen the world as the photon does. Its path to its destination would be moving relative to it at the speed of light; so the length of the path would be zero.[1605] In other words, the photon sees no path; it sees an instant transaction, a coterminous emission and absorption in a nonlocal universe. So, from its perspective, the photon has no need to know where it is going—it's already there.[a]

---

a This does not work for an electron. In any case, the question needs new physics.

101
LEADING THE CHARGE

> QCD[a] has three different kinds of charge, labeled by "color." ... [W]hen two quarks are brought together, the effective color charge is reduced by a factor of two compared to when they were separated.... That implies an attractive force. So we should consider carefully what color superconductivity can do for us.
>
> Frank Wilczek (2000)[1606]

Compared with quarks' color charges, the electric charges of the proton and electron are a simple story. But it too has its complications.

In dry air you may have had a hair-raising experience of static. When you physically brush off some electrons from hydrogen atoms on the fringes of your hair's molecules, leaving charged protons, each proton tries to move away from others. The force pushing them apart increases as the square of their separation decreases, becoming very strong if they are close.

Composite charges raise a sticky question.

Early in the particle-physics era, atom smashers showed atomic nuclei larger than hydrogen hold multiple protons. Obvious question is, what could keep them—with their strong repulsion—jammed together in the tight confines of the nucleus?

At the next level down, once it became clear that the proton itself was not elementary but rather was composed of three quarks with fractional charges,[b] the obvious next question was, what keeps *them* jammed together in the even tighter confines of the proton? At the cutting edge, forces in these confines have been computed (fig. 79).[1607] They are huge, and yet the proton is so stable it never decays.[1608]

---

a I.e., quantum chromodynamics, the theory of quarks.
b Two of the three have positive charge; see chapter 43 at page 243.

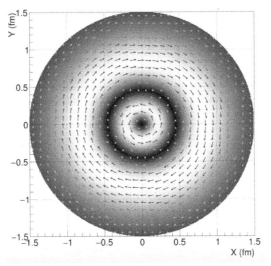

Fig. 79. Tangential forces in the proton

The answer to both questions is called the *strong nuclear force*.[1609] Siegel said,

> The strong nuclear force is responsible for a slew of incredible properties of nature, including: how protons and neutrons bind together to make atomic nuclei.[1610]

The theory of this force is called *quantum chromodynamics*.[1611] The last piece of its puzzle is what earned Wilczek (above) his Nobel Prize.[1612]

Like any good physics theory, QCD (as it's known in the trade[a]) gives a mathematical description based on symmetry. Wilczek said,

> [T]he fundamentals of QCD are simple and elegant.[1613]

In the epigram, he used an analogy to superconductivity, in which by disturbing positive ions in a solid lattice two moving negative electrons create a mutual attraction.[1614] However, descriptions of both QCD and superconductivity are based on "elementary" particles—electrons, quarks, and gluons—we are conceiving as composite. That is, though very useful, those descriptions do not tell us what is really going on.

---

a The *chromo-* in its name taps *chroma*, Greek for "color"—but it has nothing to do with color as we know it.

Our new view of the not so elementary particles suggests a simpler case that might shed light on this. We can refocus on the electron with its tweedle composition. How do *its* six negatively charged tweedles hang together?

It's a problem that may someday connect someone with another Nobel Prize.

102
THE HOLE STORY

> I only wish to make a plea for black holes to be taken seriously. . . . For who is to say . . . that they cannot play some important part in the shaping of observed phenomena?
>
> Roger Penrose (1969)[1615]

Black holes have so far had roles in maybe twenty chapters. This may be the place to pull their tale together.

The black hole is both conceived to be the smallest[a] and observed to be the largest[b] single entity in the universe. It is emergent, the ultimate instance of new properties arising from large numbers of small things.[c] Well, maybe. For the new ontology, a black hole's tweedles continue in their existence after they arrive; but the standard model's version of a black hole makes its small things—of whatever kind they may be—disappear.

> Black holes are great physics labs in the sky.

Either way, we now know the black hole plays more than "some important part"; it has a leading role.

In the nineteen thirties, Nobel Prize–winning physicist Subrahmanyan Chandrasekhar gave the first intimations that a star that's bigger than a certain size would in the end collapse.[1616]

In 1939, physicist Robert Oppenheimer (soon to be of atom-bomb fame) and a colleague showed how their black character arose from theory:

> When all thermonuclear sources of energy are exhausted a sufficiently heavy star will collapse.

---

a See particle-accelerator-created black holes; this chapter.
b See supermassive black holes; this chapter.
c See chapter 31 at page 176.

And,

> [As the star collapses] an observer comoving with the matter would not be able to send a light signal from the star.[1617]

Though black holes were born from Einstein's theory of general relativity, he did not think they were real. He said,

> Of course, these paradoxical results are not represented by anything in physical nature.[1618]

Thirty years later, when Penrose penned his plea for better understanding (above), those who thought of them at all still thought of them as theoretic figments. Penrose's Nobel Prize–winning work brought black holes back from sci-fi into sober physics journals.[1619]

We may now speak of them as really being in existence. We don't see them but we surely see their spectacular shows. However, we rely on theory (an odd mashup of general relativity, quantum theory, and thermodynamics, set in continuous space) to tell us what a black hole is[a] and much of what it does.

The biggest black holes may be the most massive hunks of matter in the universe.[b] They may also be the most consequential as astronomers see supermassive black holes (SMBHs to the cognoscenti) as responsible for organizing galaxies.[1620] In turn, the galaxies gave birth to stars.[1621] Stars synthesized the atoms that made planets, one of which gave rise to people.[1622] At least one SMBH lies in your past.

Most SMBHs we have detected lie more or less quiescent in the centers of their galaxies. But what happens when (as happens often in wide reaches of the universe[1623]) another galaxy arrives with its SMBH and—in a messy maneuver—merges? And then, when a third arrives to make things even messier?

In 2023, astrophysicist Pieter van Dokkum and colleagues described a two-hundred-thousand-light-year string of stars in the wake of a single

---

a In particular, the black hole's signature quality—its singularity of infinite density—depends on assuming space to be continuous; see chapter 93 at page 460.
b In the conventional, continuous-space view, their matter is no longer "in" the universe.

SMBH slung out of a merging galaxy at half a percent the speed of light and explained,

> There are several ways for a [SMBH] to escape from the center of a galaxy. The first step is always a galaxy merger, which leads to the formation of a binary SMBH at the center of the merger remnant. . . . [I]f a third SMBH reaches the center of the galaxy before the binary merges, a three-body interaction can impart a large velocity to one of the SMBHs leading to its escape.[1624]

Spotting evidence of three SMBHs that came so close together helps to confirm the universe has lots of them. Explaining how and when they came to exist in the observed numbers and their range of sizes is not easy; it is something of an acid test for any cosmological ontology (a test we'll see the standard model fails).

The new ontology adds a *new* source of (and three new roles for) black holes. And there are *old* sources with multiple roles. Some kinds of black holes are now clearly *real* but they all began as only *supposed*. It can get a bit confusing. To help sort it out, here is the ten-cent tour of today's black hole zoo.[a]

On the lowest rung by size, very new and barely supposed, we find miniscule black holes imagined to be made in the Large Hadron Collider. At most it might create a twenty-five-zeptogram[b] black hole—the mass of about fifteen thousand hydrogen atoms—with an event-horizon radius less than a milli-yocto-yoctometer.[c] Or maybe not; nobody knows.[1625] If this should ever happen, contrary to far-flung fears[1626] the black hole's Hawking radiation should "evaporate" it in less than a yoctosecond.[1627]

Next rung up are the old, supposed primordial black holes that Hawking said may have been "formed in the very early stages of the Universe."[d] They are (or were) "of any mass from 10^{-5} g upwards." He suggested many of them may still be around. But—as he later

---

[a] You can follow the four italicized flags in what follows.
[b] A zeptogram is 10^{-21} grams.
[c] I.e., 10^{-51} m.
[d] See chapter 68 at page 351; Hawking's proposal was based on a specific model of the early universe.

noted[1628]—if they were that small their Hawking radiation would "evaporate" them too, very quickly.[1629]

For slightly bigger ones up to, say, a billion tons,

> Any such black hole of mass less than 10^{15} g would have evaporated by now. Near the end of its life the rate of emission would be very high and about 10^{30} erg would be released in the last 0.1 s. This is a fairly small explosion by astronomical standards but it is equivalent to about 1 million 1 [mega-]ton hydrogen bombs.[1630]

That is, there may have been a lot of "fairly small" bangs soon after the Big Bang, but Hawking saw those universal–Wild West days as done.

If Hawking's supposition's real, those much larger than 10^{15} g[a] may indeed still be around but how many of them are there? And how big are they? We don't know, but some say bigger means fewer,[1631] and it's thought there are too few primordial black holes for them to get to eat each other:

> [S]ince the formation of a PBH is a rare event,[b] the probability that a small PBH is eaten by a larger one is very small.[1632]

(But, again, those primordial black holes were speculation; we will review recent observations to check them out; see below.)

Next, old and real, your common or garden stellar-mass black hole derives mostly from stars with between eight and thirty $M_\odot$. Once such a star's hydrogen has fused to helium followed by a chain of reactions that ends with iron, there's no further fusion to support its weight so it collapses (in a Type II supernova) to a neutron star (see fig. 80) or a black hole.[1633] It's no rare event. It happens roughly every ten seconds somewhere in the visible universe.[1634] We are beginning to find out more about this

Fig. 80. Radiation from a neutron star (center) after a supernova

---

a About a billion tons.
b But note the new ontology says this may have been a very common event; see this chapter, below.

kind[1635] with the new gravitational-wave observatories.[a] And, most recently, data from the Gaia satellite reveal a thirty-three-$M_\odot$ black hole nearby.[1636]

Then there is astrophysicist Priyamvada Natarajan's supposition that old big black holes might have formed from the direct collapse of big clouds of gas:

> [D]irect-collapse black holes could have been born at 10^4 or 10^5 solar masses within a few hundred million years after the big bang.[1637]

She said such an old black hole could show up as a quasar radiating such vast amounts of energy we can see it near the edge of the visible universe.

The earliest quasar observed so far was blasting out its energy some 500 million years after the universe began.[1638] It was powered by a ten-or-more-million-$M_\odot$ black hole. How it got to be so big so soon severely strains the standard model. Could it be an example of direct collapse?

There is a gap between $150M_\odot$ and $100,000M_\odot$ in which there were no clearly established black holes, and no proposed way for them to come into existence—until recently. In 2024, a team of astronomers and astrophysicists extracted data from more than five hundred Hubble images over twenty years showing multiple stars some eighteen thousand light-years away in speedy orbits around nothing that could be seen. They construe this as evidence of a real (and maybe but not likely new) black hole smack in the middle of that gap:

> From the velocities alone, we can infer a firm lower limit of the black hole mass of about $8,200M_\odot$, making this a good case for an intermediate-mass black hole in the local universe.[1639]

Next, computer simulations yield the (supposed and very old) *big* primordial black hole. If there were many, they would have set about creating galaxies immediately. If there were none, galaxies would have waited on formation of big stars, their swift burnout and col-

---

a See chapter 129 at page 617.

lapse, their stellar-sized black holes, and *their* growth or aggregation.[a] Seemingly a long wait. Thing is, simulation is a GIGO business[b] and it is not as simple as it sounds.[1640]

One indicator of (supposed and old) early big black holes may be early big galaxies. Based on simulations, the standard model picture twenty years ago was that,

> The first stars did not appear until perhaps 100 million years after the big bang, and nearly a billion years passed before galaxies proliferated across the cosmos.[1641]

Since then, more advanced simulations give somewhat sooner dates for galaxies. But they need so many assumptions their value is unclear.

These days, real and relatively new SMBHs in the up-to-ten-billion-$M_\odot$ range are a dime a dozen. One way or another, astronomers can "see" lots of them. Most (or maybe all[c]) big galaxies turn out to have one, and there are hundreds of billions of galaxies.

There were many SMBHs only a billion years after the beginning. The origin of so many so big so soon is problematic.[1642] They are so common, astronomers are searching for signs of pairs in orbit around each other (fig. 81).[1643] There is already evidence of maybe a million pairs.[1644]

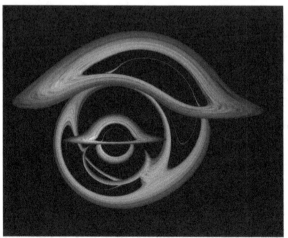

Fig. 81. NASA *simulation of two black holes in mutual orbit*

---

a Or on hypothetical direct-collapse black holes; see above.
b GIGO stands for *garbage in, garbage out*.
c Growing evidence suggests galaxy mass and black hole mass are closely related.

The largest reported SMBH to date may have a mass more than thirty billion $M_\odot$—giving it an event-horizon radius of nearly a hundred billion kilometers or three light-days—and is two billion light-years away.[1645]

Into this astronomic zoo, the new ontology's ideas bring one new source of big black holes and three new leading roles for them.

The new source is more and bigger primordial black holes. In their origin they are a bit like Hawking's. He postulated his arose from "initially large random fluctuations on all length scales."[1646]

He was speaking of events imagined to occur in the first moments after the beginning. By contrast, the new ontology offers a way for big black holes to form even sooner, based on its new view of the origin of space and matter.[a]

Neither of these views is based on quantitative physics. Suffice to say, in the new view, the many more and much bigger primordial black holes[b] could kick off galaxy formation starting immediately rather than having to wait for slow black-hole-creation processes.[c]

Indeed, new data show some ten times more and far bigger early big black holes than were expected.[1647] Astrophysicist Joseph Silk and colleagues said,

> The large population of ultracompact, dust-reddened red galaxies... implies the existence of a large population of massive black holes at very early times.[1648]

This could neatly solve an intractable problem of the standard model.[d]

It could also solve the problem of more recent black-hole-size observations.[1649] As Günther Hasinger, director of science at the European Space Agency, said,

> We don't understand how supermassive black holes could have grown so huge in the relatively short time available since the Universe existed.[1650]

---

a See chapter 94 at page 467.
b See chapter 40 at page 222.
c See chapter 69 at page 357.
d See chapter 129 at page 617.

The second new role fingers primordial black holes as the missing dark matter.[a] There is no clear limit on their size. The standard model has a problem making black holes a bit bigger than some fifty $M_\odot$.[b] Events found by LIGO included collisions of black holes having more than sixty $M_\odot$.[c] This suggests there may be many in a mass range that may point to primordial origin.[d]

Recent analysis showed the biggest black hole merger seen so far created a two-hundred-fifty-$M_\odot$ monster (far beyond the range the standard model can explain).[1651]

In the third new role, our ontology has supermassive black holes making the new space that has the universe expanding; that is, black holes create what ignorance has dubbed dark energy.[e]

Overall, the new ontology's offering of black holes—not just small but large and maybe even supermassive—in the world's initial instants suggests an entirely transformed picture of the next few hundred million years. These are the years that the James Webb space telescope[f] is, as we go to press, just starting to investigate.

And, lo, its harvest of confusion for the standard model overflows. Reporting on an August 2024, conference of astrophysicists in Santa Barbara, science journalist Rebecca Boyle said,

> [E]verything was weird during the universe's first billion years. Galaxy size, brightness, mass and shape from that period are all weird. Black holes are weird. The efficiency of star formation is weird; the correlation between brightness, astronomical power and an object's mass—essential for theoretical models—are not as astrophysicists expected. The presence of little red dots and chains of elongated galaxies: weird.[1652]

Black holes of all sizes all share the *no-hair theorem*; they have mass, charge, and angular momentum and that's all their properties

---

a See chapter 68 at page 357.
b See this chapter, above.
c See chapter 68 at page 357 and note 1170.
d But it also illustrates how they could grow by aggregation; but see note 1639.
e See chapters 67 and 69 at pages 348 and 357.
f See chapter 129 at page 617.

regardless of the diet that they ate.[1653] Another way to think of them is, they eat information. Quantum theory says there is no way that information can be lost.[1654]

In 1972, physicist Jacob Bekenstein showed (with more of that math) the information in a black hole is proportional to the area of its event horizon.[a] It's as if it stored its "bits" on Planck-area pixels at that conceptual surface, like it wears its story on its sleeve.

This shifted attention from general relativity's singularity to the event horizon. Of these contrasting crime scenes, physicist Brian Cox (who co-wrote a recent black-hole book[1655]) said,

> One of them is the description of Einstein and then there's another description, which just looks like some kind of quantum theory, some kind of building blocks of the universe that are entangled together.[1656]

The new ontology could give him an idea about building blocks; but, once again, his book was based on spacetime and the wrong kind of space. Of which, Cox acknowledged,

> [T]here is no consensus on what these calculations mean in reality.[1657]

Nonetheless, I concur with the claim of his book title, *Black Holes: The Key to Understanding the Universe*. In less than a hundred years black holes have been transformed from sci-fi curiosities to masters of the universe. They offer new laboratories for extreme physics.[b] Like Penrose in the epigram—indeed even more so—the new ontology sees them as playing an "important part in the shaping of observed phenomena."

Indeed, it suggests that understanding real black holes with real math should be central to our search for fundamental physics.

---

a The notional boundary from which even light cannot escape a black hole's gravity; see chapter 68 at page 351.
b See chapter 129 at page 617.

103
OUT OF TIME

> To make sense of the world at the Planck scale...
> we might have to give up the notion of time altogether, and learn ways to describe the world
> in atemporal terms.
>
> Carlo Rovelli (2009)[1658]

A bibliography of time would have thousands of entries. The new ontology's iterative 3-D universe opens the door for more. Stripped of its pretensions to dimensionality, time may never again be what physics has fed to our heads. But, with universal sequence as a guide, the remedy may not need to be quite as drastic as Rovelli (see above) suggested.

For readers interested in travails of time, let's take a quick look at a recent work and see where it can go. Again, I pick Rovelli—who's a leading light exploring time—to be my foil.

| Real time and clock time challenge our thinking.

In 2023, he wrote,

> It is Newton who introduced the idea that time passes "by itself" irrespectively from the events happening.... This implies that there are two distinct notions of times that are often at play. The first is the old notion of time as the simple *relative* succession of the events. The second is the Newtonian idea of a metric (measurable) time flowing *by itself.* The difference is substantial.[1659]

This started out as history—Newton's common or clock-motion time and his true or absolute time.[a] But Rovelli has blithely added that the latter is to be a *metric* or *measurable* time. (To humor him,

---

a See chapter 28 and note 635.

let's call this Newtonian time.[a]) He continued, unconsciously mimicking inexorable *Nexts*,

> For instance, if truly nothing happens, Newtonian time continues to pass; while it is meaningless to say that the counting of happenings continues if nothing happens. The traditional notion is *relational*: time is a relation between events. No events happening, no time. While the novel notion of time introduced by Newton refers to a (peculiar) *entity*, which exists by itself.

Here we can bring our clear new view to bear.

First, what does his "truly nothing happens" mean? He cannot mean the universe's iterations fail; he knows not of them (and, in any case, they won't). So, he must mean all motion stops.

At least in principle, this could occur within the new ontology; motion is made of twist-state relocations and at each iteration every one of them is random—left, right, or no move. So, what would be involved if motion should all stop?[b]

Let's narrow his imagination down to motions in a modest physics lab, say a seventy-cubic-meter room. Its space quanta would have some 10^{107} links.[c] The odds of no move of any twist-state in it for a single iteration of the universe are better than zero, but not enough to make a useful difference. Think of the chances of that many fair coin tosses being heads.

But *if* it were to happen, not for just one Planck time, but—even more bizarrely—for, say, a few minutes,[d] clocks in the lab would register no time while Newtonian time (Rovelli says) moves on. And, unwitting, beyond Newton's say-so, he is right; the universe's iterations would keep *Nexting* on.

So, (sticking with his version of the story) Rovelli—who's not really watching as he too would stop—would not notice the hiatus;

---

a But note that while Newton called it "mathematical" he did not term it *metric* or *measurable*.
b It must then stop at Planck scale.
c A reminder for those unfamiliar with scientific notation, this number is one followed by one hundred and seven zeroes.
d So, now this is like 10^{152} coin tosses being heads.

when the links resumed their moving he would check his clock and say the two times are the same, though they are not.

Thus, we see the difference—so far as he can see—is not "substantial"; there's no difference at all. When the stopping stops, his clock (having not moved) will pick up where it left off. He won't even know that his experiment occurred. The universe will offer him no evidence that it kept tocking on.[a]

The problem lies in *his* insisting that Newtonian time—*as he redefined it*—is "a metric (measurable) time." In and of themselves—stripped of those twist-state motions that connect tocks to our clocks—the iterations of the universe are not measurable.[b] Seeing evidence of Newton's time depends upon the very motions *he* decides must stop.

The universe could care less what Rovelli sees; it's heedless of his edict. *Its* version of Newton's time is also, in Rovelli's words, "a relation between events."[c] The difference is, regardless of what he may see, *its* events do not stop.

In the real world, the "traditional notion [of] time as a relation between events" and "the novel notion of time introduced by Newton" (that, like Newton, I prefer to call true time) are numerically identical because the evidence of both is given us by clocks (one by definition and the other by necessity) that measure the same motions.

The difference Rovelli focused on comes down to this. Common time works only when you have clock-apparatus and then only if it works; the clock (analog in its heart) informs the clock-watcher about how long *they* have been at work. True time's apparatus is the universe and it always works; its working is invisible but digital; even if nobody checks, the sequence registers how long *the universe* has been at work.

---

a That is, inside his lab. But we could have the entire universe stop moving, which would only make the whole experiment statistically more absurd.
b See chapter 52 at page 284.
c The events are *beings* of successive versions of the universe; the relation is succession.

What lesson can we learn? First, as we have seen, Newton's true time has no artificial metric.[a] Second, both of Newton's times are a succession of events; the difference lies in the source of the events. And, third, in the end, Rovelli's right; the difference between the two times *is* substantial. It is not a difference in what we observe; it is a difference in what we understand.

Following on the epigram above, Rovelli said,

> Time might be a useful concept only within an approximate description of the physical reality.

Time shown by our clocks is emergent, based on motion of emergent matter and emergent space. So it is approximate, although it may be quite precise.[1660] This precision can conceal the fact that it's inherently statistical, walling us off from a clear view of what Rovelli stopped just short of saying: Time is useful; indeed it's a treasured fiction but it isn't real.

Sequence number is (at least for now) less often useful; but it's real.

Writ large, the difference between Newton's two times is fundamental. Common time is a visible human artefact; true time is an invisible measure of the universe's age.

---

[a] Again, see chapter 52 at page 284.

PART VI

THE NUMBERS GAME

If mathematical facts are facts, they must be facts about something; if mathematical truths are true, something must make them true. . . . If 2 plus 2 is so definitely 4, what is it that makes it so?
Penelope Maddy (1990)[1661]

PART VI

THE
NURSERS
CRIME

104
REAL MATH

[S]omewhere deep within mathematics' big bag must lie a mathematical assemblage that [corresponds] to that of the physical world before us, even if it turns out that we will never be able to get our hands on that structure concretely.

Mark Wilson (2000)[1662]

In this part we dig deep into that big bag of math to find the "mathematical assemblage" that philosopher of mathematics Mark Wilson said must be there.

To find it we must overcome the block Muller observed stands in our way:

> All of mathematics is knowledge that is outside of physical reality. That . . . is the cause of much math phobia.[1663]

To accomplish both, we will see what the universe can tell us about understanding numbers. We'll do simple arithmetic "to get our hands on that structure." Maybe not what Wilson would call "concretely," but we'll find the assemblage is simple. And it will dispel that math phobia.

| The universe was doing real arithmetic long ago.

Though focused on the math, these chapters are integral to our worldview story; there is a close relationship between math and reality that reaches into philosophy and ontology. Issues of physical existence are intimately tied to those of mathematical existence. And, as mathematician David Hilbert observed,

> [W]e do not master a theory in natural science until we have extracted its mathematical kernel and laid it completely bare.[1664]

Though that close relationship has long been evident, it always was a puzzle.

In 1959 (before his Nobel Prize), physicist Eugene Wigner set it in full view in a now famous lecture headlined "The unreasonable effectiveness of mathematics in the natural sciences."[1665] He said,

> [T]he mathematical formulation of the physicist's often crude experience leads in an uncanny number of cases to an amazingly accurate description of a large class of phenomena. This shows that the mathematical language . . . is, in a very real sense, the correct language.[1666]

He went on to say,

> [T]he enormous usefulness of mathematics in the natural sciences is something bordering on the mysterious and . . . there is no rational explanation for it.[1667]

He was, perhaps unwittingly, echoing Einstein:

> How can it be that mathematics, being after all a product of human thought which is independent of experience, is so admirably appropriate to the objects of reality?[1668]

The new ontology offers a way to understand this. Looking over our depiction of the universe, what it is being and what it is doing—especially in the first few steps of its existence when we can grasp the whole of what it is and what it does—we see that both *are* mathematical.

That is, the universe is made of elementary quanta, which reified numbers; it takes elementary steps, which reified arithmetic.

In 1988, philosopher Alain Badiou caused a stir with his thesis (grounded in set theory[a]):

> Mathematics *is* ontology.[1669]

He threw down the ontologic gauntlet (which we will take up):

> The truth is that *there are no* mathematical objects.[1670]

To grasp the truth or otherwise of this bold negation one must grasp what mathematics is. *Britannica* says it is . . .

---

a Of which, more below.

> ... the science of structure, order, and relation that has evolved from elemental practices of counting, measuring, and describing the shapes of objects.[1671]

Objects! And counting them! We shall see that it is so.[a] Even mathematics asks *What is?*

Amid therefore some turmoil redolent of physics' ontologic issues, Maddy, who literally as well as metaphorically wrote the book on realism in mathematics, also with much reference to set theory and mathematical objects (but without reference to Badiou[b]) said,

> [T]he mathematical and physical sciences are facing the same metaphysical question.[1672]

The central question in the philosophy of mathematics through the ages is, where does math come from?

For example, philosopher of mathematics Philip Kitcher said we discover a bit of it and invent the rest:

> [A] very limited amount of our mathematical knowledge can be obtained by observations and manipulations of ordinary things. Upon this small basis we erect the powerful general theories of modern mathematics.[1673]

Another way to put the question is, is math real, a property of the universe that exists whether we discover it or not; or is math fictional, a realm of boundless variety we devise that has, like art, no existence until we create it?[1674]

The new ontology leads to a definite answer. Like Kitcher, it says *both*. Some math is real; some math is fiction. (One source of our confusion is much math that's fiction closely fits what we perceive as facts.) It goes on to say definitively which is which. This distinction has deep consequences. For example (a bit of a shocker), some of the natural numbers (like 2 and 193) exist; others (like one followed by a thousand zeroes) do not.

---

a In the next chapter.
b Maddy's definitive work did not mention Badiou's two-years-earlier book; but then it was in French, translated into English some two decades later.

When we see why, we will understand the distinction.[a] To get there we must grapple with the tangle that math logic is today. It's fortunate that we can find a clear and simple path.

As a first step, we will find that—as Wilson asserted in the epigram—there *is* a mathematical assemblage that corresponds to the physical world.

This matters since, as Maddy said (defining realism in the realm of math),

> [M]athematics is the scientific study of objectively existing mathematical entities just as physics is the study of physical entities.[1675]

The new ontology agrees with Maddy that this is a laudable (and now maybe achievable) objective.

Let's start with *natural numbers*,[b] those one counts on fingers. They lie at the root of the math hierarchy. Math is the lifeblood of modern physics. Arithmetic is the lifeblood of math. In turn, the natural numbers are the lifeblood of arithmetic. Yet the many books and learned papers on the natural numbers descend swiftly into murky controversy.

Indeed, the philosophy of mathematics is concerned—one might even say obsessed—with *What is a number?* What is its nature? Is it a real entity? Some, like Russell, say philosopher, logician, and mathematician Gottlob Frege gave a final (very messy[1676]) answer in 1884.[1677] Debate persists about it to this day.[c]

Did we invent numbers or discover them? Some still hold the view natural numbers are properties of the universe and would exist even if we did not discover them. This view builds upon the notion that physical objects we can see and handle are real; so, when we count them, their numbers too are real.[1678]

Using language of ontological commitment, Maddy set this view on a broad scientific (but logically frail) foundation:

---

[a] The reason is not just that it's too big; see chapter 111 at page 547.
[b] Also known as the positive integers.
[c] We will find a fundamental answer that differs both from Russell's and from Frege's.

> We are committed to the existence of mathematical objects because they are indispensable to our best theory of the world and we accept that theory.[1679]

At this end of the real/invented spectrum of views, Russell scornfully dismissed invention in favor of discovery:

> Arithmetic must be discovered in just the sense in which Columbus discovered the West Indies, and we no more create numbers than he created the Indians.[1680]

But this view runs into limitations (and we will show that, in this too, he was mostly wrong[a]). For example, Kitcher said,

> We require infinitely many objects as values of our variables; there are not sufficiently many physical objects to go round; hence we shall need, it seems, to posit abstract objects.[1681]

At the other end of the real/invented spectrum lies another group of views, collectively known as *antirealism*.[1682] Its reach extends far beyond mathematics.

For example, Smolin said of Bohr's worldview,

> Bohr's position is anti-realist in the extreme, in that he denies it is even possible to talk about or describe an electron as it is in itself.... Science according to this picture is not about electrons; it is about how we talk about our interactions with them.[1683]

Antirealists would agree with philosopher of mathematics Bruno Whittle (and with Badiou) that,

> Plausibly, mathematical claims are true, but the fundamental furniture of the world does not include mathematical objects.[1684]

We will challenge their view.[b]

These days, it's not physical objects but *sets*[1685] that are lead competitors as counters defining natural numbers that are real mathe-

---

[a] In chapter 111 at page 547.
[b] See below.

matical objects.[a] This is odd, because the realm of sets soon swells to include ideaistic entities, such as *the set of all sets that do not contain themselves* that undermined the footings of Whitehead's and Russell's grand program to build arithmetic on logical foundations.[b]

We'll soon see too how, in 1931, Gödel laid the concept of such programs in its grave.[c]

Not all thought all was lost. In 1937, Quine said, optimistically,

> In Whitehead and Russell's *Principia Mathematica* we have good evidence that all mathematics is translatable into logic.[1686]

And, a few pages later, more realistically,

> Gödel has shown, however, that this totality of principles can never be exactly reproduced by the theorems of a formal system.... Adequacy of our systematization must then be measured by some standard short of the totality of valid formulas.[1687]

Which is to say, Quine was willing to settle (of necessity, he thought) for half a loaf of logic. When one is bent on understanding what is real, however, it is the whole loaf or no loaf at all.

Many mathematicians and philosophers of mathematics assert that at least some numbers are real objects (though there is no clear consensus about which ones are and which are not).[1688]

Some philosophers say the natural numbers are the only numbers that are, in Whittle's words, "part of the fundamental furniture of the world" or, in our terms, real.

In 1887, philosopher John Stuart Mill said that a number corresponds to real things:

> All numbers must be numbers of something: there are no such things as numbers in the abstract. *Ten* must mean ten bodies, or ten sounds, or ten beatings of the pulse.[1689]

---

a See chapter 106 at page 528.
b They tried to codify the new field of mathematical logic; see also chapter 106; for the last hundred years a device known as Zermelo–Fraenkel set theory has provided that logical foundation; it evades paradoxes such as this set, which—as a moment's thought will show—both does and does not contain itself (assume one, you get the other).
c See chapter 114 at page 559.

And, more recently, mathematician John Bigelow reaffirmed,

> [T]he world of space and time does contain mathematical objects like numbers.[1690]

He (and others) said the natural numbers, like five and ninety-nine, are distinct from other kinds of numbers, like -5, ¾, or π. The difference is, he said,

> [T]he natural numbers can be construed at a deeper level than the rational or [so-called] real or imaginary numbers ever could be. . . . [T]here is a greater difference between one pie and three-quarters of a pie, than between one pie and three pies.[1691]

The key distinction for such numbers is they can be counted. That is, you get the natural numbers by counting things you find in the world, like fingers or pebbles. The new ontology aspires to follow. Such natural numbers are real mathematical objects; that is, they are inherent properties of the universe; *that* is, however, *if* the counted objects are, and thereby hangs a tale.[a]

It brings into figurative view a (notionally) countable thing that is ultimately well defined and indestructible—the space quantum. True, it may replicate, but only in utterly inaccessible places.[b] And, in the universe's first few moves, its replications let us track its numbers and decipher its arithmetic.

Thus, we have real mathematical objects, elemental entities that manifested their existence in the universe with no question of invention.[c] The universe began as one such quantum. It *was*. This reified the number *one*; it was the ultimate reality. When it tunneled into two it added one and reified *addition*;[d] it reified *succession* too as two was next. When the two tunneled into four, this reified *multiplication*. These were real events, without which our whole world would not exist.[e]

---

a See the next chapter.
b See chapter 66 at page 343.
c Much later we discovered them.
d Addition, not multiplication, since there was as yet no two to multiply by.
e We will have occasion to revisit them; see chapter 108 at page 535.

Of such objects, Kitcher said,

> Arithmetic truths are useful because they describe operations which we can perform on any objects.[1692]

The new ontology says certain arithmetic truths are *real* because they describe operations which the universe *does* perform on its objects. Other operations may be useful but are not thereby made real. Herein lies a cogent resolution of the long-standing ontologic issue of math's origin.

The physical inaccessibility of the universe's counting objects means we are using thought experiments. But then, that's the kind of counting philosophers and mathematicians almost invariably do.[a]

The new ontology implies we should explore in especial the math that derives from only the (really) real numbers and these real objects, *real math*, as distinct from what I call *math fiction*.[b]

Wilson's proposition (in the epigram) turns out to be prophetic. It seems ironic. We *can* see things that maybe we should *not* use for real counting;[c] but we *can't* see quanta that we are about to find we *must* use for real counting.

A persuasive reason is, the quanta have what Maddy called the key ingredient, a "determinate number property."[d]

The ultimate reason is it is not we who do real counting; it's the universe.

---

a See the next chapter.
b Math fiction of course continues to warrant studies of its own, one hopes with better clarity on what it is and what it isn't.
c See the next chapter.
d See note 1703 in the next chapter.

105
THEM APPLES

> How, we ask, can the statement "2 + 3 = 5," whose
> truth is independent of experience,
> apply to apples . . . ?
>
> Kuni Fann (2020)[1693]

Here we take a closer ontologic look into a seeming-simple question, *What are natural numbers?* As philosopher Kuni Fann's question intimates, this sends us straight to apples.

The rightful role of the philosopher includes examining minutely every aspect of the ways we acquire knowledge, prying loose hidden assumptions. There seems to be no way to know something without assumptions but philosophy does try to notice when they are in play.[1694]

> Counting needs things to count.

One that is fundamental is, objects our senses say exist, exist. We all assume this, as we must, though it is also questioned. Maddy said,

> One of the most basic ontological debates in philosophy concerns the existence of what common sense takes to be the fundamental furniture of the world: stones and trees, tables and chairs, medium-sized physical objects.[1695]

And, setting out to define realism, she said,

> [T]he assumption of objectively existing, medium-sized physical objects plays an indispensable role in our best account of experience.[1696]

Bespeaking the fundamental role of such objects as *counters*, twentieth-century philosopher Ludwig Wittgenstein (a student of the centrality of language) said,

> This is how our children learn sums . . . put down three beans and then another three beans and then count what is there. If the result at one time were 5, at another 7 . . . , then the first thing we said would be that beans were no good for teaching sums. But if the same thing happened with sticks, fingers, lines and most other things, that would be the end of all sums.[1697]

Most mathematical philosophers—not only realists—consider numbers to have meaning only by virtue of the existence of real objects they can count. Though, oddly, having insisted the objects be available, they mostly seem to count them without troubling to acquire them. And they do like apples. Some examples:

> Put two apples on a bare table . . . ; now put another two apples on the table; now count the apples that are there. You have made an experiment; the result of the counting is probably 4. (Wittgenstein[1698])

> [I]f I have two apples now, and I plan to add three apples, I will have five apples. This is knowledge gained deductively. I did not actually need to get the three other apples and place them with the first two to see that I have five. (Landauer and Rowlands[1699])

> There may be five apples on the table but the number five itself is not to be found in, on, beside or anywhere near the apples. (Paseau[1700])

You get the idea.

Maddy's "medium-sized physical objects" include apples. Her wider concept sounds like what the French call *une chose*. One might translate this as "a thing." English common law seizes on a kind of thing it calls a "*chose* in possession"[a] and, as is its wont, defines it in detail by means of lawsuits. And—shades of the philosophers' approach—one need not ever possess it but its possession must be possible for it to be adjudged a true *chose* in possession.[b] A philosopher's

---

a Much of English (and American) common law is Norman French in origin and language.
b Roughly, it is a tangible thing that can be had.

apple may (if it exists) be such a *chose*. But its legal definition cannot solve their counter problem.

Unaccountably, philosophers seem unconcerned about the quality of objects that they count, like apples; yet they bandy them about as if they were the very essence of the real world. I imagine most have never, as I have, cared for an apple tree; even a shift or two at picking apples might turn out to be enlightening. It seems they too do not need real apples; they imagine them, confident they are somewhere to be had. This (I will suggest, below) becomes significant. They too are using Einstein's favorite tool, the gedanken experiment.[a]

Imagined apples may have useful properties including certainty. That is, they seem to have sharp mental edges[b] thanks to needing little in the way of detail; in the mind's eye apples either are or they are not.

This is a lofty view; real-world apples can't live up to it. In my experience, an apple often has an air of insubstantiality. With a closer look than given by the average philosopher, one sees what one may not see in the produce store. An apple comes into existence slowly from the center of a flower or it may fail of its promise and just wrinkle into nothing, or, while only partly grown, begin to rot. A flock of Aussie parrots can transform a tree of apples into guano before breakfast. Existence, for a real apple, is uncertain in so many ways.

Not to mention there's a wide (and ever-changing) range of things that we call apples.[1701]

Simply put, *How many apples?* is a question with less meaning than maybe it seems.

The soft underbelly of the counting-medium-sized-objects business is that it assumes there is a definition of the objects one intends to count. There is and can be no such certainty. This is a problem and it is not limited to apples.

Thus, there are people who have digital extensions which one could as reasonably argue (if one could be so unkind) may be fingers or may not. If you can't count on an answer to "How many fingers

---
[a] See chapter 30 at page 173.
[b] See also chapter 130 at page 624.

are extended?" how can fingers serve as counters? And at what size do those pebbles (which some use instead of apples) become rocks or boulders or an Uluru; at what size do small pebbles merge into the sand?

Philosophers are troubled, too, by counters that are composite. For example, Frege criticized the use of aggregates.[1702] Maddy (citing Frege) thought about eggs in a carton. She said,

> The trouble is that the physical stuff in the carton has no *determinate number property*: it is three eggs, but many more molecules, even more atoms, and only a quarter of a carton of eggs. For a given mass of physical stuff, there is no predetermined way that it must be divided up, and without this, there is no determinate number property. So the physical stuff by itself cannot be three.[1703]

And she, still following Frege, continued her dissection with . . .

> . . . two boys playing in a garden [whose] physical stuff could be divided into units in various different ways—two boys, twenty boy-parts (heads, torsos, arms, hands, legs and feet), millions of cells, even more molecules, still more atoms, and so on.[1704]

Who'd have thought that counting could be such a bloody business?

One might imagine that an atom may be well defined and these days one can "see" an atom well enough to count it (almost[1705]); but physicists may split it into two or fuse it with another, while quantum physics says that at your whim that atom may become a wave.[a]

Then there was Quine, observing with serene assurance,

> [T]he terms "9" and "the number of the planets" name one and the same abstract entity.[1706]

Fifty years later the International Astronomical Union voted Pluto off the island so, as astronomer Mike Brown informed us,

> There are finally, officially, eight planets in the solar system.[1707]

---

a See chapter 73 at page 378.

Pluto didn't change; so, were there eight when Quine wrote of their number? If we can't be sure of counting fewer than two hands of *planets* maybe math should not be based on apples, eggs, or bits of boys.

This problem, the unseemly lack of real definition of our counters, is widely ignored.

There's another counter problem. Wittgenstein was not the last to say that we don't really see things; we see images of things.[1708] This leaves it open for philosophers to claim we can't count counters even if they could be well defined.

That we don't really see things is true—in spades—of counting (as I am suggesting[a]) quanta. That is, we not only cannot see them; we can't even see their images.

What we do get with the quanta is counters with none of the other failings; and we can count them through the first instants of the cosmos, as real math objects came into existence with real math operations. This, I assert, beats body parts and apples any day.

In summary of many sources, there *is* no proposed "medium-sized object"[b] that is suitable for counting. It's a bit stunning that simple arithmetic—with fundamental physics on its shoulders—sits on such a frail foundation.

Some say they solved this problem long ago with a new choice of counters: sets.

True, they are not physical objects; but since they are mental constructs at least one can conceive they are well defined.

Until one finds they're not. That's next.

---

a In chapter 104 at page 515.
b Maddy's "indispensable" requirement; see note 1696.

106
GETTING SET

> Set theory is the ultimate court of appeal on questions of what mathematical things there are, that is to say, on what philosophers call the "ontology" of mathematics.
>
> Penelope Maddy (1990)[1709]

In the nineteen twenties, philosopher of mathematics Alfred North Whitehead with his partner-in-crime Russell tried to place arithmetic upon a logical foundation based on sets.[1710]

Their logic was abstruse, but that's the way of logic. They worked long and hard to prove that $1 + 1 = 2$ (halfway there by page 360 of volume one, see fig. 82). But—built into their foundations—the *one* they used to get their counting going was based on the empty set.[a] Then, deep into their learned exposition, they defined the natural numbers as . . .

> The empty set makes one by using zero. Not really.

> . . . those that can be reached from 0 by successive additions of 1.[1711]

∗54·43. ⊢ :. α, β ∊ 1 . ⊃ : α ∩ β = Λ . ≡ . α ∪ β ∊ 2

Dem.

⊢ . ∗54·26 . ⊃ ⊢ :. α = ι'x . β = ι'y . ⊃ : α ∪ β ∊ 2 . ≡ . x ≠ y .

[∗51·231]   ≡ . ι'x ∩ ι'y = Λ .

[∗13·12]   ≡ . α ∩ β = Λ   (1)

⊢ . (1) . ∗11·11·35 . ⊃

⊢ :. (∃x, y) . α = ι'x . β = ι'y . ⊃ : α ∪ β ∊ 2 . ≡ . α ∩ β = Λ   (2)

⊢ . (2) . ∗11·54 . ∗52·1 . ⊃ ⊢ . Prop

From this proposition it will follow, when arithmetical addition has been defined, that $1 + 1 = 2$.

Fig. 82. Excerpt from *Principia Mathematica* by Whitehead and Russell

---

a That is, the set that contains nothing.

This was *something from nothing* through the magic of set theory. Worse, in natural-number world, it was *all from nothing*.[a]

Ever since, set theory has been at the heart of the program to put math on logical foundations.[1712]

For those whose schooling (like mine) came before set theory's sly move into grade-school syllabuses,[1713] a set is a collection of objects, real or imagined. Like a set of seven buttons. It sounds simple. It has its own language. It gets grander: sets of sets of sets, et cetera. It soon is far from simple.

After the Whitehead and Russell caper, sets-as-counters battle lines were drawn in the mid–nineteen hundreds.

In 1937, Quine led off the main numbers-based-on-sets foray.[1714] Leading the repulse three decades later, philosopher Paul Benacerraf showed why numbers *cannot* be sets.[1715] Neither case seems all that strong to me, but Quine held the high ground. Today, the role of sets in *number theory*[b] lies at the epicenter of an earthquake zone of problems that are mostly politely ignored in halls of math and physics.

A singular exception was mathematician Norman Wildberger, who was concerned about math teaching. He said (his italics),

> Modern mathematics *doesn't make complete sense.*

He went right on to say,

> Elementary mathematics needs to be understood in the *right way*, and the entire subject needs to be rebuilt so that it makes complete sense right from the beginning. . . . If mathematics made complete sense then the physicists wouldn't have to thrash around quite so wildly for the right mathematical theories for quantum field theory and string theory.[1716]

He railed at neglected problems with so-called real numbers and infinities and most of all with set theory. He went on to condemn lack of concern about foundational aspects . . .

---

a Shades of Mlodinow and Hawking; see chapter 91 at page 454.
b Before the twentieth century, *number theory* was called arithmetic.

> ... augmented by the twentieth century's *whole hearted and largely uncritical embrace of Set Theory*.[1717]

The new ontology says his concerns are all well founded. There's a problem built right into basing numbers upon sets: Math logic's reliance on sets arises from desire to build the natural numbers out of nothing. When one wants understanding, this is just a plan to fail.

Mathematician Robert Kaplan (whose book title *The Nothing That Is* is equally provocative) put his finger on the central problem:

> How can something as queer as the set of what doesn't exist, itself exist?[1718]

Sets' virtue is they allow you to count somethings when you have no others that are fit to count. It might be a good trick[1719]—if you could get away with it:

- You define a set that holds nothing, the set version of *zero*.
- You define a set that holds only that empty set.
- You count that set as *one*.
- You add your *one* into a set that also holds the empty set to get *two*.
- You build the other natural numbers up from there.[a]

It's like watching the magician's white-gloved hand to figure how they do it. Logic has a symbol that denotes a set: {}. This *is* the empty set, because as you can see there's nothing in it, which makes it the set-theoretic equivalent of zero. And, so, for convenience, it is also assigned the symbol Ø. (Don't watch the glove; keep your eyes on this and see if you can spot where nothing becomes something.)

Next, there's the set, {Ø}, that contains the empty set, and, as it is a different set and there is *one* such set, the number-building's underway. Kaplan (above) politely called this queer. I call it philosophic sleight of hand.

Maddy laid out how set theory can then do arithmetic "without explicit reference to number properties" and without ever needing to count a set twice:

---

a See also Whitehead and Russell, note 1711.

To say 2 + 2 = 4 is to say that if two . . . sets . . . are equi-numerous with {Ø, {Ø}}, then their union is equi-numerous with {Ø, {Ø}, {Ø, {Ø}}, {Ø, {Ø}, {Ø, {Ø}}}}.[1720]

Here, {Ø, {Ø}} is a set that contains the empty set and the set that contains that set—i.e., two distinct sets—that stands for the number two.[a] That last string of symbols is a set-theoretic version of 4.[b] As you can see without the need for detailed explanation, the whole edifice depends upon the empty set—that is, on nothing.

Its vacuity is even easier to see when we replace Ø with what it stands in for, the empty set {}, thus: 4 = {{}, {{}}, {{}, {{}}}, {{}, {{}}, {{}, {{}}}}}.

And you thought Roman numerals unwieldy?

Given the new ontology's insight that there is no nothing in the universe,[c] set theory's Ø is just as much math fiction as if it started with a zero and said there is one of them, as mathematician Giuseppe Peano is said to have done.[d] And so are all the numbers made from Ø, pure fiction.

That is, in the real world, the whole set-theoretic basis for natural numbers and arithmetic collapses.

Let's give the last word on the spreading ripples of the sorry tale of sets to Maddy (lead advocate for set theoretic realism):

> Of course, set theory is as fallible as any other science, and it could turn out that the continuum question is based on faulty presuppositions, but there is no conclusive reason to believe this now.[1721]

Maybe there was no reason then. There sure is now.[e]

The long drive to build up numbers and arithmetic from nothing using set theory was ill conceived; and (we now see) unnecessary.[f]

---

a This illustrates how far they go to ensure that they are counting only distinct counters; see chapter 108 at page 535.
b There are others (no less messy); Maddy said, "And it is this fact that generates an ontological question about numbers." See note 123 at her page 84.
c See chapter 89 at page 444.
d Not so; see chapter 108 at page 535.
e See chapter 45 at page 255.
f See chapter 108 at page 535.

107
REAL IMAGINATION

> The slogan might be coined, "Real numbers
> are not real in a causal set."
>
> Fay Dowker (2005)[1722]

Dowker's observation is more than a slogan. It may seem picky but it matters: The so-called *real numbers*[a] do not exist in the causal set we call the universe.[b]

Their existential defect goes back to roots in the imagined continuity of space and time. For hundreds of years their seductive ease of use drew physics into a condition that's akin to an addiction. Their influence on real thought in physics and philosophy is toxic.

> So-called real numbers are imaginary.

Philosopher, mathematician, and logician John Lane Bell recounted that, wrestling with the real numbers, mathematician Hermann Weyl came to believe . . .

> . . . that mathematical analysis at the beginning of the 20th century would not bear logical scrutiny, for its essential concepts and procedures involved vicious circles to such an extent that, as he says, "every cell (so to speak) of this mighty organism is permeated by contradiction."[1723]

The definition of a real number rests upon the concept of infinity (which does not exist in the world of our new ontology[c]). A shade over-simply, a real number corresponds to a point on a continuous

---

[a] In this chapter, the term *real number* has its conventional meaning, that is, "so-called real number," a number that is actually unreal.
[b] See chapter 73 at page 378.
[c] See chapters 22 and 110 at pages 133 and 543.

line (both of which also do not exist[a]). There are, says "real" mathematics, an infinity of points between any two points no matter how close they may be.

Thus, there are also infinitely many real numbers with infinitely many nonrepeating digits. Some of them might seem to have special significance, like the square root of two, 1.4142135623. . . , the length of the hypotenuse of a right triangle with two sides of length one.[b]

Kitcher explained,

> The real numbers stand to measurement as the natural numbers stand to counting.[1724]

This gives another clue to the reason underlying their unreality. There *is* no measurement (of the kind he means) in the universe, only approximations of one kind or another.[c]

Real numbers, like *pi* for example, could be truly real if the universe's space were continuous.[d] Given it's granular, real numbers are not real (yes, it's a sorry excuse for a name[e]).

Mathematician Richard Dedekind is credited with putting the real numbers on a systematic footing.[1725] His own view was not that they are real but rather,

> [N]umbers are free creations of the human mind.[1726]

His real numbers were all of that. He defined a real number by squeezing it in between the smaller and the larger numbers,[1727] thus embedding the idea in the then emerging field of sets.[f]

Not blind to its frailties, Maddy nonetheless said,

---

a There are no points (see chapter 77 at page 395); there are no lines (see note c below).
b In the nonexistent flat space of Euclidian geometry.
c Measurement of length, as on a line, determines the distance between two positions; but at Planck scale there are no such positions; see chapter 62 at page 326. Nor can we define lines at Planck scale; see chapter 110 at page 543. For the origins of the case against continuity, see chapter 12 at page 74.
d Even in continuous space, *pi* would fall afoul of the universe's curvature; see chapter 112 at page 551.
e The label traces back to Descartes who coined it to distinguish them from *imaginary* numbers, such as the square root of minus one.
f See chapter 106 at page 528.

> To this day, set theory provides our best account of mathematical analysis, which in turn plays a central role in our most successful physical theories. This achievement . . . plays an indispensable role in our best theory of the world.[1728]

This may be an accurate assessment. It is also a restatement of a basic problem. Our best theory of the world is detached from reality; it is bogged down in a swamp of contradictions and of concepts that import their innate unreality into all that they touch. Among the worst offenders are the real numbers that are the lifeblood of the continuous worldview. It is a vicious circle.

Speaking of circles, *pi* is an example of a real number,[a] the ratio of the circumference of a circle to its diameter. It is a fundamental constant in a (nonexistent) flat Euclidean space.[b] In a continuous Riemannian space, it is not a constant; it varies with both local curvature of space and the size of the circle. In a granular space, it is not well defined.[c]

In summary, real numbers are *not* real. They are fictions; we devise them. On one hand, like some other fictions, they are very useful. On the other, it's these very numbers whose imagined reality seduced millions of physicists—from Newton's time, through Riemann's and the modern era—into an almost exclusive dalliance with unreal continuous space and its affiliated math.

The cumulative consequences of those choices shape our world today.

---

a For more on *pi*, see chapter 112 at page 551.
b It finds its way far into quantum theory, with 2π appearing as divisor of Planck's constant; their combination is so widely used it gets a special symbol, **h**.
c See chapter 20 at page 122; in a granular space, the ratio has no precise definition, see chapter 112 at page 551.

108
GOT YOUR NUMBER

> Can we, in the identity 1 + 1 = 2, put for 1 in both
> places some one-and-the-same object, say
> the Moon? On the contrary, it looks as though,
> whatever we put for the first 1, we must put
> something different for the second.
>
> Gottlob Frege (1884)[1729]

Fig. 83. Evenley school

In my first school—one room (as I mistakenly recall it; see fig. 84) in an 1834 stone structure at the corner of Broad Lane and Church Lane by the village green in Evenley near Brackley in Northamptonshire, UK (see fig. 83)—Miss Care taught all six grades and we first graders studied elementary arithmetic.

We were all[a] blissfully unaware that mathematician Jacques Hadamard had just written,

> [A]rithmetic, which is the first study in elementary teaching, is one of the most difficult, if not the most difficult branch of mathematics, when one tries to penetrate it more deeply.[1730]

The universe uses numbers that are really natural.

---

a I think I may, even at this remove, speak for Miss Care in this.

Sixty years before Hadamard, Frege thought math teachers could do better. Maddy said,

> Frege was concerned to provide a foundation for ordinary arithmetic. He was scandalized by the lack of understanding, even among mathematicians, of the fundamental concepts of their subject, in particular, the concept of a natural number.[1731]

He said (above) the counters for a number must be counted only once.

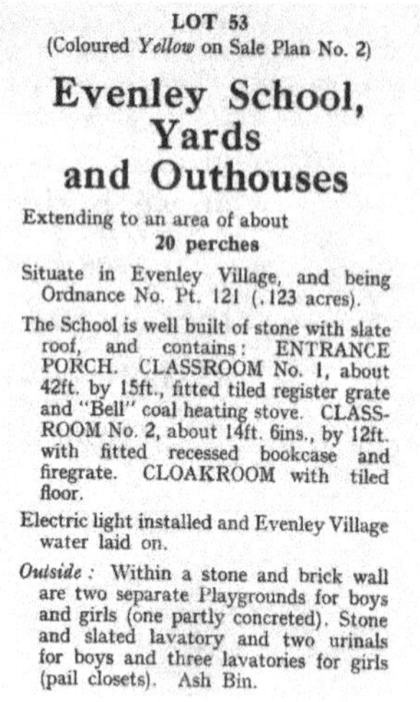

Maddy said Frege . . .

> . . . insists . . . a number is a thing, a "self-subsistent object," rather than a concept.[1732]

> LOT 53
> (Coloured *Yellow* on Sale Plan No. 2)
>
> **Evenley School, Yards and Outhouses**
>
> Extending to an area of about
> **20 perches**
>
> Situate in Evenley Village, and being Ordnance No. Pt. 121 (.123 acres).
>
> The School is well built of stone with slate roof, and contains: ENTRANCE PORCH. CLASSROOM No. 1, about 42ft. by 15ft., fitted tiled register grate and "Bell" coal heating stove. CLASSROOM No. 2, about 14ft. 6ins., by 12ft. with fitted recessed bookcase and firegrate. CLOAKROOM with tiled floor.
>
> Electric light installed and Evenley Village water laid on.
>
> *Outside*: Within a stone and brick wall are two separate Playgrounds for boys and girls (one partly concreted). Stone and slated lavatory and two urinals for boys and three lavatories for girls (pail closets). Ash Bin.

Fig. 84. School description (1938)

Our ontology is saying he was right to this extent: Some of them are.

Frege decried the lack of answers to the question what the number 1 is. He said,

> Yet is it not a scandal that our science should be so unclear about the first and foremost among its objects, and one which is apparently so simple? Small hope, then, that we shall be able to say what number is.[1733]

More than a hundred years later, Maddy was clinging to reality, backing off from sets and canvassing numbers as a property:

> [N]umbers must also be located in space-time.[1734]

Or maybe (she pursued the speculation) they could be collections (yes, it does sound quite a lot like sets), so *two* is the *collection* of all concepts of things that can be identified in twos, like Noah's list of boarding passes.

This all led into internal inconsistency, the capital offense of logic. Frege's new foundation turned out to be cracked.

Maddy summarized,

> Frege's aim was to show that arithmetic is in fact a branch of logic.... This project failed.[1735]

The deepest contradictions, as we can now see, arose from his striving to straddle incongruent, often at odds realms—reality and fiction—with a single set of objects.

The new ontology says each of these two realms has its own set of objects. The math of reality must find its objects in the universe, in what it is and in what it does—that is, in space quanta and their operations.

We've seen how the universe, within its first two iterations, reified four such objects:

- the initial condition   *1*
- *1* → *2*, succession   **S**
- *1* → *2*, addition   **+**
- *2* → *4*, multiplication   **×**

With these objects one can easily construct the natural numbers. As you read this, the universe continues work on it apace.[a]

Note, again, there is no zero.[b]

Some may wonder, *What is the successor to 9?*

True, if one wishes to express these numbers to base ten, it's *10*. But *that* zero has a different meaning; it's a place holder. The value ten arises from the *1* that's in the *tens* column; the zero serves to show the presence of the units column so the column to its left *is* the tens column.

We could choose to use the base sixteen, in which ten would be written **A**.

We could use the Roman system, in which ten is **X**.

We could just say ten.

All of which goes to show, the way we write our numbers is a matter of convenience.

The universe, with no need of convenience, expresses its numbers by making all of them. They simply are themselves.

---

a It has added and multiplied its way to a large but finite number of quanta, which has implications for the limits of real arithmetic; see chapter 111 at page 547.
b Peano's and Gödel's basic objects included these four *and* (latterly in Peano's case) the number zero.

With the universe's math objects we can create natural numbers up to some six-hundred-plus binary digits, where we reach a fundamental limit, the number of space quanta in the universe. We run out of counters.[a]

The universe's objects look to me to be the toolkit one might want for real arithmetic.

---

[a] See further, chapter 111 at page 547.

109
ZERO ZEROES

> Each number pertains to specific collections of
> things but zero to nothing at all.
>
> Robert Kaplan (1999)[1736]

Mathematician Robert Kaplan knows *nothing*. He wrote the book on it. He is not the only one to question zero. One might say it has at best a shady reputation.

Its status as a number has long been debated. For example, Barrow said,

> [T]he ancient Greeks ... demanded a logical consistency of their concepts and could not countenance the idea of "Nothing" as a something.[1737]

Like nothing, zero has no presence in the universe.

How did we manage to create the engram in our brain that is the concept zero? Psychologist Benjy Barnett and neuroscientist Stephen Fleming say,

> [T]he human ability to represent the number zero may be grounded in perceptual capacities for detecting an absence of sensory stimulation.[1738]

Like an absence of predators, Barnett suggested.[1739] If so, these days it has become even more abstract.

If one were to say that there are zero zeroes, it sounds like an oxymoron. That's okay; in the philosophy of mathematics, zero is chief source of contradictions, and some of them are much more serious than that.

To set it up, let's note that, as with other fundamental number matters, there's a strong connection to physical reality, including that the universe has no void.[a] Including too, the universe's count of manifolds began, not with a zero, but with one.

There are of course many zeroes to be seen. They do various things. For example, they can make a dollar worth much more, as in $1000. But like that bill, each zero is a fabrication; it's what we decide to think it is; it's fictional, its ovate symbol unable to bring reality to anything.

In the 600s CE, astronomer and mathematician Brahmagupta may have been the first to imbue zero (then called *cipher*) with numeric meaning.[1740]

In 1884, Frege wrote a book about arithmetic,[b] one that is influential to this day.[1741] Exploring many views on numbers, he considered at some length the number *one*.[c] He gave *zero* only a few mentions.[d]

In 1889, Peano—making no mention of Frege (with whom he later had a lengthy correspondence[1742])—set out to rebuild arithmetic upon a foundation of axioms and logic,[1743] saying,

> Questions pertaining to the foundations of mathematics ... still lack a satisfactory solution. The difficulty arises principally from the ambiguity of ordinary language.[1744]

Russell (wrongly) explained,

> [Peano's] exposition has the inestimable merit of showing that all Arithmetic can be developed from three fundamental notions (in addition to those of general Logic) and five fundamental propositions concerning these notions.... Peano's three indefinables are 0, *finite integer*, and *successor of*.[1745]

Elsewhere, he said Peano's axioms ...

> ... thus became, as it were, hostages for the whole of traditional pure mathematics. If they could be defined and proved in terms of others, so could all pure mathematics.[1746]

---

a See chapter 89 at page 444.
b See chapter 108 at page 535.
c He wrote twenty solid pages on it.
d Frege did note, "In the case of 0, we have simply no object at all from which to start our process of abstracting." *The Foundations of Arithmetic: A Logico-Mathematical Enquiry into the Concept of Number*, note 1676, his p. 57.

Zero is now generally portrayed as the source from which Peano's natural numbers spring. We are told (though it is not originally true[1747]) Peano's first axiom[a] said,

> Zero is a natural number.[1748]

His number one supposedly followed as the successor (an operation his sixth axiom established) to his axiomatic zero. And his eighth axiom is said to have said,

> There is no natural number whose successor is zero.[1749]

In fact (as the reader can see in fig. 85[1750]) his first axiom originally said *one* is a natural number.[b] None of his axioms even mentioned zero.

Fig. 85. Peano's original axioms

Like others, Russell omitted mention that Peano's original text had no zero and his numbers began with one.

This failure may seem harmless but was not. Its detrimental consequences were pervasive. Zero, an ill-defined concept (indeed a motley crew of concepts under one linguistic roof[1751]) slipped almost unchallenged into the heart of our number system.[1752] In one fell stroke it made all our arithmetic unreal.

---

a In Peano's logic language, this axiom would read: *0 ∈ N*.
b In his symbolic language, *1 ∈ N*; that is, one is among the natural numbers.

The unreality is simply this: Among the universe's math objects *there is no zero*. If there were a lack of manifolds that somehow led to the first one, it was *not* part of our universe.[a]

Another way to put this is I can see no meaningful way to posit a beginning of the universe that is no manifold.

Or, to put it a third way, the universe's mathematic objects arise from space quanta that had real existence; it has no natural expression for something that did not exist.

Or, yet another way, just as we do when counting, the universe's count began with *one*.

Of course, one might propose to obtain zero by subtraction. Peano didn't. *Semble* neither did the universe: To make (or fake) a case in favor of the universe deleting extant manifolds one needs must arbitrarily assume it. (And if it had deleted the one manifold, where would we be?)

In sum, a lesson from the new ontology is zero is not a natural number. It has no presence in the universe. It has no role in real arithmetic.

Real arithmetic requires original Peano.

This too will turn out to have consequences. (They look to be good.)

---

a This is my conclusion. I leave it to philosophers to chew the issue over.

110
INFINITE WISDOM

> The most fundamental intellectual question in all of mathematics is the nature of infinity.
>
> Reviel Netz (2013)[1753]

Historian of mathematics Reviel Netz makes a strong point that the nature of infinity needs careful study. However, there are various infinities.[1754] Are any of them real? If so, it would only be the simplest.

The question *Is there an infinity?* puzzled thinkers about that infinity at least as far back as Aristotle's day. He said,

> [I]t is incumbent on the person who specializes in physics to discuss the infinite and to inquire whether there is such a thing or not, and, if there is, what it is.[1755]

| Infinity's most important property is non-existence.

Over intervening years, some did inquire, with, it seems, at most mixed benefit.

For example, in the eighteenth century, Kant strove with the universe's size to small avail and much confusion:

> If, accordingly, we say: the world is either infinite in extension, or it is not infinite; and if the former proposition is false, its contradictory opposite—the world is not infinite—must be true. And thus I should deny the existence of an infinite, without, however, affirming the existence of a finite world.[1756]

Nineteenth-century mathematician, author of set theory, and emperor of all the infinites, Georg Cantor, claimed an infinity is real and present in this world (and he found many more of them besides).[1757]

In the late nineteenth century, Lord Kelvin weighed in, favoring an infinite universe:

> Can you suppose an end of matter or an end of space? . . . Even if you were to go millions and millions of miles the idea of coming to an end is incomprehensible.[1758]

As an argument in favor of infinity this rests upon (as we have seen[a]) a false assumption. But Kelvin was right to this extent: An end *is* incomprehensible.[b] Two centuries of tinkering with Newton's physics failed to solve the problem till Einstein, master tinkerer, said space is curved. This exposed the fallacy in Kelvin's argument; thanks to that curvature, one can go millions *of* millions of miles in a finite closed universe without ever "coming to an end."[c]

Open-ended cosmology is not the only path to an infinity. Another is by counting points on a continuous line (the so-called *real line*). Between any two points on the line (each with its number), no matter how close, there are always other points. For example, add the numbers and divide by two. Et cetera forever (aka ad infinitum).[d]

The new ontology says there's a false assumption hidden in this too: There *are* no continuous lines in this granular universe. Nor are there points.

Many mathematicians and philosophers have trouble with infinity as it can never be actualized in the real world. Even with room to do it, no one could take an infinite walk nor could they count its paces. The reality is, no matter how long one counted—say the lifetime of the universe times ten to the power of the number of atoms in the universe, just for starters—that number would be finite.

Yet physicists, on one hand, tended to blithely assume space and time are both infinite and infinitely divisible; and, on the other hand, they are vexed by infinities that in consequence crop up in their equations. Of this, Close said,

---

a Our best efforts to make sense of it say the universe is finite but endless; see chapter 22 at page 133.
b For an ancient take on it, see fig. 17 at page 133.
c Riemann made the same observation sixty years before him; see chapter 12 at page 74.
d See chapter 107 at page 532.

Part VI: The Numbers Game 545

For physicists, infinity is a code word for disaster.[1759]

Mathematicians and philosophers don't share in the equations' problems but some share in the blithe assumptions. So we find Russell casually saying in 1919,

> It is to be presumed, for example, that there are an infinite collection of trios in the world,[a] for if this were not the case the total number of things in the world would be finite, which, though possible, seems unlikely.[1760]

But, in 1925, another of the world's most influential mathematicians, David Hilbert, said scathingly,

> A careful reader will find that the literature of mathematics is glutted with inanities and absurdities which have had their source in the infinite.[1761]

Warming to his subject, he counseled,

> [T]he infinite is nowhere to be found in reality. It neither exists in nature nor provides a legitimate basis for rational thought.[1762]

Yet Maddy, notable advocate for real math, treated those who deal "with the infinite by rejecting it outright" almost as extremists, warning,

> [A] series of striking consequences follow. . . .[1763]

. . . (followed by a partial list), then saying,

> [I]ts most serious drawback is that it would curtail mathematics itself.[1764]

With due respect, not so: It would *advance* mathematics, splitting it cleanly into real math and math fiction, each with its right realm and appointed opportunities.

Maddy's stated reason was, not that she failed to see the need for a solution, but,

> My own working assumption is that the philosopher's job is to give an account of mathematics as it is practiced, not

---

a By "world" he meant the universe.

to recommend sweeping reform of the subject on philosophical grounds.[1765]

(Here I must, with much regret, part company. Philosophy's prime task in my book *is* sweeping reform.)

The new ontology sets out the ground for resolving the long-vexing question of infinity: The universe is finite[a] as it is made of a finite number of finite quanta.[b] Its history is finite.[c] It has no outside in which to find infinity.[d] Whatever future it may have is finite too.[e]

This worldview tells us straight: There's no infinity of any kind whatever in this universe.[1766]

---

a See chapter 29 at page 165.
b See chapter 45 at page 255.
c See chapter 17 at page 105.
d See chapter 22 at page 133.
e Its age may be endless, but it will always be finite; see chapter 57 at page 307.

111
THE UPPER CRUST

> In order to develop the arithmetic of natural numbers ... I need to appeal to ... the "Hypothesis of Infinity." Basically, this hypothesis says that there are infinitely many "objects."
>
> Charles Chihara (1990)[1767]

Numbers are the source and soul of math. The great mathematician Carl Gauss said,

> Mathematics is the queen of sciences and number theory is the queen of mathematics.[1768]

Number theory now brings us to a sharp but shifting boundary between what's real and what's invented.

Speaking for the views of many in philosophy and mathematics,[a] Kaplan said,

> [N]umbers cannot exceed the number of things there are.[1769]

It's obvious there is no largest natural number (and also wrong).

If, as philosopher of mathematics Charles Chihara (and many others[b]) assumed, the universe were infinite, this rule would pose no limit.

But, in our new true view, the universe is finite and holds only a finite number of things.[c] The things that are in far the largest numbers are space quanta.[d] And, once they are counted, there is nothing left over to count.

---

a Indeed, there seems to be something close to consensus about this concept, a rarity in the philosophy of mathematics.
b See, for example, Russell, quoted at endnote 1760.
c See chapter 22 at page 133.
d Or, to be technical, their links; but it is not clear they qualify as objects.

It follows that finitely many of the natural numbers—like 7 or 123456789—are real; and that infinitely many more are not. Let's take a closer look through the lens of a simple theorem of number theory. Mathematician Stephen Barker summarized,

> [I]t is a law of number theory that there is no largest natural number. In order to be able to deduce this law as a theorem of *Principia Mathematica*, Whitehead and Russell found it necessary to introduce what they called the axiom of infinity. This is an axiom asserting the existence of infinitely many entities of the lowest type. The unattractive thing about this axiom is that it does not jibe with the philosophy of realism.[1770]

More than fifty years ago, leading realist philosophers of mathematics Paul Benacerraf and Hilary Putnam traced how Frege set out to evade this problem by purporting to count points in lines:

> Another difficulty is the need for an Axiom of Infinity in deriving mathematics: in order to meet this difficulty, Frege . . . proposed to derive mathematics from geometry (where the Axiom of Infinity is true, since presumably there are infinitely many points) instead of from "logic."[1771]

Once one has infinity in view, so to speak, it seems so obvious that there is *no* largest natural number that physicist and mathematician Wolfgang Mückenheim, who taught this, could casually write,

> Of course, there is no largest natural number.[1772]

For which he might be more easily excused if the title of his paper were not "Physical Constraints of Numbers." For sure, there is that number theory theorem in support of what he said. But there's a physical constraint of numbers that makes the theorem wrong. (And so, too, the Axiom of Infinity.)

The flip proof of the theorem says, if there were a largest natural number N, just add one to it and you get $N + 1$, a larger number. QED. But this proof assumes there *is* that extra one to add. As we have noted above, Frege and most number theorists conclude it must be something extra—over and above objects already counted.

The space quanta that comprise our finite universe are plainly far more numerous than all other "objects."[a] Their number is finite. So, the proof depends upon a false assumption. The physical constraint is, when N is the number of quanta, there physically *is* no extra quantum; thus there is no 1 to add to N. Unnoticed, that purported proof slipped us over the border from reality into the realm of fiction.

This largest natural number reconciles the deep conceptual divide between the realist and antirealist; it demarks where their minds may meet. It doesn't matter what the number is at any instant, nor that it will be bigger after the next iteration. It defines in principle a boundary between two realms—below it and above it—in which the one view and the other is appropriate, respectively.

Looking to real attributes of the universe through the lens of the new ontology offers an even wider reconciliation between realism and antirealism.

The essential primitive objects of arithmetic include the natural number 1, and three operations—addition +, multiplication ×, and succession S.[1773] All four are realized in what the new ontology deduces must have happened as the multi-quanta universe burst into being.[b] By our strict criterion,[c] they are all real.

Using these four primitive objects we can construct other natural numbers, 3, 4, 5 . . . (though we cannot, please note again, construct a zero). These numbers too exist as countable quanta, up to their total number at each instant, but no more.

Thus, the realists are right. There is a body of math that is a fundamental aspect of the universe; its elements exist and we discover them. This real math includes natural numbers from one up to the number of space quanta.[d] Written to base ten, it is only a few hundred digits long.

---

a One could use a larger number by adding in objects such as atoms or apples; this would suffer from the uncertainties inherent in the definitions of such objects, see chapter 104 at page 515; it would be a negligible increase; and most significantly it would violate the rule against recycling counting objects. Or one could count links, to no advantage for the argument.
b See chapter 42 at page 236.
c The tough test set up in the Preface at page 1.
d See chapter 104 at page 515.

True, its right-hand digits change millions of times each trillionth of a trillionth of a trillionth of a second. Nonetheless, at each iteration—the very definition of complete reality—there's an exact largest natural number. That there is a larger largest natural number in the next iteration does not impair this fundamental limit.

We can do arithmetic with a number that is, say, ten times this number, but in doing so we leave math fact behind and enter the world of math fiction, where the antirealists may hold their sway.[a]

This distinction may appear pedantic. Yet it solves a problem that for a century or more has undermined the deep foundations of, not just arithmetic, but other math built on them, and the physics built upon that math. The pedantry—if such it is—disposes of a consequential problem that long seemed inscrutable.

Recognizing that our science says there *is* a largest natural number can hardly fail to start a new math revolution.

---

[a] Or of math speculation about a very distant future.

112
HUMBLED PI

> And so, in that Greek letter that looks like a shack with a corrugated tin roof, in that elusive, irrational number with which scientists try to understand the universe, I found refuge.
>
> Yann Martel (2001)[1774]

Pi shares a lifeboat with a Bengal tiger.

He's obviously fiction. Why is he called Pi? Kids can be cruel when your name's Piscine. And, as author Yann Martel has him here make clear, he is aware (in fictive fashion) his alias is tied to the said to be real number we designate with the Greek letter π.

It is less obvious why π itself is fiction.

It's such a simple concept. Pace a circle; pace across it; divide the first number of paces by the second. For better precision, pace a bigger circle or take smaller steps.

Of course, math has much better methods.[a] Indeed, you can compute π to as many digits as you might desire.[1775]

At a deeper level, we see π pretends to a precision that the cosmos can't provide. It's another of those useful fictions whose success sets up a barrier for those who seek a better understanding of the world.

Poor pi—in quantum space it loses an infinity of digits.

Mathematicians take its supposed infinite extension seriously. And there are those who memorize its digits (in decimal form). In 2015, memory trainer Suresh Sharma correctly recited some seventy thousand.[1776]

His record ventured far into the fiction. And even more of π was known; a hundred trillion digits are available out there.[1777]

---

a For example, calculate 1/1 − 1/3 + 1/5 − 1/7 + 1/9 . . . as far as you like, and multiply your answer by four.

These flourishes partake of the reality that so-called "real" numbers are all fiction.[a] But *pi* has idiosyncratic reasons for its lack of status, to do with the shape and with the granularity of space.[1778]

One reason is π's meaning runs afoul of space's quantum basis after a few hundred digits. As Planck scale approaches, the certainties of circles and diameters begin to fray.

The other is the curvature of space. At the scale of the universe, π will go as low as 2.[b] At ordinary scales it gets as close as we might want to its conventional value, but even if there is no local mass it will not be exact unless we use a vanishingly tiny circle, and there it runs into the fraying problem. So, you see, in the real world π is caught between the space-quantum devil and the universal deep blue sea. I haven't done the calculation but I bet its real digits don't get half as far as Sharma did.

Quantized space cannot support a perfect circle; but, if it would just hold still, I could trace out a good approximation. First, I would need a place where space is almost flat; that is, where there's no gravity to curve it.[c] Moving at the speed of light, as Emerson's transparent eyeball[d] I would pass from one space quantum to the next in a trajectory that is as close as possible to circular with a diameter of, let's say, one meter.

I would count the quanta as I passed, pretending they were in a line, finding 194,381,517,882,256,039,446,723,105,526,934,210 of them. This would take more than ten nanoseconds. And then I'd turn and cut through the center, counting the quanta, 61,873,558,833,333,493,606,734,952,282,288,172 of them. And there it is; π in the real world—this universe—is given by the first number divided by the second. It's a *rational number*—a ratio of natural numbers—no need for so-called "real."

If I did it again those numbers would no doubt turn out to be different. At the Planck scale, space tends to bop about all over and my

---

a See chapter 107 at page 532.
b A circle half the size of our hyperspherical universe is twice as long as its diameter; fig. 14 at page 124 may help to visualize this.
c In reality, the entire universe's tiny-but-non-zero curvature is unavoidable; see chapter 67 at page 348.
d See chapter 5 at page 43.

light-speed measurement has taken me 256,254,945,438,961,036,230,7 94,280,154,592,057 tocks of the Cosmic Clock. That's the iterations of the universe as I measured π to a mere thirty-five of Sharma's digits.[a]

Carl Sagan wrote a science-fiction novel[1779] in which aliens tip us off there is a message in π's distant digits, far more distant than even Sharma's ambition could attain.[b] Later, Sagan designed plaques that flew on the two Pioneer spacecraft that are now exiting the Solar System.[1780] Contrary to rumor the plaques do not mention π.

Just as well, I'd say. It wouldn't do to tell stray aliens our math is mythological. To send a stronger message we should figure π for the whole universe and truncate the digits at exactly the right place.[c]

---

[a] If it were possible to do this gedanken experiment, the quanta would not line up, either in a circle or straight.
[b] In the movie version, this last chapter is omitted.
[c] It would be smaller than 3.14 and larger than 2.0; see note *b* at page 552; to calculate it we would need to know the entire universe's size, or its precise curvature, which would be the point of the message.

113
SQUARING THE PLANCK CIRCLE

> The propositions of geometry [are] only of empirical certainty; they are hypotheses. We may therefore investigate their probability and inquire about the justice of their extension beyond the limits of observation, on the side both of the infinitely great and of the infinitely small.
>
> Bernhard Riemann (1854)[1781]

Geometry's frontier was once concerned with measuring the Earth. (Hence *geo+metric*.) As Riemann anticipated (above), it has become more concerned with measuring the universe at the extremes of scale.

To do common geometry you need some basic objects. For example, points and lines. With them in hand, you can construct squares, triangles, and circles. Then cubes and spheres. Or so geometers may say.

But to do geometry where things are really happening, physicists—whose discipline has become inextricably involved with it—will need to go behind those points and lines because they are not real and neither are the constructs we derive from them.

| Much depends on the kind of surface one can find to draw lines on.

As always, we should have in mind where we once were. Early geometers drew lines upon the ground for barns and houses and the like. But around 300 BCE, mathematician Euclid worked on geometry in his head.[1782] Widely translated, his books on it (collectively, the *Elementa*) became de rigueur for any decent education (e.g., see fig. 86). It was assumed there *is* only one geometry and it is Euclid's.

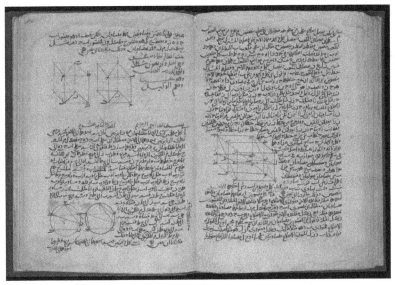

Fig. 86. Ishaq ibn Hunayn's 1270 CE Arabic Translation of Euclid's Elementa

It came to be much studied in the abstract. So, by the seventeen hundreds, philosopher Immanuel Kant could extol geometry as a creation of the mind, independent of experience.[1783] He knew Euclid's and could not imagine any other.

But in the early eighteen hundreds, Gauss espoused another view. He thought there could be more than one geometry and measurements should be the test of which of them is real.[1784]

For example, say you find angles of a small square all seem to be 90°. If you then draw a square with sides 6,225 miles and find each of its measured angles is 180°, you know you drew it on the surface of a sphere, with circumference 24,900 miles.[a] Abandon Euclid. Welcome to Earth.

In the early nineteen hundreds, Henri Poincaré was a philosopher, mathematician, and physicist. He brought all these viewpoints to bear, mocking those who found math only in their minds:

> It is by distancing themselves from reality that they acquired this perfect purity.[1785]

He concluded one can express laws of physics in many geometries.[1786] In this context, it was Einstein whose theory of general

---

a Compare with fig. 14 at page 124.

relativity—built in part on work of Poincaré—showed that a non-Euclidean geometry (one in which space itself is curved) allowed the simplest form for physics' laws of space and gravity.[1787]

Six decades earlier, in 1854, Riemann had recognized geometry *in* a space depends upon—and indeed, if one can measure well enough, reveals—properties *of* the space. He saw space could itself be curved.[a] He saw, too, it could be continuous or could be granular.[b] He said which of the two is real must "be deduced from experience."[1788]

Einstein investigated the geometry of a universe with matter in it using Riemann's math to manage space's curvature, both locally and for the entire universe.[c] But (at the time neglecting Riemann's thought on this, as others also did) he assumed space to be continuous and then embedded this assumption deep in the foundations of his math.

Today, "experience" says space is very slightly curved throughout the universe,[d] is much more curved near masses such as stars, and can be extremely curved both at the beginning of the universe and near extremely dense objects like black holes.[e] Observations show general relativity's equations describe such curvatures within the considerable limits of our measurements.[1789]

But, increasingly, "experience" was saying space is granular.[f] Einstein and a few others, reflecting on the (confused) message from Planck's quantum,[g] began to think the source of quantization must be quantized space.

Today there is a growing—but still widely disregarded—recognition that from this conclusion there is no escape.

---

a See chapter 20 at page 122.
b See chapter 12 at page 74.
c That is, in general relativity he allowed space to be curved; see chapter 20 at page 122; the theory of special relativity assumes that space is flat.
d Data from the Planck satellite show the space of the visible universe has a small positive curvature; see chapter 22 at page 133.
e Especially in black holes; see chapter 68 at page 351.
f See chapter 12 at page 74.
g See chapters 13 and 14 at pages 76 and 87.

Part VI: The Numbers Game 557

As well, a few who work on fundamentals are saying physics needs a rebuild with the quantum deep in its foundations.[a] But there is no consensus as to *Quantum of what?* We all—even the physicists; indeed, especially the physicists—require an answer.

There are three practical obstacles: minds, math, and money.

Geometry is deeply embedded in space physics. Some say Einstein's theory shows gravity *is* geometry.[1790] (It isn't.[1791]) A key role of geometry today results from our ability to see it, whether drawn or just in the mind's eye. It takes matters of serious conceptual difficulty and smoothly slips flawed visions to our minds to become part of the canonical worldview. Changing many minds on this seems likely to be difficult (one is reminded of Planck's maxim[b]). Yet for our most fundamental enterprises to make progress, one way or another they must change.

It might seem the math of space that's made of quanta should be simple.[c] It needs only simple numbers. It's devoid of zeroes and infinities. As Riemann said, it needs no arbitrary metric. Addition may be its most complex operation. But, as Einstein found,[d] discovering the math to make it simple seems to be unduly hard. It may need a breakthrough on the scale of Newton's and Leibniz's invention of the calculus. To get there swiftly calls for lots of postdocs. This brings us to obstacle three.

It's just possible the current system *might* find grant funds for this kind of research—but if so it likely would be by mistake. And grant funds are the metaphoric air that students and their supervisors breathe. As things stand, they would need to be more than a bit crazy to propose such projects. Yet that is what we need. Let's recall why.

Geometries emerged as good approximations to reality as space became much bigger than the Planck scale in the first tiny fraction of a second. Since then, points and lines and surfaces have seemed well defined even seen at subatomic scales. They aren't exact; but, if it's

---

a The alternative approach of finding more familiar ground and then quantizing it still seems to have more fans despite Einstein's disapproval; see chapter 45 and note 921 at page 255.
b See note 1991.
c See also chapter 73 at page 378.
d See note 1417.

measurements we want, they might as well be as we can't (and likely never will) detect the difference.

But if we want to understand what's going on, we must divest our minds of measurements and give up on geometries in space. We cannot comprehend what's happening in terms of points and lines and distances because, where things are happening, they don't exist. We need to find new math for quantum space; then our geometries will follow as approximations.

So don't try to think quantum geometry. Think quanta of volume and of area. Think digital math. Think causal sets.

And think grants for mad grad students.

114
O TO BE INCOMPLETE

> Gödel turned out to be an unadulterated Platonist.
>
> Bertrand Russell (1968)[1792]

Late in his life, Russell lived a while in Princeton. He had weekly chats with Pauli and Gödel and Einstein at the latter's house. Unlike others in the world of logic, he was not a Gödel fan. He said he was disappointed because,

> They all had a German bias towards metaphysics, and in spite of our utmost endeavours we never arrived at common premises from which to argue.[1793]

Do Gödel's theorems need rethinking?

Gödel's works were—and still are—widely debated. One conclusion is quite clear. He shook up logic.

Logic—right reasoning—has roots in Aristotle's *The Organon*.[1794] While modern logic rests largely on the works of nineteenth-century logician George Boole and of Frege, they embraced and enlarged upon Aristotle. American logician John Corcoran said,

> Even though Boole is thought of today as the initiator of a radical revolution that conclusively and irrevocably overthrew the "Aristotelian" paradigm then reigning in the domain of logic, he . . . accepted as valid absolutely every argument that was valid according to Aristotle— including those with "existential import."[1795]

Thus, logic seems to be (or ought to be) the relatively stable base of science. As we have seen, the nineteenth and twentieth centuries saw efforts to move math onto logic's firm foundation.[a]

---

[a] See chapters 104 and 106 at pages 515 and 528.

It was in this context that, in 1931, Gödel stunned the worlds of logic and of mathematics with his theorems.[1796] Philosopher Panu Raatikainen said,

> Gödel's incompleteness theorems are among the most important results in modern logic. These discoveries revolutionized the understanding of mathematics and logic, and had dramatic implications for the philosophy of mathematics.[1797]

They have been described in a variety of ways. One could say they showed that systems of arithmetic (such as Peano's[a]) cannot be consistent if they are complete, and vice versa. Or, no such system can prove itself to be consistent.[1798] Or one could say, as Russell seemed to think, they refuted his and Whitehead's monumental work, *Principia Mathematica*.[1799] Or, for simple clarity about the first, I like Quine's "gist" of it:

> Elementary number theory is the modest part of mathematics that is concerned with the addition and multiplication of whole numbers. Whatever sound and usable rules of proof one may devise, some truths of elementary number theory will remain unprovable; this is the gist of Gödel's theorem.[1800]

Maybe best of all is moviemaker Lutz Dammbeck's:

> The truth is superior to provability.[1801]

In the wake of these theorems there's no solid link from logic to arithmetic. Seen as a great accomplishment in logic, they were a disaster for the movement to set math (and, so, maybe physics) on a consistent logical foundation. But let's take a closer look.

Gödel showed that no matter what system of arithmetic one works with, if it has at least five fundamental objects (0, or "zero"; S, "successor to"; + "addition"; × "multiplication"; and = "equals") some true statements in that system can't be proved and some untrue statements can't be disproved. As well, one can't prove the math system is itself

---

[a] See chapter 109 at page 539.

consistent. This seemed to show the likes of Whitehead, Russell, and Hilbert were chasing a will-o'-the-wisp.

Peano and Frege began that monumental enterprise in the late eighteen hundreds; Whitehead and Russell picked it up and ran with it into the early nineteen hundreds, followed by Hilbert. As Barker described it, they sought to establish that . . .

> . . . all the laws of the mathematics of number are derivable from, or can be "reduced to," logic alone. . . . The definitions needed were definitions of all the basic nonlogical terms and symbols of number theory; that includes "zero," "immediate successor," "natural number," and "+" and "×."[1802]

Gödel relied upon this group of five math objects three decades later. In the math-philosophy world, everyone in town saw such objects—always including zero—as rock-bottom ground on which to build a system of arithmetic.

But, as we've seen, the universe's mathematics has no zero.[a]

What if Gödel had used only real objects? His proofs seem to give zero a key role;[1803] he would have needed to do something different. Could there be a workaround? If not, maybe math systems *can* be complete and consistent as the universe would seem to be—if they are entirely real.

I have no proof his theorems fail without the zero. But I have good reason to expect it: The universe *is* and it is everything; so, *it* must be consistent and complete. How, then, could its math be otherwise?

Even if the zero isn't critical, Gödel's proofs require[b] there to be infinitely many natural numbers;[1804] and there are only a finite number of them in our world.[c]

On 8 September 1930, Hilbert gave a lecture in Königsberg. He said,

> The true reason, according to my thinking, why Comte could not find an unsolvable problem lies in the fact that there is no such thing as an unsolvable problem.[1805]

---

a See chapter 109 at page 539.
b As philosopher Victor Rodych brought home to me.
c See chapter 111 at page 547.

And, (quoting Hilbert saying it) math biographer Constance Reid later said,

> At almost the same time that Hilbert was making his speech at Königsberg, a piece of work was being brought to a conclusion which was to deal a death blow to . . . the final program of Hilbert's career. On November 17, 1930, the *Monatshefte für Mathematik und Physik* received for publication a paper by a 25-year-old mathematical logician named Kurt Gödel.[1806]

Maybe Hilbert's final program is not an unsolvable problem. So maybe it can begin anew. (Without a zero, and with no infinity.)

PART VII

TESTING THE WATERS

It's difficult to speak the truth, for though *there is only one* it is alive and thus has a lively, changing face.

Franz Kafka (1920)[1807]

115
OFF THE SCALE

> The statement that a [quantum] system is at a given time and place in a certain state probably has a meaning, which the present state of our theory does not allow us to formulate.
>
> Max Born (1926)[1808]

To this day the quantum theory of which Born spoke a century ago still cannot give meaning to a certain state and time and place. We can comprehend why this is so when we see where the action is at the true quantum scale. Its time and place and state are all beyond the theory's unreal space and time's capacity for meaning.

The new ontology set out to understand the true nature of time and space, including how they fit into the quantum realm. Let's scan its scales to get a better sense of how remote they are.

For comparison, some forty years ago I co-wrote with biophysicist Don Chapman a text[a] about what radiation does to mammalian cells.[1809] It began with physics of fast events such as a light-speed particle passing an atom on a timescale of 10^{-18} s; it moved through intermediate scales of chemical events; it led to biological events that may take 10^6 s. It seemed impressive that science could gain an understanding of what we observe over a timescale range of twenty-four powers of ten.

> The frontiers of physics are now far beyond our powers of observation.

Today, twin deeply entwined disciplines—particle physics and cosmology—study physical events from the decay of a top quark in 10^{-24} s at the short end of the scale through intermediate timescales to the evolution of the universe over some 10^{18} s at the long end. This

---

[a] It was intended mainly for radiology residents.

growing picture with a timescale range of forty-two decades is a monumental human achievement. It enables science to reveal a story of our world in terms of fundamental physics based on observations.

The science is amazing and extremely useful. The story, though, is fiction. The reason is its tale is told in terms of made-up characters. We know their names; we bandy them about; but we admit we don't know what they are. In terms of understanding what is really going on, they leave us much where Planck was one hundred or so years ago—captivated and bewildered.

The new ontology tells us physics' simplest quantum systems are composed of half a dozen twist states jittering at light speed in a universe-sized Planck-scale forest of space quanta.

Science has no direct way to study twist states at the timescale at which real events occur. Planck told us what this timescale is: less than 10^{-43} seconds. To resolve an event at this timescale, we'd need measurements more than a billion-billion-fold speedier than our best. Their probings would need to be hugely energetic. Their energies would overwhelm whatever otherwise would have been going on.

The case is similarly hopeless when we think of spatial scale. Science studies objects from 10^{-58} to 10^{80} cubic meters. The low end lies in the realm of subatomic particles and the high end is the visible universe. But this impressive range of measurement—a factor of 10^{138}—falls far short of the space scale where what is really happening is happening. Again, Planck told us what this scale is: smaller than 10^{-104} cubic meters.[a] To see what's there, we need to work with sizes more than a quattuordecillion-fold[b] finer than our finest.

We might think of it like this. We seek a needle. We think it's in a haystack. But we can't detect the needle; we can't see its haystack; we can't see the field it's in or find the planet it is on or even spot its solar system. But we may yet get its galaxy into our view.

Plainly, we are not in the needle-finding business and maybe we never will be.

This leaves us with a choice.

---

[a] The Planck volume; see chapter 13 at page 76 and note *d* at page 80.
[b] A factor of 10^{45}.

We can give up on understanding what is really going on and put up with a fundamental physics that's becoming more than a bit moribund. Or we can have physics trade its links to logical positivism for a realist philosophy, rebuild a credible ontology, and use it as a guide to real physics that goes hand in hand with real understanding.

No prize for guessing my vote.

116
SEE NO EVIL

> The problem is that quantum mechanics is fundamentally about "observations." It necessarily divides the world into two parts, a part which is observed and a part which does the observing.
>
> John Bell (1987)[1810]

As Bell told us, quantum theory (and, more specifically, the brand called quantum mechanics; QM to disciples) has a thing about being observant. Not quite in the religious sense, but almost.

Consider something you set out to study as an experimental quantum system. In principle there are no limits to what it might be; but let's say it's something simple such as an electron.

Quantum mechanics describes the state of the system as a fancy math construct called its *wave function*, which sets the quantum system in a continuous space.[a] The wave function of our electron has a definite value even in far reaches of the universe[b] we can never see and quantum mechanics tells us how the wave function will change, forever. It gives us the probability of every way the state of the system could be at every instant, everywhere.[1811]

> Observing quantum events does not create reality.

QM's canonical interpretation says the quantum system is not actually *in* any real state until it is observed (and we will see why this interpretation is not true). Indeed, it says, the idea of any such reality has no meaning (and this too is untrue). When someone observes the system, its wave function is said to "collapse" throughout the

---
a There is an alternative formulation called matrix mechanics, which looks very different but is equivalent.
b Assumed to be infinite.

universe—though it's becoming clear it doesn't really[1812]—and the quantum system shows up in a certain place and state.

When we do this many times, we find QM gets all its probabilities precisely right.

Thus, backed by its success, its disciples contend there *is* no objective reality, only the results of experiments as seen by their observers.

Strangely, it is essential the observer is *outside* the quantum system. That is, the observer, and the apparatus that they use in measuring results, is *not* part of the quantum system. In physics language, the observer and their apparatus must be a classical system.

This insistence on a classical observer is a crazy problem for the theory. It's like, the quantum world's *created* by outside observers?

Why does QM need an observer at all? Let's see if this question's crazy aspect changes with the new ontology.

A key fact is quantum mechanics needs a background of continuous space and time. Indeed, this is what it is all about. Starting with the quantum system[a] at time t_0, it tells us what its wave function will be at every place at any future time, t_1.

Leave aside its need for infinite continuous space to work in (though there is no such thing). Leave aside its need to split the universe into two parts, one quantum and one classical. Leave aside its need for an observer in the latter. Leave aside as well the wave function's collapse throughout the universe the instant of the observation.[b] Leave aside too all its other craziness, such as "the only mystery" (two-slit experiments[c]) and spooky action at a distance.[d] All these observer-centric concepts of the quantum theory should fade like that Cheshire cat because its grin gives physics something new to chew on, a new causation question, future Nobel Prize alert, *How do the electron's meandering tweedles get together at their destination?*[e]

My suggestion, their arrival there is *arranged* by the "observation"—the electron's impact, for example, on an atom of a viewing

---

a That is, the state of the quantum system.
b It collapses to a value one where the particle shows up and zero everywhere else.
c See chapter 75 at page 386.
d See chapter 25 at page 144.
e See chapter 74 at page 378.

screen. But this has no need of an observer; it is just a physical event that needs the tweedles to arrive at the same place at the same time.[a]

I like to think of this in terms of tunneling. Back in the day, I used electrons tunneling across a barrier to measure the energy gap in available states that "waited" for them on the other side.[b] If there were no states that could receive them, those electrons didn't tunnel.

Over-simplistically, one might say that the electrons need somewhere to go. We have this ballistic picture in our heads of particles, like bullets, that simply follow a trajectory, arriving willy-nilly. Instead, the particle in quantum-space may be more like the crew of an airplane who take off with a flight plan that includes a place where they will land.

The confusing factor is there is neither bullet nor airplane in our quantum picture; instead, there are six twitchy tweedles navigating space according to some unknown physics law that has the tweedles get together at the end of their adventure.[c]

The virtue of quantum mechanics is it tells us all that can be known about those random-walking tweedles. Its vice is its frenetic talk of the observer. There's no need to dig up that old riddle: If a tree falls in a forest and no one is around to hear it, does it make a sound?[1813] The electron's tweedles will arrange to do electron things in the same way if the observer happens to go home before they land.

If she should, she will not "see" them all arrive but she can count on their complete cooperation.

---

a The observer's role is only to tell us that it did arrive or, we might say, to *make the measurement*. Einstein's clocks too need an observer if we want to know what times they registered and nobody makes much of that.
b See chapter 19 at page 118.
c The law is the kind of work we want from tomorrow's physics.

117
ON THE OTHER HAND

> Notice that in the braid model, the neutrino and anti-neutrino also only come in a single handedness.
>
> Niels Gresnigt (2020)[1814]

Physicist Niels Gresnigt was writing recently about Bilson-Thompson's model of the standard model particles.[a] Those particles generally come in left-handed and right-handed versions. (This handedness is also called *chirality*.) The versions are mirror images of each other that are distinguished by their opposite spins[b]—think left- and right-handed screw threads[c]—but are otherwise identical.[d]

Physicists have observed all the standard model's particles in both left- and right-handed forms. It is the basis of a fundamental symmetry (called *parity*): The laws of physics are the same for both chiralities.

But there is an exception—the neutrino. All the neutrinos that particle physicists have "observed" have been left-handed.[1815] And all antineutrinos were right-handed.[1816] Thus neutrinos are the universe's chiral oddballs; for them, the laws of physics do seem to play favorites.

> Like the universe, the tweedle model has no right-handed neutrino.

That we do observe this favoritism secured my early attention to the Bilson-Thompson explanation for the standard model's particles: His model has a left-hand but no way to make a right-hand form of the neutrino. He explained,

> [W]e create all the possible charge carrying braids in two different handedness states exactly once, but follow-

---
a This makes Gresnigt highly unusual.
b Spin in particle physics is a digital property that is loosely analogous to a spinning object.
c Roughly; a precise definition for particles with mass is messier and we won't go there.
d And compare with the two kinds of tweedles; see chapter 35 at page 190.

> ing the same procedure for the . . . neutrinos . . . would mean duplicating them, since this second pair of neutral leptons is identical to the first pair, rotated through $\pm\pi$. In other words, to avoid double-counting we can only construct the . . . neutrino in a . . . left-handed form, while all the other fermions come in both left- and right-handed forms.[1817]

Thus the Bilson-Thompson model seems to say in some sense the neutrino and the antineutrino are the selfsame particle:[a]

> Note that in this scheme we have created neutrinos, but not anti-neutrinos. This has occurred because the H_0 helon is its own anti-particle.[1818]

In particle-physics jargon this means the neutrino is a *Majorana particle*—a hypothesized kind of particle that is its own anti-particle.[1819]

Neutrinos in vast numbers are ready to hand, but famously elusive. They readily traverse the Earth quite unaffected. If you want to intercept even a few you need to put lots of mass in their way and then wait a while.

Since 2009, the IceCube Neutrino Observatory has been waiting for certain upgoing-neutrino collision-flashes in a cubic kilometer of transparent South Pole ice.[b]

One of its experiments looks for a pattern of flashes from a rare event the standard model predicts.[1820] It involves an antineutrino[c] with a specific (extremely high) energy smashing into an electron and creating a W⁻ parti-

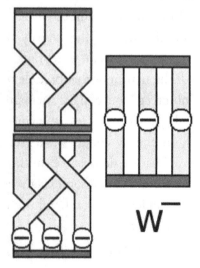

Fig. 87. *Neutrino collision with an electron makes a W⁻ particle*

---

a See note 1816.
b An upgoing neutrino has traversed the Earth and enters the ice from below.
c Neutrinos (and antineutrinos) come in three types; the prediction involved an electron antineutrino.

cle. Fig. 87 shows my cut-and-paste Bilson-Thompson–style[1821] show-and-tell for this reaction. (To follow the collision, fuse the top of the electron with the bottom of the neutrino; straighten kinks; and shrink to match the **W**⁻ on the right.)

In 2021, experimenters said they saw one such event.[1822] *Aha!* you may say—so there *is* an antineutrino.

Well, first let's note the four hundred thirteen authors of that paper called it an antineutrino because the standard model predicted an antineutrino would generate that event at that energy.[1823] Indeed, they went on to suggest this kind of event may offer a method for distinguishing between neutrinos and antineutrinos. But the standard model does not say yea or nay to the Majorana concept.[1824] That the observed event was an *anti*neutrino remains an assumption.

Next, Bilson-Thompson's model doesn't say there is no antineutrino; it says it is the same as the neutrino. This alleged identity, like all else involving the elusive neutrino, is difficult to test. But work is underway.[1825]

For example, under a mile of rock in Italy, physicists are suiting up for LEGEND-200. No, not the GPX motorbike, it's an experiment that holds two hundred kilograms of germanium crystals and shields them from background radiation.[1826] Not just any old germanium but mostly ^{76}Ge, the isotope that has forty-four neutrons, making it slightly unstable.

Mostly, the physicists will see the signature of *double beta decay*, a ^{76}Ge nucleus emitting two electrons and two neutrinos simultaneously. This decay has a half-life of some two billion-trillion years. Working that up with Avogadro's number,[a] we see they can expect to observe fewer than one such decay per minute.

Once in a long while, they may see an even rarer kind of decay called neutrinoless double beta decay.[b] It's not easy. In this modality, a ^{76}Ge nucleus would emit two simultaneous electrons and *no* neu-

---

a Avogadro's number is the number of atoms in a gram atomic weight of an element; it is 6.02214×10^{23}.
b This is sometimes denoted as $0\nu\beta\beta$ decay by contrast with $2\nu\beta\beta$ decay.

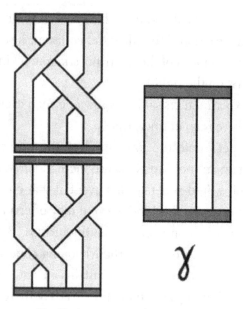

Fig. 88. A neutrino and antineutrino annihilate to form a photon

trinos. The experiment will need to show this absence of neutrinos amid blizzards of neutrinos that pervade the universe.

The reason this neutrinoless modality is seen as possible is simple. If the neutrino is its own antiparticle—that is, it is a Majorana particle—the two neutrinos, when they are created simultaneously, can instantly annihilate each other, as antiparticles are wont to do (see fig. 88 [1827]).

Showing the neutrino *is* a Majorana particle could lead us to new physics,[1828] just a small down payment on all the new physics that could arise from a real ontology.

118
NOTHING DOING

We haven't seen one.

Justin Evans (2021)[1829]

That neutrino observatory at the South Pole looks for flashes in a cubic kilometer of transparent ice.[a] More than 10^{20} neutrinos pass through it per second; most come from the Sun. A tiny fraction of them come from far beyond our galaxy, some with very high energies. Once in a long while, one of these hits a hydrogen or oxygen nucleus in the ice block and emits a flash.

Real neutrinos are mysterious enough. They rarely interact.[1830] Though billions travel through your thumbnail every second, it took a while for physicists to find them. They were invented to resolve a mystery[b] (they did this but they led to many more[1831]).

> **The tweedle model has no room for a new neutrino.**

Recently physicists invented a new version—the *sterile neutrino*—that they think they need to solve some problems; some say it might be dark matter.[1832] It is said to ignore ordinary forces.

Science writer Clara Moskowitz said,

> Such posited neutrinos are called "sterile" because they would only interact with other particles via gravity, whereas the known three flavors can do so through the weak force as well.[1833]

The weak force interacts with left-handed particles; right-handed, not so much. So (the suggestion goes) if there were right-handed neutrinos they too would just interact with gravity. Thus, the right-

---

a See the previous chapter and chapter 129 at page 617.
b Apparent violation of conservation laws.

handed neutrino (if it exists) is seen as a prime candidate to be the posited sterile neutrino.

Naturally, the IceCube crew looks for sterile neutrinos. It's in that context that physicist and founding member of the IceCube-Gen2 collaboration Justin Evans said (above), after they had taken a long look, "We haven't seen one."

I doubt they will, as Bilson-Thompson's particle menagerie has no obvious room for it. His standard model particles exhaust the simple combinations.[a] A new neutrino could call for a whole new scheme of things.

Recall he said,

> [W]e can only construct the (anti-)neutrino in a (right-)left-handed form, while all the other fermions come in both left- and right-handed forms.[b]

To me it's striking that this oddity in observations falls right out of the topology as an inevitable consequence. With typical understatement, he called it just a "pleasing result."

---

[a] See chapter 17 at page 105.
[b] See note 1817.

119
WHERE ALL THE ACTION IS

> Our task... must be to go beyond quantum mechanics to a description of the world on an atomic scale that makes sense.
>
> Lee Smolin (2019)[1834]

In his 2019 book (subtitled *The Search for What Lies Beyond the Quantum*) Smolin wrote about the challenge to "make sense" of quantum theory. Continuing, he wrote of the deep disagreement in the ontologic quantum scene,

> Physicists agree about how the quantum world behaves. We agree... that quantum mechanics works to predict some aspects of that behavior. But we don't agree about what it means that our world is a quantum world.[1835]

For this disagreement, too, the new ontology provides an answer: What it means is simply that the world was built by and of the real quantum! So I concur with Smolin's definition (above) of "our task," except that the needed "description of the world" is *not* "on an atomic scale," not even close. Here we see the heart of physics' conceptual problem.

| The quantum world is very tiny.

It's fine that physicists are seeking a solution. But they are, like Smolin, seeking at the wrong scale,[a] the atomic scale, that has held physics—and much of philosophy for that matter—in thrall for more than a hundred and twenty years.[b]

Quantum mechanics is ad hoc and abstract after the fashion of Kepler's laws of planetary motion, formulaic, brilliant but not

---
a See also chapter 115 at page 565.
b It still does; tellingly, Smolin's search "beyond the quantum" makes no mention of Planck-scale events.

insightful, ritual bereft of underlying reason. Being (as Kepler was) so successful, it begs us to find natural concepts from which its so far unexplained properties arise.

The reader might like to peek[a] at a 2011 paper by a physicist, Lucien Hardy, Smolin's colleague.[1836] A case study among other worthy efforts, it sought to address the problem Smolin was describing. At a glance, it shows to even the untutored eye how hopeless is the task without a guiding ontologic star.[b]

This example illustrates where we are at in meeting Smolin's challenge. Beyond the physics pale lies a multitrillion-dollar industry with tiny quantum workers that it does not understand—as in, no notion whatsoever, not even knowing they are there. How much better might its priesthood perform with some knowledge of its congregation!

It isn't easy. Looking to understand the quantum world at atom scale is on a par with looking through a telescope to understand bacterial infection in a globular star cluster (fig. 89). Those bugs may be all over it but it will be difficult to study them from here.

Fig. 89. A globular star cluster

---

a Search "1104.2066."
b A longer look also reveals two of the usual sources of confusion, continuity and spacetime.

The reader's study of the quantum world has had the benefit of Emerson's transparent eyeball.[a] Hopefully it has been helpful in conceiving where those quantum workers do their thing but, not to put too fine a point on it, we have been cheating. There's no way to see the scenes we've been imagining; there never will be.

But at least we have been "looking" where (as I have put it, and as Max Planck's message had it) all the action is.

The central message of our tale for quantum theorists—courtesy of Planck and his absurdly tiny number[b]—is that the action you are really working with is *not* at atom scale; it is far smaller.

Its intellectual realm—Planck-scale physics—is an almost empty field.[c]

In his book on physics' building blocks, 't Hooft said,

> There is work here to do for physicists: make a theory.[1837]

That was 1997.

About time for y'all to get with the action, *n'est-ce pas?*

---

a See chapter 5 at page 43.
b See chapter 13 at page 76.
c See, for example, note 312.

120
THE GOLDILOCKS ENIGMA

> There are literally tens of thousands of ways to curl up those six dimensions, and each one of them leads to the prediction of a different set of particles and interactions.
>
> Lee Smolin (1997)[1838]

From the early stirrings of our curiosity more than a hundred thousand years ago, odd motions of a few lights in night skies demanded explanation.[a] A reconstruction (fig. 90) of the Antikythera, an orrery dated to 205 BCE,[1839] may show how hard they worked back then to understand those cranky stars that we call planets and, to them, were gods.[b] Until the fifteen hundreds CE, the accepted stories (and their orreries) began with the fact—long seen as obvious—that Earth lies at the center of the universe.

| Physics avoiding religion is physics run by religion.

In 1543, a posthumous book by Copernicus showed it's simpler to see Earth as just another planet orbiting the Sun.[1840] It had a revolutionary impact on Western thought (including religion, as he was himself a leading figure in the Catholic Church, which accepted his worldview).[1841]

For science, the message was our place in the cosmos is not special. For example, science writer Lisa Zyga said,

> The Copernican principle states that the Earth is not the center of the universe, and that, as observers, we don't occupy a special place. First stated by Copernicus in the sixteenth century, today the idea is wholly accepted by scientists, and is an assumed concept in many astronomical theories.[1842]

---

a The Greeks knew them by their lights as the πλανηται (planetai) or wanderers.
b See, for example, the apparent motion of Mars, fig. 94 at page 612.

Part VII: Testing the Waters 581

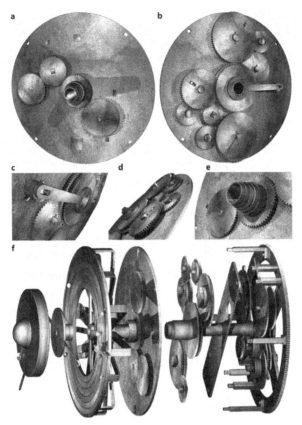

Fig. 90. A reconstruction of the Antikythera

That long, slow pendulum swung far to the not-special side. So, anything that seems to say we *are* in some way special meets with pushback from the scientific world.

Yet many physicists, writing of new data that bring new perspectives, are saying that the universe seems inconsistent with the Copernican principle in one respect. They say this universe is so good for us it's beyond improbable. Our comfortable cosmos is *less* likely to exist than crazy stuff like Neil Armstrong finding his favorite food awaiting on fine china when he stepped down on the Moon.[a]

---

a Spaghetti with meat sauce and scalloped potatoes, said NASA; that an anonymous rich astronautic jokester would put it on the Moon *and* guess the right (not as planned) landing site is—relatively speaking—much more likely than the suitability of the whole universe.

Davies said,

> To see the problem, imagine playing God with the cosmos. . . . Twiddle this knob and you make all electrons a bit lighter, twiddle that one and you make gravity a bit stronger, and so on. It happens that you need to set thirty-something knobs to fully describe the world about us. The crucial point is that some of those metaphorical knobs must be tuned very precisely, or the universe would be sterile.[1843]

Let's be clear, his "sterile" is not merely a less lovely planet. Nor even that it might be missing altogether. (Until recently, that might have seemed quite likely since a star with planets was thought to be rare. "Very few stars have planets," Russell proclaimed in 1933,[1844] uncomprehending of the scientific revolution that would say more than a hundred billion-trillion stars likely have planets.[1845]) No, Davies's issue is why there are any planets, or any stars, or even any matter whatsoever. There's a desperate need for some way to understand this without reaching for religion.[a] A literature has sprung up and he wrote the book.[1846]

Sticking closer to home, Smolin said,

> One of the things to be explained is why the whole universe from the largest scales down to the smallest produces a context that is friendly for life. This includes stable, longlived stars, needed to keep the surfaces of planets out of equilibrium for the billions of years life needs to develop, and plentiful production of carbon, oxygen, and the other chemical elements needed for life.[1847]

It does seem to leave physics with an insoluble problem. When one looks at all the ways it might be otherwise—that is, by varying adjustable parameters of physics' theories—the odds there would be stars at all are almost nonexistent. For example, Smolin estimated,

> [W]e should ask just how probable is it that a universe created by randomly choosing the parameters will con-

---

[a] Is this a reason why serious physicists are sinking to the depths of multiverses? We are coming to that.

tain stars. . . . The answer, in round numbers, comes to about one chance in 10^{229}.[1848]

In sum, when Weinberg wrote about "this . . . overwhelmingly hostile universe,"[1849] he could hardly have been more wrong. Against all odds, at every scale, the laws of physics give rise to conditions that look after us.

What's missing is an understanding of *Why these laws*? If they are the only laws they could be, if there is no way they could be other than they are, it may be a bit less of a problem. But the new ontology, which allows us to explain so much, explains these laws are all statistical; they reflect the universe's being made of quanta and the tangles of their six dimensions.[a] And, as the epigram above suggests, there are a lot of ways they could be tangled, each with its own unique universe.

Let's remember, those laws are not laws of *nature*. They are laws of physics. Physics is devised by physicists, who roll out revised versions on a roughly tridecadal basis. They don't say why their laws are what they say they are or how they came into existence. Mostly they don't even ask.

But lately physicists *are* asking *Why?* Smolin and six colleagues wrote one such paper in 2021. It's representative in that they promptly got lost in old ontological commitments, such as continuous space. This *Why?* should be a simple question; what they ended up with was,

> Why these gauge groups, why these fermion and scalar representations, why the mysteries of chirality, CP violation, and baryogenesis? Why the vast hierarchies of scale and why the particular ratios of parameters of the standard model, setting the values of the masses and mixing angles? It is sobering to contemplate that not one problem of this type has ever been solved.[1850]

Sobering indeed. Maybe there is a message there. Something about barking up the wrong ontologic tree?

But, even after simple understanding clears the ontologic air, we will still see the universe as just right amid inconceivably many bad alternatives. This will no doubt need another book.

---

a See chapter 97 at page 478.

121
MISSING MULTIVERSES

> Gradually cosmologists and physicists like myself are coming to see our ten billion light years as an infinitesimal pocket of a stupendous *megaverse*.
>
> Leonard Susskind (2006)[1851]

It would seem to be no contest: The explanatory power of one finite universe versus that of a growing (perhaps infinite) number of perhaps infinite universes. But I'm betting on the underdog. If you want to see the case for multiverses (or the pockets of the megaverse that Susskind favored, above), Greene wrote a book on it.[1852] I'm not about to spell it all out here.

A further confession; I see them as intellectual deceptions. I'll tell you their short story but excuse me if you feel I am being a bit too blunt.

Multiverses hover round a simple motive: Some say they explain away the sheer improbability of all we see;[a] but they avoid a conversation advocates don't want to have. The multiverse's secret task is not to explain the bears' situation but to do away with Goldilocks.

| The porridge is just right. Get over it.

As Ellis put it,

> Steven Weinberg, Martin Rees, Leonard Susskind and others contend that an exotic multiverse provides a tidy explanation for this apparent coincidence: if all possible values occur in a large enough collection of universes, then viable ones for life will surely be found somewhere.[1853]

---

a See the previous chapter.

We will see that multiverses suffer from two terminal conditions: They are founded on a fallacy (that they explain things); and they are not needed (as our universe provides an explanation of its own[a]).

Who are those who lend the multiverses credence?[b] Here's a quasi-random sample of respected writers who have flirted with it: Nima Arkani-Hamed; Paul Davies; Bryce DeWitt; Alan Guth; Stephen Hawking; Michio Kaku; Andrei Linde; Leonard Susskind; Max Tegmark; Alexander Vilenkin; Steven Weinberg; John Wheeler. These are names to conjure with in fundamental physics.

Physicist David Deutsch, who writes about explaining science, wrote a book on it with the doubly oxymoronic title, *The Fabric of Reality: The Science of Parallel Universes*.[1854] Some editions were even subtitled *Towards a Theory of Everything*. "Everything" indeed, and much, much more. "Theory" not so much. (His second book only compounds the problem.[1855])

First problem is, multiverses "explain" everything by assuming more than everything,[c] which is no kind of explanation.

There is still the loyal opposition. In 2004, philosopher Bede Rundle wrote,

> Postulating a million million universes, or whatever is needed to raise to an acceptable level the likelihood of this universe coming about, looks to be an expedient of unparalleled desperation.[1856]

And, five years later, Rovelli said,

> Today, many scientists do not hesitate to take seriously speculations such as . . . multiple universes, for which there isn't a wit of empirical evidence.[1857]

In 2015, Smolin wrote that it was time for physics to take out the trash, saying,

> Cosmology is in crisis. Recent experiments have given us an increasingly precise narrative of the history of our universe, but attempts to interpret the data have led to a

---

a Which is the subject of this book.
b Not all drank the Kool-Aid but they did all kiss the ring.
c And often infinitely more.

picture of a "preposterous universe" that eludes explanation in the terms familiar to scientists.[1858]

And recently, philosopher Philip Goff offered a simple way to understand the fallacy:

> You wake up with amnesia. . . . In front of you is a monkey bashing away on a typewriter, writing perfect English. This clearly requires explanation. You might think: "Maybe I'm dreaming . . . maybe it's a robot." What you would not think is "There must be lots of other monkeys around here, mostly writing nonsense."[1859]

Let's be clear here, multiple universes of any ilk—and there are several ilks—are fiction. They may sound science-like but there is no science to them whatsoever. At root, each is a trick that some suppose explains away the sheer improbability of the real universe.

Ellis said, of all of them,

> The proponents are telling us we can state in broad terms what happens 1,000 times as far as our cosmic horizon, 10^{100} times, $10^{1,000,000}$ times, an infinity—all from data we obtain within the horizon. . . . We just do not know what actually happens, for we have no information about these regions and never will.[1860]

It is, in other words, what Unger and Smolin call the cosmological fallacy[a] let loose into imagination mayhem.

Let's take a quick look at the main ways to imagine multiverses.[b]

The fad took root in math of quantum theory, which seems to say that any quantum system is simultaneously in all possible states until it's checked by an observer.[c] Think Schrödinger's famous cat in a box: The math portrays its state as both alive and dead—as long as no one looks.[1861]

In 1957 a physics student, Hugh Everett, proposed a way to solve its conceptual puzzle: Each state—live cat, dead cat—is real and your act of observation sends them on their separate ways, *each with*

---

a See chapter 132 at page 632.
b Each has various versions.
c Re the role of and need for an observer, see chapter 116 at page 568.

its own entire universe; you maybe saw the live cat and another you is in that other universe along with the dead cat. This bizarre *many worlds interpretation* of quantum physics envisages a vast multiplicity of universes (fig. 91).[1862]

Fig. 91. Everett's many-worlds multiverse

Outside a small fan club, few favor this idea. Smolin said a quarter century ago,

> To postulate an infinite multiplication of the world because one is unable to resolve a problem of interpretation is a bit like moving and purchasing a new set of kitchenware each time one doesn't feel like doing the dishes.[1863]

A second way to dream up a multiverse is via models of cosmic inflation that have the universe begin with an assumption called *false vacuum*. The *inflationary universe*[a] starts as a zone of empty space with high-energy density in the very early universe. How this math contraption came to be remains a mystery.[b] A tiny fraction of a second after the universe begins, the zone starts to expand far faster than the speed of light. It rolls down its artificial hill of energy thus turning into something—our universe.[c]

Some of these models can repeat this trick forever (fig. 92).[1864] Picture the early universe as a magician with white gloves and black hat. He pulls out of his hat not only rabbits but more well-equipped

---

a See chapters 21 and 53 at pages 126 and 291.
b See chapter 21 at page 126.
c This sketch no doubt does less than justice to the contraption; but it does deserve summary execution.

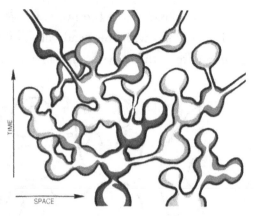

Fig. 92. *Universes spawn universes in the inflationary multiverse*

magicians with their worlds. Soon there is an endless multiplicity of universes. And, say some, like Greene . . .

> [W]e can imagine that physics varies from one universe to another.[1865]

Once one's imagination is thus cut loose to run riot, one can imagine more than that.

There is a third way. Multiverses are said to arise from the so-called *string-theory landscape* (fig. 93). Each of these supposedly may have its unique laws of physics.[1866] In 2006, Susskind said,

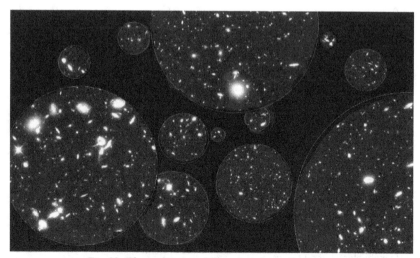

Fig. 93. *The multiverse of the string-theory landscape*

> Instead of producing a single unique elegant construct, [string theory] gives rise to a colossal landscape of Rube Goldberg machines.[1867]

If you thought his words were flipping off that landscape, think again. He said,

> The bubbling up of an infinity of pocket universes is as certain as the bubbling of an opened bottle of champagne.[1868]

It might be certain if we're certain string theory's certain but we're not.

Each magic multiverse comes gift wrapped in its magic math. They are, at best, what Penrose with fine Brit reserve calls "just nice ideas."[1869] And Ellis (sticking to the facts) said,

> None of the claims made by multiverse enthusiasts can be directly substantiated.[1870]

Such restraint is civil; but let's be clear what's going on here. Leaders in the world of fundamental physics are so desperate to fend off awkward questions about the foundations of their theories they embrace one or another of these seedy fictions. They seek to explain things that seem inexplicable; to do it they *assume* not only all the things they say they are explaining but also literally more than anybody can imagine.

Which is to say, they explain nothing.

It's not only that they don't explain the things they are intended to explain. It is not only that they cannot be substantiated. It's not only that they are (by definition) unreal. The central criticism is that something that's as simple as a new ontology allows us to explain—with just the universe we see—almost all the things that seemed so inexplicable.[a]

The multiverses earn a triply ignoble dismissal: They are ugly; they don't do their job; and they are unnecessary.

---

[a] But not our universe's Goldilocks condition, which the multiverses too do not explain.

122
THIS CHANGES EVERYTHING (ALMOST)

> Problems which were raised by Plato and Aristotle in the fourth century BC are still discussed, and the work of all the intervening centuries has brought us no nearer to finding a solution of them.
>
> Alfred Ayer (1982)[1871]

By now the persevering reader will have seen how our new worldview changes everything.[1872] Well, almost everything.

At one level, it literally throws everything out and replaces it with a new everything some 10^{43} times a second.

At another, it upends our ideas and the problems we have had with them. As Ayer observed (above), efforts to explain the world have not worked well for quite a while. Even new ideas have often tended to be new names for old problems.

| Change may come slowly in a world of rapid change.

Twenty years ago, Turner, the cosmologist who coined the term *dark energy*, said,

> [Dark energy's] deep connections to fundamental physics ... put it very high on the list of outstanding problems in particle physics.[1873]

Our new view says the solution to this problem leads to all else, because it *is* all else.[a] The entire universe is made of manifolds collectively misnamed dark energy.

In this manner and degree, the standard model of cosmology turns out to be built on a quicksand of unanswered fundamental questions. Not to put too fine a point on it, the model is an exercise in

---
[a] See part 2.

curve fitting to data that at first were bad and so the fitting was easy. Now, like my small example about cell survival,[a] it is turning out to be a bad fit to, if not good, at least better data.[1874] It is predicated on a story about what space is and why and how it is expanding that is simply wrong.

It may soon face a riptide of new data.[b] But it is a substantial structure. Replacing it with something else will—as indeed it should—face vigorous defense.

What isn't as it should be is one thing that hasn't changed: systemic roadblocks.

Fifty years ago, astronomer Gérard de Vaucouleurs was questioning the orthodox cosmology. He wrote,

> Above all I am concerned by an apparent loss of contact with empirical evidence and observational facts, and, worse, by a deliberate refusal on the part of some theorists to accept such results when they appear to be in conflict with some of the present . . . theories of the universe.[1875]

Twenty-five years later, astrophysicist and editor of astrophysics journals, Geoffrey Burbidge, wrote,

> Over the last decade or more, the vast majority of the younger astronomers have been conformists in the extreme, passionately believing what their leaders have told them, particularly in cosmology. . . . To obtain an academic position, to obtain tenure, to be successful in obtaining research funds, and to obtain observing time on major telescopes, it is necessary to conform.[1876]

Reviewing how cosmology treats new ideas, historian and philosopher of astronomy Martín López-Corredoira said, ten years ago,

> Cosmologists do not usually work within the framework of alternative cosmologies because they feel that these are not at present as competitive as the standard model. Certainly, they are not so developed, and they are not so

---
a See chapter 29 at page 165.
b See chapter 129 at page 617.

developed because cosmologists do not work on them. It is a vicious circle.[1877]

And, even more recently, Baryshev said of alternative cosmologies,

[U]sually they have small funding in modern scientific society.[1878]

Moffat may have the last word on this:

Physicists are like most other human beings; they become emotionally invested in the truth and beauty of the theoretical structures they have helped to build.[1879]

There are honorable exceptions. I have mentioned some. There is a reason why they are so few. The system grinds exceptions down.

123
ECONOMY CLASS

> We are dealing with . . . a profession that has been absorbed by theoretical constructs abstracting from human behavior. We are dealing with ingrained ways of thinking. . . . We need to pull economics back into the real world of political economy.
>
> Paul Volcker (2011)[1880]

So far, I've used the term *statistical* about two dozen times to describe emergent properties in physics. To illustrate potential ideaistic reach of our new worldview's implications, let's take a short digression to the scientific borderland.

For no clear reason, the first words of the first lecture of the first year of my minor in theory of statistics linger in my mostly missing memory:

> There are lies, damned lies, and statistics . . .[1881]
>
> Pausing, statistician Maurice Belz went on . . .
>
> . . . and then there are social statistics.[a]

Economics is not even emergent.

I never asked what "social" might include, but economics is seen as a social science.[b]

Much later, there was a time when I might have studied economics but (for friendship) I took law instead. The attraction of economics was it seemed there was not really, but should be, a *science* of economics. Like biophysics, this appeared to me to be an almost empty field, a tabula rasa calling for investigation. Easy said.

---

[a] This being no part of the well-worn quote.
[b] I limit my remarks to macroeconomics, the study of how large-scale economies work.

Any science that aspires to be quantitative[a] needs a metric and, just as with physics, the metric should be real.[b] Ontologic message: That means it should be extant, unique, digital, and indivisible so one can count it.[c] And every quantitative science other than physics has the good fortune that its real metric, if it has one, may at least be visible.

For example, pushing biology's boundaries in the nineteen seventies,[1882] ecologist Janet Dugle, a colleague from my biophysics days, devised a way to explore how relationships between the species in a natural ecosystem[d] adapted to the stress of chronic low-dose radiation.[1883] She got lots of data from a nearly million-square-meter boreal-forest plot.[1884] Each summer her crew of students would (with gamma radiation off) drop square-meter frames on random spots and count and identify all living things within them.

Her metric, in other words, was the number of each species, extant, at least supposedly unique, digital, and indivisible. If biology can do this in a boreal ecosystem, why can't economics do it in a human ecosystem? Maybe it can, but so far it hasn't.

Here too, there seem to me to be echoes of the physics story of successes that stand in the way of understanding. On Wall Street and in academe, for all its failings economics seems to get along. It has its Nobel Prize. It has its fuzzy metrics such as GDP or PPI.[e] It has its data. They are good enough only because there's nothing better. But, as economist Paul Volcker (see the epigram) suggested, at the working face they often don't work worth a damn. We can consider why.

Economics seeks to study the workings of the human economy in terms of notions like outputs and prices and express them in mathematical terms. As a guide to policy, this is a bootless enterprise. For example, central banks seek to tread a hidden path between the devil of inflation and the deep sea of recession. One need only recall recent geopolitical intrusions—such as climate change, or social media, or

---

a As distinct from merely descriptive.
b That is, to support study of reality rather than the foibles of the metric; see chapter 96 at page 475.
c This much we may intuit from Planck's experience and from Riemann's insight; see chapters 13 and 10 (at pages 76 and 63, respectively).
d Curiously, it was an essentially relational analysis.
e Gross domestic product and producer price index.

a pandemic, or a war or oil cartel—to realize that their effects on the economy routinely overwhelm those of even the most finely-tuned and timely monetary (or of fiscal, for that matter) tinkering.

Economics became sexier with the rise of the science-fantasy world of *econometrics*, which marries complex mathematical modeling to obscure statistical analysis. It has mainstream cheerleaders. For example, in 2011, International Monetary Fund economist Sam Ouliaris said,

> Econometrics uses economic theory, mathematics, and statistical inference to quantify economic phenomena. In other words, it turns theoretical economic models into useful tools for economic policymaking.[1885]

This, three decades after leading economist Sir David Hendry called econometrics out as *alchemy*.[1886] (His view may have since softened.[1887])

In the nineteen nineties, links of a sort to physics arose with the tag *econophysics*, as out-of-work physics PhDs found (or made) new markets for their skills. The American Physical Society said,

> Huge amounts of financial data suddenly become available at that time, and there were an increasing number of PhD physicists finding work on Wall Street[a] as financial analysts.[1888]

With all those PhDs and data, one might look to get something like physics' results. If so, one would be disappointed. For example, economics Nobel Prize–winner Paul Krugman said in 2009,

> [N]ot long ago economists were congratulating themselves over the success of their field. . . . Last year, everything came apart.[1889]

It's easy to blame methods that work except when they don't[1890] or squads of physics PhDs making meaningless models. But neither is the real problem.

---

a Also on Bay Street, where, incidentally, Bilson-Thompson's colleague Dr. Hackett found just such employ.

The real problem is economics isn't even trying to be real. Lacking a real metric, in the end it rests on ontological commitments that the universe refuses to support.[1891]

The notion (propped up by that pseudo–"Nobel Prize"[1892]) that economics *is* a science may continue to lose ground. People—though extant, unique, and digital by definition, and functionally indivisible like a real metric[a]—are too diverse and far too few to found emergent laws of economics. But I don't suppose that that will be the end of it.

---

[a] That they both reproduce and die (demography) would seem to be the key to anything like real economics.

124
PASSING IN THE ONTOLOGIC DARK

> Not from self, not from others, not from both,
> nor without cause. Production is not ever existing,
> anywhere of anything.
>
> Nagarjuna (c. 150 CE)[1893]

I should tip my hat to Buddhist thinking.

Though often classified as a religion, Buddhism has no deity; it abjures the supernatural; it proceeds from reason rather than authority. In all, it seems to not meet mainstream notions of what a religion is and does. It is more like a school of philosophy or, rather, group of schools.[1894]

Philosopher Mark Siderits put it this way,

> Buddhism is, then, a religion, if by this we mean that it is a set of teachings that address soteriological [i.e., salvational] concerns. But if we think of religion as a kind of faith, a commitment for which no reasons can be given, then Buddhism would not count.[1895]

Its schools argue implications of an underlying ontology (though a religion may do this too). In varying degrees, they are concerned with the nature of reality and, in particular, causality.[1896]

Thus, it might have eased my stumbling path toward the new ontology if I had studied Buddhism. Or, who knows, it might as well have turned my feet aside. In the event, I knew too little of it. Now, a bit better informed, I should note a few of what I'm told are more than a few parallels.

| Some say Buddhism taps into reality.

Here a disclaimer is in order. My purpose is not to expound or even summarize any of Buddhist thought. I seek only to acknowledge a body of works that bear some similar ideas. Though widely taught, they may be culturally inaccessible.[1897]

Buddhism's wellspring, Gautama Buddha, was, said scholar Kenneth Inada, "the Plato of his times."[1898] He lived and taught in present-day Nepal and Northern India around the time Anaximander lived and taught a half a world away. His teachings laid foundations for the schools of Buddhism.

Illustrating the broadly ontologic nature of mainstream Buddhist thought, philosopher Daisetsu Suzuki said,

> The ultimate object of the Mādhyamika school is śūnyatā [from śūnya, void, or zero] and that of the Yogācāra is dharmalaksana or ālīyavijñāna. Philosophically speaking, the former treats more of ontology and the latter chiefly of cosmogony.[1899]

Few plumb in depth what the Western philosophic worldview might learn from Buddhism's insights, which, to my Western eye, can seem (e.g., Nagarjuna, above) less concerned with *what is* and more with *what is not*.

Philosopher Peter Jones illustrated the problem of bridging their mutual divide, essaying himself . . .

> . . . to hastily sketch-out and clarify the logical relationship between the non-dual teachings of the masters and sages of the Perennial tradition and the formal metaphysics of Plato, Kant, Russell, Carnap, Wittgenstein, Chalmers and the "Academy," the stereotypically "Western" approach to metaphysics shared by . . . the majority of the world's working scientists, probably most other people and more than likely the reader.[1900]

Bridging this divide is a preoccupation of philosopher Jan Westerhoff:

> I studied Western philosophy (with a particular focus on metaphysics and the philosophy of logic) and Buddhist thought and languages (Sanskrit and Tibetan) more or less side by side, and my main motivation then and now

is to bring these two together, hopefully for the benefit of both.[1901]

The interested reader might try a recent paper that suggests to me a bridge between the two might yet be found.[1902]

From my own small sampling, certain Buddhist philosophic writing might be seen as speaking of ontology. Its foundational Sanskrit texts, such as philosopher Nagarjuna's *Mūlamadhyamakakārikā* (*The Fundamental Wisdom of the Middle Way*),[1903] consider space and time, being and becoming, in terms of a coherent worldview that is incompatible with mainstream Western thought.

Nagarjuna abjured reliance on objective reality, an approach which seems incongruent at least with the present work. Philosopher and Nagarjuna translator Jay Garfield said,

> Nagarjuna, like Western sceptics, systematically eschews the defense of positive metaphysical doctrines regarding the nature of things, arguing rather that any such positive thesis is incoherent and that, in the end, our conventions and our conceptual framework can never be justified by demonstrating their correspondence to an independent reality.[1904]

Yet, said Garfield and a colleague,[1905]

> [Nagarjuna] is telling us about the nature of ultimate reality. There are, therefore, ultimate truths. Indeed, that there is no ultimate reality is itself a truth about ultimate reality and is therefore an ultimate truth![1906]

From my perspective, this seeming incongruence is of interest because behind it lie parallels with the new ontology, which is as we have seen entirely founded upon "correspondence to an independent reality."[a]

Here are three examples of both incongruence and parallels.

Distinction between observer and observed is fundamentally enmeshed in seminal ideas we explored in part 1.[b] Thus, in Western philosophy, *a la* Descartes, my consciousness seeks to experience

---
a See note 1904.
b See chapter 2 at page 22.

and validate external reality;[1907] and, in Western science, we have the requirement for observers external to observed quantum and relativistic systems.[a] By contrast, a Buddhist worldview sees such distinctions as illusory. Rather, there is, it says (shades of indigenous worldviews) one world.[1908] Understanding it and understanding ourselves is one thing, not two, that worldview says.

So, too, for entirely other reasons, says the new ontology.[b]

In a second example, the new ontology says the locus of reality is the instant static ensemble of space quanta, with their relationships, that exists as elusive, nebulous, and uncharacterizable *Nows* and is dynamic as each *Next* creates another *Now*.[c] Considering the parallel question, Inada said,

> Buddhist reality or ontology is dynamic, and its locus is in the momentary nows, however elusive, nebulous, and uncharacterizable they may be.[1909]

For a third example, the Buddhist worldview is populated by a boundless network of causal relationships among ephemeral objects.[1910] Equivalently, the new ontology says the universe is a vast (but bounded) network of causal relationships among inaccessible space quanta.[d]

The Dalai Lama (some may be surprised to learn) studies physics as well as Buddhism. He writes of relativity and quantum theory. He too seeks after reality:

> Clearly [the current paradigm of what constitutes science] does not and cannot exhaust all aspects of reality, in particular the nature of human existence.[1911]

I do not want to overdraw the ontologic parallels but should acknowledge that they may exist.

And as we explore Planck-scale reality, we may find it provides the worldviews opportunities to kiss rather than to collide.

---

a See chapter 22 at page 133.
b See chapter 26 at page 150 for its origins.
c See chapters 53 and 57 at pages 291 and 307.
d See chapter 5 at page 43.

125
THE END OF ZEN

> That end is an intuitive realization of a single great insight—that we and the world are *one*. . . .
> Our rational intellect merely obscures this truth, and consequently we must shut it off.
>
> Thomas Hoover (1980)[1912]

Zen means meditation.[a] It is a means to an end.

In his history of Zen, writer Thomas Hoover told us the words and deeds of Zen masters through the ages all strove to the end he outlined in the epigram; what is now known as the Zen school has sought that "intuitive realization," that "single great insight," for more than two thousand years.[1913] But recently Zen's "*one*" surfaced in an unlikely place— the very essence of "rational intellect"—a new

> Knowing one is *of* rather than *in* the universe offers another path to *satori*.

ontology also based on more than two thousand years of thought. What service might it offer to Zen masters?

It shows the whole universe emerging as one single thing. The whole being; then the next whole being; with changes between successive beings causing all its doings.[b]

It shows we are *of* this universe: Our oneness with it partakes of the essence of *its* unity.

More particularly, we are clouds of braided twist relationships. Though the tale of twists and braids is ultimately physics, the paper that launched them seems a thin blend of classic *koan*, *What is this?* with a response from Lewis Carroll:

---

a In Japanese. In Mandarin or Cantonese, it is *Chan*; in Korean, *Seon*.
b As we have seen, Zeno, (no practitioner of Zen), would be ecstatic; see chapter 10 at page 63 and text of note 241.

> For convenience let us denote a twist through π as a "dum," and a twist through $-\pi$ as a "dee" (U and E for short, after Tweedledum and Tweedledee). Generically we refer to such twists by the somewhat whimsical name "tweedles." We hope to deduce the properties of quarks and leptons and their interactions from the behavior of their constituent tweedles.[1914]

It does not sound like Buddhist thought. So, what might all this offer Zen? Maybe a shortcut to their long-sought end.

Zen practitioners seek (as Hoover said, above) through lifelong meditation to suppress the intellect and gain the direct insight that observer and observed are one.

My path no doubt began with patient indigenous elders who all saw the world that way.[a] It led through no meditation but, once I had the universe's tale in view, my mind took me to its end. *Satori*[b] overcame me simply thinking about what I knew. Now, with a moment's mental focus on the universe's being-doing, my world becomes one, I am *of* it,[c] all my senses tell me it is so.

In other words, the intellect, when well informed, may not "obscure" the truth. It may rather offer a new way. It's such an easy way that even I can take it.

Is this the end of Zen? (I see this as two questions.)

---

a See chapters 26 and 130 at pages 150 and 624.
b [Japanese] *comprehension*.
c That personal limit disappears; see chapter 130 at page 624.

126
NOTHING BUT THE TRUTH

The task of a judge in the conduct of a trial
is to apply the law and to admit all evidence
that is logically probative.

Judson, J. (1971)[1915]

Law has its own concerns about the nature of reality. Its answer to the frailties of observation and their effects on deduction is the law of evidence.[a]

The central problem seems simple. As novelist Scott Turow put it,

> Something happened. Something objective but no longer verifiable.[1916]

In common-law jurisdictions, *R. v. Wray* is a leading case. Wray shot a man. He threw the rifle in a swamp. He confessed under police trickery and duress. He then showed them where to find the rifle. At trial, and on appeal, the courts excluded evidence of what he said under duress[b] and so acquitted him. On further appeal, the Supreme Court of Canada directed a new trial (at which he was convicted).

| The evidence is there.

Writing for the majority, Justice Martland said,

> [T]he learned trial judge erred in law in excluding evidence as to the facts leading up to the finding of the rifle.[1917]

The bottom line was why it was admissible. Though obtained unfairly, it was relevant; and it was relevant because it tended to show

---
a I gratefully acknowledge the support of the 1980 Cathy Turner Prize in Evidence, which is also a backhanded way of claiming I once did know a bit about it.
b Evidence obtained under duress may be excluded as it may not show the truth and so is not relevant.

"the truth." Or, one might say, to help the jury understand something that had happened, a lost episode of "reality."

There are broadly two kinds of evidence, sometimes called direct evidence (based on observation) and circumstantial evidence (based on inference). The latter is often disparaged as *"only* circumstantial" yet it may (as here) be the more probative kind.[1918]

The direct evidence showed the police found the gun in the swamp and that it fired the fatal bullet. It was of the *I was there* and *I did this* and *I saw that* variety so as to satisfy the pickiest Ayer disciple. But it could not lead to a conviction.[a]

It was the inference we all draw easily from the defendant's showing where the rifle was that clinched the case. It made simple sense.

Direct evidence takes on the appearance of a record. Consider, for example, an eyewitness saying, "I saw him shoot the victim." This "record" is a memory, expressed via thought, based on recollection. It is subject not only to the vagaries of observation but also to the processing that is inherent in remembering.[b] That is, though the thought itself may be composed of real elements,[c] this does not make it a real record. These are not minor limitations; eyewitness testimony is notoriously unreliable.[1919]

By contrast, the evidence Wray told police where to find the rifle and they found it there seems hardly susceptible to such imperfections.

The rifle was there. That it too was not in our strict sense real detracts not one whit from the inferential impact.

This too is a thought; we will pursue the question of what *that* is.[d]

Historiography shares with jurisprudence (and with me in this work) concerns about imprecision in our reconstructions of reality. For both, the new ontology asserts as a fundamental point of departure that the universe retains no record of its past. And if thousands of years' thought did not tell us what *is*, how much harder is telling us what *was*. As well, compare the resources law brings to bear on the

---

a Indeed, it was alone so ineffectual the first-trial judge directed the jury to enter an acquittal.
b See chapter 134 and note 2061 at page 655.
c See chapter 134 at page 655.
d We will see thoughts have a kind of existence; see chapter 132 at page 632.

fine foci of its reconstructions with those of the historian concerned with resurrecting, say, the life of Henry VIII.

Yet a civilization without history is like a person without memory, a terminal Alzheimer's case writ large. The aim of a history, like that of a memory, is that it should—in some real sense—be true. This may seem self-evident, but it was not (and is not) always so.[1920]

Historian and historiographer Sir Geoffrey Elton, embracing this aim, said,

> [S]ome philosophers . . . like to arrive at the conclusion that, since historical knowledge cannot, strictly speaking, exist, there is no way of establishing truth in history. . . . [I]nability to know all the truth is not the same thing as total inability to know the truth.[1921]

Elton, "a stickler for assessing the value and the provenance of a source,"[1922] sought to rely only on contemporaneous written records in his search for historical truths.

Such truths, too, when they exist at all, exist not as records but as thoughts. That's why we need historians. Or, put another way, real history is an aspect of the present; it is inseparable from its *Now*.

PART VIII

REAL PROGRESS

Out yonder there was this huge world, which exists independently of us human beings and which stands before us like a great, eternal riddle, at least partly accessible to our inspection and thinking.
Albert Einstein (1949)[1923]

127
THE DREAM OF REALITY

> Kafka ... dreamed of reality as of a goal from which he was barred by some magic obstacle, and what he asked of the dream world ... was to reveal the secret of this monstrous order of banishment and if possible the means of countermanding it.
>
> Marthe Robert (1979)[1924]

The worldview of a civilization is profoundly consequential.

Throughout history—and perhaps prehistory too—a real worldview, an ultimate understanding, a true grasp of what is going on, has been a dream. In hindsight we can see it long was unachievable.

In any case, it long was not achieved. In consequence, we lack experience of living with reality. What will a real worldview bring humankind now that (in essayist and translator Marthe Robert's words) Kafka's monstrous order may be countermanded and his goal may be attained?

| Are we ready for reality?

In one of his last lectures, "extraordinarily creative"[1925] anthropologist Marshall Sahlins said,

> [T]he state of nature is already something of a political state. It follows that, taken in its social totality and cultural reality, something like the state is the general condition of humankind.[1926]

In this lecture—and in the words of a reviewer of his last work—he pursued a central theme:

> [M]ost human beings for most of our known history inhabited cultural worlds in which nonhuman entities—spirits, gods, or Sahlins's preferred terms, "metapersons"

and "metahumans"—played a constant and decisive part in human affairs.[1927]

Sahlins's point—and mine in citing him—was, *This is where we live.*

We tell ourselves—as generations of our forebears no doubt told themselves—we are modern, we have surmounted such influences. A moment's introspective thought should tell us we are as involved with them as ever. The metapersons change;[a] *they* become modern, these days mostly mediated by the internet; *our* embrace of them into our lives continues almost unabated.

For each generation, details of the "general condition of humankind" changed but (I stress again) throughout history one thing remained constant: Humankind survived in a "reality" that wasn't real.

There is nothing inapt in this; quite the opposite. Some semblance of a common worldview—right or wrong—seems to be necessary if our ant-like genius for large-scale cooperation is to prevail.[1928] Indeed, society's survival may require it.[1929]

So, one wonders—with some trepidation—what might eventuate if that common worldview should now undergo radical change.

If such should happen, it would not be the first time. In 1953, philosopher and historian of science Alexandre Koyré gave a lecture in Baltimore that has been called "intellectual history at its finest."[1930] He began,

> It is generally admitted that the seventeenth century underwent, and accomplished, a very radical spiritual revolution of which modern science is at the same time the root and the fruit.[1931]

He attributed this revolution to man having . . .

> . . . lost his place in the world, or, more correctly perhaps, lost the very world in which he was living and about which he was thinking, and had to transform and replace not only his fundamental concepts and attributes, but even the very framework of his thought.[1932]

---

a For example, these days, for a growing portion of the population, many "metapersons" lie behind the veils of social media.

Inching it closer to our circumstance, he said,

> This scientific and philosophical revolution . . . can be described roughly as bringing forth the destruction of the Cosmos, that is, the disappearance, from philosophically and scientifically valid concepts, of the conception of the world as a finite, closed, and hierarchically ordered whole.[1933]

With this in view, let's return to Sahlins and see where we may be in his panoply and what might happen *where we live*.

There has until now been no way we could formulate a real "conception of the world" whose real elements are unseen sub-sub-subatomic entities. Physics did its best, offering its devotees an ever-finer progression of far larger fictions to absorb or venerate.

Yet somehow, we did it. By *we*, I mean all of us; believe me, if you think its causes through, you'll see it did take *all* of us. Maybe it will matter that we see that it is *ours*. Here is a mere sliver. In 1834, writing *On the Connexion of the Physical Sciences*, mathematician, astronomer, and writer Mary Somerville could say,

> The accumulated efforts of astronomers, from the earliest dawn of civilization, have been necessary to establish the mechanical theory of astronomy. The courses of the planets have been observed for ages with a degree of perseverance that is astonishing.[1934]

Be minded, then, that throughout humankind's long climb to understanding, all who were able nightly scanned the sky. Most knew constellations of the starscape in its stately unchanging progression, maybe better than they knew the faces of their friends. Untold millions of them wondered *why* five lights moved against that static backdrop in a crazy yet orderly fashion (e.g., Mars; fig. 94). In this and countless other ways we all partook of the long evolution of our worldview.

Given our humble beginnings, discovering reality was so staggering an undertaking I am ever in amaze even our insatiable curiosity could take us to it.

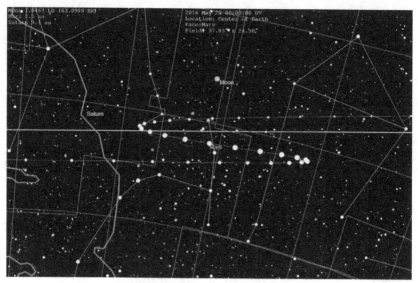

Fig. 94. *The apparent motion of Mars*

And now, it's the very pervasiveness of that "constant and decisive part in human affairs" of which Sahlins spoke that must show how much we all can achieve, how boundless the opportunity, how far the horizon, if our "general condition"[a] should move into a closer alignment with what *is*.

What *is* is a strange-seeming and mentally challenging but, in the end, supremely simple story, a modest interlocking set of well-established old—and a few new—ideas.

Its very simplicity may lead to mental change.

Or, let's say strategist James Carville was right and *It's the economy, stupid*.[1935] Freeman thought the world could monetize a Planck-scale physics revolution into something like one quadrillion dollars' worth of new production.[1936]

A wider view suggests such change in mere economic metrics such as GDP may be small beans. For example, taken as I am by the relational philosophies of long-enduring civilizations,[1937] I wonder how internalizing the insight *the world is all made of relationships* might modify entrenched power imbalances.[b]

---

a See passage cited as note 1926.
b Not only but especially the roles and influences of indigenous peoples and of women, who both tend to see the world that way.

Reality could also fundamentally change math. Gödel stripped arithmetic of its pretensions to consistency. But his proofs rested upon *zero* and an infinite supply of counting numbers.[a] The new worldview says there's no zero and no infinite supply.[b] So, can arithmetic become consistent? How not, when the world is?[c] No longer lost,[d] what might physics do with access to consistent math?

And surely, a whole new worldview offers natural philosophy a whole new lease on life.[1938]

This isn't only an emotional exuberance (though no doubt it is all of that). Reason says a real view of our world is, as Koyré's work portends, the kind of insight that could change our attitude to who we are. We might even come to understand ourselves.

Meanwhile, let this not be read as a polemic against fictions. I write fiction too, in hopes it may be useful while it entertains. We need new fictions of all kinds, more of them and even better. We should just be clear that's what they are.

Author Franz Kafka, keen observer of reality, was a fan of all our fictions, setting out to "test the truth of myths."[1939] We can overcome our obstacle. We can reveal the secret of that banishment. We can countermand the monstrous order. Did he see there was a real world just beyond it for us to explore?

The first thing we do is, let's kiss all the authors.[e]

---

a See chapter 114 at page 559.
b See chapter 104 at page 515 and chapter 111 at page 517.
c See chapter 114 at page 559.
d See chapter 40 at page 222, and Hossenfelder, "Lost in Math," note 184.
e With apologies to Shakespeare and Dick the Butcher.

128
THE MYSTERY OF MOTION

> And now I will unclasp a secret book,
> And to your quick-conceiving discontents,
> I'll read you matter deep and dangerous.
>
> William Shakespeare (1598)[1940]

Story scientist Angus Fletcher told us Shakespeare was not alone among the early modern poets to be drawn by the central cultural issue of the era, ...

> ... the predominant conceptual, scientific, metaphysical and hence philosophic power of the idea of motion.[1941]

So when, in 2001, Rovelli asked,

> What is the meaning of "moving"?[1942]

... he was asking an old and persistent question. Recall that it lies at the heart of Zeno's paradoxes, including that of the One and the Many, that stumped thinkers for thousands of years.[a]

Zeno too was not alone in this. The list of Greek philosophers who helped define the motion problem includes Thales, Anaximander, Anaximenes, Heraclitus, Parmenides, Empedocles, Anaxagoras, Gorgias, Democritus, Plato, Leucippus, and Aristotle.[1943]

| Change is one key to motion. Now we need the other. |

The latter noted an inverse consensus that came close:

> [T]ime is most usually supposed to be motion and a kind of change.[1944]

The new ontology affirms that motion is a deep and ultimate mystery and it explains why this is so.[b] It provides a key piece of the

---
a See chapter 10 at page 63.
b See also chapter 57 at page 307.

solution. Amazingly, Zeno's proposition was not paradoxical; his concept was precisely right—the universe itself *is* both one and many with its quantum space and sequence. This digital reality makes change, which manifests as motion, possible.

Even in the quantum era, the conceptual contradictions inhering in motion in continuous space and time proved to be inescapable. For example, shades of Zeno, Bohm outlined in his then leading 1951 account of quantum theory (which assumed space and time to be continuous) the clash between the *concepts* of position and motion:

> We can try to represent motion as a succession of objects at slightly different positions, as is done in a motion picture, but a succession of fixed positions does not include all the properties that are usually associated with motion. In particular, it does not seem to include the idea that a real moving object is *continuously* covering space as time passes.[1945]

This motion-picture paradigm—which Bohm advanced as faulty fiction—cleaves close to reality.

Quantum theory posited a dual relation between position and motion with Heisenberg's uncertainty principle:[a] The more that's known about one, the less that can be known of the other.[1946] But this shifted the locus of the need for explanation, all the while obscuring the core meaning.

The new ontology offers a way to reconcile this problem.[b] At each iteration, the relations of space quanta are unmoving. Motion emerges from the change in those relations from one iteration to the next. That is, the *be* and the *do* can never coincide; and (surely to the shades-of-Einstein's horror[1947]) the world *does* work in dice-driven jerks.

Compare this with Sattler's description of Zeno's stunningly insightful view of motion (which now has new Planck-scale meaning for us):

> [T]he place at which something starts its motion ceases to be its place and a new place "comes into being" as the place where this thing is now.[1948]

---

a For its Planck-scale basis, see chapter 42 at page 236.
b See chapter 42 at page 236.

Thus, the fundamental problem is the nature of inertia.[a] We are, like Einstein was, still stuck at Newton's laws,[b] which tell us that mass has (or maybe *is*) inertia with no word of why. However, the new ontology does give us a more specific focus, *How does the sequential 3-D universe encode (or conserve) motion?*[c]

Smolin offered an approach: Abandon hope of ever understanding.

> [I]f we want energy and momentum [i.e., attributes of motion] to play a role in physics, there seems to be no alternative but to put them in at the beginning.[1949]

His reasoning may seem compelling but—having the benefit of "seeing" the beginning—there is bad news for him: It also seems there is no useful way to do what he proposes. On the other hand, he may see this as good news. I doubt he likes to put things in at the beginning any more than I do.

And, if they are not there at the beginning, they needs must be emergent. Thus they are statistical in nature. Which seems to mean relations in the 3-D universe must somehow encode probabilities of twist states moving. And it will likely involve odd properties like non-Abelian physics.[1950]

This is Smolin's bailiwick. Maybe he'll figure out the way it works.

---

a See chapter 86 at page 435 and note 1443.
b E.g., his first law: A body does not change its motion until force acts on it.
c See also chapter 51 at page 279.

129
NEW-VIEWS NEWS

> I prefer to look our ignorance in the face, accept it, and seek to look just a bit further: to try to understand that which we are able to understand.
>
> Carlo Rovelli (2014)[1951]

Rovelli was speaking figuratively of looking a bit further. But looking figuratively is ultimately tied to looking literally. Here change is in the air. A whole lot of looking further—of both kinds—may be on the way.

In this book we have tapped into two and a half thousand years of evolving science. For the first two thousand, our ancestors could look only as far as they could see with their own eyes.

Four hundred years ago, optics and its lenses gave us Galileo and his telescope. Transcending the eye's size, it harvested more of light's message and it revolutionized our view. For example, he could see the Milky Way is made of stars.[1952]

New kinds of instruments will bring us new views.

Later telescopes found smudgy spots called *nebulas*. A mere ten generations ago, even better telescopes revealed a nebula to be an entire galaxy of stars.[1953] For the first time, we could explore beyond our hundred-thousand-light-year galaxy.

Thus Galileo triggered an inexorable transition. Successively, more distant sights reshaped our cosmic situation. From center of the universe, our status humbled its way down to third rock from an ordinary star parked in a side wing of one of a growing catalog of an immense number of galaxies.

A few years after nebulas showed up, in 1841, in an odd triple intersection with our story, author Edgar Allan Poe invented the detective-fiction genre with "The Murders in the Rue Morgue,"[1954] in which his detective, Dupin,[a] mused about nebular cosmogony. Poe was fascinated by cosmogony, the then obscure—one could almost say then nonexistent—science of the universe's origin.[1955]

Poe completed that odd intersection with *Eureka*, in which he said the universe began as a single "particle" that divided.[1956] That is, he anticipated Lemaître's primeval atom by almost a century and the new ontology's space quantum by almost another.[b]

As telescopes grew bigger, saw a wider spectrum, and moved outside our atmosphere, we saw vast numbers of galaxies (fig. 95).[c] And, by seeing far out into space, we saw far back in time.[d] These new views revolutionized our understanding of space and time and the evolution of the world.

Fig. 95. *Galaxies in a portion of the Hubble deep-field image*

So did the giant step we took, some sixty years ago, with the discovery of the cosmic microwave background radiation, that flash photo of the baby universe.[e] Weinberg later said,

> The most important thing accomplished by the ultimate discovery of the 3°K radiation background in 1965 was

---

a Dupin was in many ways the prototype for Sherlock Holmes, which Arthur Conan Doyle had Watson acknowledge for him in *A Study in Scarlet*.
b See chapter 17 at page 105; and see further, *Time One*.
c Fig. 95 is a portion of a Hubble Space Telescope image, from a dark part of the sky, in which even the tiniest dots are entire galaxies, each having billions of stars.
d See chapter 9 at page 57 and note *a* at page 62.
e See chapter 17 at page 105.

to force us all to take seriously the idea that there *was* an early universe.[1957]

We are on the threshold of another revolution, with new views from new kinds of instruments. They bring new capabilities, like seeing early galaxies or probing the universe's most-extreme-physics labs, black holes.

Here is my pick of three. They "see" neutrinos;[1958] and gravity waves;[1959] and light—from stars and galaxies that were the first to form after the universe began—stretched to such long wavelengths normal telescopes can't see it.[1960]

Let's start with the IceCube Neutrino Observatory. The IceCube Collaboration (more than three hundred authors) described it succinctly:

> The observatory uses [one cubic kilometer] of optically transparent glacial ice as a detection medium.[1961]

It is now strung with thousands of detectors to pick up the telltale flashes signaling a muon from a rare interaction of a high-energy neutrino with an atom in the ice. It started work in 2006.[1962]

The neutrinos of interest reach the detectors after passing, almost all of them entirely unaffected, all the way through planet Earth.[1963] The IceCube Collaboration has tracked (to a nearby galaxy, with its active galactic nucleus and central black hole) a stream of neutrinos, each of which is about a million-fold more energetic than any event we can create in earthbound particle accelerators.[1964]

Astronomer Philip Plait said,

> The neutrinos are being created outside the black hole's "point of no return"—its event horizon—although it's not clear just how.[1965]

But, just checking, I find the Collaboration *didn't* quite say where they were created. Rather, they implied they could arise from the corona of a big black hole's accretion disk according to a proposed model of the active galactic nucleus.[1966] Nobody knows for sure.

Second there is LIGO, the gravitational-wave observatory that scored its first look at a space-shaking event in 2015. Long dissed as a crazy waste of money,[a] it is two "telescopes" of sorts in two of the United States.[b] Each looks like two four-kilometer-long farm sheds built at right angles. Inside, laser beams bounce back and forth between mirrors, looking for vibration a miniscule fraction of the size of an atom.

Well, of course they get vibration! On that shaking scale, all earthly objects thrash wildly with vibrations from far-off earthquakes, distant traffic, or cows grazing in a nearby field. Its scientists used clever ways to minimize these influences. Then they hunkered down to wait for some humungous long-ago collision in a far-off galaxy to shake space with its passing wave of gravity.

It was a bit like this: Think a volcano exploding on the seafloor among distant islands.[1967] See, in your mind's eye, the shockwave moving in the water at the speed of sound. Make the intervening ocean, say, a billion miles across and try to spot the less-than-atom-high tsunami washing on a stormy shore. It was way worse; it was a wild idea;[c] it worked.[1968]

When Washington sees mirror-wiggle signals, it checks in with its Louisiana twin to see if it detected the same shake.

On 14 September 2015, that "crazy" system proved its worth (see fig. 96). The event it "saw" was two spinning black holes colliding. It shook the substance of space across the universe, a light-speed space-quake that took more than one billion years to get here.

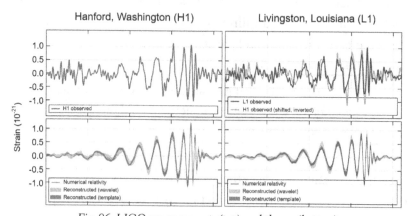

Fig. 96. LIGO measurements (top) and theory (bottom)

---

a Few expected it to achieve the necessary sensitivity and, if it did, to overcome ambient noise.
b One is in Washington State and the other is in Louisiana.
c Other attempts elsewhere had failed.

General relativity predicts this shaking and dictates its precise shape (see lower panels), so comparing with theory tells us the masses of the two black holes, how each was spinning, and the mass of the black hole that was the final product.

LIGO has since seen many such collisions, also those of merging neutron stars.[1969] Two new gravitational-wave observatories—one in Italy, the other in Japan—are now in operation. Combining data from four instruments can pin down where events occurred. Getting that word out fast tips off other telescopes to look for light or microwaves or X-rays from the same event, leading to (sometimes literal) gold mines of data from these unimaginably powerful collisions.[1970]

Images from the James Webb Space Telescope (JWST to its friends), a satellite now orbiting the Sun a million miles the other side of Earth, may be the cutting edge of the new-view revolution in the early universe thanks to its infrared sensitivity.[1971]

The reason is most stars radiate most of their light at wavelengths we can see, between about 400 and 700 nanometers.[1972] But light from stars and galaxies in the first few hundred years after the beginning of the universe arrives here stretched by the expansion of the space through which it traveled.[a] So, seeing it requires a telescope that detects longer wavelengths (in the infrared).

The first few hundred million years of the universe's life may hold the keys to the puzzle of first galaxy-formation.[b] The Hubble telescope could not observe that early era but the JWST can.[c]

Its conception, financing, design, construction, launching, in-space self-assembly, orbital positioning, and operation were all monumental undertakings and a multibillion-dollar gamble, with more than three hundred forty post-launch single points of failure, any one of which would turn it into high-priced space junk. Even for me it was a white-knuckle show. Breathtakingly, it worked.

---

a Known as the *cosmological redshift*.
b Recall that looking far out in space is looking far back in time thanks to the finite speed of light; see chapter 9 at page 57 and note *a* at page 62.
c The Euclid satellite, surveying many galaxies at high resolution, may shed new light on dark energy.

Science writer Jonathan O'Callaghan said,

> Some estimates suggest JWST could see as far as a redshift of 26, just 120 million years after the big bang, a cosmic blink of an eye.[1973]

This is an example of how new views from new instruments may have fundamental implications. The standard model can barely reconcile our finding a galaxy four hundred million years after the beginning (spotted at extreme range of the Hubble).[1974]

In other words, there will be no way to explain earlier galaxies, if such are seen,[a] by tinkering with the existing model. It would require a deep rethinking.

The standard model's picture showed the first stars as huge; they used their hydrogen fuel fast and soon collapsed in supernovas leaving black holes.[b] The model's stars and black holes formed at earliest a few hundred million years after the beginning. And those black holes were relatively small, up to fifty or so solar masses. They needed much more time to grow.

By contrast, the new ontology says many big black holes could have formed in the first fraction of a second after the beginning. Those big black holes could then have begun organizing the first galaxies much sooner than the standard model's black holes could.[c]

The JWST soon started to see very early galaxies. For example, astronomer Tom Bakx and colleagues confirmed the age of a galaxy found by JWST;[1975] it was some 370 million years[d] after the Big Bang.[1976] Another, forming stars at an astounding rate, was even younger, about 325 million years.[1977]

Of another, cosmologist Michael Boylan-Kolchin said,

> Even if you took everything that was available to form stars and snapped your fingers instantaneously, you still wouldn't be able to get that big that early.[1978]

---

a As the new ontology portends, see chapter 102 at page 499.
b See also chapter 93 at page 460.
c For more on this, see chapter 102 at page 499.
d These age estimates, converted from spectroscopically confirmed redshifts, depend on the assumed model, which adds to their uncertainties.

These observations become unsurprising when viewed through the lens of our beginning[a] and they already suggest the standard model of cosmology may need basic changes. Science writer Jonathan O'Callaghan, again quoting Boylan-Kolchin, said,

> The most startling explanation is that the canonical... cosmological model is wrong, and requires revision.... "And it's probably not a small change. We'd have to go back to the drawing board."[1979]

And, citing JWST results, Frank and Gleiser said recently,

> We may be at a point where we need a radical departure from the standard model [of cosmology], one that may even require us to change how we think of the elemental components of the universe, possibly even the nature of space and time.[1980]

Change is in the air;[b] and closing in on our beginnings is an inspiring, even an emotional, prospect. Astrophysicist Jane Rigby, operations project scientist for the JWST, described it as a feeling of...

> ... people in a broken world managing to do something right to see some of the majesty that's out there.[1981]

---

a See chapter 69 at page 357.
b Yet one should not underestimate the adaptability of multiple assumptions and adjustable parameters.

130
PERSONAL LIMITS

> "Limit" means the last point of each thing, i.e. the first point beyond which it is not possible to find any part, and the first point within which every part is; ... for this is the limit of knowledge; and if of knowledge, of the object also.
>
> Aristotle (350 BCE)[1982]

We are wont to conceive of ourselves as distinct from our surroundings, as objects (to use Aristotle's term[a]) secured within the limit of our skins. In the new light, this looks like a distinction in search of a difference.

On the complexity scale, you and I are at the far end from a subatomic particle. Its phantom flock of twists seems crisp compared with our uncertain clouds. Even if we overlook the looseness of our Planck-scale presence,[b] our casually clear distinctions between ourselves and all the rest of it conceal inherent indistinctness.

| We have no limits.

More closely considered, each of us becomes ephemeral, something more like an idea.

It's not only that we are, atomically speaking, highly complex as Feynman inelegantly said in one of his lectures,

> Is it possible that that "thing" walking back and forth in front of you, talking to you, is a great glob of these atoms in a very complex arrangement, such that the sheer complexity of it staggers the imagination as to what it can do?[1983]

---
a In the Ross translation.
b See chapter 62 at page 326.

It's also that, given what we know now about such atoms, we may view them as borrowed, maintained through more or less curated exchanges with environs, with the residue to be returned to general reserve when personal ideas run low on their telomeres.[1984]

When my mother died in Melbourne, fifteen thousand kilometers off on the far side of the world, her remains—including some 9 kg of carbon[1985]—were cremated. Nearly all her carbon atoms entered the air as carbon dioxide (CO_2).

The mass of the atmosphere is some 5×10^{18} kg.[1986] It mostly mixes on a global scale.[a] So, around the world we breathe our ancestors' carbon and, in many complex ways, those carbon atoms can exchange with ours.

Once I wondered, without meaning to be ghoulish, could one of her carbon atoms become one of mine?

This is not about the CO_2 molecules mixing in the global atmosphere; that happens more or less completely on a scale of months.[1987] Rather, it is about Avogadro's number,[b] which tells us how many of her carbon's CO_2 molecules are out there and up for grabs.

With another simplifying (and admittedly simplistic) supposition—that each carbon atom we breathe in as CO_2 becomes a part of us—the answer to my question is I may by now have taken billions of her atoms in as mine.[c]

It follows that I share her final atoms with the rest of you.

And we all share atoms from your daily exhalations.

It's not only that the atoms that comprise our physical existence are exchanged with all the world, some of them (like hydrogen and oxygen) quite quickly.[1988] It's also that the very information that keeps those atoms organized into ourselves is modified by events ranging from sub-picosecond physics and microsecond chemistry to hours-long molecular biology (as I learned working with Chapman[d]) to

---

a Technically, mostly in the troposphere rather than the stratosphere.
b Its value is 6.023×10^{23} atoms per gram atomic weight; or molecules per forty-four grams of CO_2.
c Aside, of course, from those I had at birth and managed to retain.
d See chapter 115 at page 565.

the years-long changes wrought by epigenetics.[1989] And don't get me started on our microbiomes.

Thus, day to day, we reassemble only more or less according to our parental instructions. As biologist Kenneth Walsh put it,

> The genome you are conceived with is very different from the genome you die with.[1990]

So, even at the molecular level—far more organized than Planck scale—each supposedly well-defined person can be seen on close examination to exchange with and even to merge, having no clear limit (to again use Aristotle's term), into their surroundings.

There is, then, little point in painting Planck-scale pictures of ourselves. Not only by reason of disparity of scale, but because all that is interesting about us is the consequence of serial emergences that each, alone, may be explicable,[a] while cumulatively they utterly obscure from view that we are poorly defined clouds of twists.

Nonetheless our clouds *are* our pretensions to reality if the new ontology has any traction. And there may yet be value in our thinking about them. Our twists merge seamlessly into the universe. They even partake of its nonlocality.[b] This thought falls short of "seeing" the inherent unity of everything, but it brings one to an intellectual acknowledgment: Our limits are, at every level, fuzzy fictions.

In other words, I think I know exactly who I am, but I cannot say exactly what I am.

In the literally true words of politician Kamala Harris's mother, Shyamala Gopalan,

> You think you just fell out of a coconut tree? You exist in the context of all in which you live and what came before you.[1991]

In further other words, indigenous peoples' worldview of the universe (including themselves) as a single interconnected reality[c] is more apposite than the Western world's. One might even get over it and acknowledge that they had it right all along.

---

a See chapter 94 at page 467.
b See chapter 25 at page 144.
c See chapter 26 at page 150.

Some of their civilizations have lasted thousands of years. Ours might find a more secure future if we could come to see ourselves as *of*—rather than as *in*—the universe.

That is, just as we are.

131
NEWFOUND REALISM

> Let us fix our attention out of ourselves as much as possible: Let us chase our imagination to the heavens, or to the utmost limits of the universe.
>
> David Hume (1739)[1992]

Hume was writing of *what is*—our central concern in this work—and, too, of how (poorly) we can conceive of it, "the idea of existence" as he put it, which also is our concern. For millennia these twin concerns offered a fertile field for philosophic disputation, mostly focused (as was Hume) upon the latter and its limitations. We will not likely settle all that dust but may—if our ontology delivers somewhat of its promise—shift the focus more toward the former in spite of its intrinsic difficulties.

> A real worldview has wide practical value.

The distinction may seem academic (and it was indeed a hot topic in the Academy). But it is also fundamental for more than purely philosophic reasons. When we focus on what *is* (instead of its idea) we can also focus—as we have—on what it *does*. Reality, we find, consists in both, and in their intimate relation.

Those who, like Einstein in his latter years, insist there is an objective reality are called realists[a] and their philosophy, some aspects of which we have touched upon, is realism.[b]

A hundred years ago, such views fell out of fashion. In 1931, Planck said (foolishly, but in synch with his era),

> I regard matter as derivative from consciousness.[1993]

---

a See the preface.
b See, particularly, part 6.

By 1985, polymath Freeman Dyson could truly say,

> What philosophical lessons arise from the recent discoveries in physics? . . . The old vision which Einstein maintained until the end of his life, of an objective world of space and time and matter independent of human thought and observation, is no longer ours.[1994]

As a philosophic bent, realism dates at least to Plato.[1995] It has been applied to many conceptual realms and is now graded into varying degrees.[1996] It confronts a motley opposition that is known collectively as antirealism.

Rather than canvassing the field, let's see it through the eyes of philosopher Arthur Fine, who was struck by the success of antirealism, saying,

> Physicists have learned to think about their theory in a highly nonrealist way, and doing just that has brought about the most marvelous predictive success in the history of science.[1997]

He was a former realist who, in mid-career, set out to find a middle path. He called it *natural ontological attitude*.[1998] It was, roughly speaking, a bucket of normal stuff that neither realists nor antirealists would trouble to debate. In this, he sought to find some common ground between them and move on.

In aid of this program Fine tried to shift the discussion away from any reference to actual reality. He criticized the realists as wanting to explain . . .

> . . . what is really, really the case. The full-blown version of this involves the conception of truth as correspondence with the world.[1999]

The alert reader may equate Fine's dissed "full-blown version" with the test of what is real I set up in the preface. Fine dismissed such "correspondence with the world" . . .

> . . . as superficial [decoration] that may well attract our attention but [does] not compel rational belief.[2000]

There, in those last three words, we find the nub. Like a latent infection, it lies at the root of the crisis of modern physics and its fraught relations with philosophy. These days, what, if anything, could compel rational belief?

We all—you and I, and the philosophers and physicists—are faced here with two facts.

One, there *is* no logical foundation that can compel broad rational belief. More than a century of ardent efforts failed to find a logical foundation for belief in a simple arithmetic![a]

And two, this isn't about logic; it is about choice. We can choose what compels rational belief. Indeed, we do choose (even if not always wisely) when we must. And choosing when we must choose is itself entirely rational.

Fine seemed almost to say so. In setting out his middle path, he articulated (even advocated) his own choice in this matter. He would have us . . .

. . . accept the results of science as true.[2001]

Even further, Fine sought to embrace science as a stand-in for truth. This ignores the not so hidden ontological commitments upon which our current science is dependent, not to mention all their consequential contradictions. His work serves to show (again) the awful choice: Either there is truth based on the real world; or there is only (in my language) fiction. There is not a lot of middle ground here.

This stark contrast gives rise to priorities and implications of the new ontology for physics. Its priorities require one to first look to ontology (albeit drawing upon carefully curated scientific insights) to build a consistent understanding of what is and isn't real. Its implications, once we do this, say that almost all of physics *isn't* true.[b]

This leaves lots of room for new ideas. Ideas in our minds are powerful change instruments. Once let loose, an idea knows no boundaries of language or geography; a simple idea may profoundly change the world in ways that can pass almost unnoticed. For one example where the thread can be detected, the British bright idea

---

a See part 6.
b See, e.g., part 4.

of preemptive naval air attack on the Italian battle fleet at its base in Taranto harbor informed Japan's attack at Pearl Harbor,[a] with all its consequences.[2002]

Ideas, the reader will by now have discerned, are the currency in which we seek here to deal. We assembled an ontology by choosing seminal ideas. The basis for the choosing was itself a choice, one used quite often, *It makes sense*. It was, we said, like working on a jigsaw puzzle choosing pieces from a box with many extra pieces that don't fit.

The result, so many inexplicables find explanations that I need not now essay a list.

What can compel our rational belief here is not only this ability to make sense out of many, diverse, seeming-sense-defying riddles. It is not only that each piece fits with its neighbors—size and shape and lines and colors so to speak. It's that, as with completed jigsaw puzzles, the integrity of the whole thing speaks for itself.[b]

In the wake of logical positivism, understanding—making sense of things—became estranged from physics and philosophy. Mathematics moved into the vacuum and it always explains nothing.

Therein lies the value of a real ontology for fundamental physics, a practical guide that is our best shot at, in Fine's words, *what is really, really the case*, with that being weighed and tested, not merely piecemeal against our senses or some scientific observations, but against the need to make consistent sense of the whole world.[c]

Let's be clear about this too: The main value of a real ontology is as a guide to thinking and to building better language. It should also lead to better resource allocation and to more bang for the physics buck. And to many other better things like better mental health, and better government, and better (or less bad) communication.

It's not a substitute for anybody's concept of the scientific method (on which I agree with Smolin and philosopher of science Paul Feyerabend[2003]). It is, in a word, a myth.

---

a And so led to the United States entering the war; see note 2002 above.
b This is not to say it is the entire story; plainly it is not.
c This ancient human drive seems likely to attain new prominence and value with the rise of neural networks and so-called artificial intelligence. By its nature, AI is incapable of discerning its own nonsense.

132
A HIGHER REALITY

> The creative principle resides in mathematics. In a certain sense, therefore, I hold it true that pure thought can grasp reality, as the ancients dreamed.
>
> Albert Einstein (1934)[2004]

The relation (if any) between reality and thought is both a long-standing problem of philosophy and a main thread of our story. That, as Einstein said, "thought can grasp reality" could lead us into a thicket of controversy. The which I would rather avoid, accepting Peebles's less contentious proposition,

> It is remarkable that pure thought sometimes points us in the right direction. . . . But of course it is not at all surprising that pure thought more often leads us astray.[2005]

But, in the end I am concerned with the reality of thought.

Those real counting numbers exemplify another existential issue. Distinct from a number itself is our *concept* of the number. The number two existed in the universe since the beginning,[a] while the concept of (and language for) the number two existed in the aspect of the universe we call our minds since we discovered (or, more likely, reinvented) it. So, the number and the concept of it are related but are not the same.[b]

> Thoughts are made of real things but the universe can't count them.

The number, we now see, resides (as it is real) in the universe. Where, more precisely, does the concept of it have its being (if it does)?

---

a See chapter 17 at page 105.
b This is of course no new insight (see, e.g., Maddy, note 123) but it may offer a definitive foundation.

With the old ontology, the question whether math objects like 2 or its square root exist led philosophers a merry chase. Their binary choice was, *Are they real or do we invent them?*

We have seen each answer is right in its place.[a] The universe *was* 2; it still contains 2; but it will never be or somehow fashion its square root. So, although 2 is real, 1.414213562373.... is only a notion we invented.[b]

However, we can think of both as concepts, the concept of the number and of its square root. We can see what they are: We think them; they are thoughts. Can we say such thoughts exist? In efforts to make sense of this, serious philosophic minds have bent themselves into cerebral pretzels.

A case in point. Russell said abstract ideas (here called *universals*) *subsist* but they don't *exist*:

> We shall find it convenient only to speak of things "existing" when . . . we can point to some time at which they exist. . . . Thus thoughts and feelings, minds and physical objects exist. But universals do not exist in this sense; we shall say that they *"subsist"* or *"have being,"* where "being" is opposed to "existence" as being timeless. The world of universals, therefore, may also be described as the world of being. [Russell's italics.][2006]

Philosopher Arnold Cusmariu said this gives two meanings to the word *exist*, that Russell used subsistence as a work-around of Plato's thesis that his abstract forms had more reality than real things.[2007] Indeed, the long shadow of Plato's perfect forms[2008]—which truly existed only as concepts—looms darkly across the issue here.[c]

Quine said, somewhat dismissively,[d]

> The questions as to what subsists evidently struck [Russell] as less substantial, more idly verbal perhaps, than questions as to what exists.[2009]

---

a In part 6.
b See chapter 107 at page 105.
c One might see the never-to-be-seen Calabi-Yau manifold as the ultimate Platonist form.
d And Quine went on to note, "The fact is that Russell has stopped talking of subsistence . . . by 1914."

Quantum space allows us to descry two kinds or levels of existence. Lest this appear to agree with Russell, let us first be clear it doesn't. Russell's two kinds would have your thought of *two* exist because he thinks you think it at a moment, while *two* itself he thought merely subsists since it is timeless.[a]

The two kinds of existence quantum space supports are almost opposite. *Two* meets our test[b] for something to be real—that is, it must be an accurately described objective aspect of the universe—and, in its creation, was a bit like Russell's "existing" at a time.[c] Here, *existence* is the totality of space quanta and their relations at an iteration of the universe.

Thus, the infant universe in each of its first few iterations existed; and we can understand it was the entire universe.[d] Similarly of the current iteration: we understand it *is*;[e] our phrase *the universe* describes it now as it did then. We don't need to specify the number of quanta or what are their relations to say they, in their entirety, exist. By our test then, they are real in totality.

It gets a little messy if one tries to specify a subset of them, such as an electron. But we can enumerate its real ingredients[f] and at least circumscribe all their locations[g] so, though composite, at each iteration *the electron* seems to qualify as real.

It gets much messier for more emergent things, like atoms. But an atom, too, involves an enumerable number of twist states within some spatial zone.

We are each composed of atoms. But even at a given iteration they are not specifiable. We have already canvassed how utterly each of us is, at every scale, physically dispersed, a fuzzy zone with no clear

---

a By timeless, Russell means subsisting at all times.
b See the preface at page 1.
c In the sense the universe exists only Now and is timeless in not having a property of time.
d See chapter 42 at page 236.
e Though of course it is long gone before our thought can reach out to it.
f Links between space quanta; see chapter 35 at page 190 and fig. 27 at page 195.
g They are within n links of their last interaction as a particle, where n is the (increasing) number of iterations since then; and then within a similar (decreasing) number of links from their next interaction; see chapter 100 at page 492.

boundary and so no clean list of constituents.*ᵃ* In real terms, each of us is not only ill defined but also indefinable.

Amid that fuzzy zone we find the mind, thinking its often abstract thoughts. For no given reason, Russell picked on *north of London* as a universal; let's think of this as a thought. We can conceive that thought is confined in the brain, and at each instant we can circumscribe the limits of the brain, ensconced within its skull.[2010] But if it were no more than what is there, it would carry no meaning.

The difficulty of accurately describing it arises not so much from where it is in space as when it is in time.

Russell blithely said (above) "we can point to some time at which" a thought *exists*. I beg to differ, if by *some time* he meant some instant and, if not, what time might he have meant—some railway time like ten past three?*ᵇ* The frozen tweedle pattern at a given iteration holds no thought: A thought *takes* time, or rather, as we now see, iterations.[2011] The symphonic evolution of that pattern with its indeterminate but large array of atoms*ᶜ* moving for vast but uncertain numbers of the universe's iterations is the essence of a thought.*ᵈ*

Might we then think of such a thought as real? It too is made of real elements. But, in contrast with examples of the first kind of existence, there is no way to say which elements (notwithstanding that we may say generally where they are). And, in further contrast, by its nature thought must extend over very many iterations with, again, no way of saying which.

On the one hand, that thought (I think of it now, *north of London*; not the mere words, but the concept; for me it evokes where I started school*ᵉ*) plainly exists there in my brain. On the other, I am unwilling

---

a See chapter 130 at page 624.

b The action potential is the element of thought; each one takes milliseconds, without clear definition of beginning or of end; see fig. 97 at page 648. A fleeting thought partakes of billions of action potentials—with their relations in brain-connection-space over time—and lasts far longer, with *its* beginning and end being both indefinable and unconfinable.

c The action potential consists, roughly speaking, of an influx of sodium ions into the neuron followed by an efflux of potassium ions.

d See further, chapter 134 at page 647.

e See chapter 108 at page 535.

to dilute my test for what is real with this kind of existence based on Russell's reasoning (or, for that matter, mine). But, surely, we should not ignore it. So, somewhat like him (but far from the same), I end up with two kinds of existence, with the second falling short of being clearly real but being, going forward, maybe the more interesting for us of the two.

Let's take a tour through these existences with logic's language. It includes an assertion often represented by $\exists$. Seeing it, one thinks, *There is*. . . .

Like a badge of reality, it says something exists; the something is whatever follows.

So, I could assert there is a number *n*, among the natural numbers **N**, that is greater than one, using this bit of logic language (we won't dive in; it just shows you how one finds one's way in Russell's world):

$$\exists n \in N : (n > 1)$$

This is true; take, for example, two. Our Russell could proceed (no doubt at length) to prove it and his proof would then, for him, also exist.

But I could also assert there is, among the natural numbers **N**, a number *n* that is less than one:

$$\exists n \in N : (n < 1)$$

This we know is not true. Zero is the only possibility and it is *not* a natural number.[a] But Russell bought into the notion that it is.[b] With zero on his natural number list, he could prove this false proposition. The false proof would have the same quality of existence as the true one.

There is too the logic of nonexistence. The symbol $\nexists$ asserts something (that follows it) does not exist, seeming to set that something up only to knock it down. Philosopher Maria Reicher said,

> Some important philosophers have thought that the very concept of a nonexistent object is contradictory (Hume) or logically ill-formed (Kant, Frege), while others (Leib-

---
a See chapter 109 at page 539.
b See chapter 108 at page 535.

niz, Meinong, the Russell of *Principles of Mathematics*) have embraced it wholeheartedly.²⁰¹²

Logic is a perfect GIGO gadget. It yields no more than what follows from its input assumptions. Yet there is a curious power in bandying ∃ about. Of course, saying something exists does not make it so. And then ∄ involves thinking of the nonexistent. That thought then has existence, and *its* existence (think of all those thoughts of *zero*) has real consequences.

Contradictory it surely seems. But once the concept takes on form in thought it takes on an existence in the world. Thus, for example, though its actuality is nonexistent in this universe, the concept of *infinity* formed a lasting Earth-wide fog of innumerable tweedles whose being and doing came to shape our world profoundly. The reader can no doubt envision more.

Cogito ergo est. I think, therefore it is.

A thought's existence has a meaning only in the mutual relations of our neurons and synapses.[a] So, at least initially, we do know where it is. But we cannot count its tweedles—not merely as in it's impossible but as in they are undefined and undefinable. All the more so if I should succeed in telling you of my idea and it enters into our relations. And even more so if lots of us tag it with a name that ends up on the dictionary shelf, endlessly infecting other minds.

This thought has brought us into a short skirmish with the nature of consciousness as a physical phenomenon, a relatively recent issue in the history of philosophy.²⁰¹³ The first question would seem to be *What is the question?* Nonetheless, there is a large interdisciplinary literature on questions about the nature of consciousness, with some of it purported to be linked to physics.²⁰¹⁴

Our small excursion into Russell's realm with tweedles showed our thoughts have a kind of existence embedded (one might rather say disseminated) in the world.[b] Though less than real as entities, they shape the universe in no less physical degree than their real relatives.

---

a See further, chapter 134 at page 647.
b For the role of thoughts relating to past truths in legal evidence and history, see chapter 126 at page 603.

This needs much more than the tag end of a chapter but that's all it can get here, so I will jump to a few obvious conclusions.

The universe now has a consciousness and it is ours.[a]

Thoughts *are* actions, no less than acts that are more overt.[b]

Our false thoughts are as existent as the true, and as consequential. Self-curating what we think may be an existential duty.[c]

---

a I do not mean by this to dally with Erwin Schrödinger's views or endorse panpsychism.
b A literature sees them as opposed; Arendt, for example, spoke "of the mind's withdrawal as the necessary condition of all mental activities"; see my note 1033 (Book One), at her page 96.
c Here the failure to think—or what Arendt called "this absence of thinking," speaking of Eichmann and "the banality of evil"—is kin to the failure to act; see my note 1033 (Book One), at her page 4.

133
A FUTURE FOR PHYSICS

> Fundamental ideas play the most essential role in forming a physical theory. Books on physics are full of complicated mathematical formulae. But thought and ideas, not formulae, are the beginning of every physical theory.
>
> Albert Einstein and Leopold Infeld (1938)[2015]

The idea of real quanta offers an entirely new career path for the few boldest practitioners of physics. Thus far (notwithstanding Planck) it is an ontologic, and not yet a physics, concept. Yet already it suggests experiments. It offers inroads into physics even without the new math that it cries out for. It is hard, maybe impossible, to foresee even outlines of the shapes of things—the theories and experiments—to come. As it ever was.

| The prescription for new physics is: Get real.

Einstein and Infeld were describing the evolution of the idea of quanta as the basis of reality (above). They used distance from New York to rail stations ("quantized," compared with places accessible by car, "continuous") to illustrate the idea as they understood it. They were embedding their quanta in a conceptual map—a *continuous* space. As we have seen, deriving discontinuous quanta by quantizing in continuous space—here using the device of trains they said don't stop between the stations (their experience of trains was different from mine)—troubled Einstein. Even as he wrote this,[2016] for at least two decades his reason had been telling him this idea is not real.[a]

Yet his life's work—along with essentially all of physics—remained stuck in the cerebral swamp of continuous space.

---

a See chapters 2 and 45 at pages 22 and 255, and note 921.

Consistently with what he said about the role of fundamental ideas, the new ontology offers physics a new opportunity: Leave the swamp of sense experiences; build anew upon the rock of making sense.

The idea observation (rather than belief) would reveal truth about our world and people's place in it has ancient roots.[2017] Even religious thought tapped tentatively into it; for example, in 1733, Catholic poet Alexander Pope wrote,

> See worlds on worlds compose one universe,
> Observe how system into system runs,
> What other planets circle other suns,
> What varied Being peoples every star,
> May tell why Heav'n has made us as we are.[2018]

Those ancient roots—that would ask *why* about what one observes, and about who we are—have been a-withering of late. To regain its role in a reason-based approach, physics requires fundamental change.[a] First and foremost, it needs to return to a concern about reality.[b] This will not be easy. Observing need for far less fundamental change,[c] Planck said,

> A new scientific truth does not triumph by convincing its opponents and making them see the light, but rather because its opponents eventually die, and a new generation grows up that is familiar with it.[2019]

Must physics continue to eschew reality? Will Planck's forecast keep on being right—will physics keep advancing, as the saying goes, one funeral at a time?[2020]

Physics always has opportunity to refocus on reality. On its face it looks to be a simple choice. Yet almost to a man (and too few women), physicists work within walls of an ontology that's obviously wrong. For this reason, few could claim to spend their days on real physics.

---

a See chapter 4 at page 34.
b Febrile voices may seize on and promote my repeated message, physics is almost all fiction. Let me say again: It works! Only those who want what doesn't work should espouse anti-science. The issue is, can we do better with a senior science that becomes grounded in reality?
c Ludwig Boltzmann's battle to establish the atomic theory of gases.

This poverty of vision is ingrained by the time students get to grad school. But grade-school students still imagine physics will enable them to understand. For example, science student Daniel, aspiring to teach it, saw,

> It's how everything works, why the universe is [what] it is, why this does that, why that does this. It's just sort of why everything does everything.[2021]

Some months later, he said,

> Physics is the senior science, that's the problem. . . . [I]t probably sits, like it's at the heart of all the other sciences, physics leads to chemistry, chemistry leads to biology. Science, it's about understanding the universe we live in, trying to make sense of it.[2022]

How much better might all science meet such expectations if physics for its part could come to truly be "about understanding"? Meanwhile, systematic disenchantment of inquiring students perpetuates the problem.

Smolin outlined why it's hard to break out and take new directions at the research-program level:

> Even if I'm convinced that something very new is needed, I have little idea how to search for scientific truth except by building on an existing research program, using a well-honed tool kit and methodology. This is research as it is taught, recognized, funded, and rewarded by the academic community.[2023]

The physics forum's ongoing rift with philosophy, especially its neglect of or even disdain for ontology, plays a part in the paralysis. It was not always so. As Rovelli noted (in an appendix titled, "On method and truth"),

> [A] dialog between physics and philosophy . . . has played a major role during the other periods in which science has faced fundamental problems.[2024]

He was thinking of "the time of Newton, Faraday, Heisenberg and Einstein."

Amid all this, physics has a genuine quandary. How should it pick its paths?[a] Is a blind leap the only way for physics to renew?

And with physics working beyond boundaries of observation, the elephant in this room is the so-called scientific method.[2025] Roughly, wild theorizing must be constrained by experimental data.[b] But the constraint is itself under increasing pressure. For example, in 2009, Amelino-Camelia wrote of the problem at Planck scale,

> Unfortunately it is not unlikely that experimenters might never give us any clear lead toward Quantum Gravity, especially if our intuition concerning the role of the tiny Planck length . . . turns out to be correct. . . . [B]ut we must try.[2026]

Controversy over the perceived impossibility of testing theory at Planck scale has seen the rise of post-empiricism,[c] especially in string theory circles.[2027] A key part of the issue here is—or should be—the role of ontology. Whether or not they make it explicit, physicists make research choices that are guided by their ontological commitments. Their ontological commitments can cause problems; they could also point toward solutions.

My solution to physics' pick-a-path and scientific-method quandaries is, resurrect the attitude that guided fundamental physics in its heyday. Rebuild real ontology. Let *it*, not mathematics, blaze the trails. Test ontological commitments and companion theories for their fit with *all* we know about our world, not just some siloed data set seen through a narrow lens.

That is, make sense.

This is no new proposal. In 1949, Einstein said,

> The justification of the constructs, which represent "reality" for us, lies alone in their quality of making intelligible what is sensorily given (the vague character of this

---

a Here we revisit Ayer's question from chapter 4 at page 34; see note 136.
b See the preface at page 1.
c See chapter 15 at page 95.

expression is here forced upon me by my striving for brevity).[2028]

As we saw, Smolin, maybe more than many leaders in the fundamental physics business, has long been engaged with ontologic issues.[a] Setting his face against the trend, he declared for realism, and said,

> [R]ealists are interested in ontology, which is the study of what exists.[2029]

Elsewhere, he said we need a new ontology.[2030] Some may wonder why *he* did not write this book.[b]

Between the lines one can sense Smolin's angst about the future of physics. It's not just an urgency to make that breakthrough he can almost taste; it's a responsibility he's not as blunt about as I was with Sundance:

> You and I both got a very expensive education and the public paid for it. Don't you think the public is entitled to a return on investment?[c]

No one seems to track what the world spends on physics each year but it may be in the order of a hundred billion U.S. dollars.[d] Moving a mere hundred million to engage with reality could bring big benefits for paying publics.[e]

And, in my mind's eye, even greater value may lie in humankind having a real-world vision of itself, one that is concordant with its science. No doubt my tapestry should be critiqued, torn down and—one hopes—rewoven in a better form. Meantime I have this (no doubt naive) notion that, armed with a better worldview, we could be better at being and do better at doing.

---

a See the preface and chapter 1.
b As do I. Maybe he would not want to cite himself more than a hundred times.
c In the garden; see page 201.
d This is no more than a rough estimate (starting with a million physicists) that begs the question, what *spending on physics* is.
e I am again reminded of Freeman's eyeball estimate of economic value of a Planck-scale breakthrough, some six or seven orders of magnitude higher than the fancied funding level; see note 1936.

I have too this sense of where we are today: Social media disseminate untruth faster than truth,[2031] making technology of mass addiction and a culture of fear[2032] into core strategy for trillion-dollar businesses. The very concept of reality is crumbling.[2033]

On both sides—opportunities and problems—we have a growing need for physics that relates to a real worldview. Even sixty years ago, Planck saw this need:

> Man wants . . . a standard, a measure of his actions, a criterion of what is valuable and what is worthless. He wants an ideology and philosophy of life.[2034]

He looked to science and found it wanting:

> [S]cience is not built on any principle of such universal validity, and at the same time of such portentous meaning, as to be fit to support the edifice properly.

He spoke of the individual worldview,

> [T]his practical world picture which every human carries within himself is not a directly given notion, but an idea elaborated gradually on the basis of facts of experience. . . . The same principle applies to the scientific world picture.

With this last assertion, I must beg to differ. When science seeks a real world-picture, the principle cannot be the same as that for individuals. The difference relates in large part to the state of today's physics. Its roots lie in Unger's and Smolin's *cosmological fallacy* . . .

> . . . the mistake of taking a scientific methodology . . . outside the domain where it can make contact with experiment and observations.[2035]

By this they meant, specifically, the error of extending, to parts of the universe we cannot see, science based upon observing some part of the universe we can see.[a] They were conflicted. On one hand they set up their fallacy as trapping science in the inescapable domain of

---

a One might apply it also to extending to the Planck scale science built by observing atoms.

our experience while, on the other, their research agenda aimed to help cosmology escape this trap.[2036]

Like Planck, they asked by implication, *What is the right relation between science and ontology?*

Our answer should recognize the prime role of ontology in guiding scientific enterprise: One way or another, scientists elaborate[a] a worldview that influences how they think and work.[b]

We've seen how physicists' ontological commitments—to space as continuous, to time as a dimension, to atom scale as fundamental, to particles as elementary, to spacetime as their stage—imprison physics in a falsely circumscribed inquiry.[c] If science is to preserve its truth-seeking value while widening its purview to the universe, it will need physicists to work with a more real ontology. That may also be a worldview we can all begin to understand.[d]

The new ontology—as far as I've been able to decipher and unfold it here—gives rise to new questions and provides glimpses of entirely new fields for new physics.

Getting from here to there will require—as all increase of knowledge does—an iterative approach. For example, if experience shows doing physics with an ontological commitment to a *first law*, $1 \to 2$, leads physicists to fertile fields, we may with greater confidence extend Einstein's speed-of-light law out to the whole universe on one hand and to Planck-scale physics on the other.[e]

Physicists could then cultivate two reasons to embrace their new commitments, two indicia of truth: They work *and* they make sense.

Is it too much to hope that in this way *what works* could resume its romance with *what is*?

---

a Adopting Planck's verb.
b And, regardless of its inbuilt contradictions, their worldview spills over into ours.
c See part 4.
d Here I am mindful of Arendt's cautionary analysis of Descartes's doubt, wherein she said, "Intelligibility to human understanding does not at all constitute a demonstration of truth, just as visibility did not at all constitute a proof of reality." See p. 275 of note 102.
e This may be current practice with no sound foundation for a few who work with or think of those two inaccessible realms.

Worst case, we might find we're fabricating physics upon ontological foundations that turn out to *not* be real. But that would merely be more of the same.

134
BEING

> I know of no more moving story . . . than that of intelligence struggling step by step from the beginning . . . to master the dimensions and the relief of the universe.
>
> Teilhard de Chardin (1955)[2037]

In review, we see the story of the universe is a marathon road show that is a play in two short acts. Likewise its existential language has two verbs: to *be* and to *do*.

It is surely not some kind of grand cosmic coincidence that, in our evolution of the English language, they should turn out to be two of its three biliteral verbs. Nor that the third should speak to motion that those two make happen.

The same two verbs lie at the heart of the philosophy of existentialism. For example, toward the end of his classic existential essay, Sartre settled in on the interplay between being and doing. Therein, he said, we find the essential elements.[2038]

> The universe has two verbs: be and do (so we can go).

The new ontology agrees; it takes his thesis to a deeper level. It says the world uniquely reifies these verbs and their successive interaction.

Each *Now* iteration *is*.[a] (It does nothing.)

Each *Next* iteration *does*.[b] (It has no existence.)

The iterated *do-be-do-be-do* sequence[c] is also the source of what we call causation.[d] Their successive interplay causes everything. As each causes the other.

---

a See chapter 56 at page 301.
b See chapter 57 at page 307.
c With a nod to songwriters Charles Singleton and Eddie Snyder.
d It is also what got bastardized as spacetime; see chapter 83 at page 420.

So in a related question—when it comes to *our* being and *our* doing, which is the cause (and which is the effect): Who we are or what we do?

Do we choose what we do because we are who we are? Or do our choices to do what we do create who we are?

After trying the apparent truth of each of these seeming mutually exclusive views, I conclude they must, like the *be* and *do* of the world, rather be mutually synthetic.

One thing is clear. In each moment we are who we are and choose what we do. Or, as jurist Michelle O'Bonsawin said,

> Our actions are reflections of ourselves.[2039]

In 1967, new PhD in hand,[a] I moved to a small town set in ancient rocks and waters of the Canadian Shield. Nearby, at a new (and the world's only organic-cooled) nuclear reactor, an eclectic band of biophysicists was getting underway. They seemed to lack discernible relation to the business of their employer, Atomic Energy of Canada Limited; rather, they were a small center of excellence that helped attract the scientists and engineers whose skills that business needed.

As a postdoc with the National Research Council of Canada (which seemed to care not what I did) I had more freedom than tenured professors. And from my late teens I had wanted answers to three questions:

- How does the nerve membrane do its seemingly unphysical tricks (see fig. 97)?[b]

- How does the human brain work, as in, what are memory and thinking?

- How did the universe begin?

While I was drawn into other interesting studies with colleagues, for eight years[c] I delved into these three questions.

---

[a] Figuratively speaking; I moved to Canada so suddenly I never got to pick it up.
[b] It generates a large reversal of its electric field known as the *action potential* after a relatively small depolarizing stimulus—smaller even than the 27-millivolt scale of thermal energy at body temperature. In physics and chemistry this is (and in biology was coming to be seen as) a fundamental voltage scale factor, **kT/e** (the Boltzmann constant times the absolute temperature divided by the charge on the electron).
[c] The last five, employed by AECL.

Part VIII: Real Progress   649

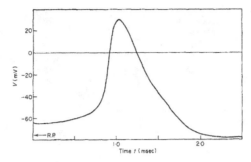

Fig. 97. Nerve action potential after a 10-mV drop from the resting potential

I found answers for the first two, sufficient for my curiosity.[a] In those days I had a cavalier attitude to the public's return on investment.[b] I thought the world unready for the way the brain works.[c]

The brain is a network (fig. 98[2040]) of some hundred billion nerve cells (*neurons*) and about the same number of neuroglia.[2041] The neurons are linked by maybe a quadrillion connections (called *synapses*) each of which can either stimulate or inhibit its neuron, which

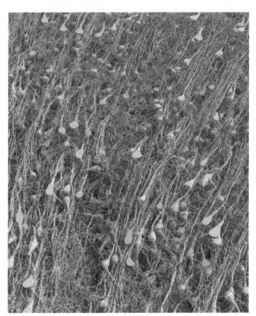

Fig. 98. A *tiny speck of human brain*

---

a The third had to wait.
b See note c and related text at page 643.
c It still is; but what is now arising from neural networks and artificial intelligence has me reconsidering.

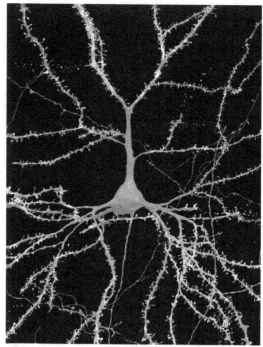

Fig. 99. A human-brain neuron with 5,600 synapses

integrates its inputs and may then itself fire nerve impulses (action potentials)[2042] many times per second. These impulses travel in one direction, from the cell body down its axon, which leads to synaptic inputs of other, often distant, neurons.

Various specialized parts of the brain have various kinds of neurons[2043] that make connections with idiosyncratic patterns of complexity (see a human brain neuron with 5,600 synapses,[2044] fig. 99); but the brain works as an integrated whole.[2045] It is so highly connected the shortest path from any one neuron to any other among all its billions is only a few synaptic hops.[2046]

We have long known how complex the brain can be; see, for example, Nobel Prize–winning physiologist

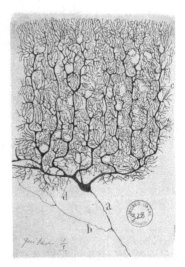

Fig. 100. Physiologist Ramón y Cajal's drawing of a Purkinje-fiber cell

Ramón y Cajal's celebrated 1899 drawing of a single brain cell (fig. 100; the axon extends lower right).[2047]

The human brain is a seemingly simple input-output system that has evolved into stunning complexity. Science journalist Thomas Lewton asked if it is truly the most complex object in the universe and reported it's a poorly defined question.[2048] But molecular biologist and Nobel laureate James Watson, called it . . .

> . . . the last and grandest biological frontier, the most complex thing we have yet discovered in our universe.[2049]

At various rates through our lives, depending on age and usage, existing synapses are deleted and new synapses are formed. And each synapse exhibits plasticity—past activity affects the strength of future activity.[2050] These changes are the basis of memory and, one might say, their *doing* is our *being*.

Though the brain has areas of specialization, like vision or language processing, our minds and memories are essentially nonlocal.[2051] And loss of one area may be largely compensated by plasticity of other areas.[2052] Those who find the concept hard to grasp might

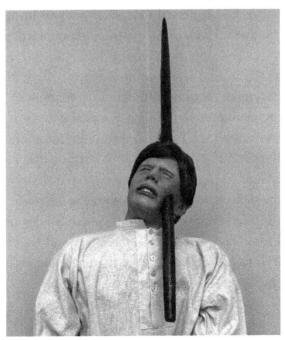

Fig. 101. The rail-spike brain injury

check out the case of Phineas Gage, who in 1848 survived a railway spike through his head (fig. 101) with . . .

> . . . his intellectual faculties being decidedly impaired, but not totally lost; nothing like dementia, but they were enfeebled in their manifestations, his mental operations being perfect in kind, but not in degrees or quantity.[2053]

Thus, *engrams* are elements of memory encoded in insignificantly tiny changes in the numbers and strengths of synapses and firing patterns of neurons. They are nonlocally encoded; each engram consists in relationships among a large part of the brain's neurons and synapses (or perhaps the entirety) that also encode all else we remember. That is, memory is relational.

As psychologist Karl Pribram said,

> [E]xperiments and observations made it clear that engrams, memory traces, could not be localized and that perceptual images and motor patterns displayed constancies and equivalences for which it was difficult to conceive any permanent "wiring diagram."[2054]

Yet scientists have detected individual neurons firing in relation to specific words.[2055] The connected structure, nonlocal functioning, and plasticity provide the neurological basis for learning, memory, and thought.[2056]

So, for example, your mind's concept (or engram) of "chair" consists in commonalities of the firing patterns of billions of your brain's neurons arising over time from the diverse objects designed or used for sitting as reported by your senses.[2057] It is, if you like, the embodied *concept* of a chair.

Thus, the mind is, for want of more precise language, a 4-D hologram,[2058] three of its "dimensions" being physical interconnectedness of neurons (now known as the *connectome*) and the fourth being time.[a] That is, its information content is distributed in relationships among the firing propensities of all its synapses.[b]

---

[a] They are not dimensions of the universe.
[b] This is a simplified description; parts of the brain are also more or less specialized.

Compare this conceptually with the more familiar hologram where relationships embedded in the patterns of a 2-D image encode a record of a 3-D scene. If you cut it in two, each piece still bears a record of the whole scene, but with poorer resolution.

My reason for digressing into human neurophysiology is to bring out its intriguing intersection with the universe viewed as a hologram.[a]

That is, if the universe and your worldview of it are each some kind of hologram, your lifelong efforts to grasp the world around you may take on a new perspective.

For instance, the existence and activity (being and doing) of your hologram give rise to consciousness.

The mediator between holograms we call *language*, concepts encoded in words. It is of the essence that our languages, unlike a computer language, must be inexact. Linguist Jila Ghomeshi said,

> There is no such thing as ambiguity-free language.[2059]

With our pseudo-4-D holograms in gear, sitting pretty in our succession of 3-D quantum spaces of our maybe 2-D holographic universes, we can begin to see why. And we see why in the end the word *real* is inescapably ambiguous, though I have tried my best to pin it down.

We can see too why sense perception—logical positivism's sole road to reality[b]—has both fundamental functions and fundamental limitations.[2060]

Your hologram preserves a highly processed memory of past events.[c] It has long been thought the memory and what is remembered are different (though related) things. Around 420 CE, Augustine of Hippo said,

> [W]hat is hidden and retained in the memory is one thing, and what is impressed by it in the thought of the

---

a See chapter 64 at page 340.
b See the preface.
c It is highly processed in part because soon after the event its engram becomes less the product of action-potential patterns caused by the event and more the product of later action-potential patterns caused by remembering the event. Long term, we can remember because we did remember but, each time, the engram inevitably changes. See, for example, "The science behind Brian Williams's mortifying memory flub" by Amy Ellis Nutt in the *Washington Post* on 5 February 2015.

one remembering is another thing, even though they appear to be one and the same thing when they are joined together.[2061]

Memory is a feat the universe is unable to match without you—not only because it has nowhere to keep it but also because, to the extent some shadow of its pasts does linger in its ever-present state there is no process for it to recall them and apply them.[a] However, like mine your hologram is an integral aspect of the universe; so the existence of our and others' holograms bring an emergent property—memory—into the world.

Likewise, sharing the conceit that we do understand at least a bit of how things work, we all think of the future. Any small sampling of editorial pages reveals that varied views of how things work provide the basis for most of our disagreements. Nonetheless, some of our future-thoughts (like plans) do come to pass. This present pseudo-future is another property the world would lack without us.

One way to summarize this is the universe evolved intelligence, and we are it.[2062] Mayhap this thought could serve us well.

One could write a book about all this, or many books, but they are not this book.

---

[a] Thus history is a creative enterprise; see also chapter 126 at page 603.

135
DOING

> Last, but not least, it cannot be otherwise than important to a teacher of metaphysics, to be able to say with universal assent, that what he expounds is Science, and that thereby genuine services will be rendered to the commonweal.
>
> Immanuel Kant (1783)[2063]

One might say Kant had the first word[a] on our subject with his *Prolegomena to Any Future Metaphysics That Will Be Able to Present Itself as a Science*. The epigram then becomes his last word on his first word. It was a word on words as actions with consequences.

For the many ardent seekers of the real over the years—some looking into ever-larger spaces; others smashing ever-smaller structures; some observing seemingly distinctly different things; many plumbing their own minds; each doing their own bit to render service to the commonweal—it's been a back-and-forth and often sideways journey. And so, theirs is as we have seen a bits-and-pieces, stitched-together kind of story.

> Being has no separate existence from doing.

Amazingly, as their smallest and largest, and their earliest and latest stories diverged to their limits, they turned out to meld: The one is many; the many are one.

Amazingly too, our many *Why?*s and all their fragmentary answers can now be seen to stitch together in a single picture with two faces: This is what it is; this is what it does.

---

a *Prolegomena* means "introductory remarks."

Looking back, we see that Einstein has been our essential guide to finding out the universe's doing.[a] His genius lay not so much in finding the right answers—he was often wrong[b]—but rather, it resided in his posing the right questions in pursuit of what is real.

It was not quantum theory's challenge to his concept of causality (the universe's doing) that troubled him the most; it was its rejection of reality (its being):

> The sore point lies less in the renunciation of causality than in the renunciation of the representation of a reality thought of as independent of observation.[2064]

Too late for him, the new ontology hands us such a reality. What shall we do with it?

Its insight into ourselves, our world, and our place in it comes at a time when much looks grim. But then, it often seemed that way. For example, early in the last century, Bertrand Arthur William Russell, 3rd Earl Russell of Kingston Russell, Viscount Amberley of Amberley and of Ardsalla, wrote to a friend,

> [N]othing seems worth doing or worth having done. . . . These times have to be lived through: there is nothing to be done with them.[2065]

Few mathematicians with a philosophic bent have had more cause to fall in with the mordant view of the aristocratic Russell than did Stephen Hawking. Yet Hawking found things worth doing and against all odds found ways to do them. He foresaw new doings based on his, and wide enlightenment beyond his own:

> [I]f we do discover a complete theory, it should in time be understandable in broad principle by everyone, not just a few scientists. Then we shall all, philosophers, scientists, and just ordinary people, be able to take part in the discussion of the question of why it is that we and the universe exist.[2066]

---

a In token measure of our debt: Einstein is quoted, cited, or referenced here about five hundred times.
b See note 655.

Part VIII: Real Progress

If there is to be "a complete theory" that meets Hawking's prescription, it should be a theory of a way to understand our world. Understanding needs to be set into its foundations; or, as my friend transportation planner Manfred Rehbock put it,

> If it's not there for the takeoff it won't be there for the landing.[2067]

There may well be better ways to stitch understanding together but now we have a way that does at least set out to be real. Such as it is, it embodies thoughts of many thinkers—some of whom I've quoted—each reaching for the real world from their place in Plato's cave. Until we find a better way, maybe the kind of theory it needs and the physicists inclined to do something about it could be allotted a small slice of the next billion physics dollars. If I seem to harp upon this theme, it expresses my concern we may at this point need "a complete theory" to conceive a better human action plan.[a]

Pseudo-Zen fashion I may contemplate the beings/doings of far reaches whose vast messages will reach us ten or twenty billion years from now and feel our oneness with them. The same universal vista harbors teeming human billions sleeping, eating, anxiously pursuing their life choices. At all scales—from Planck's to the universe's, including ours—the new ontology depicts doing as the flip side of the being coin.

---

[a] In this, I concur with Frank and his colleagues; see notes 108 and 109.

136
YOU AND I AND WE

> Know then thyself... The proper study of mankind is Man.
>
> Alexander Pope (1734)[2068]

Above all, this is a story about us. About our long striving to rise. About our pursuit of the advice wisely inscribed upon the Temple of Appollo at Delphi, Γνωθι σαυτον (*Gnothi sauton*; "know thyself"). And maybe, too, about "touching a certain place in the soul" as songwriter and singer Robbie Robertson expressed it.[2069]

My main muse, Einstein, saw all this; he sought to act upon it:

> One cannot help but be in awe when he contemplates the mysteries of eternity, of life, of the marvelous structure of reality. It is enough if one tries merely to comprehend a little of this mystery every day.[2070]

We have a growing understanding of our deepest human mysteries thanks to thousands, millions really, who, as Einstein said, tried "to comprehend a little" of it every day. We see it's an amazingly simple but surpassingly strange world that, while explaining much, also invokes *You can't make this stuff up.*

| The new ontology's personal message is one of hope.

Today we have easy access to much more knowledge than previous generations. Yet many find themselves adrift, drowning in disinformation, much of it state sponsored.[2071]

The degree to which we've lost our compass is reflected in our language, for example in the adjectives we use describing supposed facts; language that not so long ago was utterly derogatory—like *incredible*[2072] or *fantastic*[2073]—now conveys praise.[a] By the nineteen nineties, *unbelievable* was beginning to mean "extraordinary," edging

---

a For examples, see notes 1610 and 756.

out "too dubious or improbable to be believed."[2074] To seize on one example of this linguistic upheaval, eighty or so years ago popular (and fashionably loquacious) author Leslie Charteris wrote,

> I have become . . . used to seeing the adjective "incredible" regularly used even in the most flattering reviews of the Saint's adventures.[2075]

The critics meant his stories were not credible.

We have daily need for some of these lost meanings.

Have in mind here how our language forms foundations for our thought.[a] Are we en masse performing amateur brain surgery?

At the same time, we are largely cut off from what once were our shared sense-experiences. We seek synthetic substitutes for real nature, real dangers, real emotions, even real skies.[2076] Our strivings for lost links to reality show in our linguistic nomination of *authentic* as 2023's most-increasingly used word.[2077]

Are we losing our place?

For millennia, this was a managed problem at least in the occidental world. A formulation called the *great chain of being* played a central role in its worldview.[2078] It put every entity in its place, from God at the top to minerals at the bottom. (That's us, second row, just below angels, in fig. 102.) It imposed an understanding that was almost universally accepted. We can hardly grasp the steady influence it exercised on the whole Western human enterprise for some two thousand years.

Philosopher and historian of ideas, Arthur Lovejoy, said of it,

> The Chain of Being . . . was a perfect example of an absolutely rigid and static scheme of things.[2079]

Evolutionary biologist and ecologist Sean Nee said, simply,

> For centuries the "great chain of being" held a central place in Western thought.[2080]

No longer immersed in this scheme, we are embarked (we think) on seas of individual adventure, each left to devise or find our own personal identity and ethic.

---

a See the introduction and chapter 72 at page 373.

Fig. 102. A sixteenth-century drawing of the great chain of being

Arendt, whose worldview began—as one might think is right—with birth (being) saw initiative (doing) as our high purpose. She said,

> With word and deed we insert ourselves into the human world.[2081]

Our *be-do* ontology seems to somehow resonate.

In this way the many authors of the new worldview offer a new view of our chain of being. No longer actors playing set roles on an ordained stage, we are integral to it and, in real and practical degree, create it. We conceive purpose; we have agency; we choose; we do the things that make the world. In so saying, I am taking a position in the debate on free will. Two recent books bookend it well;[2082] our worldview offers a new window into an old conversation.

Einstein's Nietzschean view of a strictly causal world[a] allowed for no such agency.[2083] He saw our future as already written like our past.[2084] In 1931, he said,

> If the moon were gifted with consciousness . . . it would be quite sure it was moving on its way of its own accord based on its own definitive decision. So would one with higher insight and with perfect perception see man and his deeds and smile at the illusion he was acting through free will.[2085]

Taking this view, it's odd he would suggest (above[b]) there's any point in trying if we have no choice.

We can now understand how, totally engaged but mired in old ontology, he would have missed this: Our purpose is the universe's—to be and to do. Yet he did both and, with what he was dealt, he did both well.

For my part, I take it as an observable that life has purpose and it is to be the best that we can be and do the best that we can do. Both aspects involve choices (one immediate, the other longer term). This twin purpose has a consonance with being/doing as the ontologic objects that the universe insistently makes real, its existential yin and yang.[c]

Our purpose calls for us to endure being and to essay doing. It sets us up for lifelong striving: At each moment, we are who we are and we can be no other; the *Next* iterations open opportunity to choose. Then being who we are, we choose and do; and what we choose to do creates who we become.

The many hundreds whose ideas I have drawn upon—and untold others—strove, each by their lights, to do. It would be surprising if any *didn't* try to do their best. (Even, clearly from his writings, Einstein the determinist.)

In their being and their doing, they helped us to get to where— and become who—we are. Change any one of them and we would all be somewhere—and somebody—else, or more likely not be at all.

---

a See chapter 74 at page 381.
b Note 2070.
c See chapter 42 at page 236.

The new ontology that underlies my notion of life's purpose is surely far from the whole story. And what there is of it is surely not all right. Yet, it is fashioned from utmost simplicity. If the universe is simple,[a] it seems likely that this kind of view of it is mostly right.

One reason for thinking so is any deeply different understanding of our world will almost surely call for many more assumptions.

A for me more cogent reason can be seen by setting it beside the old ontology, which we in varying degree create, accept, adjust, and live by, thinking it helps us to understand the world. Notwithstanding its strange aspect, the new may help us understand much more much better than the old.

And in doing so it offers us a clear alternative to what a favorite author called . . .

> . . . life which slips gently toward death without fuss or stirring, without even asking Why?[2086]

Let's revisit what and where we are.

For some number of universal iterations,[b] we each borrow a uniquely ordered randomly changing loose cloud of a mind-boggling number of tweedles—every one of which is real—that collectively express emergent properties, like sense perceptions, knowledge, reason, purpose, pain, and consciousness, that, among other extraordinary things, enable us to look at the night sky and contemplate the entire universe—though we *see* almost none of it—and how it works, all the way from tweedles betwixt tiny quanta, each a replica of the beginning, through layers of emergent entities like quarks, atoms, cells, and plants and planets to black holes and galaxies, each and every one of them essential for our own personal cloud to be.[c]

It is indeed, as Baez foresaw, . . .

> . . . a world more strange, more beautiful, but ultimately more reasonable than our current theories of it.[d]

---

a See chapter 9 at page 57.
b The average lifetime is 72.6 years or about forty-two trillion-trillion-trillion-trillion iterations of the universe; of course, neither beginning nor end is well defined.
c I advance this as a proposition for some student of causation to prove.
d See note 46.

Your cloud of tweedles could not exist alone, without the rest of us. Indeed, as brief reflection on life's daily incidents and fine-tuned random aspects of human reproduction should tell you, with a slight change in almost any choice by almost anyone a hundred or more years ago, you would almost surely not have been conceived.[a]

Each of our tweedles is inseparable from the rest of the universe. My mantra, yet again, *We are not in but of the universe.* We are intrinsic to the whole of it. Our reality is its.

So, with a little practice, you can close your eyes and become in your imagination as you really are, an integral aspect of its vast cloud of swift-changing twists. If this sounds like getting lost, it's not. In that unity is ultimate identity.

Thus, we are the products of much more than genes that we inherit, air we breathe, and food we eat; we all partake of a chain of causation.[b]

This chain runs unbroken back through thousands of generations of our species. Once we grasp the real nature of causation, we confront, for example, the plain reality that we owe all our lives—that is, not only the station to which we were born and the conditions in which we exist but also our very existences—to slaves and the institution of slavery. With that single insight, Henry Clay's self-made man[2087] is exposed as shabby fiction.

This personal embodiment of causation runs back much further—through all the chances in the chain of life back to the first surviving reproducing organism arising from random photochemistry more than four billion years ago.[2088] And yet further, all the way to the ultimate cause.[c]

It's not easy to perceive amid "the clashing discords and the din of life"[2089] how each of us matters. Yet we each create the world by what we do (including what we might do but do not) every minute every day. Our choices spread like ripples, some slowly, some with light speed. At the scale where this is happening, their effects are undetectable. Yet, suffice here to say, each small choice each of us

---

a To-be theorem number two.
b See chapter 27 at page 154.
c See chapter 57 at page 307.

makes a hundred times a day resets the entire population of the Earth a hundred years from now.[a] Be sure your choices have consequences far beyond what with seeming good reason poet Linda Besner called . . .

> . . . our small and separate lives. [2090]

One consequential choice is to in some wise embrace the new worldview. It depends upon a few bizarre assumptions. The old depends on many more, no less bizarre (though more familiar). The new would seem to be a more rational choice, at least as long as it brings better consequences.

The foremost consequence, I think, will be pervasive change arising as we find our place in what nuclear engineer, submariner, farmer, and politician Jimmy Carter called,

> . . . a vast and awesome universe. [2091]

I do not say that we have mastered its reality. But we are at last coming to close grips. Being and doing is our nature and the nature of our world. And, in answer to Piet Hein's appeal[b], we can now know some of who we are and what we are doing before we are done.

I hope your mental journey with those many seekers after what is real will sustain you as they lead us from the cave into the sunlight.

This is our almost fourteen billion years' heritage. I wonder what our children and their children and theirs too will do with it.

And who they will *be*.

---

[a] Let this be causation theorem three.
[b] See note 1.

ACKNOWLEDGMENTS

As always, indeed far more than as always, I owe far more to far more than I can say here.

First—their contributions being first—those whose works and words I have quoted, drawn upon, and cited, the authors of the many threads from which the tapestry of my tale is contrived. *My tale,* I say, owning responsibility mostly for the snipping and the stitching; almost all of it is theirs.

Closer to home, my special thanks to Sundance Bilson-Thompson, Norm Brandson, David Cherniack, Marilu Chiofalo, Ron Coke, Gord Collis, Doug Davison, Moira Eyjolfsson, Alan Freeman, Susan Guindon, Howard Gurevich, DeLloyd Guth, Tony Harwood-Jones, Craig Henderson, Daryl Jantzen, Fred Kelly, Taras Luchak, Dwight MacAulay, Simone Mahrenholz, Beverly McCaffrey, Mike McKernan, David Miller, Thomas Monias, John Myers, William Osborne, Gloria Phares, Victor Rodych, Kyle Shiells, David Topper, Devan Towers, Sloane Waldie, and Simon West. And to Nirdosh Ganske, who declined the job of editing the manuscript and then just did it anyway.

None of whom are in the least responsible for those errors that will escape even the sharpest eyes in a work of this kind. They are all mine.

IMAGE CREDITS

Fig. 1	Particle Data Group.
Fig. 2	Petrus Apianus.
Fig. 4	Claes Jansz Visscher.
Fig. 6	René Descartes.
Fig. 7	Albert Einstein, reproduced with permission of Fondation Martin Bodmer.
Fig. 8	National Institute of Standards and Technology.
Fig. 9	Omni Calculator.
Fig. 10	Andrew J. Hanson, reproduced with his permission.
Fig. 14	ESO, Non-Euclidean Triangle, supernova.eso.org/exhibition/images/1120_noneuclid-4K/, CCA 4.0 International License (creativecommons.org/licenses/by/4.0/).
Fig. 15	MDPI, © 2015 Pierre-Henri Chavanis, A Cosmological Model Describing the Early Inflation, the Intermediate Decelerating Expansion, and the Late Accelerating Expansion of the Universe by a Quadratic Equation of State, doi:10.3390/universe1030357, CCA 4.0 International License.
Fig. 16	Textile Mercury.
Fig. 17	Camille Flammarion, Hachette.
Fig. 18	The Globe of Death Chronicles.
Fig. 19	Joseph Streator.
Fig. 20	Steven J. Carlip, Causal sets and an emerging continuum, doi.org/10.1007/s10714-024-03281-1, © S. Carlip 2024, CCA 4.0 International License.
Fig. 22	Jürgen Goertz/Colin Gillespie.
Fig. 23	A. Einstein, image reproduced courtesy of the Leo Baeck Institute, New York.
Fig. 24	NASA/CXC/U. Texas.

Figs. 25, 27 to 29	Sundance Bilson-Thompson, reproduced with his permission.
Fig. 30	Yana Lehey, reproduced with her permission.
Fig. 32	John M. Sullivan, reproduced with his permission.
Fig. 33	Peter Prevos, Topological Magic Tricks, horizonofreason.com/magic/topological-magic-tricks/, CCA 4.0 International License.
Figs. 35 to 37 and 40 to 43	Sundance Bilson-Thompson, reproduced with his permission.
Figs. 45 and 46	From REALITY IS NOT WHAT IT SEEMS by Carlo Rovelli, published by Penguin Press. Copyright © Raffaello Cortina Editora SpA, 2014. Reprinted by permission of Penguin Books Limited.
Fig. 47	Sakurambo, commons.wikimedia.org/wiki/File:3D_coordinate_system.svg, CCA-SA 3.0 Unported license (creativecommons.org/licenses/by-sa/3.0/deed.en).
Figs. 48 and 49	Sundance Bilson-Thompson, reproduced with his permission.
Fig. 50	The Royal Society.
Fig. 51	Sundance Bilson-Thompson, reproduced with his permission.
Fig. 52	Sundance Bilson-Thompson/Colin Gillespie, modified and reproduced with permission from Sundance Bilson-Thompson.
Fig. 53	Rice University, Open Stax, College Physics, Simultaneity and Time Dilation, © 1999-2025 Rice University, openstax.org/books/college-physics/pages/28-2-simultaneity-and-time-dilation, CCA 4.0 International License.
Fig. 54	Albert Einstein/Christie's, autograph letter (fragment), www.christies.com/en/lot/lot-6089302.
Fig. 56	Rube Goldberg.
Fig. 57	Eugene Antipov, 2D-projection of 3D-projection of Hypersphere of 4D-space, https://en.wikipedia.org/wiki/3-sphere#/media/File:Hypersphere.png, CC BY-SA 3.0 license (https://creativecommons.org/licenses/by-sa/3.0/).
Fig. 58	Unknown/Academic, Japanese addressing system (detail), © Academic, 2000-2025, https://en-academic.com/dic.nsf/enwiki/252711.

Image Credits

Fig. 59	Fandom, Cubic honeycomb, verse-and-dimensions.fandom.com/wiki/Cubic_honeycomb?file=Yellow_Cubic_Honeycomb_against_Black_Background.png, CC BY-SA 3.0 license.
Fig. 61	NASA/Swift/Cruz deWilde.
Fig. 62	Photo File 6.93 Cooper Rubin, Archives and Special Collections, Vassar College Library.
Fig. 63	NASA/LAMBDA Archive/WMAP Science Team.
Fig. 64	An Illustration of The Allegory of the Cave from Plato's Republic (rendered in b&w), commons.wikimedia.org/wiki/File:An_Illustration_of_The_Allegory_of_the_Cave,_from_Plato%E2%80%99s_Republic.jpg, CCA 4.0 International License.
Fig. 65	Stib, World line (rendered in b&w), commons.wikimedia.org/wiki/File:World_line.png, CCA BY-SA 3.0 license.
Fig. 66	Thomas Young.
Fig. 69	NASA/LAMBDA Archive/WMAP Science Team.
Fig. 70	CERN.
Fig. 72	Elias Niederwieser, reproduced with his permission.
Fig. 73	Newton Henry Black.
Fig. 74	University of St Andrews.
Fig. 75	NASA/CXC.
Figs. 77 and 78	CERN.
Fig. 79	V.D. Burkert, reproduced with bis permission.
Fig. 80	NASA/ESA/CS/STScI/C. Fransson/M. Matsuura/M. J. Barlow/P. J. Kavanagh/J. Larsson.
Fig. 81	Goddard Space Flight Center/Jeremy Schnittman and Brian P. Powell.
Fig. 82	Alfred North Whitehead and Bertrand Russell.
Fig. 83	Courtesy Evenley Village Hall Management Committee.
Fig. 84	Messrs. Fox & Sons.
Fig. 85	Iosef Peano.
Fig. 86	Ishaq Ibn Hunayn/Chester Beatty Library.

Figs. 87 and 88	Sundance Bilson-Thompson/Louis K. Kauffman, reproduced with permission from Sundance Bilson-Thompson.
Fig. 89	NASA/ ESA/Giampaulo Piotto/Ata Sarajedini.
Fig. 90	Tony Freeth, The Antikythera Mechanism, CCA 4.0 International License.
Fig. 92	Andrei Linde, reproduced from "Particle Physics and Inflationary Cosmology," *Physics Today*, **40** (1987), p. 61, with the permission of the American Institute of Physics.
Fig. 94	Tomruen, Retrograde motion of Mars during opposition, https://commons.wikimedia.org/wiki/File:Mars_motion_2016.png, CCA BY-SA 4.0 International License.
Fig. 95	NASA/ESA/H. Teplitz/M. Rafelski/A. Koekemoer/R. Windhorst/Z. Levay.
Fig. 96	First observation of gravitational waves by LIGO (signal GW150914), B. P. Abbott et al., (LIGO Scientific Collaboration and Virgo Collaboration), CCA 3.0 Unported license.
Figs. 98 and 99	Viren Jain, reproduced with his permission.
Fig. 100	Ramón y Cajal/Museo Cajal.
Fig. 101	S. Kattam/St Albans museum.
Fig. 102	Didacus Valades.

[Other figures are courtesy of the author or are of unknown source and in the public domain.]

ENDNOTES

NATIONAL SKEPTICS CONVENTION

"I don't believe you don't believe we've met."

The natural object of endnotes is to cite sources of works quoted. Here my aim is to provide original sources of authoritative works and, where available, an online near-equivalent.

A further object is to offer interested readers entry points to topics this work brings together so they can dig deeper if they wish. My aim here is to point to serious nontechnical examples that are informative, though they may not always be accurate.

A third object—maybe this should be the first—is to help the world's skeptics perform their public service: checking it out.[2092]

1 P. Hein, *Grooks* (New York: Doubleday, 1969), #49.
2 For example, summing up his popular exploration of reality, science writer Jim Baggott wrote, "So, what is real? We have to admit that we don't know." J. Baggott, *A Beginner's Guide to Reality: Exploring Our Everyday Adventures in Wonderland* (New York: Pegasus Books, 2005), p. 226.
3 See, e.g., P. L. Berger and T. Luckman, *The Social Construction of Reality: A Treatise in the Sociology of Knowledge* (New York: Penguin Books, 1966).
4 Y. Harari, "Yuval Noah Harari Argues That AI Has Hacked the Operating System of Human Civilisation," *The Economist*, 28 April 2023.
5 See, e.g., M. H. Cárdenas, "How Does Language Affect Our World View?," thesis, Universidad de Jaén, (2016); https://crea.ujaen.es/bitstream/10953.1/3377/1/Herrera_Crdenas_Mara_TFG_EstudiosIngleses.pdf.

6   C. Rovelli, "Unfinished Revolution." in D. Oriti, ed., *Approaches to Quantum Gravity: Toward a New Understanding of Space, Time and Matter* (Cambridge: Cambridge University Press, 2009), p. 3.

7   C. Rovelli, *La realtà non è come ci appare: La struttura elementare delle cose* (Milan, Rafaello Cortina, 2014), trans. S. Carnell and E. Segre, *Reality Is Not What It Seems: The Journey to Quantum Gravity* (New York: Riverhead Books, 2017), p. 13; his italics.

8   S. Iyengar, *Think Bigger: How to Innovate* (New York: Columbia University Press, 2023).

9   Following Einstein's insights can be challenging; there are twists and turns because he changed his mind. He wrote to a colleague, with the *Schalkheit* (as such wry wit was then known) for which Berliners were noted: "*Es ist bequem mit dem Einstein. Jedes Jahr widerruft er, was er das vorige Jahr geschrieben hat.*" ("Einstein has it easy. Every year he retracts what he wrote in the preceding year.") A. Einstein, letter to Paul Ehrenfest, 26 December 1915, in A. M. Hentschel, transl., *The Collected Papers of Albert Einstein, Vol. 8, The Berlin Years: Correspondence, 1914–1918*, (Princeton: Princeton University Press, 1995), doc. 173, p. 167.

10   A. Einstein, "Das Raum-, Äther- und Feld-Problem der Physik," in *Mein Weltbild* (Amsterdam: Querido Verlag, 1934), p. 138; "The Problem of Space, Ether, and the Field in Physics," in A. Harris, transl., *Essays in Science* (New York: Philosophical Library, 1934), p. 61.

11   Ibid., p. 11; Smolin's emphasis.

12   For a review of the relationship between realism and truth, see H. Putnam, "What Is Realism?" in J. Leplin, ed., *Scientific Realism* (Berkeley: University of California Press, 1984), p. 140.

13   W. V. Quine, *Theories and Things* (Cambridge MA: Belknap Press, 1981), p. 90.

14   I was not alone in my misapprehension. For example, physicist and science communicator Jim Al-Khalili said, "If I wanted satisfactory answers to the many deep questions about the nature of the universe and the meaning of existence bubbling up in my teenage mind, then physics was the subject I had to study." J. Al-Khalili, *The World According to Physics* (Oxford: Princeton University Press, 2020), p. vii.

15   S. Weinberg, in R. Evans and B. Clegg, *Ten Physicists Who Transformed Our Understanding of Reality* (Philadelphia: Running Press, 2015), p. xi.

16   From Book One of this work, *Time One: Discover How the Universe Began* (New York: Rosetta Books, 2013): "Quantum theory's a loose term. Its meaning has evolved. It begins a hundred years ago as the idea of energy that comes in tiny pieces known as quanta. Quanta do strange things. One might say QT calculates the ways small things are strange."

17   Many serious scientists not only accept the "shut up and calculate" ethic but also expound it as philosophy; see, e.g., T. Andersen, "Quantum Wittgenstein," *Aeon*, 12 May 2022, https://aeon.co/essays/how-wittgenstein-might-solve-both-philosophy-and-quantum-physics. Its origins lie as far back as the nineteen twenties, as shown, for example, by a 1921 bicycle-touring conversation between to-be-quantum-physicists Werner Heisenberg, who "did not really know what was meant by 'understanding'" relativity theory, and Wolfgang Pauli, who told him, "But once you have grasped the mathematical framework you can surely predict what an observer at rest and a moving observer ought to observe or measure. . . . What more can you ask?" See W. Heisenberg, *Physics and Beyond: Encounters and Conversations* (New York: Harper & Row, 1972), p. 29. Thirty years later, Heisenberg attributed this view to logical positivism (and still disagreed with it); ibid., p. 206. In 1994, Stephen Hawking, describing himself as "a positivist" in a debate with Roger Penrose at Cambridge University, said, "I don't demand that a theory correspond to reality because

I don't know what it is. . . . All I'm concerned with is that the theory should predict the results of experiments." S. W. Hawking, "Quantum Cosmology," lecture in Cambridge, 1994; repr. in S. W. Hawking and R. Penrose, *The Nature of Space and Time* (Oxford: Princeton University Press, 1996), p. 121.

18 See, e.g., S. Carroll, "Even Physicists Don't Understand Quantum Mechanics: Worse, They Don't Seem to Want to Understand It," *The New York Times*, 7 September 2019, https://www.nytimes.com/2019/09/07/opinion/sunday/quantum-physics.html.

19 S. W. Hawking, *A Briefer History of Time*, (New York: Bantam Dell, 1988), p. 68.

20 F. Herbert, *Dune* (New York: Ace Books, 1965), p. 60.

21 Plato, Πολιτεία; G. M. A. Grube, transl., *The Republic*, in *Plato's Republic*, (Indianapolis: Hackett Publishing Company, 1974), Book 1, p. 165.

22 The decline is not confined to physics: E.g., M. Park et al., "Papers and Patents Are Becoming Less Disruptive over Time," *Nature*, (2023) **613**, p. 138; https://www.nature.com/articles/s41586-022-05543-x. See also chapter 2 at page 22.

23 A. Einstein, "Ernst Mach," *Phys. Zeitschrift*, (1916) **17**, p. 101; repr. in A. Engel, transl., *The Collected Papers of Albert Einstein, Vol. 6, The Berlin Years: Writings, 1914–1917*, (Princeton: Princeton University Press, 1995), p. 141.

24 J. H. Jeans, *Physics and Philosophy* (Cambridge: Cambridge University Press, 1943), p. 15.

25 G. Holton, "The Increasing Coalescence of Scientific Disciplines," *Edge*, https://www.edge.org/response-detail/10180.

26 E. O. Wilson, *Consilience: The Unity of Knowledge* (New York: Alfred A. Knopf, 1998), p. 4.

27 See, e.g., E. K. Chen, *Laws of Physics* (Cambridge: Cambridge University Press, 2024).

28 A. Einstein, address to the *Deutsche Physikalische Gesellschaft* (German Physical Society), Berlin, 1918, for Max Planck's sixtieth birthday; A. Einstein, "Prinzipien der Forschung," in *Mein Weltbild*, n. 10, p. 107; A. Einstein, "Principles of Research," repr. in S. Bargmann, transl., C. Seelig, ed., *Ideas and Opinions* (New York: Three Rivers Press, 1954), p. 225.

29 The meaning of *particle* is fraught and not well understood even by physicists; see, e.g., N. Wolchover, "What Is a Particle?," *Quanta Magazine*, 12 November 2020, https://www.quantamagazine.org/what-is-a-particle-20201112/.

30 C. Rovelli, *Sette Brevi Lezioni di Fisica* (Milan, Adelphi Edizioni, 2015); S. Carnell and E. Segre, transl., *Seven Brief Lessons on Physics* (New York: Riverhead Books, 2016), p. 20.

31 See, e.g., A. Pais, "Einstein and the Quantum Theory," *Rev. Mod. Phys.*, (1979) **51**, p. 863; https://eclass.aegean.gr/modules/document/file.php/511165/projects/einstein_quantum.pais.pdf.

32 W. Walsh and D. DaGradi, *Mary Poppins*, Walt Disney Productions, (1964); https://www.imdb.com/title/tt0058331/characters/nm0000267.

33 This is a widely held view of his later philosophy; he earlier had somewhat positivist opinions. But there are many takes on this complex question; see, e.g., D. Howard, "Was Einstein Really a Realist?," *Persp. Sci.*, (1993) **1**, p. 204; https://direct.mit.edu/posc/article-abstract/1/2/204/111551/Was-Einstein-Really-a-Realist.

34   For a review of the relation between realism and truth, see J. Asay, "Realism and Theories of Truth," in J. Saatsi, ed., *The Routledge Handbook of Scientific Realism* (London: Routledge, 2017), p. 383.

35   See, e.g., M. Proietti et al., "Experimental Test of Local Observer Independence," *Sci. Adv.*, (2019) **5**, aaw9832; https://www.science.org/doi/10.1126/sciadv.aaw9832.

36   G. Sparrow, *What Shape Is Space?* (London: Thames & Hudson Ltd., 2018), p. 111.

37   H. Reichenbach, in M. Reichenbach, ed., *The Direction of Time* (Berkeley: University of California Press, 1956), p. 1.

38   W. V. Quine, "Two Dogmas of Empiricism," *Phil. Rev.*, (1951) **60**, p. 20; repr. in W. V. Quine, *From a Logical Point of View: Nine Logico-Philosophical Essays* (New York: Harper & Row, 1961), p. 44; https://www.jstor.org/stable/4106675.

39   See, e.g., T. Bateman, "Scientists Tend to Superspecialize—But There Are Ways They Can Change," *The Conversation*, 8 December 2015, https://theconversation.com/scientists-tend-to-superspecialize-but-there-are-ways-they-can-change-51644.

40   E. O. Wilson, *Consilience*, n. 26, p. 39.

41   C. Rovelli, *Quantum Gravity* (New York: Cambridge University Press, 2004), p. 420.

42   R. P. Feynman, *The Meaning of It All: Thoughts of a Citizen-Scientist* (New York: Basic Books, 1998), p. 3.

43   Philosopher Gottfried Leibniz, himself a leading linguistic theorist, said, "I venture to say that no European language is better suited than German for this testing and examination of philosophical doctrines by a living tongue." G. W. Leibniz, "Preface to an Edition of Nizolius," (1670); repr. in L. E. Loemker, transl. and ed., *Philosophical Papers and Letters*, (Chicago: University of Chicago Press, 1956), Vol. I, p. 193.

44   A. Pais, *Subtle Is the Lord: The Science and the Life of Albert Einstein* (Oxford: Oxford University Press, 1982), p. vii.

45   C. Rovelli, *Seven Brief Lessons on Physics*, n. 30, p. 24.

46   J. C. Baez, "Higher-Dimensional Algebra and Planck-Scale Physics," (1999), in C. Callender and N. Huggett, eds., *Physics Meets Philosophy at the Planck Scale: Contemporary Theories in Quantum Gravity* (Cambridge: Cambridge University Press, 2001), p. 177; https://www.researchgate.net/publication/2418811_Higher-Dimensional_Algebra_and_Planck-Scale_Physics.

47   F. Herbert, *Dune*, n. 20, p. 69.

48   J. Baez, "Can We Understand the Standard Model Using Octonions?," *The n-Category Café*, 31 March 2021; https://pdf.pirsa.org/files/21040005.pdf.

49   A. Becker, *What Is Real? The Unfinished Quest for the Meaning of Quantum Physics* (New York: Basic Books, 2018).

50   Ibid., p. 287.

51   A. Becker, "The Origins of Space and Time," *Sci. Am.*, (2022) **326**, p. 26.

52   Ibidem.

53   S. Firestein, *Ignorance: How It Drives Science* (Oxford: Oxford University Press, 2012), p. 111.

54   B. Greene, *The Fabric of the Cosmos: Space, Time and the Texture of Reality* (New York: Alfred A. Knopf, 2004), p. 272.

55   Ibid., p. 23.

56   E.g., B.Greene, *The Fabric of the Cosmos*, (2011); https://www.youtube.com/watch?v=8C2fipfdwZw.

57   S. Davies, "This Physics Pioneer Walked Away from It All, *Nautilus*, **38**, 19 July 2016; https://nautil.us/this-physics-pioneer-walked-away-from-it-all-4942/.

58   L. Smolin, quoted by S. Davies, ibidem.

59   Quoted by S. Davies, "This Physics Pioneer Walked Away from It All," n. 57.

60   A concept about which she was skeptical; see "Fotini Markopoulou—"What Is a Theory of Everything?," https://www.youtube.com/watch?v=i3B7xC45DMU. Peebles was of like mind, saying, "Physicists have a regrettable tendency to declare that we are at last approaching the final theory, the discovery of the ultimate nature of reality." P. J. E. Peebles, *The Whole Truth: A Cosmologist's Reflections on the Search for Objective Reality* (Princeton: Princeton University Press, 2022), p. 29.

61   L. Smolin, *Three Roads to Quantum Gravity* (London: Weidenfeld & Nicolson, 2000), p. vii.

62   R. M. Unger in R. M. Unger and L. Smolin, *The Singular Universe and the Reality of Time: A Proposal in Natural Philosophy* (Cambridge: Cambridge University Press, 2015), p. 232.

63   See, e.g., S. Davies, "Her Second Act: Fotini Markopoulou on Life Beyond Physics," *Nautilus*, 19 July 2016, https://nautil.us/her-second-act-fotini-markopoulou-on-life-beyond-physics-4945/.

64   R. Penrose, *The Road to Reality: A Complete Guide to the Laws of the Universe* (London: Jonathan Cape, 2004).

65   Ibid., p. 730.

66   R. Penrose, *Cycles of Time: An Extraordinary New View of the Universe* (London: Bodley Head, 2010).

67   He can be more poetic than precise; see, e.g., a rather rough but apposite review by physicist Lisa Randall, "Unpeeling the Universe," *The New York Times*, 3 March 2017, book review p. 15; https://www.nytimes.com/2017/03/03/books/review/reality-is-now-what-it-seems-carlo-rovelli.html.

68   C. Rovelli, "The Relational Interpretation of Quantum Physics," in *The Oxford Handbook of the History of Quantum Interpretations* (Oxford: Oxford University Press, 2022).

69   C. Rovelli, *The Order of Time*, transl. E. Segre and S. Carnell (New York: Riverhead Books, 2018), p. 137.

70   L. Smolin, *The Trouble with Physics: The Rise of String Theory, the Fall of a Science, and What Comes Next* (New York: Houghton Mifflin Company, 2007), p. xxiii.

71   L. Smolin, "The Place of Qualia in a Relational Universe," *PhilArchive* (London ON: Centre for Digital Philosophy, 2020), https://philarchive.org/rec/SMOTPO-3.

72   R. M. Unger in R. M. Unger and L. Smolin, *The Singular Universe and the Reality of Time*, n. 62, p. 239 ff.

73   See L. Smolin, interview by A. Gefter, "How to Understand the Universe When You're Stuck Inside of It," *Quanta Magazine*, 27 June 2019, https://www.quantamagazine.org/were-stuck-inside-the-universe-lee-smolin-has-an-idea-for-how-to-study-it-anyway-20190627/.

74   L. Smolin, *Einstein's Unfinished Revolution: The Search for What Lies Beyond the Quantum* (New York: Penguin Press, 2019), p. 259.

75   Ibid., pp. 229 ff.

76   See, e.g., T. Lewton, "Neil Turok on the Case for a Parallel Universe Going Backwards in Time," *New Scientist*, 25 January 2023, https://www.newscientist.com/article/mg25734230-100-neil-turok-on-the-case-for-a-parallel-universe-going-backwards-in-time/.

77   N. Turok, quoted in S. Davies, "This Physics Pioneer Walked Away from It All," n. 57.

78   J. A. Wheeler, *Geons, Black Holes & Quantum Foam: A Life in Physics* (New York: W.W. Norton & Company, 1998), p. 247.

79   F. Wilczek, *Fundamentals*, n. 100, p. 227.

80   Ibid., p. 228.

81   For a short summary see H. Murayama and K. Riesselman, "DOE Explains . . . the Standard Model of Particle Physics," https://www.energy.gov/science/doe-explains-the-standard-model-particle-physics. Note that its status is more volatile than its name might seem to imply; its ingredients at any moment are part of what Teilhard de Chardin called "these complicated and fragile edifices"; *La Phénomene Humain* (Paris: Editions du Seuil, 1955); B. Wall, transl., *The Phenomenon of Man* (New York: Harper & Row, 1975), p. 39.

82   See P. J. E. Peebles, *Principles of Physical Cosmology* (Princeton: Princeton University Press, 1993), p. 3 ff. For a more recent review of the Standard Model of Cosmology see D. Scott, "The Standard Model of Cosmology: A Skeptic's Guide," *Proceedings* of the *International School of Physics Enrico Fermi, Varenna*, 2018, https://ned.ipac.caltech.edu/level5/March18/Scott/paper.pdf.

83   In turn, Einstein built on others; see, e.g., J. Renn and H. Gutfreund, *The Einsteinian Revolution: The Historical Roots of His Breakthroughs* (Princeton: Princeton University Press, 2023).

84   Principally in The Collected Papers of Albert Einstein, *Princeton University*, https://einsteinpapers.press.princeton.edu/, and Albert Einstein Archives, The Hebrew University of Jerusalem, https://albert-einstein.huji.ac.il/.

85   E. A. Poe, *Eureka: An Essay on the Material and Spiritual Universe* (New York: Geo. P. Putnam, 1848); repr. in S. Levine and S. F. Levine, eds., *Eureka* (Urbana IL: University of Illinois Press, 2004), p. 8; Poe's emphasis.

86   *Webster's Encyclopedic Unabridged Dictionary of the English Language* (New York: Gramercy Books, 1996), cosmos 1.

87   S. F. Teiser, "The Spirits of Chinese Religion," in D. S. Lopez, Jr., ed., *Religions of China in Practice* (Princeton: Princeton University Press, 1996), p. 29; and see J. Needham, *Science and Civilisation in China*, vol. 2: History of Scientific Thought (Cambridge: Cambridge University Press, 1956).

88   D. L. Couprie, "Anaximander (c. 610–546 BCE)," *Internet Encyclopedia of Philosophy*, https://iep.utm.edu/anaximan/.

89   Ibidem.

90   C. H. Kahn, *Anaximander and the Origins of Greek Cosmology* (Indianapolis: Hackett Publishing Company, 1994), p. 199.

91   J. Evans, "Anaximander," n. 96, https://www.britannica.com/biography/Anaximander.

92   C. Rovelli, *Reality Is Not What It Seems*, n. 7, p. 17.

93   Rovelli thinks so too; see C. Rovelli, M. L. Rosenberg, transl., *The First Scientist: Anaximander and His Legacy*, (Yardley PA: Westholme, 2007).

94   C. Rovelli, *Reality Is Not What It Seems*, n. 7, p. 16.

95   E.g., "Plato's Divided Line," The Information Philosopher, https://www.information philosopher.com/knowledge/divided_line.html.

96   See, e.g., D. W. Zimmerman, "universal," *Britannica* (Chicago: Encyclopaedia Britannica Inc., 2022), https://www.britannica.com/topic/universal; and see B. Russell, *The Problems of Philosophy* (New York: H. Holt & Co., 1912), p. 145; https://gutenberg.org/files/5827/5827-8.txt.

97   E.g., A. N. P. Stevens and J. R. Stevens, "Animal Cognition," Nature Education Knowledge, (2012) **3**, p. 1; https://www.nature.com/scitable/knowledge/library/animal-cognition-96639212/.

98   See the classic: J. W. F. Piaget, *La construction du réel chez l'enfant* (Paris: Delachaux & Niestle, 1937); M. Cook, transl., *The Construction of Reality in the Child* (New York: Basic Books, 1954).

99   See, e.g., S. Grimm, "Understanding," (2021), *Stanford Encyclopedia of Philosophy*, E. N. Zalta, ed., (Stanford, Center for the Study of Language and Information, Stanford University, 2020), https://plato.stanford.edu/entries/understanding/#ExplUnde.

100 F. Wilczek, *Fundamentals: Ten Keys to Reality* (New York: Penguin Press, 2021), p. xviii.

101 R. Descartes, *Meditationes de Prima Philosophia in Qua Dei Existentia et Animæ Immortalitas Demonstratur* (Paris: Michael Soly, 1641), Meditation 1; E. S. Haldane, transl., *Meditations on First Philosophy* (Cambridge: Cambridge University Press, 1911), pp. 1–8.

102 H. Arendt, *The Human Condition* (Chicago: University of Chicago Press, 1958), 2nd ed., 1998, p. 277.

103 L. Smolin in R. M. Unger and L. Smolin, *The Singular Universe and the Reality of Time*, n. 62, p. 359.

104 P. J. E. Peebles, *The Whole Truth*, (Princeton: Princeton University Press, 2022), n. 60, p. 208.

105 See, e.g., N. Ben-Yehuda, "The European Witch-Craze of the 14th to 17th Centuries," *Am. J. Sociol.*, **86**, p. 1.

106 Physics intruded into the worldview of philosophy extensively by mid–twentieth century; see, e.g., H. Margenau, *The Nature of Physical Reality: A Philosophy of Modern Physics* (New York: McGraw-Hill, 1950); and see a short review by G. Gamow, *Science*, (1950) **112**, p. 155, https://www.science.org/doi/10.1126/science.112.2901.155.c.

107 J. Baggott, *Farewell to Reality: How Modern Physics Has Betrayed the Search for Scientific Truth* (New York: Pegasus Books, 2013), p. x.

108 A. Frank and M. Gleiser, "The Story of Our Universe May Be Starting to Unravel," *The New York Times*, 2 September 2023; https://www.nytimes.com/2023/09/02/opinion/cosmology-crisis-webb-telescope.html.

109 A. Frank et al., *The Blind Spot: Why Science Cannot Ignore Human Experience* (Cambridge: MIT Press, 2024), p. ix.

110 L. Smolin, *Einstein's Unfinished Revolution*, n. 74, p. xx.

111 A. Aspect, in J. S. Bell, *Speakable and Unspeakable in Quantum Mechanics* (Cambridge: Cambridge University Press, 1987), 2nd ed., p. xvii.

112 M. Beller, *Quantum Dialogue: The Making of a Revolution* (Chicago: University of Chicago Press, 1999), p. 2.

113 Ibidem.

114 A. Einstein, lecture at Oxford, 10 June 1933, "Zur Methodik der theoretischen Physik," in *Mein Weltbild*, n. 10, p. 113; repr. as "On the Method of Theoretical Physics," in A. Harris, transl., *Essays in Science*, n. 10, p. 20; my translation.

115 See, e.g., W. L. Craig and Q. Smith, *Einstein, Relativity and Absolute Simultaneity* (London: Routledge, 2008), p. 3: "Most historians of science now recognize that Einstein's rejection of Newtonian absolute time and space was predicated upon a positivist philosophy of science."

116 E.g., D. Deutsch, "Einstein the Realist," *Project Syndicate*, 2 September 2011, https://www.project-syndicate.org/commentary/einstein-the-realist; but see also n. 33.

117 A. Einstein, "Space and Time in Pre-Relativity Physics," (1921), the Stafford Little Lectures, Princeton University, in E. P. Adams, transl., *The Meaning of Relativity: Including the Relativistic Theory of the Non-Symmetric Field* (Princeton: Princeton University Press, 1922), p. 3.

118 See, e.g., a little-known dialog with Nobel Prize–winning philosopher and poet Rabindranath Tagore, "On the Nature of Reality," *Modern Review*, (1931) 49, p. 72; https://mast.queensu.ca/~murty/einstein_tagore.pdf; and see G. Holton, "Einstein's Search for the 'Weltbild'," *Proc. Am. Phil. Soc.*, (1981) 125, p. 2.

119 G. Holton, ibid., p. 15.

120 L. Crane, "Categorical Geometry and the Mathematical Foundations of Quantum Gravity," in D. Oriti, ed., *Approaches to Quantum Gravity*, n. 6, p. 84.

121 P. Bricker, "Ontological Commitment," (2014), *Stanford Encyclopedia of Philosophy*, n. 99, https://plato.stanford.edu/entries/ontological-commitment/.

122 Ibidem.

123 P. Maddy, *Realism in Mathematics* (Oxford: Clarendon Press, 1990), p. 4.

124 E.g., M. Beller, "The Sokal Hoax: At Whom Are We Laughing?," *Phys. Today*, (1998) 51, p. 29: "Bohr was notorious for the obscurity of his writing. Yet . . . Bohr's obscurity is attributed, time and again, to a 'depth and subtlety' that mere mortals are not equipped to comprehend." See also the text of n. 1683.

125 A. Einstein in conversation with N. Bohr and A. Sommerfeld in 1923, quoted by Bohr, cited in L. Gilder, *The Age of Entanglement: When Quantum Physics Was Reborn* (New York: Vintage Books, 2008), p. xv.

126 J. von Neumann, *Mathematische Grundlagen der Quantenmechanik* (Berlin: Julius Springer, 1932); R. T. Beyer, transl., *Mathematical Foundations of Quantum Mechanics* (Princeton: Princeton University Press, 1955), p. 3.

127 Quoted by L. Kruesi, "Clashing Cosmic Numbers Challenge Our Best Theory of the Universe," *Quanta Magazine*, 19 January 2024, https://www.quantamagazine.org/clashing-cosmic-numbers-challenge-our-best-theory-of-the-universe-20240119/.

128 L. Smolin, *Three Roads to Quantum Gravity*, n. 61, p. 6.

129 A. Einstein, "On the Method of Theoretical Physics," n. 114, p. 15.

130 A. Einstein and L. Infeld, *The Evolution of Physics: The Growth of Ideas from Early Concepts to Relativity and Quanta* (New York: Cambridge University Press, 1938), p. 295.

131 R. M. Unger in R. M. Unger and L. Smolin, *The Singular Universe and the Reality of Time*, n. 62, p. 232.

132 D. Gross, "Einstein and the Quest for a Unified Theory," in P. Galison et al., eds., *Einstein for the 21st Century: His Legacy in Science, Art and Modern Culture* (Princeton: Princeton University Press, 2008), p. 286; my emphasis.

133 L. Smolin, *Three Roads to Quantum Gravity*, n. 61, p. 6.

134 J. D. Barrow, *New Theories of Everything: The Quest for Ultimate Explanation* (Oxford: Oxford University Press, 2007), p. 1.

135 D. Bohm and B. J. Hiley, *The Undivided Universe: An Ontological Interpretation of Quantum Theory* (London: Routledge, 1993).

136 A. J. Ayer, *The Central Questions of Philosophy* (London: Weidenfeld & Nicholson, 1973), repr. (Harmondsworth: Penguin Books, 1976), p. 11.

137 A. Einstein, "Physik und realität," *J. Franklin Inst.*, (1936) **221**, pp. 313 and 349 resp.; repr. in A. Einstein, "Physics and reality", *Ideas and Opinions*, n. 28, p. 322.

138 G. Hinton, interviewed by Ian Brown, *The Globe and Mail*, 13 June 2024, p. A8; https://www.theglobeandmail.com/business/article-geoffrey-hinton-artificial-intelligence-machines-feelings/; for role of ontology in psychology, see, e.g., J. Sugarman, "Historical Ontology and Psychological Description," *J. Theor. Philos. Psychol.*, (2009) **29**, p. 5.

139 T. R. Gruber, "A Translation Approach to Portable Ontology Specifications," *Knowledge Acquisition*, (1993) **5**, p. 199.

140 A. Einstein, address at UCLA, in *Builders of the Universe: From the Bible to the Theory of Relativity* (Los Angeles: U. S. Library Association, 1932), p. 94.

141 Nicola Guarino et al., "What Is an *Ontology?*," in S. Staab and R. Studer, eds., *Handbook on Ontologies* (Berlin: Springer-Verlag, 2009), p. 1; the meaning in computer science, given here, is stricter than that in philosophy and so may be apter to the stringency of the present venture.

142 I. Hacking, *Historical Ontology* (Cambridge MA: Harvard University Press, 2002).

143 J. Sugarman, "Historical Ontology and Psychological Description," n. 138.

144 See, e.g., "Metaphysics, or alternatively ontology, is that branch of philosophy whose special concern is to answer the question 'What is there?' These expressions derive from Aristotle, Plato's student." A. Silverman, "Plato's Middle Period Metaphysics and Epistemology," (2014), *Stanford Encyclopedia of Philosophy*, n. 99, https://plato.stanford.edu/entries/plato-metaphysics/.

145 See, e.g., A. P. Martinich, "Epistemology," n. 96, https://www.britannica.com/topic/epistemology.

146 "The whole effect of Einstein's work is to make physics more philosophical (in a good sense). . . ." H. A. Lorentz, *The Einstein Theory of Relativity: A Concise Statement* (New York: Brentano's, 1920), p. 24; https://www.lorentz.leidenuniv.nl/IL-publications/sources/Lorentz_Einstein_1920.pdf.

147 In the language of philosophy, physics is now about epistemology, the study of what we know (or think we do). See, e.g., M. Steup and R. Neta, "Epistemology," *Stanford Encyclopedia of Philosophy*, n. 99, https://plato.stanford.edu/entries/epistemology/.

148 S. Critchley, "Being and Time, part 1: Why Heidegger Matters," *Guardian*, 8 June 2009, https://www.theguardian.com/commentisfree/belief/2009/jun/05/heidegger-philosophy.

149 M. Heidegger, *Der Begriff der Zeit* (Frankfurt am Main: Klostermann, 2004); W. MacNeill, transl., *The Concept of Time* (Oxford: Blackwell, 1992), p. 1E.

150 M. Heidegger, *Sein und Zeit* (Tübingen, Max Niemeyer, 1927), p. 18; transl. J. Stambaugh, *Being and Time* (New York: State University of New York Press, 1996), p. 16.

151 E.g., E. Anderson, *The Problem of Time: Quantum Mechanics Versus General Relativity* (Cham: Springer, 2017).

152 C. Rovelli, *The Order of Time*, n. 69, p. 43.

153 Quoting Alain Aspect; in J. S. Bell, *Speakable and Unspeakable in Quantum Mechanics*, n. 111, p. xix. In explanation, Aspect said, "Before the realization of the importance of Bell's theorem, which happened only in the nineteen-seventies, the conventional wisdom among physicists was that the 'founding fathers' of quantum mechanics had settled all the conceptual questions."

154 J. S. Bell, "Quantum Mechanics for Cosmologists," in C. Isham, R. Penrose and D. Sciama, eds., *Quantum Gravity 2* (Oxford: Clarendon Press, 1981), p. 611; repr. in J. S. Bell, *Speakable and Unspeakable in Quantum Mechanics*, n. 111, p. 117.

155 Even eighty years after its conception, this view of reality continued to sow growing confusion; see, e.g., B. Rosenblum and F. Kuttner, "The Observer in the Quantum Experiment," *Found. Phys.*, (2002) **32**, p. 1273; https://arxiv.org/ftp/quant-ph/papers/0011/0011086.pdf.

156 A. Fine, "The Einstein-Podolsky-Rosen Argument in Quantum Theory," *Stanford Encyclopedia of Philosophy*, 31 October 2017, n. 99, https://plato.stanford.edu/entries/qt-epr/#ArguText.

157 See, e.g., S. Hossenfelder, *Lost in Math: How Beauty Leads Physics Astray* (New York: Basic Books, 2018).

158 Plato, *The Republic* (c. 375 BCE); B. Jowett, transl., https://gutenberg.org/ebooks/1497.

159 W. Shakespeare, *As You Like It*, in *Mr. William Shakespeare's Comedies, Histories & Tragedies* (London: Isaac Jaggard & Edward Blount, 1623).

160 R. W. Emerson, *Nature* (New York: James Munroe & Co., 1836), ch.1.

161 Strictly, general relativity is not entirely background-free; see, e.g., J. Baez, "What Is Background-free Theory?," *John Baez's Stuff*, 5 May 2000, https://math.ucr.edu/home/baez/background.html; and see L. Smolin, "The Case for Background Independence," in D. Rickles et al., eds., *The Structural Foundations of Quantum Gravity*, p. 196; https://arxiv.org/abs/hep-th/0507235.

162 L. Smolin, "The Case for Background Independence," ibidem.

163 L. Smolin, *The Trouble with Physics*, n. 70, p. 239; Smolin's italics.

164 Ibid., p. 240; my emphasis.

165 See, e.g., "Relational Theory," *Scholarly Community Encyclopedia*, 21 October 2022, https://encyclopedia.pub/entry/30586.

166 E.g., P. Martin-Dussaud, "Relational Structures of Fundamental Theories," *Found. Phys.*, (2021) **51**, p. 1; https://arxiv.org/pdf/2012.05584v1.pdf.

167 L. Smolin, "The Case for Background Independence," in D. Rickles et al., eds., *The Structural Foundations of Quantum Gravity* (Oxford: Oxford University Press, 2006), p. 196.

168 Ibidem.

169 L. Smolin, *Einstein's Unfinished Revolution*, n. 74, p. 232.

170 A. Einstein, letter to Arnold Sommerfeld, 14 January 1908; repr. in A. Beck, transl., *The Collected Papers of Albert Einstein, Vol. 5, The Swiss Years, Correspondence 1902–1914* (Princeton: Princeton University Press, 1995), p. 50.

171 C. Rovelli, "A Dialog on Quantum Gravity," *Int. J. Mod. Phys.*, (2003) **D 12**, p. 1509; https://arxiv.org/abs/hep-th/0310077v2.

172 C. Rovelli, "Unfinished Revolution," n. 6, p. 4.

173 D. Hoffman, "Spacetime is not fundamental," *IAI NEWS*, Institute of Art and Ideas, 27 October 2022, https://iai.tv/articles/donald-hoffman-spacetime-is-not-fundamental-auid-2281.

174 N. Huggett & C. Wüthrich, "Emergent spacetime and empirical (In)coherence," *Stud. Hist. Philos. Sci. B - Stud. Hist. Philos. Mod. Phys.*, (2012) **44**, p. 276.

175 C. Rovelli, *Reality Is Not What It Seems: The Journey to Quantum Gravity*, n. 7, p. 20.

176 Ibid., p. 25; his emphasis.

177 W. Heisenberg, lecture in Zürich on 9 July 1948, repr. in W. Heisenberg, *Philosophic Problems of Nuclear Science* (Greenwich CT: Fawcett Publications, 1952), p. 109.

178 A. Einstein, "Über die von der molekularkinetischen Theorie der Wärme geforderte Bewegung von in ruhenden Flüssigkeiten suspendierten Teilchen" *Ann. d. Phys.*, (1905) **322**, p. 549; A. D. Cowper, transl., Investigations on the Theory of the Brownian Movement (London: Methuen & Company Limited, 1926); https://archive.org/details/investigationont0000albe/page/n5/mode/2up.

179 Experimental confirmations used molecules rather than atoms but the results established the atomic principle; see, e.g., C. O'Raifeartaigh, "Einstein and the Atomic Theory," *Eng. J.*, (2005) **59**, p. 386; https://www.researchgate.net/publication/279646461_Einstein_and_the_Atomic_Theory.

180 Protons and neutrons are made of quarks; see, e.g., M. Riordan, "The Discovery of Quarks," *Science*, (1992) **256**, p. 1287; https://www.science.org/doi/10.1126/science.256.5061.1287.

181 J. Swift, "On Poetry: A Rhapsody," (1733), repr. in T. Sheridan et al., eds., *The Works of the Rev. Jonathan Swift* (London: J. Johnson, 1801), **8**, p. 166; https://www.online-literature.com/swift/3515.

182 B. Hobbs, "The Standard Model of particle physics is brilliant and completely flawed," *ABC News*, 14 July 2017, https://www.abc.net.au/news/science/2017-07-15/the-standard-model-of-particle-physics-explained/7670338.

183 J. Hackett, "Locality and Translations in Braided Ribbon Networks," *Class. Quant. Grav.*, (2007) **24**, p. 5757; https://arxiv.org/pdf/hep-th/0702198.

184 See, e.g., J. M. Sonneveld, "Searches for Physics Beyond the Standard Model at the LHC," *Afr. Rev. Phys.*, (2018) **13**, p. 17; https://arxiv.org/abs/1905.06239.

185 Physicist Sabine Hossenfelder said we don't need a new big accelerator: S. Hossenfelder, "The World Doesn't Need a New Gigantic Particle Collider," *Sci. Am.*, 19 June 2020, https://www.scientificamerican.com/article/the-world-doesnt-need-a-new-gigantic-particle-collider/.

186 T. de Chardin, *The Phenomenon of Man*, n. 81, p. 41.

187 R. Southey, "What All the World Is Made Of," manuscript, (c. 1820); see P. Opie and I. Opie, *The Oxford Dictionary of Nursery Rhymes* (Oxford: Oxford University Press, 1997), 2nd ed., p. 100.

188 See, e.g., n. 91.

189 G. W. Leibniz, "Principes de la nature et de la grâce fondés en raison," (1714) in C. I. Gerhardt, ed., *Die Philosophische Schriften von Gottfried Wilhelm Leibniz Vol. VI* (Berlin: Weidmannsche Buchhandlung, 1875); L. E. Loemker, ed. and transl., "Principles of Nature and of Grace, Based on Reason," *Philosophical Papers and Letters* (Chicago: University of Chicago Press, 1956), vol. 2, p. 1033; my revised translation.

190 G. W. Leibniz, "The Monadology," (1714); repr. in L. E. Loemker, transl. and ed., *Philosophical Papers and Letters*, (Chicago: University of Chicago Press, 1956), Vol. II, p. 1044.

191 W. Heisenberg, *Physics and Philosophy: The Revolution in Modern Science* (New York: Harper & Row, 1958), p. 61.

192 F. Wilczek, *The Lightness of Being: Mass, Ether and the Unification of Forces* (New York: Basic Books, 2008), p. 74.

193 G. 't Hooft, *In Search of the Ultimate Building Blocks* (Cambridge: Cambridge University Press, 1996), p. 59.

194 I. Newton, *Opticks* (London: Samuel Smith & Benjamin Walford, 1704); and see A. Chalmers, *The Scientist's Atom and the Philosopher's Stone: How Science Succeeded and Philosophy Failed to Gain Knowledge of Atoms*, (Cham: Springer Link, 2009), pp. 123 ff.; https://link.springer.com/chapter/10.1007/978-90-481-2362-9_7.

195 The benchmark publication was A. Lavoisier, R. Kerr, transl., *Elements of Chemistry in a New Systematic Order Containing All the Modern Discoveries*, (Edinburgh: William Creech, 1790).

196 See, e.g., "Periodic Table of Elements," *PubChem* (Bethesda, National Center for Biotechnology Information), https://pubchem.ncbi.nlm.nih.gov/periodic-table/.

197 See, e.g., "Who Discovered Electrons, Protons and Neutrons?," *UCSB ScienceLine*, University of California Santa Barbara, (2003), http://scienceline.ucsb.edu/getkey.php?key=408.

198 J. W. Moffat, *Cracking the Particle Code of the Universe: The Hunt for the Higgs Boson* (Oxford: Oxford University Press, 2014), p. 2.

199 Physicist Murray Gell-Mann tells the tale of quarks succinctly in M. Gell-Mann, *The Quark and the Jaguar: Adventures in the Simple and the Complex* (New York: W.H. Freeman & Company, 1994), pp. 180 ff.

200 E.g., A. Finkbeiner, "Looking for Neutrinos, Nature's Ghost Particles," *Smithsonian Magazine*, November 2010, https://www.smithsonianmag.com/science-nature/looking-for-neutrinos-natures-ghost-particles-64200742/.

201 L. Lederman and D. Teresi, *The God Particle: If the Universe Is the Answer, What Is the Question?* (New York: Delta, 1993), p. 2.

202 See, e.g., H. Murayama and K. Riesselmann, "DOE Explains . . . the Standard Model of Particle Physics," Office of Science, U. S. Department of Energy, https://www.energy.gov/science/doe-explainsthe-standard-model-particle-physics. Ignore the Higgs boson for now.

203 E.g., V. T. Toth, "The Parameters of the Standard Model," *Spinor Info*, 4 November 2014, https://spinor.info/weblog/?p=6355.

204 R. N. Oerter, *The Theory of Almost Everything: The Standard Model, the Unsung Triumph of Modern Physics* (New York: Plume, 2006), p. 12.

205 See, e.g., E. A. Thompson, "MIT's Wilczek Wins 2004 Nobel Prize in Physics," *MIT News*, 5 October 2004, https://news.mit.edu/2004/mits-wilczek-wins-2004-nobel-prize-physics.

206 M. L. Perl et al., "Searches for Fractionally Charged Particles," *Ann. Rev. Nucl. & Part. Sci.*, (2009) **59**, p. 47; https://www.annualreviews.org/doi/10.1146/annurev-nucl-121908-122035; (for consistency, I've replaced his *q* with my *e*).

207 R. Penrose, *The Road to Reality*, n. 64, p. 1034.

208 A. Einstein, "On the Method of Theoretical Physics," n. 114, p. 15.

209 A. Einstein, (1955), advice to Pat Miller, in W. Miller, (Ed.), Old Man's Advice to Youth: "Never Lose a Holy Curiosity." *Life*, 2 May 1955.

210 G. Holton, "Einstein's Search for the 'Weltbild,'" n. 118, p. 2.

211 One can dissect the concept of *simplicity* at length, showing it to be susceptible of erudite analysis; see e.g., K. Popper, *The Logic of Scientific Discovery* (New York: Basic Books, 1959) pp. 136 ff.; the present work adopts a less technical and more intuitive notion of simplicity that will become evident as it unfolds.

212 The rule predates William of Ockham's use of it. Sixteen centuries earlier, Aristotle said, "Other things being equal, we may assume the superiority of a demonstration that uses fewer postulates or hypotheses." Aristotle, *Posterior Analytics*, (c. 330 BCE), Part 25; cited in H. G. Gauche Jr., *Scientific Method in Brief* (New York: Cambridge University Press, 2012), p. 175.

213 See, e.g., S. Kaye, "William of Ockham (Occam, c. 1280–c. 1349)", *Internet Encyclopedia of Philosophy*, https://iep.utm.edu/ockham/.

214 I. Newton, *Philosophiæ Naturalis Principia Mathematica* (London: Joseph Streater, 1687); I. B. and A. Whitman, transls., *The Principia: Mathematical Principles of Natural Philosophy* (Oakland: University of California Press, 1999), p. 440; Motte's early translation was "To this purpose the philosophers say that Nature does nothing in vain, and more is in vain when less will serve; for Nature is pleased with simplicity, and affects not the pomp of superfluous causes." See n. 634; https://redlightrobber.com/red/links_pdf/Isaac-Newton-Principia-English-1846.pdf.

215 A. Einstein, "Physics and Reality," *J. Franklin Inst.*, **221**, p. 349; repr. in A. Einstein, *Ideas and Opinions*, n. 28, p. 293; Einstein's emphasis.

216 A. Einstein, "Autobiographical Notes," in P. A. Schilpp, trans. and ed., *Albert Einstein: Philosopher-Scientist* (London: Cambridge University Press, 1949), p. 3; repr. in S. W. Hawking, ed., *A Stubborn and Persistent Illusion: The Essential Scientific Works of Albert Einstein* (London: Running Press, 2007), p. 339.

217 A. Einstein, "Remarks on Bertrand Russell's Theory of Knowledge," in P. A. Schilpp, ed., *The Philosophy of Bertrand Russell* (Chicago: Open Court Publishing, 1944), p. 18.

218 See, e.g., A. J. Ayer, *Language, Truth and Logic* (London: Victor Gollancz, 1936); repr., (London: Penguin, 1990), pp. 32 ff.; https://archive.org/stream/AlfredAyer/LanguageTruthAndLogic_djvu.txt.

219 D. Hume, *Philosophical Essays: Concerning Human Understanding* (London: A. Millar, 1748), Essay 12, Part 3; *Eighteenth Century Collections Online*, https://quod.lib.umich.edu/cgi/t/text/text-idx?c=ecco;idno=004806472.0001.000.

220 See, e.g., A. E. Blumberg and H. Feigl, "Logical Positivism: A New Movement in European Philosophy," *J. Phil.*, (1931) **28**, p. 281; https://www.pdcnet.org/jphil/content/jphil_1931_0028_0011_0281_0296.

221 See, e.g., A. J. Ayer, *Language, Truth and Logic*, n. 218, p. 24: "All metaphysical assertions are nonsensical."

222 M. Friedman, "Philosophy and the Exact Sciences: Logical Positivism as a Case Study," in J. Earman, ed., *Inference, Explanation, and Other Frustrations: Essays in*

the Philosophy of Science (Berkeley: University of California Press, 1992), p. 91; https://publishing.cdlib.org/ucpressebooks/view.

223 J. Passmore, "Logical Positivism," in P. Edwards, ed., *The Encyclopedia of Philosophy*, vol. 5, (New York: Macmillan, 1967), p. 52.

224 See also n. 17.

225 W. V. Quine, "On What There Is," *Rev. Metaphys.*, (1948) **2**, p. 21; repr. in *From a Logical Point of View: Logico-Philosophical Essays*, n. 38, p. 1.

226 Ibid., p. 16.

227 Ibid., p. 19.

228 A. Einstein, letter to Cornelius Lanczos, 24 January 1938; *Albert Einstein Archives*, # 15–268, The Hebrew University of Jerusalem; quoted in G. Holton, "Mach, Einstein, and the Search for Reality," *Daedalus*, (1968), p. 636; repr. in G. Holton, *Thematic Origins of Scientific Thought: Kepler to Einstein* (Cambridge MA: Harvard University Press, 1973), p. 219.

229 J. Eisenstaedt, *Einstein et la relativité générale: Les chemins de l'espace-temps* (Paris: CRNS Editions, 2002); A. Sangalli, transl., *The Curious History of Relativity: How Einstein's Theory of Gravity Was Lost and Found Again* (Princeton: Princeton University Press, 2006), p. 25.

230 J. Al-Khalili, *The World According to Physics*, n. 14, p. 14.

231 L. Smolin, *Einstein's Unfinished Revolution*, n. 74, p. 272.

232 Ibid., p. 226.

233 KISS is a design principle attributed to Rear Admiral Paul Stroop, head of the U.S. Navy Weapons Bureau; *Chicago Daily Tribune*, 4 December 1960, p. 43.

234 See, e.g., S. L. Bonatto and F. M. Salzano, "A Single and Early Migration for the Peopling of the Americas Supported by Mitochondrial Sequence Data," *Proc. Natl. Acad. Sci. U.S.A.*, (1997) **94**, p. 1866.

235 W. James, *Pragmatism: A New Name for Some Old Ways of Thinking* (London: Longmans Green & Co., 1907), Lecture IV, "The One and the Many."

236 See, e.g., J. Palmer, "Zeno of Elea," *Stanford Encyclopedia of Philosophy*, 9 January 2008, n. 99, https://plato.stanford.edu/entries/zeno-elea/.

237 B. Russell, *The Principles of Mathematics* (Cambridge: The University Press, 1903), p. 327; edited and republished, K. C. Klement, https://people.umass.edu/klement/pom/pom-portrait.pdf, p. 506.

238 Zeno of Elea, in H. A. Diels, *Die Fragmente der Vorsokratiker: griechisch und deutsch* (*The Fragments of the Pre-Socratics: Ancient Greek and German*) (Berlin: Wiedmann, 1903), fragment 3; for original Greek, see A. Haas, *Hegel and the Problem of Multiplicity* (Evanston: Northwestern University Press, 2000), p. 293.

239 B. M. Sattler, *The Concept of Motion in Ancient Greek Thought* (Cambridge: Cambridge University Press, 2021), p. 128.

240 Ibid., p. 136.

241 Ibid., p. 127.

242 Ibid., p. 21.

243 "... erscheint diese Deduktion als das widersprechende Beginnen, aus der Einheit die Mannigfaltigkeit ..." G. W. F. Hegel, *Differenz des Fichteschen und Schellingschen*

Systems der Philosophie (Jena: Seidler, 1801), p. 36; quoted in A. Haas, *Hegel and the Problem of Multiplicity*, n. 238, p. 307; my translation.

244 For which happenstance I acknowledge with gratitude the guidance of my thesis advisor, Bill Rachinger; https://www.monash.edu/vale/home/articles/vale-emeritus-professor-william-bill-rachinger.

245 A. Einstein, *Über die spezielle und die allgemeine Relativitätstheorie: (Gemeinverständlich)* (Brunswick: Vieweg Verlag, 1917); R. W. Lawson, transl., *Relativity: The Special and the General Theory* (New York: Crown Trade Paperbacks, 1961), p. 156; the translated third edition can be found online at https://www.gutenberg.org/files/36114/36114-pdf.pdf; note the excerpt is in Appendix 5, "Relativity and the Problem of Space," which was added in the fifteenth edition in 1952.

246 R. Bondar, quoted by science journalist Ivan Semeniuk, "Bondar Is Cultivating Curiosity on Earth," *The Globe and Mail*, 22 January 2022, p. A16.

247 See W. J. Slater, *Lexicon to Pindar* (Berlin: Walter de Gruyter & Co., 1969); https://www.perseus.tufts.edu/hopper/text?doc=Perseus:text:1999.04.0072.

248 Aristotle, *Phusike akroasis* (c. 360 BCE); transl. R. P. Hardie and R. K. Gaye, *Physics*, in J. Barnes, ed., *Complete Works of Aristotle, Volume 1: The Revised Oxford Translation* (Princeton: Princeton University Press, 1984), book IV, part 7, 213b31; http://classics.mit.edu/Aristotle/physics.4.iv.html.

249 Ibid., book IV, part 2.

250 R. Descartes, *Le traité du monde et de la lumière* (Paris: Michel Bobin and Nicolas le Gras, 1664); M. S. Mahoney, transl., *The World, or Treatise on the Light* (New York: Abaris Books, 1979). Newton was at pains to dispose of the vortices; see *The Principia*, n. 214, p. 436.

251 I. Newton, *Opticks: or, a Treatise of the Reflexions, Refractions, Inflexions and Colours of Light* (London: Smith & Walford, 1704), Query 20.

252 I. Kant, (1770), collected writings; the original text is: "Spatiem non est aliquid objective et realis, nec substantia, nec accidens, nec relatio; sed subjectivum et ideale et a natura mentis."; in *Kant's gesammelte Schriften* (Berlin: Königlich-Preussische Akademie der Wissenschaften, 1902), **2**, p. 403; https://korpora.zim.uni-duisburg-essen.de/Kant/aa02/403.html.

253 An outline of the standard ontology can be found in J. Al-Khalili, *The World According to Physics*, n. 14.

254 A. Einstein, "Autobiographical Notes," n. 216, p. 3, (p. 339 in Hawking).

255 H. A. Lorentz, *The Theory of Electrons: And Its Applications to the Phenomena of Light and Radiant Heat*, n 319, p. 11.

256 A. Einstein, "Zur Elektrodynamik bewegter Körper," *Annalen der Physik*, (1905) 17, p. 891; A. Beck, transl., "On the Electrodynamics of Moving Bodies," *The Collected Papers of Albert Einstein, Vol. 2, The Swiss Years, Correspondence 1900–1909* (Princeton: Princeton University Press, 1989), p. 140.

257 L. Infeld, *Albert Einstein: His Work and Influence on Our World* (New York: Charles Scribner's Sons, 1950), p. 24.

258 A. Einstein, lecture at University of Leiden on May 5, 1920, "Äther und Relativitätstheorie," (Berlin: Julius Springer, 1920); A. Einstein, W. Perrett and G. Jeffery, transls., "Ether and the Theory of Relativity," in *Sidelights on Relativity* (London: Methuen & Co., 1922), p. 3.

259 Ibidem.

260 Ibidem.

261 Ibidem.

262 A. Einstein, "Aether und Relativitätstheorie," my transcription and translation of his handwritten verbatim lecture notes, 5 May 1920 (see fig. 7); https://www.e-codices.unifr.ch/en/fmb/ms-Einstein-E-004-001/10v/0/; "Zusammenfassend können wir sagen. Nach der allgemeinen Relativitätstheorie ist der Raum mit physikalischen Qualitäten ausgestaltet; es existiert also in diesem Sinne ein Aether. Dieser Aether darf aber nicht mit der für ponderable Medium characteristisches Eigenschaft ausgestaltet gedacht werden aus durch die Zeit verfolgbaren Teilen zu bestehen. Bewegungstgriff darf auf ihn nicht angewendet werden."

263 A. A. Michelson and E. W. Morley, "On the Relative Motion of the Earth and the Luminiferous Ether," *Am. J. Sci.*, (1887) **34**, p. 333.

264 It seems unfortunate that Einstein did not explicitly address the experiment in his 1920 lecture. However, he always maintained it had no role in his work on relativity. For example, in a 9 February 1954, letter to J. C. Davenport, he said, "In my own development, Michelson's result has not had a considerable influence. I even do not remember if I knew of it at all when I wrote my first paper on the subject." See also, M. Nauenberg, "Millikan, Einstein, and the Birth of Relativity," *APS News* (2004) **13**, letters; https://www.aps.org/publications/apsnews/200403/letters.cfm.

265 A. Einstein, "Über den Äther," *Verh. Schweiz. Naturforsch. Ges.*, (1924) **105**, p. 85; repr. as S. Saunders, transl., "On the Ether" in S. Saunders and H. R. Brown, eds., *The Philosophy of Vacuum* (Oxford: Clarendon Press, 1991), p. 13.

266 H. Gutfreund, notes in *Einstein's Masterpiece: The Foundation of General Relativity* (Jerusalem: Israel Academy of Sciences and Humanities, 2011), p. 83.

267 See n. 261.

268 E. Whittaker, *A History of the Theories of Aether and Electricity: The Modern Theories 1900–1926* (London: Thomas Nelson & Sons, 1953).

269 R. P. Feynman, *The Character of Physical Law* (New York: Modern Library, 1994), p. 166.

270 E.g., see K. Brown, *Physics in Space and Time* (Morrisville NC: Lulu Press, 2018), p. 707; https://www.mathpages.com/home/kmath526/kmath526.htm.

271 See, e.g., J. Bennett, "Space and Subtle Matter in Descartes's Metaphysics," in R. Gennaro and C. Heunemann, eds., *New Essays on the Rationalists* (Oxford: Oxford University Press, 1999), p. 3.

272 P. J. E. Peebles, *Principles of Physical Cosmology*, n. 82.

273 As Riemann modestly put it, "I could make use of no previous labors." G. F. B. Riemann, "Über die Hypothesen welche der Geometrie zu Grunde liegen," habilitation thesis, Göttingen University, (1854); W. K. Clifford, transl., "On the Hypotheses Which Lie at the Foundation of Geometry," *Nature*, (1867) **8**, p. 14; https://www.maths.tcd.ie/pub/HistMath/People/Riemann/Geom/WKCGeom.html.

274 The reader who wants a peek into the world of non-Euclidean geometries might look at philosopher of science Hans Reichenbach's classic work *Philosophie der Raum-Zeit-Lehre* (Berlin: De Gruyter, 1928); M. Reichenbach and J. Freund, transls. and eds., *The Philosophy of Space & Time* (New York: Dover Publications Inc., 1957), with the caution that he was a strong proponent of logical positivism, of which more will be said below; see especially chapter 36.

275 Ibid., pp. 14 and 36.

276   G. F. B. Riemann, n. 273. Riemann summed up this part of his thesis with a heading, "Bestimmte Theile einer Mannigfaltigkeit heissen Quanta"—which I render as "Definite Parts of a Multiplicity Are Called Quanta." See G. F. B. Riemann, "Über die Hypothesen welche der Geometrie zu Grunde liegen," *Abh. Konigl. Ges. Wiss. Gottingen*, (1868) **13**, p. 133; https://www.deutschestextarchiv.de/book/download_txt/riemann_hypothesen_1867. Clifford's translation said, "Definite portions of a manifold, distinguished by a mark or by a boundary, are called Quanta. Their comparison with regard to quantity is accomplished in the case of discrete magnitudes by counting."; n. 272.

277   Beyond this one work, Riemann sought wholeheartedly to exploit the mathematical conveniences of continuous space; see C. Jungnickel and R. McCormmach, *Intellectual Mastery of Nature: Theoretical Physics from Ohm to Einstein, Vol. 1, The Torch of Mathematics 1800–1870* (Chicago: University of Chicago Press, 1986), pp. 176 ff.

278   Einstein did not, as physicists are often taught—e.g., P. K. McKeown, "Gravity Is Geometry," *The Physics Teacher*, (1984) **22**, p. 557; https://doi.org/10.1119/1.2341671—show gravity *is* geometry; see, e.g., D. Lehmkuhl, "Why Einstein Did Not Believe That General Relativity Geometrizes Gravity," *Stud. Hist. Philos. Sci. Part B*, (2014) **46**, p. 316; https://www.sciencedirect.com/science/article/pii/S1355219813000695/pdfft?md5=f2d7dbd87a18845978cb24f7660d1f60&pid=1-s2.0-S1355219813000695-main.pdf

279   L. Susskind, *The Black Hole War: My Battle with Stephen Hawking to Make the World Safe for Quantum Mechanics* (New York: Little, Brown and Company, 2008), p. 335.

280   For a valiant effort to sustain it, see A. Hagar, *Discrete or Continuous: The Search for Fundamental Length in Modern Physics* (New York: Cambridge University Press, 2014).

281   M. Planck, *Wissenschaftliche Selbstbiographie* (Leipzig: Barth, 1948); M. Planck, "A Scientific Autobiography," in F. Gaynor, transl., M. von Laue, ed., *Scientific Autobiography and Other Papers* (New York: Philosophical Library, 1949), p. 13.

282   Curiously, by the time of Planck's death von Laue was working on superconductivity: M. von Laue, *Theorie der Supraleitung* (Berlin: Springer, 1947); L. Meyer and W. Band, transl., *Theory of Superconductivity* (New York: Academic Press, 1952).

283   M. von Laue, memorial address, Göttingen, 7 October 1947; repr. in *Max Planck: Scientific Autobiography and Other Papers*, n. 281, p. 7.

284   For a detailed (and unorthodox, but my modest researches suggest accurate) analysis of Planck's program and his evolving views, see T. S. Kuhn, *Black-Body Theory and the Quantum Discontinuity, 1894–1912* (Oxford: Oxford University Press, 1978), repr. with new 1986 Afterword (Chicago: University of Chicago Press, 1987).

285   Ibid., p. 350.

286   Ibid., p. viii.

287   I. Duck and E. C. G. Sudarshan, *100 Years of Planck's Quantum* (Singapore: World Scientific, 2000), p. 9; https://www.worldscientific.com/doi/suppl/10.1142/4426/suppl_file/4426_chap01.pdf.

288   M. von Laue, memorial address, n. 283.

289   M. Planck, "Über das Gesetz der Energieverteilung im Normalspectrum," *Verhandl. Dtsch. phys. Ges.*, (1900) **2**, p. 237; transl. and repr. "On the Theory of the Energy Distribution Law of the Normal Spectrum," in D. ter Haar, ed., *The Old Quantum Theory* (Oxford: Pergamon Press, 1967), p. 82.

290   M. Planck, letter to Robert W. Wood in 1931, repr. in A. Hermann, C. W. Nash, transl., *The Genesis of the Quantum Theory (1899–1913)* (Cambridge MA: MIT Press, 1971), p. 23; https://mitpress.mit.edu/books/genesis-quantum-theory-1899-1913.

291 H. Kragh, "Max Planck: The Reluctant Revolutionary," *Physics World*, (2000) **13**, p. 31; https://physicsworld.com/a/max-planck-the-reluctant-revolutionary/.

292 L. Smolin, *Einstein's Unfinished Revolution*, n. 74, p. 71.

293 L. J. Boya, "The Thermal Radiation Formula of Planck (1900)," (2004), http://arxiv.org/pdf/physics/0402064.pdf.

294 M. Planck, "Über irreversible Strahlungsvorgänge," *Ann. d. Phys.*, (1900) **1**, p. 69; https://uni-tuebingen.de/fileadmin/Uni_Tuebingen/Fakultaeten/MathePhysik/Institute/IAP/Forschung/MOettel/Geburt_QM/planck_AnnPhys_306_69_1900.pdf; transl. and repr., "On Irreversible Radiation Processes," in I. Duck and E. C. G. Sudarshan, transl. and eds., *100 Years of Planck's Quantum*, n. 287, p. 18.

295 Ibidem.

296 Ibid., p. 23; original italics.

297 For example, he told Boltzmann his radiation law had an "atomistic foundation"; M. Planck, "A Scientific Autobiography," n. 281, p. 33.

298 M. Planck, "Über irreversible Strahlungsvorgänge," *Sitzungsberichte der Königlich Preussischen Akademie der Wissenschaften zu Berlin* (Sessional Reports of the Royal Prussian Academy of Science), (1899) Pt. 1, p. 440, my translation; https://www.biodiversitylibrary.org/item/93034#page/498/mode/1up.

299 M. Planck, "A Scientific Autobiography," n. 281, p. 41.

300 M. Planck, "Entropie und Temperatur strahlender Wärme," *Ann. d. Phys.*, (1900) **1**, p. 719; repr. and transl., "Entropy and Temperature of Radiant Heat" in I. Duck and E. C. G. Sudarshan, eds., *100 Years of Planck's Quantum*, n. 287, p. 27.

301 Ibid., p. 28.

302 See, e.g., J. Stein, "Planck's Constant: The Number That Rules Technology, Reality, and Life," *The Nature of Reality*, Nova, 24 October 2011, https://www.pbs.org/wgbh/nova/article/plancks-constant/; almost all accounts get the story wrong.

303 M. Planck, "Über das Gesetz der Energieverteilung im Normalspectrum," *Ann. d. Phys.*, (1901) **4**, p. 553; repr. and transl., "On the Energy Distribution in the Blackbody Spectrum," in I. Duck and E. C. G. Sudarshan, eds., *100 Years of Planck's Quantum*, n. 287, p. 31.

304 M. Planck, "Zur Geschichte der Auffindung des physikalischen Wirkungsquantums," *Naturwissenschaften*, (1943) **31**, p. 153; repr. and transl., "Reflections on the Discovery of the Quantum of Action," in I. Duck and E. C. G. Sudarshan, eds., *100 Years of Planck's Quantum*, n. 287, p. 48.

305 E. Planck, undated third-hand report, cited in T. S. Kuhn, *Black-Body Theory and the Quantum Discontinuity, 1894–1912*, n. 284, p. 113; see also A. Hermann, *Max Planck in Selbstzeugnissen und Bilddokumenten* (Hamburg: Rowohlt, 1973), p. 29.

306 M. Planck, "A Scientific Autobiography," n. 281, p. 44.

307 Ibidem; my emphasis.

308 For more on this unique story, see M. Nauenberg, "Max Planck and the Birth of the Quantum Hypothesis," *Am. J. Phys.*, (2016) **84**, p. 709; https://doi.org/10.1119/1.4955146; and chapter 1 of M. Kumar, *Quantum: Einstein, Bohr and the Great Debate About the Nature of Reality* (New York: W.W. Norton & Co., 2008); https://dokumen.pub/quantum-einstein-bohr-and-the-great-debate-about-the-nature-of-reality-9781848311039-1848311036.html.

309 M. Planck, quoted in I. Duck and E. C. G. Sudarshan, transl. and eds., "Reflections on the Discovery of the Quantum of Action," n. 304, p. 48.

310 M. Planck, "A Scientific Autobiography," n. 281, p. 45.

311 I. Duck and E. C. G. Sudarshan, *100 Years of Planck's Quantum*, n. 287, p. 9.

312 E.g., "He guessed, therefore, that he should recombine these two expressions in the simplest possible way."; "October 1900: Planck's Formula for Black-Body Radiation," *APS News*, 11, October 2002, https://www.aps.org/publications/apsnews/200210/history.cfm.

313 Y. J. Ng, "Selected Topics in Planck-scale Physics," *Mod. Phys. Lett.*, (2003) **A18**, p. 1073; https://arxiv.org/pdf/gr-qc/0305019.pdf.

314 Planck's life descended into tragedy; see, e.g., B. R. Brown, *Planck: Driven by Vision, Broken by War* (Oxford: Oxford University Press, 2015).

315 A. G. Ekstrand, award ceremony speech, Nobel Prize, 1 June 1920; https://www.nobelprize.org/prizes/physics/1918/ceremony-speech/.

316 G. R. Kirchhoff, "Über den Zusammenhang zwischen Emission und Absorption von Licht and Wärme," *Monatsber. Akad. Wiss. Berlin*, (1859), p. 783, and "Über das Verhältnis zwischen dem Emissionsvermögen und dem Absorptionsvermögen der Körper für Wärme und Licht," *Ann. d. Phys.*, (1860) **109**, p. 275; repr. in, *Gesammelte Abhandlungen von G. Kirchhoff* (Leipzig: Johann Ambrosius Barth, 1882), p. 566 and p. 571; https://archive.org/details/gesammelteabhan01unkngoog.

317 Indeed, until 1906, "he [Einstein] had nothing whatsoever to say about Planck's theory"; T. S. Kuhn, *Black-Body Theory and the Quantum Discontinuity, 1894–1912*, n. 284, p. 182.

318 H. A. Lorentz, "The Theory of Radiation and the Second Law of Thermodynamics," *Proc. K. Ned. Akad. Wet., Ser. B*, (1901) **3**, p. 436; https://adsabs.harvard.edu/full/1900KNAB....3..436L.

319 H. A. Lorentz, *The Theory of Electrons: And Its Applications to the Phenomena of Light and Radiant Heat* (Leipzig: B. G. Teubner, 1909); and see, e.g., R. McCormmach, "Lorentz, Henrik Antoon," *Complete Dictionary of Scientific Biography*, 27 June 2018, https://www.encyclopedia.com/people/science-and-technology/physics-biographies/hendrik-antoon-lorentz.

320 F. Wilczek, "Scaling Mount Planck 1: A View from the Bottom," in F. Wilczek, ed., *Fantastic Realities: 49 Mind Journeys and a Trip to Stockholm* (Singapore: World Scientific, 2006), p. 125.

321 A. Pais, *Subtle Is the Lord: The Science and the Life of Albert Einstein*, n. 44, p. 358.

322 B. Hoffmann, *The Strange Story of the Quantum* (New York: Dover Publications, 1947), p. ix; https://archive.org/stream/the-strange-story-of-the-quantum-hoffmann/The%20Strange%20Story%20of%20the%20Quantum%20-%20Hoffmann_djvu.txt.

323 A. Einstein, "Autobiographical Notes," n. 216, p. 45, (p. 359 in Hawking).

324 P. G. Bergmann, *The Riddle of Gravitation* (New York: Dover Publications, 1992), p. x.

325 M. Planck et al., Proposal for Einstein's Membership in the Prussian Academy of Sciences, *The Collected Papers of Albert Einstein, Vol. 5, The Swiss Years: Correspondence, 1902–1914*, n. 170, Doc. 445, p. 526; repr. in https://einsteinpapers.press.princeton.edu/vol5-doc/576.

326 M. Born, lecture at Massachusetts Institute of Technology, 14 November 1925; repr. in M. Born, *Problems of Atomic Dynamics* (Mineola NY: Dover Publications, 1926), p. 2; his emphasis.

327 H. Kragh, "Max Planck: The Reluctant Revolutionary," n. 291.

328 E.g., P. Taylor, "Number of smartphone mobile network subscriptions worldwide from 2016 to 2023, with forecasts from 2023 to 2028", 22 May 2024, https://www.statista.com/statistics/330695/number-of-smartphone-users-worldwide/.

329 M. Planck, lecture at the 1911 Solvay Conference, "La Loi du Rayonnement Noir et l'Hypothése des Quantités Élémentaires d'Action," *Solvay Conference Proceedings*, p. 93; http://www.solvayinstitutes.be/pdf/Proceedings_Physics/1911.pdf; quoted by M. J. Klein in "Thermodynamics and Quanta in Planck's Work," *Phys. Today*, (1966) **19**, p. 23.

330 In the nineteen twenties it was thought all atoms might have the same size; see, e.g., R. W. G. Wyckoff, "On the Hypothesis of Constant Atomic Radii," *Proc. Nat. Acad. Sci. U.S.A.*, (1923) **9**, p. 33.

331 Skepticism about the real significance of Planck's tiny numbers extended into the current century; see, e.g., D. Meschini, "Planck-scale Physics: Facts and Beliefs," *Found. Sci.*, (2007) **12**, p. 277; https://arxiv.org/abs/gr-qc/0601097.

332 C. A. Mead, "Possible Connection Between Gravitation and Fundamental Length," *Phys. Rev.*, (1964) **135**, p. B849; https://harvest.aps.org/v2/journals/articles/10.1103/PhysRev.135.B849/fulltext.

333 C. A. Mead, "Walking the Planck Length Through History," *Phys. Today*, (2001) **54**, p. 15; https://physicstoday.scitation.org/doi/10.1063/1.1428424.

334 The phrase "Shut up and calculate" seems to originate in an article by physicist David Mermin, "What's Wrong with This Pillow?," *Phys. Today*, (1989) **42**, p. 9; https://physicstoday.scitation.org/doi/10.1063/1.2810963; repr. in N. D. Mermin, *Boojums All the Way Through: Communicating Science in a Prosaic Age* (New York: Cambridge University Press, 1990), p. 199.

335 E.g., W. Myrvold, "Philosophical Issues in Quantum Theory," *Stanford Encyclopedia of Philosophy*, n. 99, 23 March 2022, https://plato.stanford.edu/entries/qt-issues/#OntoIssu.

336 See, e.g., O. Darrigol, "The Historians' Disagreements over the Meaning of Planck's Quantum," in J. Büttner et al., *Revisiting the Quantum Discontinuity* (Berlin: Max Planck Institute for the History of Science, 2000); https://scholar.google.com/citations?view_op=view_citation&hl=de&user=CO_1lBAAAAAJ&citation_for_view=CO_1lBAAAAAJ:u5HHmVD_uO8C.

337 H. Reichenbach, *The Rise of Scientific Philosophy* (Los Angeles: University of California Press, 1951), p. 170.

338 The Editors of Encyclopaedia Britannica, "Planck's Radiation Law," n. 96, https://www.britannica.com/science/Plancks-radiation-law.

339 H. Poincaré, *Dernières Pensées* (Paris: Flammarion, 1913); J. W. Bolduc, transl., *Last Thoughts, Mathematics and Science: Last Essays*, (New York: Dover, 1963) p. 80; https://henripoincarepapers.univ-nantes.fr/chp/hp-pdf/hp1963ms.pdf.

340 G. Kramm and N. Molders, "Planck's Blackbody Radiation Law: Presentation in Different Domains and Determination of the Related Dimensional Constants," *J. Calcutta Math. Soc.*, (2009) **5**, p. 27; https://arxiv.org/pdf/0901.1863.

341 S. Bose, A. Einstein transl. to German, "Plancks Gesetz und Lichtquantenhypothese," *Zeits. Physik*, (1924) **26**, p. 178; repr. and transl., "Planck's Law and the Light Quantum Hypothesis" in O. Theimer and B. Ram, "The Beginning of Quantum Statistics," *Am. J. Phys.*, (1976) **44**, p. 1057; https://pubs.aip.org/aapt/ajp/article-abstract/44/11/1056/1050235/The-beginning-of-quantum-statistics-A-translation.

342 A. Einstein, "On the Ether," n. 264, p. 13.

343 A. Einstein, "Relativistic Theory of the Non-Symmetric Field," (1954), in *The Meaning of Relativity, Fifth Edition* (New York: MJF Books, 1956), Appendix II, p. 165.

344 J. Wheeler, *Einstein's Vision: Vie steht es heute mit Einstein's Vision, alles als Geometrie aufzufassen* (New York: Springer Verlag, 1968); quoted in V.I. Konushko, "Granular Space and the Problem of Large Numbers," *J. Mod. Phys.*, (2011) **2**, p. 289, (transl. unknown); https://www.researchgate.net/publication/228779936_Granular_Space_and_the_Problem_of_Large_Numbers.

345 See J. A. Wheeler, *Geons, Black Holes & Quantum Foam: A Life in Physics*, n. 78.

346 G. 't Hooft, *In Search of the Ultimate Building Blocks*, n. 193, p. 149.

347 C. Rovelli, *Quantum Gravity*, n. 41, p. 367.

348 L. Smolin, *Einstein's Unfinished Revolution*, n. 74.

349 By 1999, it was creeping into the purview of the *The New York Times*: G. Johnson, "How Is the Universe Built? Grain by Grain," https://faculty.washington.edu/smcohen/320/GrainySpace.html.

350 For an analysis of weaknesses of the case made for Planck-scale physics, see D. Meschini, "Planck-scale Physics: Facts and Beliefs," n. 331.

351 L. Euler, *Methodus Inveniendi Lineas Curvas Maximi Minive Proprietate Gaudentes* (Geneve: Marc-Michel Bousquet, 1744) Addentiamentum 2; J. S. D. Glaus, transl., *A Method for Finding Curved Lines Enjoying Properties of Maximum or Minimum*, The Euler Archive, https://scholarlycommons.pacific.edu/euler-works/65/.

352 See W. Dittrich, *The Development of the Action Principle: A Didactic History from Euler-Lagrange to Schwinger* (Cham: Springer, 2021); https://link.springer.com/book/10.1007/978-3-030-69105-9.

353 R. P. Feynman et al., "The Principle of Least Action," *The Feynman Lectures on Physics, Vol. II* (Reading MA: Addison-Wesley, 1964), ch. 19; https://www.feynmanlectures.caltech.edu/II_19.html.

354 See, e.g., H. Tributsch, "On the Fundamental Meaning of the Principle of Least Action and Consequences for a 'Dynamic' Quantum Physics," *J. Mod. Phys.*, (2016) 7, p. 365, https://www.scirp.org/journal/paperinformation.aspx?paperid=63922.

355 S. W. Hawking, "Classical Theory," lecture in Cambridge in 1994; repr. in S. W. Hawking and R. Penrose, *The Nature of Space and Time*, n. 17, p. 4.

356 G. 't Hooft, "The fundamental nature of space and time," in D. Oriti, (ed.), *Approaches to Quantum Gravity*, n. 6, p. 13.

357 M. Kumar, *Quantum*, n. 308, p. 360.

358 D. Oriti, quoted in G. Musser, "What Is Spacetime?," *Nature*, (2018) **557**, p. S2; https://media.nature.com/original/magazine-assets/d41586-018-05095-z/d41586-018-05095-z.pdf.

359 G. Musser, "What Is Spacetime?," ibidem.

360 E.g., C. Wood and V. Stein, "What Is String Theory?," *Space*, 20 January 2022, https://www.space.com/17594-string-theory.html.

361 See, e.g., S. Weinberg, *Dreams of a Final Theory* (New York: Pantheon Books, 1992).

362 Ibid., p. 8.

363 C. Rovelli, "Unfinished Revolution," n. 6, p. 5.

364 G. Ellis, "Theories Beyond Testability?," *Science*, (2013) **342**, p. 934; https://www.science.org/doi/10.1126/science.1246302.

365 Two works may bookend the issue: Smolin's *The Trouble with Physics*, n. 70; and R. Dawid, *String Theory and the Scientific Method* (Cambridge: Cambridge University Press, 2013).

366 J. Butterfield and C. Isham, "Spacetime and the Philosophical Challenge of Quantum Gravity," in C. Callender and N. Huggett, eds., *Physics Meets Philosophy at the Planck Scale: Contemporary Theories in Quantum Gravity* (Cambridge: Cambridge University Press, 2001), p. 177; https://arxiv.org/abs/gr-qc/9903072.

367 Ibid., p. 34.

368 Generated by A. J. Hanson, "A Construction for Computer Visualization of Certain Complex Curves," *Not. Amer. Math. Soc.*, (1994) **41**, p. 1156.

369 To be a tad technical, the space these string theories need is called a *Calabi-Yau threefold*. It has six dimensions, three of which are denoted as real and three as imaginary after the fashion of complex numbers; for a short introduction to the latter, see, e.g., "Complex Number," *Cuemath*, https://www.cuemath.com/numbers/complex-numbers/.

370 See, e.g., B. Greene, "String Theory on Calabi-Yau Manifolds," 23 February 1997, https://arxiv.org/abs/hep-th/9702155.

371 It may be infinite; see Hilbert space in Wikipedia.

372 E.g., A. Einstein and P. Bergmann, "On a Generalization of Kaluza's Theory of Electricity," *Ann. Math.*, (1938) **39**, p. 683.

373 D. Gross, "Einstein and the Search for Unification," *Current Sci.*, (2005) **89**, p. 2035.

374 For an explanation of Euclidean space, see, e.g., W. L. Hosch, "Euclidean space," n. 96, https://www.britannica.com/science/Euclidean-space.

375 See S-T. Yau and S. Nadis, *The Shape of Inner Space: String Theory and the Geometry of the Universe's Hidden Dimensions* (New York: Basic Books, 2010), p. 169.

376 For an easy onramp to these ideas, see *Time One*, n. 16.

377 For the methodology of making such 2-D projections, see, e.g., "Images of Mathematical Physics" on *Andrew J. Hanson's Homepage*, https://legacy.cs.indiana.edu/~hanson/.

378 L. Susskind, *The Black Hole War: My Battle with Stephen Hawking to Make the World Safe for Quantum Mechanics*, n. 279, p. 343.

379 Ibid., p. 339.

380 For more on these strange critters, see, e.g., S.-T. Yau, "Calabi-Yau Manifold," *Scholarpedia*, (2009) **4**, 6524; http://www.scholarpedia.org/article/Calabi-Yau_manifold.

381 For those who, unafraid of math, want a sense of the work already done on it decades ago, see T. Hübsch, *Calabi-Yau Manifolds: A Bestiary for Physicists* (Singapore: World Scientific, 1992).

382 R. L. Amoroso et al., "Exploring Novel Cyclic Extensions of Hamilton's Dual-Quaternion Algebra," in R. L. Amoroso et al., eds., *The Physics of Reality: Space, Time, Matter, Cosmos* (Singapore: World Scientific, 2012), p. 81.

383 T. Kaluza, "Zum Unitätsproblem der Physik," *Sitzungsberichte der Königlich Preußischen Akademie der Wissenschaften zu Berlin*, (1921), p. 966; https://ia800905.us.archive.org/1/items/sitzungsberichte1921preussi/sitzungsberichte1921preussi_bw.pdf.

384 E.g., D. Berman, "Kaluza, Klein and Their Story of a Fifth Dimension," *Plus*, 10 October 2012, https://plus.maths.org/content/kaluza-klein-and-their-story-fifth-dimension.

385 E.g., R. Garisto, "Curling Up Extra Dimensions in String Theory," *Phys. Rev. Focus*, (1998) **1**, p. 7; https://physics.aps.org/story/v1/st7.

386 L. Susskind, *The Black Hole War*, n. 279, p. 279.

387 L. Smolin, *The Trouble with Physics*, n. 70.

388 S. Hossenfelder, *Lost in Math*, n. 157, p. 188.

389 R. Altman et al., "New Large-Volume Calabi-Yau Threefolds," *Phys. Rev.*, (2018) **D97**, 046003, original emphasis; https://journals.aps.org/prd/pdf/10.1103/PhysRevD.97.046003.

390 I. Duck and E. C. G. Sudarshan, *100 Years of Planck's Quantum*, n. 313, p. 17.

391 A. Einstein, letter to Michele Besso, 10 August 1954, quoted and transl. in J. Stachel, "Einstein and the Quantum," in R. G. Colodny, ed., *From Quarks to Quasars*, n. 758, p. 380; Einstein's italics.

392 See, e.g., "Konrad Zuse: The First Programmable, Digital Computer," *The German Way*, https://www.german-way.com/notable-people/featured-bios/konrad-zuse/.

393 K. Zuse, *Rechnender Raum* (Wiesbaden: Springer, 1969); and see K. Zuse, "Calculating Space" in H. Zenil, ed., *A Computable Universe: Understanding & Exploring Nature as Computation* (Singapore: World Scientific, 2012), p. 729; https://link.springer.com/book/10.1007/978-3-663-02723-2.

394 Ibid., p. 786.

395 H. R. Pagels, *Perfect Symmetry: The Search for the Beginning of Time* (New York: Bantam Books, 1986), p. 371 ff.

396 Lloyd asserted "the universe is observationally indistinguishable from a giant quantum computer."; S. Lloyd, *Programming the Universe: A Quantum Computer Scientist Takes On the Cosmos* (New York: Alfred A. Knopf, 2006); and see S. Lloyd, "The Universe as a Computer," (2013), https://arxiv.org/pdf/1312.4455.pdf.

397 M. Tegmark, *Our Mathematical Universe: My Quest for the Ultimate Nature of Reality* (New York: Alfred A. Knopf, 2014).

398 N. Bostrom, "Are You Living in a Computer Simulation?," *Phil. Quart.*, (2003) **53**, p. 243; https://simulation-argument.com/simulation.pdf.

399 M. Gleiser, "Reality Is Not a Simulation and Why It Matters," Institute of Art and Ideas, 4 January 2023, https://iai.tv/articles/reality-is-not-a-simulation-and-why-it-matters-auid-2343.

400 See, e.g., L. Zyga, "Physicists Investigate the Structure of Time, with Implications for Quantum Mechanics and Philosophy," Phys Org, 1 February 2016, https://phys.org/news/2016-02-physicists-implications-quantum-mechanics-philosophy.html.

401 P. Davies and J. Gribbin, *The Matter Myth: Dramatic Discoveries That Challenge Our Understanding of Physical Reality* (New York: Simon & Schuster, 1992), p. 140.

402 A. A. Milne, *Now We Are Six* (London: Methuen & Co., 1927).

403 B. Russell, *Why I Am Not a Christian and Other Essays on Religion and Related Subjects* (London: George Allen & Unwin, 1957), p. 4.

404 I. Kant, *Kritik der reinen Vernunft* (Riga: Johann Friedrich Hartknoch, 1781); P. Guyer and A. W. Wood, eds. and transl., *Critique of Pure Reason* (Cambridge: Cambridge University Press, 1999), Div. 2, Sec. 1.

405 P. J. E. Peebles, lecture in Winnipeg, 5 July 2022. Certainly, events showed Lemaître understood the implications of the general relativity equations better than Einstein did.

406 G. Lemaître, "Un Univers homogène de masse constante et de rayon croissant rendant compte de la vitesse radiale des nébuleuses extra-galactiques," *Annales de la Société Scientifique de Bruxelles*, (1927) A47, p. 49; transl., *Gen. Rel. Grav.*, (2013) 45, p. 1635.

407 G. Lemaître, "The Expanding Universe," *Mon. Not. R. Astron. Soc.*, (1931) 41, p. 491.

408 To this day, his work is underrated. Serious physicists, writing of the Big Bang theory's origins, fail to even mention Lemaître; see, e.g., M. Kaku and J. T. Thompson, *Beyond Einstein: The Cosmic Quest for the Theory of the Universe* (New York: Anchor Books, 1995), pp. 130 ff.

409 W. de Sitter, letter to Harlow Shapley, 17 April 1930, Widner Library, Harvard University Archives; quoted in R. W. Smith, *The Expanding Universe, Astronomy's Great Debate, 1900–1931* (Cambridge: Cambridge University Press, 1982), p. 187.

410 E. Hubble, "A Relation Between Distance and Radial Velocity Among Extra-Galactic Nebulae," *Proc. Nat. Acad. Sci. USA*, (1929) 15, p. 168; https://www.pnas.org/doi/10.1073/pnas.15.3.168.

411 For example, in his 1989 preface to the Dover edition of his 1965 history of cosmology, science historian John (J. D.) North wrote of it: "Like most writers thirty years ago I did not properly appreciate [Lemaître's] very considerable physical intuition. A letter he wrote to *Nature* in 1939 [presumably he meant 1931] . . . now has a prophetic look about it." J. D. North, *The Measure of the Universe: A History of Modern Cosmology* (New York: Dover Publications, 1990), p. i.

412 G. Lemaître, "The Beginning of the World from the Point of View of Quantum Theory," *Nature*, (1931) 127, p. 706; https://www.nature.com/articles/127706b0.

413 In 1848, Edgar Allan Poe published *Eureka*, n. 85, a cosmogony in the form of an epic prose poem in which he foreshadowed the one-particle origin. Seeking the "extreme of *Simplicity*," he described the universe beginning as "*one* particle—a particle of *one* kind—of *one* character—of *one* nature—of *one* size" that divided, "*forcing* the originally and therefore normally *One* into the abnormal condition of *Many*." See note 85 at §§ 45 to 48; his italics. Poe was a serious student of the infant science of cosmology. In 1934 Einstein read *Eureka* and wrote to Poe scholar Richard Gimbel (Shelby White and Leon Levy Archives Center, Institute for Advanced Study, Princeton, Office of the Director, Faculty Files, Box 10; https://albert.ias.edu/bitstream/handle/20.500.12111/2901/54424.pdf) that it was "eine sehr schöne Leistung eines ungewöhnlich selbständigen Geistes" ("a very beautiful achievement of an unusually original mind"; my transl.) though he later seemed to change his view. For a modern astrophysicist's appreciation, see Alberto Cappi, http://www.bo.astro.it/~cappi/poe.html. There are those who would credit Poe with much more, such as originating general relativity or a competing theory, to which I do not mean to give credence here.

414 W. D. MacMillan, "Some Mathematical Aspects of Cosmology," *Science*, (1925) 62, pp. 63, 96, and 121; https://www.science.org/doi/epdf/10.1126/science.62.1595.63 ff.

415 See "Continuous Creation," BBC Third Programme, 28 March 1949, *Radio Times*, Issue 1328, p. 17; https://genome.ch.bbc.co.uk/schedules/service_third_programme/1949-03-28.

416 G. Lemaître, *Hypothèse de l'Atome Primitif* (Neuchâtel: Éditions du Griffon, 1946); B. H. Korff and S. A. Korff, transl., *The Primeval Atom: An Essay on Cosmogony* (New York: Van Nostrand Company, 1950), p. 78.

417 In this model, matter was assumed to be created steadily at the rate needed to compensate the expansion of the universe. See, e.g., M. K. Munitz, "Creation and the 'New' Cosmology," *Brit. J. Phil. Soc.*, (1954) **5**, p. 32.

418 G. Lemaître, interview on Vlaamse Radio-en Televisieomroeporganisatie, 14 February 1964; transcr. in S. G. A. Gontcho et al., eds. and transl., "Resurfaced 1964 VRT video interview of Georges Lemaître," https://arxiv.org/pdf/2301.07198.pdf.

419 For a Cook's tour of the main universe-origin stories, see J. D. Barrow, *The Book of Universes: Exploring the Limits of the Cosmos* (New York: W. W. Norton & Co., 2011), pp. 54 ff.

420 Einstein is quoted as saying, "I want to know how God created this world." R. W. Clark, *Einstein: The Life and Times* (New York: World Publishing Company, 1971), p. 19.

421 S. W. Hawking, "Origins of the Universe," lecture at the University of California, Berkeley, 13 March 2007, https://newsarchive.berkeley.edu/news/media/releases/2007/03/16_hawking_text.shtml.

422 B. Clegg, *Before the Big Bang: The Prehistory of Our Universe* (New York: St. Martin's Press, 2009), p. 136.

423 E.g., V. J. Vicent and V. Trimble, "Cosmologists in the Dark," J. A. Rubiño-Martín et al., eds., *ASP Conference Series*, (2009) **409**, p. 47.

424 E.g., E. Howell and D. Dobrijevic, "What Is the Cosmic Microwave Background?," *Space*, 28 January 2022, https://www.space.com/33892-cosmic-microwave-background.html.

425 This does not mean he always thought time had a beginning. In 1983, physicist James Hartle and he proposed a mathematical no-boundary boundary from which the universe emerged: J. B. Hartle and S. W. Hawking, "Wave Function of the Universe," *Phys. Rev.*, (1983) **28**, p. 2960. He later changed his mind.

426 See K. Knox and R. Noakes, *From Newton to Hawking: A History of Cambridge University's Lucasian Professors of Mathematics* (Cambridge: Cambridge University Press, 2003); rev. by L. Pyenson, "In Newton's long shadow," *Nature*, (2004) **428**, p. 258; https://www.nature.com/articles/428258b.pdf.

427 S. W. Hawking, "Is the End in Sight for Theoretical Physics?," inaugural lecture at Cambridge University, 29 April 1980, (Cambridge: Cambridge University Press, 1980), p. 3.

428 S. W. Hawking, "Quantum Cosmology," lecture in Cambridge, 1994; repr. in S. W. Hawking and R. Penrose, *The Nature of Space and Time*, n. 17, p. 76.

429 S. W. Hawking, "The Beginning of Time," lecture in 1996, https://www.hawking.org.uk/in-words/lectures/the-beginning-of-time.

430 S. W. Hawking and L. Mlodinow, *The Grand Design* (New York: Bantam Books, 2010).

431 J. Horgan, "Cosmic Clowning: Stephen Hawking's 'New' Theory of Everything Is the Same Old CRAP," *Sci. Am.*, 13 September 2010, https://blogs.scientificamerican.com/cross-check/cosmic-clowning-stephen-hawkings-new-theory-of-everything-is-the-same-old-crap/.

432 T. Hertog, *On the Origin of Time: Stephen Hawking's Final Theory* (New York: Bantam, 2023).

433 R. P. Crease, "The Never-Ending Quest for a Beginning," *Nature*, (2023) **616**, p. 243.

434 See, e.g.: H. Kragh, *Matter and Spirit in the Universe: Scientific and Religious Preludes to Modern Cosmology* (London: Imperial College Press, 2004).

435 B. Greene, *Until the End of Time: Mind, Matter, and Our Search for Meaning in an Evolving Universe* (New York: Vintage Books, 2020), p. 45.

436 J. S. Bell, "La Nouvelle Cuisine," in *Between Science and Technology*, A. Sarlemijn and P. Kroes, eds., (Amsterdam: Elsevier, 1990), p. 97; repr. in *Speakable and Unspeakable in Quantum Mechanics*, n. 111, p. 234.

437 H. Arendt, *The Human Condition*, n. 102, p. 178.

438 *Supra*, n. 416.

439 B. Gamow and G. Gamow, "Said Ryle to Hoyle," from G. Gamow, *Mr. Tompkins in Wonderland* (Cambridge: Cambridge University Press, 1940), original emphasis; https://history.aip.org/exhibits/cosmology/ideas/ryle-vs-hoyle.htm.

440 See his website: https://marcuschown.com/review/the-nashville-tennessean/.

441 M. Chown, "7 Questions About the Expansion of the Universe," *Sky at Night Magazine*, BBC Studios, 3 March 2022, https://www.skyatnightmagazine.com/space-science/expansion-universe/.

442 See, e.g., P. J. E. Peebles, "The Standard Model of Cosmology," in M. Greco, ed., *Les Rencontres de Physique de la Vallee d'Aosta 1998* (Gif-sur-Yvette: Ed. Frontières, 1998); https://ned.ipac.caltech.edu/level5/Peebles1/Peeb_contents.html.

443 R. A. Muller, *Now: The Physics of Time* (New York: W.W. Norton & Co., 2016), p. 130.

444 T. M. Davis and C. H. Lineweaver, "Expanding Confusion: Common Misconceptions of Cosmological Horizons and the Superluminal Expansion of the Universe," in R. Durrer et al., eds., *Cosmology and Particle Physics CAPP 2000* (New York: American Institute of Physics, 2001), p. 443; https://www.cambridge.org/core/services/aop-cambridge-core/content/view/EFEEEFD8D71E59F86DDA82FDF576EFD3/S132335800000607Xa.pdf/expanding_confusion_common_misconceptions_of_cosmological_horizons_and_the_superluminal_expansion_of_the_universe.pdf.

445 E.g., Steven Weinberg is reported to have said, "Space does not expand. Cosmologists sometimes talk about expanding space—but they should know better." M. Chown, "All You Ever Wanted to Know About the Big Bang," *New Scientist*, (1993) **137**, p. 32; and see Appendix B of n. 444, which gives twenty-five "Examples of Misconceptions or Easily Misinterpreted Statements in the Literature."

446 Regrettably, few speak of this number (known as the Hubble constant) in terms of sensible units like percent per annum; for reasons almost as old and obscure as measuring length by somebody's foot, it is usually given in km/s/Mpc, a unit that is not worth the effort to explain; see, e.g., J. P. Huchra, "The Hubble Constant," The Center for Astrophysics, (2008); https://lweb.cfa.harvard.edu/~dfabricant/huchra/hubble/.

447 See, e.g., T. Lewton, "What Might Be Speeding Up the Universe's Expansion?," *Quanta Magazine*, 27 April 2020, https://www.quantamagazine.org/why-is-the-universe-expanding-so-fast-20200427/.

448 E. Gates, *Einstein's Telescope: The Hunt for Dark Matter and Dark Energy in the Universe* (New York: W. W. Norton & Company, 2009), p. 3.

449 P. J. E. Peebles, lecture in Winnipeg, n. 405.

450 See, e.g., E. L. Wright, "Brief History of the Universe," 15 July 2004, https://www.astro.ucla.edu/~wright/BBhistory.html.

451 V. M. Slipher, "Radial Velocity Observations of Spiral Nebulae," *Observatory*, (1917) **40**, p. 304; see also V. M. Slipher, "Spectrographic Observations of Nebulae," *Pop. Astron.*, (1923) **15**, p. 21; https://ui.adsabs.harvard.edu/link_gateway/1915PA.....23...21S/ADS_PDF.

452 A. A. Friedmann, "Über die Krümmung des Raumes," *Zs. f. Phys.*, (1922) **10**, p. 377; B. Doyle, transl., "On the Curvature of Space," General Relativity and Gravitation, (1999) **31**, p. 1991; http://wwwphy.princeton.edu/~steinh/ph563/friedmann.pdf.

453 A. Einstein, "Remark on the Work of A. Friedmann (Friedmann 1922) 'On the Curvature of Space'," *Zs. f. Phys.*, (1922) **11**, p. 326.

454 A. Einstein, "A Note on the Work of A. Friedmann 'On the Curvature of Space,'" *Zs. f. Phys.*, (1923) **16**, p. 228.

455 I. Steer, "Who Discovered Universe Expansion?," *Nature*, (2012) **490**, p. 176; https://www.nature.com/articles/490176c.

456 G. Lemaître, "Un univers homogène de masse constante et de rayon croissant, rendant compte de la vitesse radiale des nébuleuses extra-galactiques," *Ann. Soc. Sci. Bruxelles*, (1927) **A 47**, p. 49.

457 G. Lemaître, "A Homogeneous Universe of Constant Mass and Increasing Radius Accounting for the Radial Velocity of Extra-Galactic Nebulae," *Mon. Not. R. Astron. Soc.*, (1931) **91**, p. 483.

458 M. Livio, "Mystery of the Missing Text Solved," *Nature*, (2011) **479**, p. 171.

459 E.g., H. Nussbaumer and L. Bieri, "Who Discovered the Expanding Universe?," https://arxiv.org/abs/1107.2281.

460 M. Livio, "Mystery of the Missing Text Solved," n. 458.

461 J. Farrell, *The Day Without Yesterday: Lemaître, Einstein, and the Birth of Modern Cosmology* (New York: Thunder's Mouth Press, 2005), p. 85; his emphasis.

462 E. Hubble, "A Relation Between Distance and Radial Velocity Among Extra-Galactic Nebulae," *Proc. Natl. Acad. Sci. U.S.A.*, (1929) **15**, p. 168; https://doi.org/10.1073/pnas.15.3.168.

463 For example, a new technique uses change in wavelength of delayed and gravitationally-lensed light from supernovas: P. L. Kelly et al., "Constraints on the Hubble Constant from Supernova Refsdal's Reappearance," *Science*, (2023) **380**, p. 1322; see also K. Cooper, "Gravitational Lensing of Supernova Yields New value for Hubble Constant," *Physics World*, 23 May 2023, https://physicsworld.com/a/gravitational-lensing-of-supernova-yields-new-value-for-hubble-constant/.

464 E.g., P. M. Sutter, "Is the Hubble constant not . . . Constant?", *Universe Today*, 30 May 2021, https://www.universetoday.com/151325/is-the-hubble-constant-notconstant/. See also M. G. Dainotti et al., "On the Hubble Constant Tension in the SNe Ia Pantheon Sample," *Astrophys. J.*, (2021) **912**, p. 150; https://iopscience.iop.org/article/10.3847/1538-4357/abeb73/pdf.

465 W. L. Freedman et al., "Status Report on the Chicago-Carnegie Hubble Program (CCHP): Three Independent Astrophysical Determinations of the Hubble Constant Using the James Webb Space Telescope," (2024), https://www.arxiv.org/abs/2408.06153.

466 W. Freedman, interviewed by B. Turner, *Live Science*, 22 December 2023, https://www.livescience.com/space/it-could-be-profound-how-astronomer-wendy-freedman-is-trying-to-fix-the-universe; see also D. Clery, "New results intensify debate over cosmic expansion rate," *Science*, (2024) **385**, p. 698; https://www.science.org/doi/epdf/10.1126/science.ads4465.

467 J. D. Barrow, *New Theories of Everything*, n. 134, p. 90.

468 P. M. Sutter, "Do We Live in a Quantum World?," *Space*, 27 November 2019, https://www.space.com/do-we-live-in-quantum-world.html.

469 For helpful works that bear upon our present inquiry, see, e.g., J. D. Barrow, *Pi in the Sky: Counting, Thinking and Being* (Oxford: Oxford University Press, 1992); *The Origin of the Universe* (New York: Basic Books, 1994); and *New Theories of Everything*, n. 134.

470 See figure from T. Henning, *Charging Effects in Niobium Nanostructures*, thesis, Chalmers Tekniska Högskola AB och Göteborgs Universitet, Göteborg, 1999, Fig. 2.1; https://www.arxiv.org/abs/cond-mat/9901308.

471 See, e.g., J. Schombert, "An Introduction to Quantum Tunneling," *AZO Quantum*, 15 May 2012, https://www.azoquantum.com/Article.aspx?ArticleID=12.

472 E.g., G. Nimtz, "On Superluminal Tunneling," *Prog. Quantum Electron.*, (2003) **27**, p. 417; and P. Eckle et al., "Attosecond Ionization and Tunneling Delay Time Measurements in Helium," *Science*, (2008) **322**, p. 1525; https://www.science.org/doi/epdf/10.1126/sciadv.adl6078.

473 And see P. Schach and E. Giese, "A Unified Theory of Tunneling Times Promoted by Ramsey Clocks," *Sci. Adv.*, (2024) **10**, eadl6078; https://www.science.org/doi/epdf/10.1126/sciadv.adl6078.

474 See, e.g., "A Decay: Tunnel Effect," *Radioactivity*, Institute of Nuclear Physics and Particle Physics, Dresden, https://radioactivity.eu.com/articles/phenomenon/tunnel_effect.

475 See, e.g., E. Merzbacher, "The Early History of Quantum Tunneling," *Phys. Today*, (2002) **55**, p. 44; https://doi.org/10.1063/1.1510281.

476 For a clear description, see hhttps://radioactivity.eu.com/articles/phenomenon/tunnel_effect.

477 E.g.: N. Moulin et al., "Tunnel Junction I(V) Characteristics: Review and a New Model for p-n Homojunctions," *HAL* (2020), https://hal.science/hal-03035269/document. For a somewhat broader review see S-D. Liang, *Quantum Tunneling and Field Electron Emission Theories* (Singapore: World Scientific Publishing, 2014); https://www.worldscientific.com/worldscibooks/10.1142/8663.

478 S. W. Hawking, "Quantum Black Holes," lecture in Cambridge, 1994, repr. in S. W. Hawking and R. Penrose, *The Nature of Space and Time*, n. 17, p. 57 and his figure 3.16; https://spacetimemodel.blogspot.com/2020/06/quantum-black-hole.html.

479 C. J. Gillespie, *Study of Superconductors by Electron Tunneling*, thesis, Monash University, 1967; https://www.academia.edu/78950186/Study_of_Superconductors_by_Electron_Tunneling.

480 S. Alexander et al., "Cosmological Bardeen–Cooper–Schrieffer Condensate as Dark Energy," *Phys. Rev. D*, (2010) **81**, 043511; https://arxiv.org/pdf/0906.5161.

481 S. M. O'Mahony et al., "On the Electron Pairing Mechanism of Copper-oxide High Temperature Superconductivity," *Proc. Natl. Acad. Sci. U.S.A.*, (2022) **119**, e2207449119; https://europepmc.org/article/med/36067325.

482 E.g., "Quantum Tunneling of Particles Through Potential Barriers," *LibreTexts*, 12 September 2022, https://phys.libretexts.org/Bookshelves/University_Physics/Book%3A_University_Physics_(OpenStax)/University_Physics_III_-_Optics_and_Modern_Physics_(OpenStax)/07%3A_Quantum_Mechanics/7.07%3A_Quantum_Tunneling_of_Particles_through_Potential_Barriers.

483 E.g., A. Marais et al., "The Future of Quantum Biology," *J. R. Soc. Interface*, (2018) **15**, p. 0640; https://royalsocietypublishing.org/doi/10.1098/rsif.2018.0640.

484 E.g., A. Jain et al., "Wireless Electrical-Molecular Quantum Signaling for Cancer Cell Apoptosis," *Nat. Nanotechnol.*, (2023), https://doi.org/10.1038/s41565-023-01496-y.

485 "The Nobel Prize in Physics 1986," Nobel Prize, https://www.nobelprize.org/prizes/physics/1986/summary/.

486 "Ivar Giaever," Nobel Prize, https://www.nobelprize.org/prizes/physics/1973/giaever/facts/.

487 C. J. Gillespie, *Study of Superconductors by Electron Tunneling*, n. 479, p. 111.

488 R. P. Feynman et al., "Curved Space," *The Feynman Lectures on Physics, Vol. II*, n. 353, ch. 42; https://www.feynmanlectures.caltech.edu/II_42.html.

489 J. D. Norton, "Gravity Near a Massive Body," *Einstein for Everyone*, 22 February 2017, https://sites.pitt.edu/~jdnorton/teaching/HPS_0410/chapters/general_relativity_massive/index.html.

490 As it was soon shown it could be: O. H. L. Heckmann, "Über die Metrik des sich ausdehnenden Universums," *Nachr. Gesell. Wiss. Göttingen, Math.-Phys. Klasse*, (1931) 2, p. 126.

491 As Einstein and de Sitter said in 1932 we could determine; A. Einstein and W. de Sitter, "On the Relation Between the Expansion and the Mean Density of the Universe," *Proc. Nat. Acad. Sci. USA*, (1932) 18, p. 213.

492 E. Di Valento et al., "Planck Evidence for a Closed Universe and a Possible Crisis for Cosmology," *Nature Astr.*, (2020) 4, p. 196; https://www.nature.com/articles/s41550-019-0906-9.

493 E.g., W. Hu, "Spatial Curvature," http://background.uchicago.edu/~whu/beginners/curvature.html.

494 The West Antarctic Ice Sheet's gravity raises the nearby sea level significantly: J. Bamber et al., "Reassessment of the Potential Sea-Level Rise from a Collapse of the West Antarctic Ice Sheet," *Science*, (2009) 324, p. 901; http://www.sciencemag.org/content/324/5929/901.full.pdf.

495 J. Michell, "On the Means of Discovering the Distance, Magnitude, &c. of the Fixed Stars, in Consequence of the Diminution of the Velocity of Their Light, in Case Such a Diminution Should Be Found to Take Place in Any of Them, and Such Other Data Should Be Procured from Observations, as Would Be Farther Necessary for That Purpose," *Phil. Trans. Roy. Soc.*, (1784) 74, p. 35.

496 See, e.g., A. May, "8 Ways We Know That Black Holes Really Do Exist," *Live Science*, 25 August 2021, https://www.livescience.com/how-we-know-black-holes-exist.html.

497 See, e.g., P. M. Sutter, "What Happens at the Center of a Black Hole?," Expert Voices, *Space*, 9 February 2022, https://www.space.com/what-happens-black-hole-center.

498 E.g., W. Clavin, "LIGO-Virgo Network Catches Another Neutron Star Collision," LIGO, January 6, 2020, https://www.ligo.caltech.edu/news/ligo20200106.

499 E.g., "Neutron Stars," *Imagine the Universe!*, March 2017, https://imagine.gsfc.nasa.gov/science/objects/neutron_stars1.html.

500 High Energy Advisory Panel, Quantum Universe Committee, *Quantum Universe: The Revolution in 21st Century Particle Physics* (Washington: National Science Foundation, 2004), p. v; https://science.osti.gov/-/media/hep/pdf/files/pdfs/quantum_universe_gr.pdf.

501 A. D. Linde, "A New Inflationary Universe Scenario: A Possible Solution of the Horizon, Flatness, Homogeneity, Isotropy and Primordial Monopole Problems," *Phys. Lett.*

B, (1982) **108**, p. 389; https://link.springer.com/chapter/10.1007/978-1-4613-2701-1_13; see chapter 21.

502 J. Updike, in an interview by J. Holt, c. 2013, quoted in J. Holt, *Why Does the World Exist?*, n. 752, p. 249.

503 P. M. Sutter, "How Did Inflation Happen?—And Why Do We Care?," Expert Voices, *Space*, 26 October 2018, https://www.space.com/42261-how-did-inflation-happen-anyway.html.

504 F. Wilczek, *Fundamentals*, n. 100, p. 157.

505 Figure from V. Ullman, "Gravitation and the Global Structure of the Universe," https://astronuclphysics.info/Gravitace5-5.htm.

506 Ibidem.

507 D. Scott, "The Standard Model of Cosmology: A Skeptic's Guide," n. 82.

508 See A. Guth, *The Inflationary Universe: The Quest for a New Theory of Cosmic Origins* (Reading MA: Helix Books, 1997); Russian cosmologist Alexei Starobinsky proposed the same idea a year or so earlier, but his paper went unnoticed at the time: A. A. Starobinsky, "Spectrum of Relict Gravitational Radiation and the Early State of the universe," *Pis'ma Zh. Eksp. Teor. Fiz.*, (1979) **30**, p. 719; *J. E. T. P. Lett.*, (1979) **30**, p. 682. He too was trying to solve another problem—the universe's origin as a mathematical point in the equations of the standard model.

509 Inflation was invented to explain why we see no magnetic monopoles, an esoteric entity predicted by theories some physicists would like to preserve. Actually, the theories don't need a lot of monopoles, and we might have just missed them. For a comprehensive but technical account, see S. Coleman, "The Magnetic Monopole Fifty Years After," in M. K. Gaillard and R. Stora, eds., *Meeting on Gauge Theories in High Energy Physics, Les Houches*, (Amsterdam: North-Holland Publishing Co., 1983), p. 461.

510 A. H. Guth, "Inflationary Universe: A Possible Solution to the Horizon and Flatness Problems," *Phys. Rev. D*, (1981) **23**, p. 347; https://journals.aps.org/prd/issues/23/2.

511 J. Gribbin, "Inflation for Beginners," (1996), *Cosmology for Beginners*.

512 P. J. E. Peebles, *The Whole Truth*, n. 60, p. 145.

513 P. M. Sutter, "How Did Inflation Happen?—And Why Do We Care?," n. 503.

514 Corrected for Earth's motion, the variation between two points in any direction is a few parts in ten thousand; see, e.g., E. L. Wright, "Cosmic Microwave Background Anisotropy," (2013), https://www.astro.ucla.edu/~wright/CMB-DT.html.

515 See, e.g., C. Cottier, "What Shape Is the Universe? As Far as Cosmologists Can Tell, Space Is Almost Perfectly Flat. But What Does This Mean?," *Astronomy*, 23 February 2021, https://astronomy.com/news/2021/02/what-shape-is-the-universe.

516 E.g., D. Clery, "Into the Dark," *Science*, (2023) **380**, p. 1212; https://www.science.org/doi/epdf/10.1126/science.add3433.

517 A. D. Miller, "A Measurement of the Cosmic Microwave Background from the High Chilean Andes," thesis, Princeton University, (2000), p. 100; https://dornsife.usc.edu/assets/sites/1/docs/about/Miller_dissertation.pdf.

518 C. L. Bennett et al., "Nine-Year Wilkinson Microwave Anisotropy Probe (WMAP) Observations: Final Maps and Results," *Astrophys. J. Supp.*, (2013) **208**, p. 20; https://iopscience.iop.org/article/10.1088/0067-0049/208/2/19/pdf.

519 J. D. Barrow, *The Origin of the Universe*, n. 465, p. 8.

520 E. J. Wollack, (2010), "Inflation," *WMAP's Universe*, National Aeronautics and Space Administration, https://wmap.gsfc.nasa.gov/universe/bb_cosmo_infl.html.

521 For a short summary, see A. Z. Jones and D. Robbins, "The Inflation Theory: Solving the Universe's Problems of Flatness and Horizon," *Dummies*, (Hoboken: John Wiley & Sons, 2021); https://www.dummies.com/education/science/physics/the-inflation-theory-solving-the-universes-problems-of-flatness-and-horizon/.

522 P. J. Steinhardt, "Paul Steinhardt disowns inflation, the theory he helped create: Is the theory at the heart of modern cosmology deeply flawed?," *Scientific American*, (2014) **304**, p. 36.

523 A. Ijjas, A. Loeb, and P. Steinhardt, "Inflationary Paradigm in Trouble after Planck 2013," *Phys. Lett. B.*, (2013) **723**, p. 261.

524 G. Lemaître, *The Primeval Atom*, n. 416, p. 38.

525 See, e.g., "The Greek Worldview," *Ideas of Cosmology*, Center for History of Physics, American Institute of Physics, (2022), https://history.aip.org/exhibits/cosmology/ideas/greekworldview.htm.

526 D. J. Furley, "The Greek Theory of the Infinite Universe," *J. Hist. Ideas*, (1981) **42**, p. 571; https://www.jstor.org/stable/2709119.

527 See, e.g., K. Easwaren et al., "Infinity," *Stanford Encyclopedia of Philosophy*, 29 April 2021, n. 99, https://plato.stanford.edu/entries/infinity/.

528 See, e.g., Giovanni Aquilecchia, "Giordano Bruno: Italian Philosopher," n. 96, https://www.britannica.com/biography/Giordano-Bruno.

529 I. Newton, *De Gravitatione et æquipondio fluidorum*, unpublished (c. 1666), MS Add. 4003, Cambridge University Library; J. Bennet, transl. and ed., "Descartes, Space and Body," (2013), § 8; https://www.earlymoderntexts.com/assets/pdfs/newton1666.pdf; and see, e.g., E. Harrison "Newton and the Infinite Universe," *Phys. Today*, (1986) **39**, p. 24; https://physicstoday.scitation.org/doi/abs/10.1063/1.881049.

530 S. W. Hawking, *A Brief History of Time* (New York: Bantam Books, 1998), p. 5.

531 J. D. Norton, "The Cosmological Woes of Newtonian Gravitation Theory," in H. Goenner et al., eds., *The Expanding Worlds of General Relativity (Einstein Studies Vol. 7)* (Boston: Birkhäuser, 1999), p. 271.

532 G. F. B. Riemann, "On the Hypotheses Which Lie at the Foundation of Geometry," n. 273, p. 36.

533 A. Einstein, "Kosmologische Betrachtungen zur allgemeinen Relativitatstheorie," (1917), *Sitzungsberichte der Königlichen Preussischen Akademie der Wissenschaften*, Berlin, p. 142; W. Perrett and G. B. Jeffery, transls., "Cosmological Considerations on the Theory of General Relativity," in H. Lorenz, H. Weyl and H. Minkowski, transls., *The Principle of Relativity* (New York: Dover Publications, 1952), p. 179.

534 Physicist Franz Selety challenged Einstein's reasoning; F. Selety, "Beiträge zum kosmologischen Problem," *Annal. d. Phys.*, (1922) **68**, 281. "Simply put, he provided a recipe for an infinitely large hierarchical universe, which contained an infinite mass of clustered stars filling the whole of space, yet with a zero average density and no special center." J. D. Barrow, *The Book of Universes* (London: Vintage Books, 2012), p. 89.

535 D. R. Topper, *How Einstein Created Relativity Out of Physics and Astronomy* (New York: Springer, 2013), p. 162.

536 A. Einstein, "On the Theory of Relativity," lecture at King's College, London, 13 June 1921; repr. in A. Einstein, *Essays in Science* (New York: Philosophical Library, 1934), p. 52.

537 E. Di Valento et al., "Planck Evidence for a Closed Universe and a Possible Crisis for Cosmology," n. 492.

538 Technically, the hypersurface of a hypersphere; see, e.g., J. G. Cramer, "Is the Universe a Hypersphere?," *Alternate View*, 17 January 2020, https://www.npl.washington.edu/av/altvw206.html.

539 P. C. W. Davies, *The Cosmic Jackpot: Why Our Universe Is Just Right for Life* (New York: Houghton Mifflin, 2007), p. 39.

540 For help in grasping this unintuitive idea, see P. M. Sutter, "Is There Anything Outside the Universe?," *Expert Voices*, 25 February 2022, https://www.space.com/whats-beyond-universe-edge.

541 C. O'Raifeartaigh et al., "Einstein's 1917 Static Model of the Universe: A Centennial Review," *Eur. Phys, J.*, (2017) **42**, p. 431.

542 R. Penrose, *The Road to Reality*, n. 64, p. 754.

543 See, e.g., K. Malley et al., "2nd Law of Thermodynamics," *Chemistry LibreTexts* (2020); https://chem.libretexts.org/Bookshelves.

544 Entropy is a fundamental concept for understanding the universe; there are many explanations out there. Here is a reasonably good one for the novice: "Entropy: The Hidden Force That Complicates Life," *FS*, (Ottawa: Farnam Street Media Inc., 2022), https://fs.blog/entropy/.

545 E.g., D. A. Leeming, *Creation Myths of the World: An Encyclopedia* (Santa Barbara: ABC-CLIO, 2010), p. 9.

546 C. Lineweaver, "The Entropy of the Universe and the Maximum Entropy Production Principle," in R. C. Dewar et al., eds., *Beyond the Second Law: Understanding Complex Systems* (Berlin: Springer-Verlag, 2014), p. 415; https://www.mso.anu.edu.au/~charley/papers/Chapter22Lineweaver.pdf.

547 J. D. Barrow, *New Theories of Everything*, n. 134, p. 189.

548 R. Penrose, *The Road to Reality*, n. 64, p. 726.

549 S. Carroll, *From Eternity to Here: The Quest for the Ultimate Theory of Time* (New York: Dutton, 2010), p. 301.

550 C. Rovelli, *The Order of Time*, n. 69, p. 31.

551 Ibid., p. 143.

552 Ibid., pp. 143, 156.

553 B. Greene, *The Fabric of the Cosmos: Space, Time and the Texture of Reality* (New York: Alfred A. Knopf, 2004), p. 174.

554 Ibidem; his italics.

555 G. Lemaître, "The Beginning of the World from the Point of View of Quantum Theory," n. 412.

556 J. Farrell, *The Day Without Yesterday*, n. 461, p. 106.

557 Ibid., p. 107; his italics.

558 R. Penrose, *Cycles of Time*, n. 66.

559 In case this religious impediment might not attract wide notice, he wrote his book (n. 416) *sub nom* Canon Georges Lemaître. Then, in 1951, the pope sealed the fate of this seminal idea as far as physics was concerned, by seeming to embrace it into Catholic dogma; Pius XII, "Discorso di Sua Santità Pio XII ai Cardinali, ai Legate delle Nazione

Estere e ai Soci della Pontificia Accademia delle Scienze," *Acta Apostolicae Sedis*, (Roma: Tipografia Poliglotta Vaticana, (1951) **44**, p. 31; http://www.vatican.va/holy_father/pius_xii/speeches/1951/documents/hf_p-xii_spe_19511122_di-serena_it.html.

560 F. Markopoulou, "The Internal Description of a Causal Set: What the Universe Looks Like from the Inside," *Comm. Math. Phys.*, (2000) **211**, p. 559.

561 L. Smolin, "A Theory of the Whole Universe," *Conversation: Universe* (New York: Edge, 1996), https://www.edge.org/conversation/lee_smolin-chapter-17-a-theory-of-the-whole-universe.

562 L. Smolin, *Three Roads to Quantum Gravity*, n. 61, p. 40.

563 It's ironic that a magic trick may be the only other happening that must have an observer; without them (or it) there *is* no illusion; see, e.g., J. Cami and L. M. Martinez, *The Illusionist Brain: The Neuroscience of Magic* (Princeton: Princeton University Press, 2022); also J. Cami et al., "On the Cognitive Bases for Illusionism: An Untapped Tool for Brain and Behavioural Research," https://www.preprints.org/manuscript/202001.0011/v1.

564 L. Gilder, *The Age of Entanglement: When Quantum Physics Was Reborn* (New York: Random House, 2008), p. 5.

565 J. S. Bell, "Quantum Mechanics for Cosmologists," n. 154.

566 L. Smolin, *The Life of the Cosmos* (Oxford: Oxford University Press, 1997), p. 244.

567 J. S. Bell, "Quantum Mechanics for Cosmologists," n. 154.

568 L. Smolin, *Three Roads to Quantum Gravity*, n. 61, p. 40.

569 A. Gefter, "How to Understand the Universe When You're Stuck Inside of It," n. 73.

570 L. Smolin, in A. Gefter, ibidem.

571 L. Smolin in R. M. Unger and L. Smolin, *The Singular Universe and the Reality of Time*, n. 62, p 371.

572 J. Baez, "Quantum Quandaries," in D. Rickles et al., eds., *The Structural Foundations of Quantum Gravity* (Oxford: Oxford University Press, 2006), p. 262.

573 M. Gleiser, "Consciousness, Quantum Physics and Reality," ICE, 11 January 2018, https://ice.dartmouth.edu/blog/consciousness-quantum-physics-and-reality.

574 M. Gleiser, "Could a Hidden Variable Explain the Weirdness of Quantum Physics?," Big Think, 1 February 2023, https://bigthink.com/13-8/quantum-entanglement-hidden-variable/.

575 I. Newton, letter to Richard Bentley, 25 February 1692 (or 93), Trinity College Library, Cambridge UK, 189.R.4.47, ff. 7-8; repr. in *The Newton Project*, October 2007, https://www.newtonproject.ox.ac.uk/view/texts/normalized/THEM00258.

576 See, e.g., G. Musser, "How Einstein Revealed the Universe's Strange 'Nonlocality,'" *Sci. Am.*, 1 November 2015, https://www.scientificamerican.com/article/how-einstein-revealed-the-universe-s-strange-nonlocality/.

577 For an accessible and more complete account, see G. Lowther, "Quantum Entanglement," *Almost Sure: A Random Mathematical Blog*, 19 September 2020, https://almostsuremath.com/2020/09/19/quantum-entanglement/.

578 R. Sorkin, "Does Locality Fail at Intermediate Length Scales?," in D. Oriti, (ed.), *Approaches to Quantum Gravity*, n. 6, p. 26; https://arxiv.org/abs/gr-qc/0703099.

579 J. Emspak, "Quantum Entanglement: A Simple Explanation," *Space*, 16 March 2022, https://www.space.com/31933-quantum-entanglement-action-at-a-distance.html.

580 L. Smolin, *Einstein's Unfinished Revolution*, n. 74, p. 48.

581 A. Aczel, *Entanglement: The Greatest Mystery in Physics* (Vancouver: Raincoast Books, 2002), p. xii.

582 A. Einstein et al., "Can Quantum-Mechanical Description of Physical Reality Be Considered Complete?," *Phys. Rev.*, (1935) **47**, p. 777; http://journals.aps.org/pr/pdf/10.1103/PhysRev.47.777.

583 See, e.g., B. Brubaker, "How Bell's Theorem Proved Spooky Action at a Distance Is Real," *Quanta Magazine*, 20 July 2021, https://www.quantamagazine.org/how-bells-theorem-proved-spooky-action-at-a-distance-is-real-20210720/.

584 The name *hidden variables* is unfortunate; Bell said it was absurd; J. S. Bell, "Are There Quantum Jumps?," in *Schrodinger: Centenary of a polymath* (Cambridge: Cambridge University Press, 1987), p. 172; repr. in *Speakable and Unspeakable in Quantum Mechanics*, n. 111, p. 201.

585 See, e.g., A. Muller, "What Is Quantum Entanglement? A Physicist Explains Einstein's 'Spooky Action at a Distance,'" *SciTech Daily*, 19 December 2022, https://scitechdaily.com/what-is-quantum-entanglement-a-physicist-explains-einsteins-spooky-action-at-a-distance/.

586 The gentle ridicule is even better in the original German: *spukhafte fernwirkungen*; A. Einstein, letter to M. Born, 3 March 1947; repr. in I. Born, transl., *The Born-Einstein Letters* (New York: Walker, 1971).

587 J. S. Bell, "Bertlmann's Socks and the Nature of Reality," *J. Physique*, (1981) **42**, C2 suppl. 3, p. 41, his emphasis; repr. in *Speakable and Unspeakable in Quantum Mechanics*, n. 111, p. 144.

588 E.g., A. Fine, "The Einstein-Podolsky-Rosen Argument in Quantum Theory," *Stanford Encyclopedia of Philosophy*, n. 99, https://plato.stanford.edu/entries/qt-epr/.

589 C. S. Wu and I. Shaknov, "Angular Correlation of Scattered Annihilation Radiation," *Phys. Rev.*, (1950) **77**, p. 136; https://journals.aps.org/pr/issues/77/1.

590 J. S. Bell, "On the Einstein-Podolsky-Rosen Paradox," *Physics*, (1964) **1**, p. 195; repr. in *Speakable and Unspeakable in Quantum Mechanics*, n. 111, p. 14.

591 A. Aspect et al., "Experimental Realization of Einstein-Podolsky-Rosen-Bohm Gedankenexperiment: A New Violation of Bell's Inequalities," *Phys. Rev. Lett.*, (1982) **49**, p. 91; https://journals.aps.org/prl/pdf/10.1103/PhysRevLett.49.91; see also A. Aspect et. al., "Experimental Test of Bell's Inequalities Using Time-varying Analyzers," *Phys. Rev. Lett.*, (1982) **49**, p. 1804; https://journals.aps.org/prl/pdf/10.1103/PhysRevLett.49.1804.

592 J. Handsteiner et al., "Cosmic Bell Test: Measurement Settings from Milky Way Stars," *Phys. Rev. Lett.*, (2017) **188**, 060401; https://journals.aps.org/prl/pdf/10.1103/PhysRevLett.118.060401.

593 S-R. Zhao et al., "Loophole-Free Test of Local Realism via Hardy's Violation," *Phys. Rev. Lett.*, (2024) **133**, 060201; https://journals.aps.org/prl/abstract/10.1103/PhysRevLett.133.060201.

594 J. J. Stachel, "Einstein and Bose," lecture in Calcutta, 1994, repr. in J. Stachel, *Einstein from "B" to "Z"* (Boston: Birkhäuser, 2002), p. 519.

595 E.g., D. Garisto, "The Universe Is Not Locally Real," *Sci. Am.*, (2022) **328**, p. 48, https://www.scientificamerican.com/article/the-universe-is-not-locally-real-and-the-physics-nobel-prize-winners-proved-it/.

596 A. Davour, C. Barnes, transl., "How Entanglement Has Become a Powerful Tool," (2022), Nobel Prize Outreach, https://www.nobelprize.org/prizes/physics/2022/popular-information/.

597 P. Ball, "The Science of the Inconceivable: A Different Kind of Logic," *Prospect*, 14 July 2016, https://www.prospectmagazine.co.uk/magazine/the-science-of-the-inconceivable.

598 L. Susskind, letter to colleagues, "Dear Qubitzers, GR=QM," (2017), https://arxiv.org/pdf/1708.03040.pdf.

599 R. Webb, "Quantum Physics: Our Best Picture of How Particles Interact to Make the World," *New Scientist*, https://www.newscientist.com/definition/quantum-physics/.

600 R. V. Buniy and S. D. H. Hsu, "Everything Is Entangled," *Phys. Lett. B*, (2012) **718**, p. 233; https://www.sciencedirect.com/science/article/pii/S037026931200994X#.

601 B. Desjarlais, quoted in *The Globe and Mail*, Toronto, 22 November 2021, p. A4.

602 There is a far-flung literature on indigenous worldviews. E.g., "A common strand . . . is the holism of Māori cosmology. That is, the inherent connectedness and interdependence of all things." M. Cheung, "The Reductionist-Holistic Worldview Dilemma," *MAI Review*, (2000) **3**, p. 2; archived at http://www.review.mai.ac.nz/mrindex/MR/article/view/186/196.html.

603 This is the essential method of modern physics. It now has many pieces and can't understand any of them. Tolkien had his wizard Gandalf say, "He that breaks a thing to find out what it is has left the path of wisdom." J. R. R. Tolkien, *Lord of the Rings* (London: HarperCollins, 2007), vol. 1, p. 337.

604 L. Smolin, *The Life of the Cosmos*, n. 566, p. 33.

605 J. Newsum, "Disaggregation," *Stratechi*, (2022), https://www.stratechi.com/disaggregation/.

606 There is a tantalizing link between the math of one kind of spacetime in relativity's equations (anti–de-Sitter space) and a quantum theory (called conformal field theory); they can flip back and forth with the same math (called AdS/CFT): J. M. Maldacena, "The Large N Limit of Superconformal Field Theories and Supergravity," *Int. J. Theor. Phys.*, (1999) **38**, p. 1113; https://arxiv.org/pdf/hep-th/9711200.

607 D. Bohm, "Hidden Variables and the Implicate Order," in B. J. Hiley and E. D. Pitt, eds., *Quantum Implications: Essays in Honour of David Bohm* (London: Routledge & Kegan Paul, 1987), p. 33.

608 C. Rovelli, *Reality Is Not What It Seems*, n. 7, p. 148.

609 L. Smolin, *The Trouble with Physics*, n. 70, p. 5.

610 A. Einstein, "On the Method of Theoretical Physics," n. 114, p. 12.

611 He was speaking of gravity and electrodynamics; A. Einstein, Nobel Prize lecture, *Nobel Lectures, Physics 1901–1921* (Amsterdam: Elsevier Publishing Company, 1967), p. 489; https://www.nobelprize.org/uploads/2018/06/einstein-lecture.pdf.

612 A. Einstein, "Considerations Concerning the Fundaments of Theoretical Physics," *Science*, (1940) **91**, p. 487; https://www.science.org/doi/epdf/10.1126/science.91.2369.487.

613 L. Smolin, "Space and Time in the Quantum Universe," paper for the 1988 Osgood Hill Conference, North Andover, in A. Ashtekar and J. Stachel, eds., *Conceptual Problems of Quantum Gravity* (Basel: Birkhauser, 1991), p. 228.

614 L. Susskind, quoted in A. Becker, "The Origins of Space and Time," n. 51.

615 The reader who wants to pursue this theme might try physicist Heinrich Päs's recent book, *The One: How an Ancient Idea Holds the Future of Physics* (New York: Basic Books, 2023).

616 C. Rovelli, *The Order of Time*, n. 69, p. 45.

617 See, e.g., "There Is No Now," PBS, 30 June 2014, https://www.pbs.org/wgbh/nova/article/there-is-no-now/; excerpted from M. Gleiser, *The Island of Knowledge: The Limits of Science and the Search for Meaning* (New York: Basic Books, 2014).

618 A. Einstein, *Relativity*, n. 245, p. 170.

619 L. Gilder, *The Age of Entanglement* (New York: Alfred A. Knopf, 2008), p. 19.

620 He defined time as what is measured by a clock; see A. Einstein, *Relativity*, n. 245, chapter 8, "On the Idea of Time in Physics."

621 C. Rovelli, *The Order of Time*, n. 69, p. 197.

622 Proponents conflate special relativity with reality; e.g., V. Petkov, "Is There an Alternative to the Block Universe?," https://philsci-archive.pitt.edu/2408/1/Petkov-BlockUniverse.pdf.

623 See, e.g., K. Miller, "The Block Universe Theory, Where Time Travel Is Possible but Time Passing Is an Illusion," *ABC News*, 1 September 2018, https://www.abc.net.au/news/science/2018-09-02/block-universe-theory-time-past-present-future-travel/10178386.

624 H. Arendt, *Between Past and Future* (London: Penguin, 2006), p. 10.

625 L. Infeld, *Albert Einstein: His Work and Its Influence on Our World* (New York: Charles Scribner's Sons, 1950), p. 23.

626 G. Holton, "Einstein's Search for the 'Weltbild,'" n. 118, p. 6.

627 G. Amelino-Camelia, "Testable Scenario for Relativity with Minimum-Length," *Phys. Lett.*, (2001) **B510**, p. 255; https://arxiv.org/abs/hep-th/0012238; and see J. Kowalski-Glikman, "Introduction to Doubly Special Relativity," *Lect. Notes Phys.*, (2005) **669**, p. 131; https://arxiv.org/pdf/hep-th/0405273.pdf.

628 When rightly understood in terms of observing moving rods and clocks, the paradoxes make sense; see, e.g., J. D. Norton, "Is Special Relativity Paradoxical?," *Einstein for Everyone*, 20 January 2022, https://sites.pitt.edu/~jdnorton/teaching/HPS_0410/chapters/.

629 The corresponding school of philosophy is *presentism*; see M. Hinchliff, "The Puzzle of Change," *Phil. Perspec.*, (1996) **10**, p. 119; https://www.jstor.org/stable/2216239.

630 The movie took seven Oscars at the 95th Academy Awards in 2023, but does not seek to literally live up to its name; https://www.imdb.com/title/tt6710474/.

631 L. Smolin, *Einstein's Unfinished Revolution*, n. 74, p. 265.

632 Rovelli listed eight notions of time in C. Rovelli, "Halfway Through the Woods: Contemporary Research on Space and Time," n. 755.

633 E.g., B. Cotterell et al., "Ancient Egyptian Water-Clocks: A Reappraisal," *J. Archaeol. Sci.*, (1986) **13**, p. 31; https://www.sciencedirect.com/science/article/abs/pii/0305440386900257.

634 I. B. Cohen, "A Brief History of the Principia," in I. Newton, *The Principia*, n. 214, p. 2.

635 I. Newton, *The Principia*, n. 214; A. Motte, transl., *The Mathematical Principles of Natural Philosophy* (London: Benjamin Motte, 1729), p. 77; https://redlightrobber.com/red/links_pdf/Isaac-Newton-Principia-English-1846.pdf.

636 J. C. Hafele and R. E. Keating, "Around-the-World Atomic Clocks: Predicted Time Gains," *Science*, (1972) **177**, p. 166; https://www.science.org/doi/pdf/10.1126/science.177.4044.166.

637 B. Fullerton, "Time Is Not Absolute," *Chemistry LibreTexts*, 9 September 2020, https://phys.libretexts.org/Bookshelves/Conceptual_Physics/Book%3A_Conceptual_Physics_(Crowell)/08%3A_Relativity/8.01%3A_Time_Is_Not_Absolute.

638 See, e.g., n. 151.

639 Most notably in R. M. Unger and L. Smolin, *The Singular Universe and the Reality of Time*, n. 62.

640 S. J. Carlip, "Causal Sets and an Emerging Continuum," *General Relativity and Gravitation*, (in press, 2024); https://link.springer.com/article/10.1007/s10714-024-03281-1.

641 R. Sorkin, "Does Locality Fail at Intermediate Length-Scales," n. 578, p. 26; and see S. J. Carlip, "Causal Sets and an Emerging Continuum," *General Relativity and Gravitation*, (in press); https://arxiv.org/pdf/2405.14059.

642 L. Smolin, *Einstein's Unfinished Revolution*, n. 74, p. 258.

643 Ibid., p. 265.

644 Ibid., p. 266.

645 M. Cortes and L. Smolin, "The Universe as a Process of Unique Events," *Phys. Rev. D*, (2014) **90**, p. 044035; https://arxiv.org/abs/1307.6167.

646 R. M. Unger and L. Smolin, *The Singular Universe and the Reality of Time*, n. 62.

647 He got there seeming to conflate time with causation. L. Smolin, *Einstein's Unfinished Revolution*, n. 74, p. 204.

648 Rovelli tabulated ten "notions" of time in C. Rovelli, *Quantum Gravity*, n. 41, Table 2.1, p. 86. At p. 82 he said, "The main point I intend to emphasize is that a single, clear and pure notion of 'time' does not exist."

649 L. Smolin, *Einstein's Unfinished Revolution*, n. 74, p. 260; see also D. A. Meyer, "The Dimension of Causal Sets," thesis, M.I.T., (1989), "The proposal is that the substance, or structure, underlying spacetime is what Riemann might have called an ordered discrete manifold; we will refer to it as a causal set."; https://www.researchgate.net/publication/37601708_The_dimension_of_causal_sets.

650 See, e.g., The Editors of Encyclopaedia Britannica, "speed of light," *Encyclopaedia Britannica*, n. 96, https://www.britannica.com/science/speed-of-light.

651 See, e.g., R. Rynasiewicz, "Newton's Views on Space, Time and Motion," *Stanford Encyclopedia of Philosophy*, n. 99, https://plato.stanford.edu/entries/newton-stm/.

652 G. Gamow, "The Evolutionary Universe," *Sci. Am.*, (1956) **195**, p. 136. Regarding controversy over the origin (Einstein or Gamow) of the words, see D. R. Topper, *How Einstein Created Relativity Out of Physics and Astronomy*, n. 535, p. 165; and the *The New York Review*, 8 May 2014, letters; https://www.nybooks.com/articles/2014/05/08/einsteins-blunder/.

653 A. Einstein, "Cosmological Considerations on the Theory of General Relativity," n. 533, p. 179.

654 C. Rovelli, "Making Mistakes Is a Sign of Intelligence—and Einstein's Errors Prove It: Einstein's Flip-flopping Was a Vital Part of His Learning Process," *i News*, 12 November 2020, https://inews.co.uk/news/science/carlo-rovelli-making-mistakes-sign-of-intelligence-einsteins-errors-757427.

655 A. Einstein, letter to Paul Ehrenfest, 26 December 1915, transl. Ann M. Hentschel, *Collected Papers of Albert Einstein, Vol. 8, The Berlin Years: Correspondence, 1914–1918* (Princeton: Princeton University Press, 1998), p. 167; http://einsteinpapers.press.princeton.edu/vol8-trans/195.

656 A. Einstein, "Cosmological Considerations on the Theory of General Relativity," n. 533, p. 179.

657 E.g., A. Mann, "What Is the Cosmological Constant? Not a Blunder Anymore," *Live Science*, 16 February 2021, https://www.livescience.com/cosmological-constant.html.

658 E.g., S. Carroll, "The Cosmological Constant," *Living Rev. Relativity*, (2001) **4**, p. 1: "The cosmological constant turns out to be a measure of the energy density of the vacuum—the state of lowest energy."

659 A. Einstein, letter to Georges Lemaître, 26 September 1947; *Albert Einstein Archives*, # 15–85, The Hebrew University of Jerusalem.

660 Quoted in D. E. Vincent and D. R. Topper, "Einstein's Oxford Cosmology Blackboards: Open Portals to 1931," *Eur. Phys. J. H*, (2022) **47**, p. 14.

661 See, e.g., B. Greene, *The Hidden Reality: Parallel Universes and the Deep Laws of the Cosmos* (New York: Alfred A. Knopf, 2011), p. 129; and J. Earman, "Lambda: The Constant That Refuses to Die," *Arch. Hist. Exact Sci.*, (2001) **55**, p. 189; https://www.jstor.org/stable/41134106.

662 E.g., S. Weinberg, "The Cosmological Constant Problem," *Rev. Mod. Phys.*, (1989) **61**, p. 1; https://journals.aps.org/rmp/abstract/10.1103/RevModPhys.61.1.

663 See, e.g., comments in P. J. Steinhardt and N. Turok, "Why the Cosmological Constant Is Small and Positive," *Science*, (2006) **312**, p. 1180; https://www.science.org/doi/10.1126/science.1126231.

664 R. Lea, "A New Generation Takes on the Cosmological Constant," *Physics World*, (2021) **34**, p. 42.

665 See, e.g., G. Mariani, "Henrietta Leavitt—Celebrating the Forgotten Astronomer," American Association of Variable Star Observers, https://www.aavso.org/henrietta-leavitt-%E2%80%93-celebrating-forgotten-astronomer.

666 See, e.g., U. S. Department of Energy, "DOE Explains...Supernovae," https://www.energy.gov/science/doe-explainssupernovae.

667 Two teams arrived at this result independently: S. Perlmutter et al., "Measurements of Ω and Λ from 42 High-Redshift Supernovae," *Astrophys. J.*, (1999) **517**, p. 565, and A.G. Riess et al., "Observational Evidence from Supernovae for an Accelerating Universe and a Cosmological Constant," *Astronom. J.*, (1998) **116**, p. 1009.

668 A. Wright, "Nobel Prize 2011: Perlmutter, Schmidt and Riess," *Nat. Phys.*, (2011) **7**, p. 833; https://www.nature.com/articles/nphys2131. The prize was for showing the expansion is accelerating.

669 P. J. E. Peebles and B. Ratra, "The Cosmological Constant and Dark Energy," *Rev. Mod. Phys.*, (2003) **75**, p. 559; https://arxiv.org/pdf/astro-ph/0207347.

670 P. J. E. Peebles, lecture in Winnipeg, n. 405; see also P. J. E. Peebles, "Growth of the Nonbaryonic Dark Matter Theory," *Nature Astron.*, (2017) **1**, 0057, and P. J. E. Peebles, "How the Nonbaryonic Dark Matter Theory Grew," 20 January 2017, https://arxiv.org/pdf/1701.05837.pdf.

671 See, e.g., "Planck Mission Brings Universe into Sharp Focus," News Features and Press Releases, NASA, 21 March 2013, https://www.nasa.gov/mission_pages/planck/news/planck20130321.html.

672   A. Einstein, in conversation with Werner Heisenberg—W. Heisenberg, *Der Teil und das Ganze: Gespräche im Umkreis der Atomphysik* (München: R. Piper & Co. Verlag, 1926), p. 91—said, "Erst die Theorie entscheider darüber, was man beobachten kann."; A. J. Pomerans, transl., *Physics and Beyond: Encounters and Conversations* (New York: Harper Torchbooks, 1972), p. 63.

673   J. D. Chapman and C. J. Gillespie, "Radiation-Induced Events and Their Time Scale in Mammalian Cells," *Advances in Radiation Biology* (New York: Academic Press, 1981), Vol. 9, p. 143; https://www.sciencedirect.com/science/article/abs/pii/B9780120354092500102.

674   D. L. Dugle et al., "DNA Strand Breaks, Repair and Survival in X-Irradiated Mammalian Cells," *Proc. Natl. Acad. Sci. U.S.A.*, (1976) **73**, p. 809; https://www.pnas.org/doi/abs/10.1073/pnas.73.3.809.

675   E.g., C. J. Gillespie et al., "The Inactivation of Chinese Hamster Cells by X Rays: Synchronized and Exponential Cell Populations," *Rad. Res.*, (1975) **64**, p. 353; https://www.jstor.org/stable/3574271.

676   E.g., C. J. Gillespie et al., "DNA Damage and Repair in Relation to Mammalian Cell Survival: Implications for Microdosimetry," in J. Booz et al., eds., *Fifth Symposium on Microdosimetry*, Verbania Pallanza, 1975, (Brussels: Euratom, 1976), p. 799.

677   E.g., E. J. Hall and A. J. Giaccia, *Radiobiology for the Radiologist* (Philadelphia: Lippincott Williams & Wilkins, 2018), 8th ed., ch. 19.

678   J. D. Chapman and C. J. Gillespie, "The Power of Radiation Biophysics—Let's Use It," *Int. J. Radiat. Oncol. Biol. Phys.*, (2012) **84**, p. 309; https://www.redjournal.org/article/S0360-3016(12)00567-6/fulltext.

679   T. Freeman, "Hypofractionated Radiotherapy: Faster, Simpler and Equally Effective," *Physics World*, 5 October 2023, https://physicsworld.com/a/hypofractionated-radiation-therapy-faster-simpler-and-equally-effective/.

680   See, e.g., D. N. Spergel, "The Dark Side of Cosmology: Dark Energy and Dark Matter," *Science*, (2015) **347**, p. 1100; https://www.science.org/doi/10.1126/science.aaa0980.

681   A. Einstein, "On a Stationary System with Spherical Symmetry Consisting of Many Gravitating Masses," *Ann. Math.*, (1939) **40**, p. 922; https://www.jstor.org/stable/1968902.

682   W. S. Churchill, *The Second World War* (Boston: Houghton Mifflin Company, 1950), Vol. VI, p. 445.

683   For the story of the missing-mass mystery, see, e.g., R. Panek, *The 4% Universe: Dark Matter, Dark Energy, and the Race to Discover the Rest of Reality* (Boston: Houghton Mifflin Harcourt, 2011).

684   For the dark-matter story, see K. Freese, *The Cosmic Cocktail: Three Parts Dark Matter* (Princeton: Princeton University Press, 2014).

685   For the dark-energy story, see H. Kragh and J. M. Overduin, *The Weight of the Vacuum: A Scientific History of Dark Energy* (Heidelberg: Springer, 2014).

686   N. Aghanim et al., "Planck 2018 Results: VI. Cosmological Parameters," *Astron. Astrophys.*, (2020) **641**, p. A6; https://www.researchgate.net/publication/344382189_Planck_2018_results_VI_Cosmological_parameters; Table 2, column 2, shows dark energy (Ω_Λ) is some 67.9% of the universe.

687   S. Perkowitz, "Gedankenexperiment," n. 96, https://www.britannica.com/science/Gedankenexperiment.

688   See, e.g., G. Segre, "What Scientific Concept Would Improve Everybody's Cognitive Toolkit?," (2011), https://www.edge.org/response-detail/10157.

689   In somewhat different style, it was known in ancient Greece; see, e.g., N. Rescher, "Thought Experiment in Pre-Socratic Philosophy," in T. Horowitz and G. J. Massey, eds., *Thought Experiments in Science and Philosophy* (Lanham MD: Rowman & Littlefield, 1991), p. 31.

690   J. R. Brown, "Thought Experiment," *Stanford Encyclopedia of Philosophy*, n. 99, https://plato.stanford.edu/entries/thought-experiment/.

691   For examples, see M. Cohen, *Wittgenstein's Beetle and Other Classic Thought Experiments*, (Hoboken NJ: Wiley-Blackwell, 2005).

692   J. Witt-Hansen, "H. C. Ørsted, Immanuel Kant and the Thought Experiment," *Danish Yearbook of Philosophy*, (1976) **13**, p. 48.

693   E. Helms, "Oersted, Mach, and the History of 'Thought Experiment,'" *Brit. J. Hist. Phil.*, (2022) **30**, p. 837; https://www.academia.edu/72142163/Orsted_Mach_and_the_History_of_Thought_Experiment.

694   See, e.g., J. Mehra, *The Golden Age of Theoretical Physics*, (2 vols.), (Singapore: World Scientific Publishing Co., 2001).

695   In his 1952 autobiography, Max von Laue wrote, "The freedom of science which we lost in 1933 we have not yet regained."; in H. Hartmann, *Schöpfer des neuen Weltbildes* (Bonn: Athenäum Verlag, 1952); P. P. Ewald, ed., P. P. Ewald and R. Bethe, transls., *50 Years of X-ray Diffraction*, p. 303; http://ww1.iucr.org/iucr-top/publ/50YearsOfXrayDiffraction/von_laue.pdf. The degree of German dominance in physics thought and practice before that time is hard to imagine now; it declined in the wake of World War I, see, e.g., C. Jungnickel and R. McCormmach, *Intellectual Mastery of Nature: Theoretical Physics from Ohm to Einstein, Vol. 2, The Now Mighty Theoretical Physics 1870–1925* (Chicago: University of Chicago Press, 1986), pp. 348 ff.

696   It is not always at its best, particularly as used in philosophy; for a consideration of the difference, see J. Peijnenburg and D. Atkinson, "When Are Thought Experiments Poor Ones?," *J. Gen. Phil. Sci.*, (2003) **34**, p. 305; https://sites.ualberta.ca/~francisp/Phil488/PeijenburgAtkinsonPoorThoughtExps03.pdf.

697   See, e.g., J. Baggini, "Plato Got Virtually Everything Wrong," *Prospect*, 20 September 2018, https://www.prospectmagazine.co.uk/magazine/plato-got-virtually-everything-wrong.

698   Much of Plato's greater influence compared with other early philosophers may be due to written records of his works surviving; see, e.g., J. M. Cooper, ed., *Plato: Complete Works* (Indianapolis: Hackett Publishing Company, 1997).

699   The single exception on record is his verification—working with Lorentz's son-in-law, Wander de Haas—of the Einstein-de Haas effect; A. Einstein, "Experimenteller Nachweis der Ampèreschen Molekularströme" (Experimental Proof of Ampère's Molecular Currents), *Naturwissenschaften*, (1915) **3**, p. 237; see also D. R. Topper, *Quirky Sides of Scientists: True Tales of Ingenuity and Error from Physics and Astronomy* (New York: Springer, 2007), p. 10.

700   See, e.g. (re direct economic value), Center for Economics and Business Research, "The Importance of Physics to the Economies of Europe," European Physical Society, September 2019, https://cdn.ymaws.com/www.eps.org/resource/resmgr/policy/eps_pp_physics_ecov5_full.pdf.

701   See, e.g., B. Hepburn, "Scientific Method," n. 99.

702   E.g., N. Rescher, "Thought Experiment in Pre-Socratic Philosophy," n. 689, p. 31.

703   Inconsistency or antinomy is a tool in philosophy that may be useful in a gedanken experiment, though it does not guarantee the product is real; see, e.g., I. Kant, *Critique of Pure Reason*, n. 404, Ch. 2.

704   D. Frauchiger and R. Renner, "Quantum Theory Cannot Consistently Describe the Use of Itself," *Nature Comm.*, (2018) 9, p. 3711; https://www.nature.com/articles/s41586-023-06839-2. (The paper evoked a bunch of nonsense I shall not go into here.)

705   S. W. Hawking and L. Mlodinow, *The Grand Design*, n. 430, p. 5.

706   F. Bacon, *Novum Organum, sive Indicia Vera de Interpretatione Naturae* (1620); unknown transl., *New Way of Reasoning, or True Direction for the Interpretation of Nature, Book I, On the Interpretation of Nature and the Empire of Man* (New York: P. F. Collier & Son, 1902), aph. VI; https://gutenberg.org/cache/epub/45988/pg45988.txt

707   S. Weinstein, "Quantum Gravity," (2019), *Stanford Encyclopedia of Philosophy*, n. 99, https://plato.stanford.edu/entries/quantum-gravity/.

708   For more on the first two see L. Smolin, *Three Roads to Quantum Gravity*, n. 61.

709   Even compactifying the math is problematic; see, e.g., R. Garisto, "Curling Up Extra Dimensions in String Theory," *Phys. Rev. Focus*, (1998) 1, p. 7; https://physics.aps.org/story/v1/st7.

710   For an anguished critique by a former leading practitioner, see L. Smolin, *The Trouble with Physics: The Rise of String Theory, the Fall of a Science, and What Comes Next*, n. 70.

711   Or spacetime; see, e.g., N. Huggett and C. Wüthrich, in N. Huggett and C. Wüthrich, eds., *Out of Nowhere: The Emergence of Spacetime in Quantum Theories of Gravity* (Oxford: Oxford University Press, 2025, in press), ch. 9; https://philsci-archive.pitt.edu/17204/1/HuggettOON9.pdf.

712   See, e.g., C. Rovelli, *Reality Is Not What It Seems: The Journey to Quantum Gravity*, n. 7, p. 159.

713   C. Rovelli, "Loop Quantum Gravity," *Living Rev. Relativ.*, (1998) 1, p. 1; https://pubmed.ncbi.nlm.nih.gov/28937180/.

714   L. Smolin, *Time Reborn: From the Crisis in Physics to the Future of the Universe* (New York: Mariner Books, 2014), p. 82.

715   R. D. Sorkin, "Causal Sets: Discrete Gravity," in A. Gomberoff and D. Marolf, eds., *Proceedings of the Summer School*, Valdivia, Chile, (2002), (New York: Springer, 2003), p. 305; https://arxiv.org/abs/gr-qc/0309009.

716   F. Dowker, "Causal sets and the deep structure of spacetime," in A. Abhay, ed., *100 Years of Relativity: Space-Time Structure—Einstein and Beyond* (Singapore: World Scientific Publishing Co., 2005), p. 445; https://arxiv.org/pdf/gr-qc/0508109.pdf.

717   Ibidem.

718   Ibid., at p. 448; original emphasis.

719   A. Einstein, letter to John Moffat, 4 June 1953; dup. in Mudd Library, Princeton University, Item 17-390; quoted in R. B. Salgado, *Toward a Quantum Dynamics for Causal Sets*, thesis, University of Chicago, (2008), p. 190, referring to attempts to quantize gravity.

720   G. W. Leibniz, "Principles of Nature and of Grace, Based on Reason," n. 188, p. 1038.

721 Y. Y. Melamed and M. Lin, "Principle of Sufficient Reason," 7 September 2017, *Stanford Encyclopedia of Philosophy*, n. 99, https://plato.stanford.edu/entries/sufficient-reason/.

722 G. W. Leibniz, "The Monadology," n. 189, §§ 31 and 32.

723 Oddly, the connection is not widely noted; but see R. J. Johnson, "The Problem: The Theory of Ideas in Ancient Atomism and Gilles Deleuze," thesis, Duquesne University, (2013); https://dsc.duq.edu/etd/706/.

724 T. Carus, *De Rerum Natura*, (c. 60 BC), W. E. Leonard, transl., *Of the Nature of Things* (New York: E.P. Dutton & Co., 1916), Book 1.

725 See, e.g., A. Haynes, "Essence, Existence, and Necessity: Spinoza's Modal Metaphysics," (2012), p. 36, University of Rhode Island, Senior Honors Projects, paper 345; https://digitalcommons.uri.edu/srhonorsprog/345.

726 L. Smolin, *Einstein's Unfinished Revolution*, n. 74, p. 233.

727 Ibid., p. 229 ff.

728 See, e.g., D. Yates, "Emergence, Downward Causation and the Completeness of Physics," *Phil. Q.*, (2009) **59**, p. 110.

729 See, e.g., P. Forrest, "The Identity of Indiscernibles," (2010), *Stanford Encyclopedia of Philosophy*, n. 99, https://plato.stanford.edu/entries/identity-indiscernible/.

730 J. Jeans, *Physics and Philosophy* (Cambridge: Cambridge University Press, 1943), p. 81.

731 There is a literature around this idea; see, e.g., H. Pās, *The One: How an Ancient Idea Holds the Future of Physics* (New York: Basic Books, 2023).

732 T. de Chardin, *The Phenomenon of Man*, n. 81, p. 44; original capitalization.

733 R. M. Unger in R. M. Unger and L. Smolin, *The Singular Universe and the Reality of Time*, n. 62, p. 359.

734 See, e.g., C. Rovelli, "The Strange Equation of Quantum Gravity," (2015), https://arxiv.org/pdf/1506.00927.

735 R. M. Unger in R. M. Unger and L. Smolin, *The Singular Universe and the Reality of Time*, n. 62, p. 194.

736 L. Smolin, *The Life of the Cosmos*, n. 566, p. 14.

737 Ibid., p. 19.

738 A. Guth, "Was Cosmic Inflation the Bang of the Big Bang?," *Beamline*, (1997) **27**, p. 14; https://ned.ipac.caltech.edu/level5/Guth/Guth_contents.html.

739 N. Aghanim et al., "Planck 2018 Results: VI. Cosmological Parameters," n. 686; Dark Energy (Ω_Λ) is some 67.9% of the universe.

740 J. C. Smuts, "The Scientific World Picture of Today," address to the British Association for the Advancement of Science upon taking office as its president, 24 October 1931; repr. in "Evolution of the Universe," Supplement to *Nature*, (1931) **128**, p. 718.

741 A. Einstein and L. Infeld, *The Evolution of Physics: The Growth of Ideas from Early Concepts to Relativity and Quanta*, n. 130, p. 3.

742 Ibid., p. 76; my emphasis.

743 Ibid., p. 3.

744 A. Einstein, *Relativity*, n. 245, p. 178; my emphasis.

745 See, e.g., T. Folger, "Einstein's Grand Quest for a Unified Theory," *Discover*, 29 September 2004, https://www.discovermagazine.com/the-sciences/einsteins-grand-quest-for-a-unified-theory.

746 G. Simenon, *Le Voleur de Maigret* (Paris: Presses de la Cité, 1967); N. Ryan, transl., *Maigret and the Pickpocket* (San Diego: Harcourt Brace, 1967), p. 4.

747 G. Simenon, *Félicie est Là* (Paris: Gallimard, 1944); E. Ellenbogen, transl., *Maigret and the Toy Village* (San Diego: Harcourt Brace, 1978), p. 33.

748 G. Simenon, *L'Inspecteur Cadavre* (Paris: Gallimard, 1944); H. Thomson, transl., *Maigret's Rival* (San Diego: Harcourt Brace, 1979), p. 133.

749 This investigative strategy is developed in perhaps the most explicit detail in *Maigret à New York* (Paris: Presses de la Cité, 1947); Adrienne Foulke, transl., *Maigret in New York's Underworld* (Garden City, NY: Doubleday, 1955)—a story that also seeks simplicity—with passages like, "All this added up to nothing. These were not even thoughts." And, "No good to run after the truths. The only thing is to become permeated by the pure and simple truth."

750 J. S. Bell and M. Nauenberg, "The Moral Aspect of Quantum Mechanics," in A. De Shalit et al., eds., *Preludes in Theoretical Physics*, (1966), p. 279; repr. in J. S. Bell, *Speakable and Unspeakable in Quantum Mechanics*, n. 111, p. 22.

751 H. Reichenbach, *The Rise of Scientific Philosophy*, n. 337, p. vii.

752 J. Updike, in an interview by J. Holt, c. 2013, quoted in J. Holt, *Why Does the World Exist? An Existential Detective Story* (London: Profile Books, 2012), p. 249.

753 W. V. Quine, "What Is It All About?," lecture at Mount Holyoke College, April 1980; repr. as "Things and Their Place in Theories" in W. V. Quine, *Theories and Things*, n. 13, p. 17.

754 L. Susskind, *The Black Hole War*, n. 279, p. 314.

755 C. Rovelli, "Halfway Through the Woods: Contemporary Research on Space and Time," in J. Earman and J. D. Norton, eds., *The Cosmos of Science: Essays of Exploration* (Pittsburgh: University of Pittsburgh Press, 1997), p. 180; his emphasis.

756 R. P. Feynman, *The Character of Physical Law*, n. 269, p. 166.

757 A. Einstein, letter to Walter Dällenbach, after 15 February 1917, in A. M. Hentschel, transl., *The Collected Papers of Albert Einstein, Vol. 8, The Berlin Years: Correspondence, 1914–1918* (Princeton: Princeton University Press, 1995), p. 285.

758 A. Einstein, letter to Paul Langevin, 3 October 1935, quoted in and transl. by J. Stachel, "Einstein and the Quantum: Fifty Years of Struggle," in R. G. Colodny, ed., *From Quarks to Quasars: Philosophical Problems in Modern Physics* (Pittsburgh: University of Pittsburgh Press, 1986), p. 380.

759 A. Einstein, letter to David Bohm, 28 October 1954, repr. in J. Stachel, "Einstein and the Quantum: Fifty Years of Struggle," ibid., p. 349.

760 The idea came (he later told me) as he read a book by James P. Hogan, *Voyage from Yesteryear* (Harmondsworth: Penguin Books, 1984), pp. 232 ff. (The idea's clearly his, not Hogan's.)

761 See "arXiv.org," Cornell University Library, https://engineering.library.cornell.edu/database/arxiv-org/.

762 S. O. Bilson-Thompson, "A topological model of composite preons," (2005), https://arxiv.org/abs/hep-ph/0503213.

763 L. Smolin, *The Trouble with Physics*, n. 70, p. 254.

764 E.g., S. O. Bilson-Thompson, F. Markopoulou, and L. Smolin, "Quantum Gravity and the Standard Model," *Class. Quantum Gravity*, (2007) **24**, p. 3975; https://arxiv.org/abs/hep-th/0603022.

765 Figure is a detail from Sundance Bilson-Thompson's Figure 1 in n. 762.

766 Figure from Sundance Bilson-Thompson, n. 878.

767 S. Bilson-Thompson, personal communication, 4 September 2024.

768 L. Smolin, *The Trouble with Physics*, n. 70, p. 254.

769 M. Lazaridis, "Address by Mike Lazaridis, PI Chair, to the Public Policy Forum," 8 April 2009, *Perimeter Institute*, https://perimeterinstitute.ca/news/address-mike-lazaridis-pi-board-chair-public-policy-forum.

770 L. Smolin, *The Trouble with Physics*, n. 70, p. 254.

771 The notion that the standard model's particles are elementary is being consigned to the pages of science history; see, e.g., K. Gavroglu, "Simplicity and Observability: When Are Particles Elementary?," *Synthese*, (1989) **79**, p. 543.

772 T. S. Eliot, *The Hollow Men* (London: Faber & Faber, 1925); https://allpoetry.com/the-hollow-men.

773 R. Leverton, "Spoon Fed Advanced Physics," Amazon, 17 January 2014, https://www.amazon.ca/Time-One-Discover-Universe-Began/product-reviews/B072877TR8/.

774 An empty universe is not quite as useless as it may sound. The de Sitter universe is a solution to the equations of general relativity that is seen as a key concept in cosmology, and it is "empty" space. See, e.g., M. Trodden, "de Sitter Space and Cosmology," *Discover*, 15 April 2012, https://www.discovermagazine.com/the-sciences/de-sitter-space-and-cosmology.

775 While writing this I came on Smolin's account of the impact of his first reading. It mirrored mine. He said, "I knew this was the missing idea." L. Smolin, *The Trouble with Physics: The Rise of String Theory, the Fall of a Science, and What Comes Next*, n. 70, p. 254.

776 A. Ashtekar et al., "Weaving a Classical Geometry with Quantum Threads," *Phys. Rev. Lett.*, (1992) **69**, p. 237.

777 "Meandering Marsupials," https://meanderingmarsupials.blogspot.com/.

778 S. Bilson-Thompson, "Braided Topology and the Emergence of Matter," *J. Phys.: Conf. Ser.*, (2012) **360**, 012056; https://iopscience.iop.org/article/10.1088/1742-6596/360/1/012056/pdf.

779 S. W. Hawking, interviewed by David Cherniack for *Stephen Hawking's Universe*, 20 June 1986, https://www.allinonefilms.com/transcripts/hawking.htm.

780 More irony: The acronym PI drove the name, not vice versa. PI's founding director wrote: "You might as well think about . . . the acronym you want. Perimeter Institute formed the acronym 'PI,' which easily reflects the Greek letter 'PI,' one of the first transcendental numbers to be discovered and a clear reference to the long and glorious history of fundamental inquiry through the ages." H. Burton, *First Principles: The Crazy Business of Doing Serious Science* (Toronto: Key Porter Books, 2009), p. 50.

781 L. Hardy, "Reformulating and Reconstructing Quantum Theory," 25 August 2011, p. 152; https://arxiv.org/pdf/1104.2066.pdf.

782 A. Einstein, "Principles of Research," A. Einstein, *Ideas and Opinions*, n. 28, p. 224.

783 J. D. Barrow, *New Theories of Everything*, n. 134, p. 67.

784 A. Einstein, "On the Electrodynamics of Moving Bodies," n. 256.

785 A. Einstein, "Über einen die Erzeugung und Verwandlung des Lichtes betreffenden heuristischen Gesichttspunkt," *Annalen der Physik*, (1905) **17**, p. 132; "On a Heuristic Point of View Concerning the Production and Transformation of Light," *The Collected Papers of Albert Einstein*, Vol. 2, n. 256, p. 86.

786 "Albert Einstein," Nobel Prize, https://www.nobelprize.org/prizes/physics/1921/einstein/facts/.

787 P. J. E. Peebles, *The Whole Truth*, n. 60, p. 3.

788 L. Smolin, *The Life of the Cosmos*, n. 566, p. 178.

789 See, especially: L. Smolin, *The Trouble with Physics*, n. 70.

790 S. Hossenfelder, *Lost in Math*, n. 157.

791 S. Hossenfelder, "Book Update," *BackReAction*, 25 October 2017; https://backreaction.blogspot.com/2017/10/book-update.html.

792 S. Hossenfelder, *Lost in Math*, n. 157, p. 52.

793 There is an extensive literature on such distinctions in the philosophy of mathematics. See, for example, J. Bigelow, *The Reality of Numbers: A Physicalist's Philosophy of Mathematics* (Oxford: Clarendon Press, 1988), saying, at p. 1, ". . . the world of space and time does contain mathematical objects like numbers."

794 A. Einstein, A. Harris, transl., *Essays in Science*, n. 10, p. 18.

795 It has become fashionable to say philosophy is not a science. No doubt, like other sciences, it could do better. But the front door of the physics department of my first alma mater (then well into its second hundred years) entitled it Department of Natural Philosophy. And, too, why do a million physicists have PhD degrees?

796 E.g., S. de Haro, "Science and Philosophy: A Love–Hate Relationship," *Found. Sci.*, (2020) **25**, p. 297; https://doi.org/10.1007/s10699-019-09619-2.

797 C. Rovelli, "Physics Needs Philosophy/ Philosophy Needs Physics," *Observations*, *Sci. Am.*, 18 July 2018, https://blogs.scientificamerican.com/observations/physics-needs-philosophy-philosophy-needs-physics/.

798 Ibidem.

799 See S. Hossenfelder, *Lost in Math*, n. 157; and see *Time One*, n. 16, chapter "Faith in Math."

800 C. Rovelli, *Reality Is Not What It Seems*, n. 7, p. 211.

801 E.g., J. Norman, "Russell and Whitehead's Principia Mathematica," *History of Information*, https://www.historyofinformation.com/detail.php?entryid=2067.

802 K. Gödel, "Über formal unentscheidbare Sätze der *Principia Mathematica* und verwandter Systeme," *Mon. Math. Phys.*, (1931) **38**, p. 173; B. Meltzer, transl., *On Formally Undecidable Propositions of Principia Mathematica and Related Systems* (New York: Dover Publications, 1992).

803 The precise meaning of Gödel's achievement defies brief description; readers looking for more may find it in E. Nagel and J. R. Newman, *Gödel's Proof* (New York: New York University Press, 1958).

804 J. Baez, "The Inconsistency of Arithmetic," *The n-Category Café*, 27 September 2011.

805 J. D. Barrow, *Pi in the Sky*, n. 465, p. 296.

806   M. Frisch, *Inconsistency, Asymmetry, and Non-Locality: A Philosophical Investigation of Classical Electrodynamics* (Oxford: Oxford University Press, 2005), abstract.

807   Euclid, *The Thirteen Books of Euclid's Elements*, T. L. Heath, transl. (Cambridge: The University Press, 1908). See, e.g., N. Swartz, "Axioms and Postulates of Euclid," https://www.sfu.ca/~swartz/euclid.htm.

808   J. Polchinski, lecture in honour of Paul Dirac, Tallahassee, 6 December 2002, repr. in "Monopoles, Duality, and String Theory," *Int. J. Mod. Phys. A*, (2004) **19**, p. 145.

809   E.g., J. Silk, "Physics: The Impulse of Beauty," *Nature*, (2015) **523**, p. 156; https://www.nature.com/articles/523156a.

810   E.g., B. Duignan, "empiricism," n. 96, https://www.britannica.com/topic/empiricism; there are kinds and degrees of empiricism but we have no need to go into them here.

811   "Logical positivism remains the tacit philosophy of many scientists." M. Bunge, *Finding Philosophy in Social Science* (New Haven: Yale University Press, 1996), p. 317.

812   A. J. Ayer, *Philosophy in the Twentieth Century* (London: George Weidenfeld & Nicholson Ltd., 1982), repr. (London: Unwin, 1984), p. 3.

813   A. J. Ayer, *The Foundations of Empirical Knowledge* (London: Macmillan, 1940), p. 220 in 1969 edition; https://archive.org/stream/in.ernet.dli.2015.46395/2015.46395.Foundations-Of-Empirical-Knowledge_djvu.txt.

814   Just as there are many views about the nature of dark energy and dark matter, there are as many, or even more, ideas of ways to detect them; see, e.g., "Dark Energy, Dark Matter," NASA Science: Astrophysics, https://science.nasa.gov/astrophysics/focus-areas/what-is-dark-energy.

815   A. J. Ayer, *Language, Truth and Logic*, n. 218, p. 21.

816   G. Greene, *The End of the Affair* (London: Heinemann, 1951), p. 66.

817   A. J. Ayer, *Language, Truth and Logic*, n. 218, p. 146.

818   A. J. Ayer, in G. Unwin, ed., *What I Believe* (London: George Allen & Unwin Ltd., 1966), p. 13.

819   E.g., B. Duignan, "empiricism," n. 96; see heading, "Contemporary Philosophy."

820   John F. Kennedy has been blamed for inventing this trope, now widely used by politicians, but this too is not quite true: see, e.g., B. Zimmer, "Crisis = Danger + Opportunity: The Plot Thickens," *Language Log*, 27 March 2007, http://itre.cis.upenn.edu/~myl/languagelog/archives/004343.html.

821   For one of many views, see S. Nagel, "Physics in Crisis," *Phys. Today*, (2002) **55**, p. 55.

822   We see one octave of photon frequencies, out of a range of more than sixty octaves other apparatus can detect; see, e.g., R. White, "How Do Submarines Communicate with the Outside World?," *Naval Post*, 3 May 2021, https://web.archive.org/web/20241003160739/https://navalpost.com/how-do-submarines-communicate-with-the-outside-world/, and R. Yirka, "Highest Energy Photons Ever Coming from the Crab Nebula," Phys.org, 26 June 2019, https://phys.org/news/2019-06-highest-energy-photons-crab-nebula.html.

823   E. H. Land, "Color Vision and the Natural Image: Part 1," *Proc. Natl. Acad. Sci. USA*, (1959) **45**, p. 115; https://www.pnas.org/doi/pdf/10.1073/pnas.45.1.115.

824   See also, e.g., P. Tacikowski *et al.*, "Human hippocampal and entorhinal neurons encode the temporal structure of experience," *Nature*, 25 September 2024, https://www.nature.com/articles/s41586-024-07973-1.

825 There is evidence the process begins at least three months before birth; see, e.g., M. McElroy, "While in Womb, Babies Begin Learning Language from Their Mother," *UW News*, 2 January 2013, https://www.washington.edu/news/2013/01/02/while-in-womb-babies-begin-learning-language-from-their-mothers/.

826 The adult human brain has about one hundred billion neurons, surrounded by about the same number of neuroglia, and interconnected by more than one hundred trillion synapses; see, e.g., "Scale of the Human Brian," *AI Impacts*, https://aiimpacts.org/scale-of-the-human-brain/.

827 In the nineteen seventies, I had an opportunity to study how neurons could create and transmit nerve impulses with exquisite (and seemingly unphysical) sensitivity to input signals: C. J. Gillespie, "Towards a Molecular Theory of the Nerve Membrane: The Significance of the Quasithreshold Behaviour," *J. theor. Biol.*, (1973) **42**, p. 519; https://www.sciencedirect.com/science/article/abs/pii/0022519373902440.

828 See, e.g., D. Shohamy and A. D. Wagner, "Integrating Memories in the Human Brain: Hippocampal–Midbrain Encoding of Overlapping Events," *Neuron*, (2008) **60**, p. 378; https://www.ncbi.nlm.nih.gov/pmc/articles/PMC2628634/.

829 See, e.g., D. I. Slobin, ed., *The Cross-Linguistic Study of Language Acquisition Vol. 5: Expanding the Contexts* (New York: Psychology Press, 2014); https://www.taylorfrancis.com/books/mono/10.4324/9781315805825/crosslinguistic-study-language-acquisition-dan-isaac-slobin.

830 J. Polchinski, "String Theory to the Rescue," paper for the meeting "Why Trust a Theory? Reconsidering Scientific Methodology in Light of Modern Physics," Munich, 7-9 December 2015; https://arxiv.org/abs/1512.02477.

831 R. Webb, "Quantum Physics: Our Best View of How Particles Interact to Make the World," *New Scientist*, online, https://www.newscientist.com/definition/quantum-physics/.

832 K. E. Boulding, quoted by Mancur Olsen in "The No-Growth Society," *Daedelus*, (1973) **102**, p. 3.

833 Ibidem.

834 This is typical (and, in this case, anonymous) chatter; "Calibi [sic] Yau Spaces," *Act for Libraries*, http://www.actforlibraries.org/calibi-yau-spaces/.

835 J. Baez, "Quantum Gravity and the Algebra of Tangles," 4 May 1992, https://arxiv.org/pdf/hep-th/9205007.pdf.

836 Ibid., p. 13.

837 D. Scott, "The Standard Model of Cosmology: A Sceptic's Guide," in E. Coccia et al., eds., *Gravitational Waves and Cosmology*, (2022), p. 133; https://www.semanticscholar.org/reader/ab9dd99283fb367fa36c898e0b25bb66c67de330.

838 Ibidem. Nobel Prize–winning physicist Enrico Fermi quoted mathematician John von Neumann as saying, "With four parameters I can fit an elephant, and with five I can make him wiggle his trunk." And, just for fun, see also, J. Mayer et al., "Drawing an Elephant with Four Complex Parameters," *Am. J. Phys.*, (2010) **78**, p. 648; https://sci-hub.se/10.1119/1.3254017.

839 G. Lemaître, *The Primeval Atom: An Essay on Cosmogony*, n. 416, p. 78.

840 See, e.g., A. Knapp, "The Seduction of the Exponential Curve," *Forbes*, 17 November 2011, https://www.forbes.com/sites/alexknapp/2011/11/17/the-seduction-of-the-exponential-curve/.

841 D. Gross, "Einstein and the Quest for a Unified Theory," n. 132, p. 295.

842 A. C. Clarke, "Second Dawn," in *Expedition to Earth* (New York: Ballantine Books, 1953), p. 33.

843 Smolin seems to have experienced a somewhat similar reaction upon seeing Bilson-Thompson's paper; see n. 763, p. 253.

844 C. Rovelli and L. Smolin, "Discreteness of Area and Volume in Quantum Gravity," *Nucl. Phys. B*, (1995) **442**, p. 593; https://arxiv.org/pdf/gr-qc/9411005.pdf; see also L. Smolin, "Atoms of Space and Time," *Sci. Am.*, (2004) **290**, p. 66; https://www.scientificamerican.com/article/atoms-of-space-and-time-2006-02/.

845 See, e.g., T. Thiemann, "The Fabric of Space: Spin Networks," Einstein Online, (2005), https://www.einstein-online.info/en/spotlight/spin_networks/.

846 L. Smolin, "General Predictions of Quantum Theories of Gravity," in D. Oriti, (ed.), *Approaches to Quantum Gravity*, n. 6, p. 548.

847 S. O. Bilson-Thompson, "A Topological Model of Composite Preons," n. 762, p. 1.

848 See, e.g., D. Gunderman and R. Gunderman, "The Mathematical Madness of Möbius Strips and Other One-Sided Objects," *Smithsonian Magazine*, 25 September 2018, https://www.smithsonianmag.com/science-nature/mathematical-madness-mobius-strips-and-other-one-sided-objects-180970394/.

849 S. O. Bilson-Thompson, F. Markopoulou and L. Smolin, "Quantum Gravity and the Standard Model," n. 764.

850 For more on C-Y manifolds, dimensions, size, and string theory see, e.g., M. Freiberger, "Hidden Dimensions," *Plus*, 21 December 2010, https://plus.maths.org/content/hidden-dimensions.

851 A. Einstein, "The Problem of Space, Ether, and the Field in Physics," n. 10, p. 68.

852 L. Smolin, *Einstein's Unfinished Revolution*, n. 74, p. 236.

853 A. Becker, "The Origins of Space and Time," n. 51.

854 See, e.g., A. Becker, ibidem.

855 See A. Einstein, *Relativity*, n. 245, his chapter 31, "The Possibility of a 'Finite' and Yet 'Unbounded' Universe"; https://www.gutenberg.org/ebooks/36114.

856 S. O. Bilson-Thompson, F. Markopoulou, and L. Smolin, "Quantum Gravity and the Standard Model," n. 764.

857 This is not what those words mean in English but rather is an example of a common linguistic strategy called word meaning extension; see S. J. Greenhill, "A Shared Foundation of Language Change: Short-term Development and Long-term Evolution of Language Share Mechanisms," *Science*, (2023) **381**, p. 374.

858 With a distant salute to J. R. R. Tolkien, who would likely have not liked my taking this liberty; see also http://www.timeone.ca/debt-stream/.

859 F. Markopoulou, "The Internal Description of a Causal Set: What the Universe Looks Like from the Inside," n. 560.

860 C. Dickens, *A Christmas Carol* (London: Chapman & Hall, 1843).

861 J. von Neumann, *Mathematical Foundations of Quantum Mechanics*, n. 126, pp. 417–8.

862 D. Hume, *A Treatise of Human Nature: Being an Attempt to Introduce the Experimental Method of Reasoning into Moral Subjects* (London: John Noon, 1739), Book 1, Part II, Sect. VI; https://oll.libertyfund.org/title/bigge-a-treatise-of-human-nature.

863   R. M. Unger in R. M. Unger and L. Smolin, *The Singular Universe and the Reality of Time*, n. 62, p. 34.

864   L. Smolin, *Einstein's Unfinished Revolution*, n. 74, p. 232.

865   D. D. Murphey and D. Miller, "An Intellectual Mind-Twister for Our Readers: Going to the Outer Limits of Cosmology," *J. Soc. Pol. Ec. Sci.*, (2013) **38**, p. 235; https://specialcollections.wichita.edu/collections/ms/2007-03/2007-3-a.html.

866   K. Vonnegut, *Bluebeard* (New York: Dell Publishing, 1987), p. 207.

867   See also F. Markopoulou, "An Insider's Guide to Quantum Causal Histories," *Nucl. Phys. B Proc. Suppl.*, (2000) **88**, p. 308; https://core.ac.uk/download/pdf/25274228.pdf; and see F. Markopoulou, "Quantum Causal Histories," *Class. Quant. Grav.*, (2000) **17**, p. 2059.

868   F. Markopoulou, "The Internal Description of a Causal Set: What the Universe Looks Like from the Inside," n. 560.

869   H. Reichenbach, M. Reichenbach, ed., *The Direction of Time*, n. 37, p. 24; his emphasis.

870   S. Surya, "The Causal Set Approach to Quantum Gravity," *Living Rev. Relativ.*, (2019) **22**, # 5; https://arxiv.org/abs/1903.11544.

871   See *Time One*, chapter "The Copenhagen Hegemony," http://www.timeone.ca/wp-content/uploads/2015/07/the-copenhagen-hegemony.pdf.

872   J. T. Cushing, *Quantum Mechanics: Historical Contingency and the Copenhagen Hegemony* (Chicago: University of Chicago Press, 1994), p. 117.

873   A. Einstein, letter to Max Born, 4 December 1926, J. N. James et al., transls., *The Collected Papers of Albert Einstein, Vol. 15, The Berlin Years: Writings & Correspondence, June 1925–May 1927* (English Translation Supplement), p. 403. Original emphasis.

874   E.g., M. Frisch, "Causation in Physics," *Stanford Encyclopedia of Philosophy*, n. 99, https://plato.stanford.edu/entries/causation-physics/.

875   E.g., Y. Ben-Menahem, "Struggling with Causality: Einstein's Case," *Science in Context*, (2008) **6**, p. 291; Cambridge University Press, 26 September 2008, https://www.cambridge.org/core/journals/science-in-context/article/abs/struggling-with-causality-einsteins-case/2DB574036CED542BFEA325230D217290.

876   For key steps, see L. Smolin, *Three Roads to Quantum Gravity*, n. 61, pp. 184 ff, p. 228.

877   See, e.g., "A Step Towards Quantum Gravity: Resolving the Problem of Time," *Science Daily*, 12 August 2022, https://www.sciencedaily.com/releases/2022/08/220812130806.htm.

878   S. O. Bilson-Thompson, "Braids, loops, and the emergence of the standard model," lecture at the International Conference on Quantum Gravity, Morelia, Mexico, 28 June 2007; https://www.matmor.unam.mx/eventos/loops07/talks/5B/Bilson-Thompson.pdf.

879   L. Smolin, "An Invitation to Loop Quantum Gravity," in P. C. Argyres et al., eds., *Quantum Theory and Symmetries* (Singapore: World Scientific, 2004), p. 655; https://arxiv.org/abs/hep-th/0408048.

880   Smolin also gave a paper at the conference and it did plug Bilson-Thompson's model. L. Smolin, "Chiral excitations of quantum geometry as elementary particles," lecture at the International Conference on Quantum Gravity, Morelia, Mexico, 29 June 2007; https://www.matmor.unam.mx/eventos/loops07/talks/PL5/Smolin.pdf.

881   Ibidem.

882 S. O. Bilson-Thompson, n. 878.

883 L. Smolin, "Chiral excitations of quantum geometry as elementary particles," n. 880.

884 Note Bilson-Thompson posed his volume and area designations in a question; ibidem.

885 Ibidem.

886 E.g., "Neutrinos from Beta Decay," All Things Neutrino, https://neutrinos.fnal.gov/sources/beta-decay/.

887 The ups and downs of quarks do not complete their complex story: for example, they also have a certain charm; The NNPDF Collaboration, "Evidence for Intrinsic Charm Quarks in the Proton," *Nature*, (2022) **608**, p. 483; https://doi.org/10.1038/s41586-022-04998-2.

888 The leptons are the electron, muon, and tau, and their neutrinos; https://www.britannica.com/science/lepton.

889 Color charge has nothing to do with electrical charge; see, e.g., E. Siegel, "Quarks Don't Actually Have Colors," *Forbes*, 18 April 2019, https://www.forbes.com/sites/startswithabang/2019/04/18/quarks-dont-actually-have-colors/.

890 E.g., F. Fernflores, "The Equivalence of Mass and Energy," *Stanford Encyclopedia of Philosophy*, n. 99, 15 August 2019, https://plato.stanford.edu/entries/equivME/#InteEoMc2HypoConcNatuMatt.

891 This fundamental virtue is mentioned only fleetingly in his main paper: n. 762, p. 6.

892 See, e.g., R. Nave, "Quarks," *HyperPhysics*, http://hyperphysics.phy-astr.gsu.edu/hbase/Particles/quark.html.

893 See, e.g., M. Williams, "What Are Leptons?," *Universe Today*, 1 December 2016; https://www.universetoday.com/46935/leptons/.

894 See, e.g., "The Standard Model," SLAC, https://www-project.slac.stanford.edu/e158/StandardModel.html.

895 See, e.g., "Are All Neutrinos Left-handed?," All Things Neutrino, https://neutrinos.fnal.gov/mysteries/handedness/.

896 J. C. Zorn et al., "Experimental Limits for the Electron-Proton Charge Difference and for the Charge of the Neutron," *Phys. Rev.*, (1963) **129**, p. 2566; https://journals.aps.org/pr/abstract/10.1103/PhysRev.129.2566.

897 E.g., M. Strassler, "Particle/Anti-Particle Annihilation," *Of Particular Significance*, 25 March 2012, https://profmattstrassler.com/articles-and-posts/particle-physics-basics/particleanti-particle-annihilation/.

898 E.g., "Collisions of Light Produce Matter/Antimatter from Pure Energy," Brookhaven National Laboratory, 28 July 2021, https://www.bnl.gov/newsroom/news.php?a=119023.

899 S. O. Bilson-Thompson, "Topological Preon Models: A Braid New World," PIRSA, 16 November 2005, https://pirsa.org/05110009.

900 See, e.g., "The Atom Builder Guide to Elementary Particles," PBS, https://www.pbs.org/wgbh/aso/tryit/atom/elempartp.html.

901 S. O. Bilson-Thompson, "A Topological Model of Composite Preons," n. 762.

902 S. O. Bilson-Thompson, "A Topological Model of Composite Preons," n. 762; S. O. Bilson-Thompson, F. Markopoulou, and L. Smolin, "Quantum Gravity and the Standard Model," n. 764; S. O. Bilson-Thompson, "Braided Topology and the Emergence of Matter," n. 778.

903 S. O. Bilson-Thompson et al., "Particle Identifications from Symmetries of Braided Ribbon Network Invariants," https://arxiv.org/pdf/0804.0037.pdf; S. O. Bilson-Thompson et al., "Particle Topology, Braids, and Braided Belts," J. Math. Phys., (2009) **50**, 113505; https://arxiv.org/pdf/0903.1376.pdf; S. O. Bilson-Thompson et al., "Emergent Braided Matter of Quantum Geometry," *SIGMA*, (2012) **8**, 014; https://arxiv.org/pdf/1109.0080.pdf. A more recent published work on another topic is S. Bilson-Thompson et al., "Tachyonic Media in Analog Models of Special Relativity," *Phys. Rev. D*, (2023) **108**, p. 1.

904 See, e.g., N. Gresnigt, "The Standard Model Particle Content with Complete Gauge Symmetries from the Minimal Ideals of Two Clifford Algebras," *Eur. Phys. J. C*, (2020) **80**, p. 583; https://link.springer.com/content/pdf/10.1140/epjc/s10052-020-8141-1.pdf. He has authored or co-authored more than fifty papers relating to the Bilson-Thompson model in the last eighteen years.

905 B. Greene, *The Elegant Universe: Superstrings, Hidden Dimensions, and the Quest for the Ultimate Theory*, n. 932, p. 87.

906 M. Arndt et al., "Wave-Particle Duality of C60 Molecules," *Nature*, (1999) **401**, p. 680.

907 See, e.g., "Richard E. Smalley, Robert F. Curl, and Harold W. Kroto," Science History Institute, https://www.sciencehistory.org/education/scientific-biographies/richard-smalley-robert-curl-harold-kroto/.

908 L. Hackermüller et al., "Decoherence of Matter Waves by Thermal Emission of Radiation," *Nature*, (2004) **427**, p. 711.

909 C. J. Gillespie, *Study of Superconductors by Electron Tunneling*, n. 479, p. 29.

910 See, e.g., F. Sonnemann, *Resistive Transition and Protection of LHC Superconducting Cables and Magnets*, Rheinisch-Westfälischen Technischen Hochschule Aachen, thesis (2001), p. 23; https://cds.cern.ch/record/499591/files/thesis-2001-004.pdf

911 The Editors of Encyclopaedia Britannica, "quantum," n. 96, https://www.britannica.com/science/quantum.

912 Online Etymology Dictionary, https://www.etymonline.com/word/quantum.

913 J. Cresswell, "quantum," *Little Oxford Dictionary of Word Origins* (Oxford: Oxford University Press, 2014), p. 290.

914 Google Books Ngram Viewer, English (2019) corpus; and see J.-M. Michel et al., "Quantitative Analysis of Culture Using Millions of Digitized Books," *Science*, (2010) **331**, p. 176; https://www.science.org/doi/10.1126/science.1199644.

915 E. A. Poe, "A Decided Loss," *Philadelphia Saturday Courier*, 10 November 1832, repr. as "Loss of Breath: A Tale Neither In Nor Out of 'Blackwood,'" in *The Works of the Late Edgar Allan Poe, Vol. IV, Arthur Gordon Pym, &c* (New York: Redfield, 1859), p. 304; https://www.eapoe.org/works/tales/lssbthe.htm.

916 E. A. Poe, *Eureka*, n. 85, § 45.

917 G. 't Hooft, *In Search of the Ultimate Building Blocks*, n. 192, p. 11.

918 C. Rovelli, *Reality Is Not What It Seems*, n. 7, p. 172.

919 E.g., B. C. Hall, *Quantum Theory for Mathematicians* (New York: Springer Verlag, 2013), *Graduate Texts in Mathematics*, vol. 267, chs. 22 and 23.

920 B. Holm, "Dear Mr. Einstein: Waterloo Scholar Corresponded with Legendary Physicist in 1950s," *Record*, 2 October 2010, https://www.therecord.com/life/2010/10/02/dear-mr-einstein-waterloo-scholar-corresponded-with-legendary-physicist-in-1950s.html.

921 A. Einstein, letter to John Moffat, n. 719.

922 A. Einstein, "Physics and Reality," n. 215, p. 319.

923 A. Einstein, letter to Michele Besso, 10 August 1954, quoted by and transl. in J. Stachel, "Einstein and the Quantum," in R. G. Colodny, ed., *From Quarks to Quasars*, n. 758, p. 380; Einstein's italics.

924 Even the length contraction of special relativity may fail to shrink them: G. Amelino-Camelia, "Relativity in Space-Times with Short-Distance Structure Governed by an Observer-Independent (Planckian) Length Scale," *Int. J. Mod. Phys. D*, (2002) **11**, p. 35. In any event, the quanta do not possess the 1-D property of linear size.

925 C. Rovelli, *Reality Is Not What It Seems*, n. 7, p. 165.

926 Ibid., p. 164.

927 L. Smolin, in R. M. Unger and L. Smolin, *The Singular Universe and the Reality of Time*, n. 62, p. 517.

928 J. A. Wheeler, interview by Ken Ford, "Update on John Wheeler," *Princeton Physics News*, (2006) **2** (1), p. 5.

929 See, e.g., M. Shirber, "Why Only Three Dimensions?," *Science*, 4 October 2005, https://www.science.org/content/article/why-only-three-dimensions.

930 L. Smolin, *Einstein's Unfinished Revolution*, n. 74, p. 233.

931 E.g., L. Randall, *Warped Passages: Unraveling the Mysteries of the Universe's Hidden Dimensions* (New York: Ecco, 2005); https://www.amazon.ca/Warped-Passages-Unraveling-Mysteries-Dimensions/dp/0060531096.

932 See, e.g., B. Greene, *The Elegant Universe: Superstrings, Hidden Dimensions, and the Quest for the Ultimate Theory* (New York: Vintage Books, 2000), p. 207.

933 Ibid., p. 141; also S. Sahoo, "String Theory: Big Problem for Small Size," *Eur. J. Phys.*, (2009) **30**, p 901; https://arxiv.org/ftp/arxiv/papers/1209/1209.5498.pdf.

934 F. Dowker, "Causal Sets and the Deep Structure of Spacetime," n. 716.

935 E.g., Z. Lu and X. Sun, "On the Weil-Petersson Volume and the First Chern Class of the Moduli Space of Calabi-Yau Manifolds," *Comm. Math. Phys.*, (2006) **261**, p. 297; https://arxiv.org/abs/math/0510021.

936 C. Rovelli, *Quantum Gravity*, n. 41, p. 19.

937 G. Lemaître, "The Beginning of the World from the Point of View of Quantum Theory," n. 412.

938 E.g., M. Williams, "What Is the Big Bang Theory?," Phys.org, 18 December 2015, https://phys.org/news/2015-12-big-theory.html.

939 R. Foot et al., "Electric Charge Quantization," *J. Phys. G: Nucl. Part. Phys.*, (1993) **19**, p. 361; https://arxiv.org/pdf/hep-ph/9209259.pdf.

940 E.g., E. Siegel, "Ask Ethan: Why Is the Universe Electrically Neutral?," Big Think, 15 April 2022, https://bigthink.com/starts-with-a-bang/universe-neutral/.

941 "The Nobel Prize in Physics 1923: Robert A. Millikan," Nobel Prize, https://www.nobelprize.org/prizes/physics/1923/millikan/facts/.

942 In 2021, layperson and Reddit user "oldendude" asked, "I understand that a proton's charge is due to quarks, two up and one down, and their charges sum up to +1. But an electron is elementary, doesn't have quarks, so the charge of -1 doesn't come from quarks. So protons and electrons acquire their charges in different ways, AND these charges happen to have exactly the same magnitude. Is that right? This seems like an amazing coincidence. Is there an explanation of this fact? If not, do physicists believe that an

explanation is important? Or do they just not worry about it?" The responses he received to his astute the-emperor-is-not-well-dressed observation and question were trite at best; https://www.reddit.com/r/AskPhysics/comments/st6zj3/why_do_electrons_and_protons_have_charge_of/.

943 R. A. Millikan, Nobel lecture, "The Electron and the Light-quant from the Experimental Point of View," Nobel Prize, 23 May 1924, https://www.nobelprize.org/uploads/2018/06/millikan-lecture.pdf.

944 M. Gell-Mann "A Schematic Model of Baryons and Mesons," *Phys. Lett.*, (1964) 8, p. 214.

945 No one has observed an isolated quark. A review by Nobel Prize–winning physicist Martin Perl and colleagues said, "There have been a very large number of searches but there is no confirmed evidence for existence of isolatable fractional charge particles. It may be that they do not exist.," M. L. Perl et al., "A Brief Review of the Search for Isolatable Fractional Charge Elementary Particles," *Mod. Phys. Lett.* A, (2004) 19, p. 2595.

946 Y. Fang et al., "Structured Electrons with Chiral Mass and Charge," *Science*, (2024) 385, p. 183; https://www.science.org/doi/10.1126/science.adp9143.

947 E.g., R. Nave, "The Equality of the Number of Protons and Electrons," *HyperPhysics Concepts*, Georgia State University, http://hyperphysics.phy-astr.gsu.edu/hbase/Astro/wcep.html.

948 See, e.g., E. M. Purcell and D. J. Morin, *Electricity and Magnetism* (Cambridge: Cambridge University Press, 1969), p. 4.

949 J. D. Barrow, *New Theories of Everything*, n. 134, p. 95.

950 R. P. Feynman, Nobel lecture, "Quantum Electrodynamics," Nobel Prize, 11 December 1965, https://www.nobelprize.org/prizes/physics/1965/feynman/lecture/.

951 F. Wilczek, "The Surprise of Splitting Electrons," *Wall Street Journal*, 24 June 2023; https://www.wsj.com/articles/the-surprise-of-splitting-electrons-8078568d.

952 S. Arrhenius, "The Nobel Prize in Physics 1921," presentation speech, Nobel Prize, The Royal Swedish Academy of Sciences, 10 December 1922; for the story behind this, see D. R. Topper, *How Einstein Created Relativity Out of Physics and Astronomy*, n. 535, pp. 132 ff.

953 M. Niaz et al., "Reconstruction of the History of the Photoelectric Effect and Its Implications for General Physics Textbooks," Wiley Online Library, 14 January 2010, https://onlinelibrary.wiley.com/doi/full/10.1002/sce.20389.

954 A. Einstein, letter to Paul Bonofield, 18 September 1939, dup. in Mudd Library, Princeton University, Item 6-118-1.

955 Strictly, accelerating charges. E.g., P. Walorski, "Why Is That Electrons Radiate Electromagnetic Energy When They Are Accelerated?," PhysLink, https://www.physlink.com/education/askexperts/ae436.cfm.

956 A. Einstein and L. Infeld, *The Evolution of Physics*, n. 130, p. 276.

957 E.g., D. Lincoln, "Weinberg's Angle," *FermiLab at Work*, 11 November 2011, https://news.fnal.gov/2011/11/weinberg-s-angle/. Weinberg mixing is an integral aspect of the so-called weak interaction in particle physics, an inescapably technical topic. It was the subject of the 1979 Nobel Prize for physics; https://www.nobelprize.org/prizes/physics/1979/press-release/.

958 S. O. Bilson-Thompson, "A Topological Model of Composite Preons," n. 762, p 3.

959 "Nature's Laws," *The Economist*, 31 July 2021, p. 74.

960 R. Laughlin, *A Different Universe: Reinventing Physics from the Bottom Down* (New York: Basic Books, 2005), p. 15.

961 It has been so since 1983, when the 17th *Conférence Générale des Poids et Mesures* in Sévres, passed its Resolution 1: "The metre is the length of the path travelled by light in vacuum during a time interval of 1/299792458 of a second." That is, the metre is now defined, not by a bar of platinum, but so as to make the speed of light *exactly* 299,792,458 meters per second. See Bureau International des Poids et Mesures, "Resolution 1 of the 17th CGPM (1983)," https://www.bipm.org/en/committees/cg/cgpm/17-1983/resolution-1. Of course, this definition does not *explain* the value of the speed of light.

962 I. Newton, *Opticks*, n. 250, question 30, ". . . and may not Bodies receive much of their Activity from the Particles of Light which enter their Composition?"

963 A. Einstein, "On a Heuristic Point of View Concerning the Production and Transformation of Light," n. 785.

964 A. Einstein, "On the Electrodynamics of Moving Bodies," n. 256.

965 M. Planck, "A Scientific Autobiography," n. 281, p. 47.

966 A. Einstein, "On the Electrodynamics of Moving Bodies," n. 256; and see A. Einstein, *Relativity*, n. 245, p. 21 ff.

967 S. O. Bilson-Thompson, "A Topological Model of Composite Preons," n. 762.

968 E.g., A. Berger, "Positron Emission Tomography," *Br. Med. J.*, (2003) **326**, p. 1449, https://www.ncbi.nlm.nih.gov/labs/pmc/articles/PMC1126321/.

969 S. O. Bilson-Thompson, "Braided Topology and the Emergence of Matter," n. 778.

970 See, e.g., "Kinematics," n. 96, https://www.britannica.com/science/kinematics.

971 S. O. Bilson-Thompson, "Braided Topology and the Emergence of Matter," n. 778.

972 J. Hackett, "Locality and Translations in Braided Ribbon Networks," n. 182.

973 Ibidem.

974 He worked for several years at the Special Research Centre for the Subatomic Structure of Matter in Adelaide, https://sciences.adelaide.edu.au/physical-sciences/research/physics-research/cssm.

975 E.g., F. Christensen, "The Problem of Inertia," *Phil. Sci.*, (1981) **48**, p. 232; https://www.jstor.org/stable/187183.

976 See B. M. Sattler, *The Concept of Motion in Ancient Greek Thought*, n. 239.

977 S. O. Bilson-Thompson, F. Markopoulou, and L. Smolin, "Quantum Gravity and the Standard Model," n. 764.

978 Ibidem.

979 R. M. Unger, in R. M. Unger and L. Smolin, *The Singular Universe and the Reality of Time*, n. 62, p. 233.

980 Ibid., p. 234.

981 L. Smolin, *Time Reborn*, n. 714, p. 83.

982 Ibidem.

983 See, e.g., A. Los, "Trits Instead of Bits—A Short Introduction to Balanced Ternary," Code Project, 19 December 2014, https://www.codeproject.com/Articles/855365/Trits-Instead-of-bits-A-Short-Introduction-to-Bala.

984 See, e.g., A. Broshar, "Introduction to Synchronous and Asynchronous Processing," Koyeb, 15 March 2021, https://www.koyeb.com/blog/introduction-to-synchronous-and-asynchronous-processing.

985 See, e.g., M. Hawthorne, "What Is Sequential Processing?," *TechniPages*, 12 September 2022, https://www.technipages.com/what-is-sequential-processing/.

986 For the meaning of a *global state*, see, e.g., V. K. Garg, *Principles of Distributed Systems* (Boston: Springer, 1996), p. 71; https://link.springer.com/chapter/10.1007/978-1-4613-1321-2_4.

987 See, e.g., *Merriam-Webster* dictionary, "massively parallel," https://www.merriam-webster.com/dictionary/massively%20parallel.

988 See, e.g., "What Is Distributed Computing?," *TXSeries for Multiplatforms*, IBM, 19 April 2021, https://www.ibm.com/docs/en/txseries/8.2?topic=overview-what-is-distributed-computing.

989 U. Coope, *Time for Aristotle*, (Oxford: Oxford University Press, 2006); https://global.oup.com/academic/product/time-for-aristotle-9780199247905?cc=ca&lang=en&..

990 At least two learned authors other than Smolin have invoked the extra-universal perspective: Bell asked, "When the 'system' in question is the whole world where is the measurer to be found?," J. S. Bell, "Quantum Mechanics for Cosmologists," n. 154. Fotini Markopoulou set up a paper on causal sets with: "In general, an entire spacetime . . . can only be seen by an observer either in the infinite future or outside the universe." F. Markopoulou, "The Internal Description of a Causal Set: What the Universe Looks Like from the Inside," n. 560. Then, too, there was my detective, Frank, who, being terminally fictitious, found a way to check it from the outside (which, as Markopoulou said, "is unphysical"): see *Time One*, n. 16.

991 Current accuracy is about one second in the lifetime of the universe; see M. J. Martin et al., "Sr Lattice Clock at 1x10^{-16} Fractional Uncertainty by Remote Optical Evaluation with a Ca Clock," *Science*, (2008) **319**, p. 1805; https://arxiv.org/ftp/arxiv/papers/0801/0801.4344.pdf.

992 A. Einstein, "On the Electrodynamics of Moving Bodies," n. 256.

993 For his simple explanation with a railway, see A. Einstein, R. W. Lawson, transl., *Relativity*, n. 245, p. 29.

994 See "o'clock (adj.)," *Online Etymology Dictionary*, https://www.etymonline.com/word/o%27clock.

995 But not general relativity, for which there is no observable time variable; see, e.g., C. Rovelli, "Unfinished Revolution," n. 6, p. 7.

996 Figure by OpenStax College; https://courses.lumenlearning.com/physics/chapter/28-2-simultaneity-and-time-dilation/.

997 C. Rovelli, "The Layers That Build Up the Notion of Time," in R. Lestienne and P. Harris, eds, *Time and Science* (Singapore: World Scientific, 2023); https://arxiv.org/ftp/arxiv/papers/2105/2105.00540.pdf.

998 C. Rovelli, "Halfway Through the Woods: Contemporary Research on Space and Time," n. 755, p. 214, my emphasis.

999 P. Forrest, "Relativity, the Passage of Time and the Cosmic Clock," in D. Dieks, ed., *The Ontology of Spacetime II* (Amsterdam: Elsevier, 2008), p. 245.

1000 C. Rovelli, *The Order of Time*, n. 69.

1001 L. Smolin, in R. M. Unger and L. Smolin, *The Singular Universe and the Reality of Time*, n. 62, p. 491.

1002 Ibid., p. 522.

1003 E.g., J. Butterfield and C. Isham, "Spacetime and the Philosophical Challenge of Quantum Gravity," n. 366, p. 76.

1004 For a somewhat technical review of preferred foliations, see D. Delphenich, "Proper Time Foliations of Lorentz Manifolds," (2002), https://arxiv.org/abs/gr-qc/0211066.

1005 A. Valentini, "Hidden Variables and the Large-scale Structure of Space-time," in W. L. Craig and Q. Smith, eds., *Einstein, Relativity and Absolute Simultaneity*, n. 115, p. 125.

1006 B. Greene, *Until the End of Time*, n. 435, p. 15.

1007 This is the reciprocal of the Planck time; see, e.g., "Planck Time," COSMOS—*The SAO Encyclopedia of Astronomy*, https://astronomy.swin.edu.au/cosmos/p/Planck+Time.

1008 Yu. V. Baryshev, "Expanding Space: The Root of Conceptual Problems of Cosmology," *Pract. Cosmol.*, (2008) **2**, p. 20; https://arxiv.org/abs/0810.0153.

1009 M. Chown, "All You Ever Wanted to Know about the Big Bang," *New Scientist*, 17 April 1993.

1010 Definitively: see A. Einstein, lecture at University of Leiden on 5 May 1920, "Ether and the Theory of Relativity," n. 258.

1011 P. J. E. Peebles, lecture in Winnipeg, n. 405.

1012 C. Rovelli, *Reality Is Not What It Seems*, n. 7, p. 237.

1013 See, e.g., A. Knapp, "The Seduction of the Exponential Curve," *Forbes*, 17 November 2011, https://www.forbes.com/sites/alexknapp/2011/11/17/the-seduction-of-the-exponential-curve/.

1014 A. A. Friedmann, "On the Curvature of Space," n. 454.

1015 See, e.g., G. Pitruzzello, "A Bright Future for Attosecond Physics," *Nat. Photonics*, (2022) **16**, p. 550.

1016 This was his famous paper about "Hawking radiation" from black holes and he was referring to the radius near them; S. W. Hawking, "Black Hole Explosions?," *Nature*, (1974) **248**, p. 30; https://www.nature.com/articles/248030a0.

1017 P. J. E. Peebles, *The Whole Truth*, n. 60, p. 144.

1018 S. Weinberg, *The First Three Minutes: A Modern View of the Origin of the Universe* (New York: Basic Books, 1977), p. 5.

1019 K. Vonnegut, *Bluebeard*, n. 866, p. 191.

1020 S. Weinberg, *Gravitation and Cosmology: Principles and Applications of the General Theory of Relativity* (New York: Wiley, 1972).

1021 S. Weinberg, *The First Three Minutes*, n. 1018, p. vii.

1022 Ibid., p. 8.

1023 Ibid., p. 9.

1024 "The Big Bang," CERN: *The Heart of the Matter*, Origins, (2000), https://www.exploratorium.edu/origins/cern/ideas/bang.html.

1025 A detonation wavefront moves at about the speed of sound in the medium; for example, in an enhanced TNT, at about 7 mm/μs (or 7 km/s), see H. Dorsett and

M. D. Cliff, "Detonation Front Curvature Measurements and Aquarium Tests of Tritonal Variants," Defence Science and Technology Organization, 1 April 2003, Table 1; https://apps.dtic.mil/sti/pdfs/ADA484053.pdf.

1026   E.g., "Clusters and Groups of Galaxies," Max Planck Institute for Extraterrestrial Physics, https://www.mpe.mpg.de/2040034/clusters_and_groups_of_galaxies.

1027   E.g., J. S. Kartaltepe et al., "Probing the Large-scale Structure around the Most Distant Galaxy Clusters from the Massive Cluster Survey," *Mon. Not. R. Astron. Soc.*, (2008) **389**, p. 1240; https://academic.oup.com/mnras/article/389/3/1240/1018173.

1028   S. Weinberg, *The First Three Minutes*, n. 1018, p. 8.

1029   C. Rovelli, *The Order of Time*, n. 69, p. 45.

1030   J.-P. Sartre, *La Nausée* (Paris: Librairie Gallimard, 1938); L. Alexander, transl., *Nausea* (New York: New Directions Publishing Corporation, 1959), p. 95.

1031   J.-P. Sartre, *L'être et le néant: Essai d'ontologie phénoménologique* (Paris: Librairie Gallimard, 1943); H. E. Barnes, transl., *Being and Nothingness: An Essay on Phenomenological Ontology* (New York: Washington Square Press, 1966).

1032   H. E. Barnes in *Being and Nothingness*; ibidem.

1033   H. Arendt, lecture at the University of Aberdeen, 1973, repr. in M. McCarthy, ed., *The Life of the Mind* (Orlando: Harcourt Brace & Co., 1981), Book Two, p. 12.

1034   A. Einstein, letter to Vero Besso and Bice Rusconi, 21 March 1955; my transcription and translation. It is variously quoted without attribution. The letter was sold by Christie's on 12 July 2017, and a facsimile (fig. 54) is shown at https://www.christies.com/en/lot/lot-6089302. The relevant sentence reads: "Für uns gläubige Physiker hat die Scheidung zwischen Vergangenheit, Gegenwart und Zukunft nur die Bedeutung einer wenn auch hartnäckige Illusion."

1035   A. Becker, interview with M. Marshall and M. Hogenboom, "Physics suggests that the future has already happened," REEL, BBC, 5 February 2019; https://www.bbc.com/reel/video/p04s223f/physics-suggests-that-the-future-has-already-happened.

1036   A. Becker, *What Is Real?*, n. 49.

1037   See n. 1035.

1038   B. M. Sattler, *The Concept of Motion in Ancient Greek Thought*, n. 239, p. 91, her note 56.

1039   Ibid., p. 104.

1040   J. D. Norton, "Special Theory of Relativity: The Principles," *Einstein for Everyone*, 14 January 2022, https://sites.pitt.edu/~jdnorton/teaching/HPS_0410/chapters/Special_relativity_principles/index.html.

1041   A. Einstein, *Relativity*, n. 245, p. 18.

1042   For a mostly non-mathematical history of early conceptual issues see J. D. North, *The Measure of the Universe: A History of Modern Cosmology* (Oxford: Oxford University Press, 1965), pp. 349 ff.

1043   See, e.g., "Cosmic Microwave Background Dipole," COSMOS—*The SAO Encyclopedia of Astronomy*, https://astronomy.swin.edu.au/cosmos/c/Cosmic+Microwave+Background+Dipole.

1044   E. K. Conklin, "Velocity of the Earth with Respect to the Cosmic Background Radiation," *Nature*, (1969) **222**, p. 971.

1045   Ibidem.

1046 E.g., A. Notari and M. Quartin, "On the Proper Kinetic Quadrupole CMB Removal and the Quadrupole Anomalies," https://arxiv.org/pdf/1504.02076.

1047 T. Maudlin, *Quantum Non-Locality and Relativity: Metaphysical Intimations of Modern Physics* (Hoboken NJ: Blackwell Publishing Ltd., 2002), p. 202.

1048 P.-S. Laplace, *Essai philosophique sur les probabilités* (Paris: Mme. Ve. Courcier, 1814); F. W. Truscott and F. W. Emory, transl., *A Philosophical Essay on Probabilities* (London: John Wiley & Sons, 1902), p. 10; https://www.gutenberg.org/ebooks/58881.

1049 M. Merleau-Ponty, *Phénoménologie de la Perception* (Paris: Librairie Gallimard, 1945); C. Smith transl., *Phenomenology of Perception* (London: Routledge & Kegan Paul Ltd., 1962), p. 268; https://archive.org/stream/merleaupontyphenomenologyofperception/Merleau-Ponty%20-%20Phenomenology%20of%20Perception_djvu.txt.

1050 C. Rovelli, *The Order of Time*, n. 69, p. 94. Marcus Aurelius said it too.

1051 A. J. Ayer, *The Central Questions of Philosophy*, n. 136, p. 16; my emphasis.

1052 J. S. Bell, "On the Problem of Hidden Variables in Quantum Mechanics," *Rev. Mod. Phys.*, (1966) 38, p. 447; repr. in *Speakable and Unspeakable in Quantum Mechanics*, n. 111, p. 10.

1053 See, e.g., N. Huggett, "Zeno's Paradoxes," *Stanford Encyclopedia of Philosophy*, n. 99, 11 June 2018, § 3.3, https://plato.stanford.edu/entries/paradox-zeno/#Arr.

1054 Aristotle, *Physics*, n. 248, book VI, part 9.

1055 F. Wilczek, "Provoked by Zeno's Paradoxes," *Wall Street Journal*, 1 April 2021; https://www.wsj.com/articles/provoked-by-zenos-paradoxes-11617293384.

1056 S. Weinberg, *The First Three Minutes*, n. 1018, p. 149.

1057 Aristotle, *Metaphysics*, (350 BCE); J. H. McMahon, transl., *The Metaphysics*, Book I, Chapter I.

1058 L. Smolin, in R. M. Unger and L. Smolin, *The Singular Universe and the Reality of Time*, n. 62, p. 358.

1059 See, e.g., G. Klempner, "What's So Bad about an Infinite Regress?," *Ask a Philosopher*, 3 October 2016, https://askaphilosopher.org/2016/10/03/whats-so-bad-about-an-infinite-regress/.

1060 F. Close, *The Void* (Oxford: Oxford University Press, 2007), p. 5.

1061 Ibid., p. 156.

1062 All sources I can find trace to Holt, n. 752 at p. 24; he cites no source.

1063 J. Holt, *Why Does the World Exist?*, n. 752, p. 275.

1064 Especially *L'être et le néant*, n. 1031.

1065 Though ultimate cause of some kind is essential in any finite causal universe, some serious authors call it brute fact to disdain it; e.g., F. H. Bradley, *Essays on Truth and Reality* (Oxford: Clarendon Press, 1914), p. 314: "Your ultimate brute fact is in brief your own half-thought-out theory."

1066 T. Hübsch, *Calabi-Yau Manifolds*, n. 381, p. vii.

1067 J. W. Gibbs, *Elementary Principles in Statistical Mechanics: Developed with a Special Reference to the Rational Foundation of Thermodynamics* (New York: Charles Scribner's Sons, 1902), p. iv; https://www.gutenberg.org/files/50992/50992-pdf.pdf.

1068 W. J. M. Rankine, quoted by L. D. B. Gordon, "Obituary Notice of Professor Rankine," *Proc. Roy. Soc. Edinburgh*, (1873–74) 6, p. 296; https://www.cambridge.org/core/

services/aop-cambridge-core/content/view/51F0BCF1216E8A1D538E610F6B033312/ S0370164600029618a.pdf/div-class-title-2-obituary-notice-of-professor-rankine-div.pdf.

1069 S. Neamati, "How Do Different Definitions of Entropy Connect with Each Other?," *Physics, Stack Exchange*, 10 January 2021, https://physics.stackexchange.com/questions/606722/how-do-different-definitions-of-entropy-connect-with-each-other.

1070 See, e.g., G. Parisi, S. Carnell, transl., *In a Flight of Starlings: The Wonder of Complex Systems* (London: Penguin, 2023).

1071 G. [Parisi], "How Do Different Definitions of Entropy Connect with Each Other?," *Physics, Stack Exchange*, 13 January 2021, https://physics.stackexchange.com/questions/606722/how-do-different-definitions-of-entropy-connect-with-each-other.

1072 Ibidem.

1073 H. Reichenbach, *The Direction of Time* (Mineola NY: Dover Publications, 1956), p.117.

1074 A. Eddington, Gifford lectures 1927, *The Nature of the Physical World* (Cambridge: Cambridge University Press, 1928), p. 69.

1075 M. Bronstein, "On the Expanding Universe," *Phys. Z. Sowjetunion*, (1933) **3**, p. 73.

1076 R. M. Unger and L. Smolin, *The Singular Universe and the Reality of Time*, n. 62.

1077 Ibid., p. 233.

1078 See, e.g., A. Eddington, *The Nature of the Physical World*, n. 1061.

1079 S. Carroll, *From Eternity to Here: The Quest for the Ultimate Theory of Time* (New York: Dutton, 2009), p. 3.

1080 P. C. W. Davies, *About Time: Einstein's Unfinished Revolution* (New York: Simon & Schuster, 2005), p. 278.

1081 J. Barbour, *The End of Time: The Next Revolution in Physics* (New York: Oxford University Press, 1999), p. 18.

1082 L. Smolin, *Einstein's Unfinished Revolution*, n. 74, p. 202.

1083 J. Barbour, *The End of Time: The Next Revolution in Physics*, n. 1081, p. 16.

1084 For a quick look at dimensions in physics, see D. D. Nolte, "A Short History of Multiple Dimensions," *Galileo Unbound*, 8 March 2023, https://galileo-unbound.blog/2023/03/08/a-short-history-of-hyperspace/.

1085 N. Copernicus, *De revolutionibus orbium coelestium libri vi* (Nuremberg: Johannes Petreius, 1543); E. Rosen, transl., *Six Books on the Revolutions of the Heavenly Spheres* (Warsaw: Polish Scientific Publications, 1978), p. 24; https://www.reed.edu/math/wieting/mathematics537/DeRevolutionibus.pdf.

1086 N. Wolchover, "What Shape Is the Universe? A New Study Suggests We've Got It All Wrong," *Quanta Magazine*, 4 November 2019, https://www.quantamagazine.org/what-shape-is-the-universe-closed-or-flat-20191104/.

1087 See, e.g., G. F. Lewis and P. van Oirschot, "How Does the Hubble Sphere Limit Our View of the Universe?," *Mon. Not. R. Astron. Soc.*, (2012) **423**, p. L26; http://academic.oup.com/mnrasl/article-pdf/423/1/L26/9454201/423-1-L26.pdf.

1088 Precisely what one concludes about how expansion affects observation depends on the model that one uses; see T. M. Davis and C. H. Lineweaver, "Expanding Confusion: Common Misconceptions of Cosmological Horizons and the Superluminal Expansion of the Universe," n. 444.

1089   L. Smolin in R. M. Unger and L. Smolin, *The Singular Universe and the Reality of Time: A Proposal in Natural Philosophy*, n. 62, p. 410.

1090   C. A. Mead, "Possible Connection Between Gravitation and Fundamental Length," n. 332.

1091   P. G. Bergmann, *The Riddle of Gravitation*, n. 324, p. ix.

1092   One might say, there is no way to build a Planck-scale version of an inukshuk, the stone cairn that Inuit peoples use to aid navigation in featureless tundra landscapes; see, e.g., N. Hallendy, "Inuksuk (Inukshuk)," *The Canadian Encyclopedia*, 8 December 2020, https://www.thecanadianencyclopedia.ca/en/article/inuksuk-inukshuk.

1093   See B. Bullock, "How Does the Japanese Addressing System Work?," https://www.sljfaq.org/afaq/addresses.html; see also: I. Singh, "How the Crazy Japanese Address System Works, Explained," *GEO Awesome*, 28 March 2019, https://geoawesomeness.com/how-the-crazy-japanese-addressing-system-works-explained/.

1094   See https://en-academic.com/pictures/enwiki/65/Area_Guide_Board-Japan1.png.

1095   For a review from an ontological perspective, see J. Hilgevoord and J. Uffink, "The Uncertainty Principle," (2016), n. 99, https://plato.stanford.edu/archives/win2016/entries/qt-uncertainty/.

1096   See, e.g., G. Kane, "Are Virtual Particles Really Popping In and Out of Existence? Or Are They Merely a Bookkeeping Device for Quantum Mechanics?," *Sci. Am.*, 9 October 2006, https://www.scientificamerican.com/article/are-virtual-particles-rea/.

1097   For more on such efforts, see N. Huggett & C. Rovelli, "Quantum Spacetime," *Sci. Am.*, (2024) **331** (2), p. 65; https://www.scientificamerican.com/article/do-space-and-time-follow-quantum-rules-these-mind-bending-experiments-aim-to-find-out/.

1098   C. Hogan, "Random Twists of Place: How Quiet Is Quantum Space-Time at the Planck Scale?," Fermilab News, 12 February 2021, https://news.fnal.gov/2021/02/random-twists-of-place-how-quiet-is-quantum-space-time-at-the-planck-scale/.

1099   S. O. Bilson-Thompson, F. Markopoulou and L. Smolin, "Quantum Gravity and the Standard Model," n. 764.

1100   A. Einstein, "On the Method of Theoretical Physics," n. 114, p. 20.

1101   Ibid., p. 21.

1102   The concept of particle in QFT leads to conceptual inconsistencies; D. Krause and O. Bueno, "Ontological Issues in Quantum Theory," *Manuscrito—Rev. Int. Fil.*, (2010) **33**, p. 269; https://web.as.miami.edu/personal/obueno/Site/Online_Papers_files/KraBue2008_Manuscrito_Final.pdf.

1103   C. Rovelli, "Halfway Through the Woods: Contemporary Research on Space and Time," n. 755, p. 208.

1104   E. Shaghoulian, "A Tale of Two Horizons," *Sci. Am.*, (2022) **327**, p. 42.

1105   L. Smolin, *Three Roads to Quantum Gravity*, n. 61, p. 13.

1106   L. Susskind, quoted in J. Stromberg, "Some Physicists Believe We're Living in a Giant Hologram—and It's Not That Far-fetched," *Vox*, 29 June 2015, https://www.vox.com/2015/6/29/8847863/holographic-principle-universe-theory-physics.

1107   For more on versions of the holographic principle, its origins, and its relationship to quantum-gravity physics, see L. Smolin, *Three Roads to Quantum Gravity*, n. 61, pp. 175 ff.

1108   Ibidem.

1109 G. 't Hooft, "Dimensional Reduction in Quantum Gravity," (1993), https://arxiv.org/pdf/gr-qc/9310026.pdf.

1110 L. Susskind, *The Black Hole War*, n. 279, p. 298; my emphasis.

1111 Ibid., p. 299.

1112 Ibid., p. 301.

1113 J. Maldacena, "The Large N Limit of Superconformal Field Theories and Supergravity," *Adv. Theor. Math. Phys.*, (1998) **2**, p. 231; https://arxiv.org/pdf/hep-th/9711200.pdf.

1114 B. Greene, *The Hidden Reality*, n. 661, p. 263.

1115 P. W. Phillips et al., "Stranger Than Metals," *Science*, (2022) **377**, p. 169; https://www.science.org/doi/10.1126/science.abh4273.

1116 T. M. Davis and C. H. Lineweaver, "Superluminal Recession Velocities," in R. Durrer et al., eds., *Cosmology and Particle Physics 2000*, American Institute of Physics, Conference Proceedings, (2001) **555**, p. 348; https://arxiv.org/abs/astro-ph/0011070.

1117 H. van Till, "Relativity, Special Theory of," *Encyclopedia of Science and Religion*, Tucows Inc., 17 May 2018, https://www.encyclopedia.com/science-and-technology/physics/physics/special-relativity-physics.

1118 T. M. Davis and C. H. Lineweaver, "Superluminal Recession Velocities," n. 1102.

1119 T. M. Davis and C. H. Lineweaver, "Expanding Confusion: Common Misconceptions of Cosmological Horizons and the Superluminal Expansion of the Universe," n. 444; their emphasis.

1120 Ibidem.

1121 D. Miller in D. D. Murphey and D. Miller, "An Intellectual Mind-Twister for Our Readers: Going to the Outer Limits of Cosmology," n. 854.

1122 D. Clery, "X-ray Telescope to Study How Magnetic Objects Sculpt Light," *Science*, (2021) **374**, p. 1310; https://www.science.org/content/article/nasa-telescope-study-how-extreme-cosmic-objects-sculpt-x-ray-light.

1123 E.g., D. Clery, "What Powers a Black Hole's Mighty Jets?," *Science*, 19 November 2014, https://www.science.org/content/article/what-powers-black-holes-mighty-jets.

1124 Some are about thirty billion degrees Kelvin; see D. C. Homan et al., "Intrinsic Brightness Temperatures of AGN Jets," *Astrophys. J.*, (2006) **642**, p. L115; https://iopscience.iop.org/article/10.1086/504715.

1125 D. Miller in D. D. Murphey and D. Miller, "An Intellectual Mind-Twister for Our Readers: Going to the Outer Limits of Cosmology," n. 854, p. 255.

1126 See, for example, data used in J-M. Marti, "Numerical Simulations of Jets from Active Galactic Nuclei," *Galaxies*, (2019) 7, p. 24; https://www.mdpi.com/2075-4434/7/1/24.

1127 R. Antonucci, "Active Galactic Nuclei and Quasars: Why Still a Puzzle after 50 Years?," (2105), http://arxiv.org/abs/1501.02001.

1128 A. Lobanov, "Overview of Active Galactic Nuclei," Max Planck Institut fur Radioastronomie; https://www3.mpifr-bonn.mpg.de/staff/sbritzen/lobanov.pdf.

1129 E.g., K. Nyland et al., "Revolutionizing Our Understanding of AGN Feedback and Its Importance to Galaxy Evolution in the Era of the Next Generation Very Large Array," *Astrophys. J.*, (2018) **859**, p. 23; https://iopscience.iop.org/article/10.3847/1538-4357/aab3d1.

1130 M. Young, "A Mad New Way to Make Black Hole Jets," *Sky & Telescope*, 12 June 2014, https://skyandtelescope.org/astronomy-news/new-way-make-black-hole-jets/.

1131   D. C. Homan et al., "Mojave XII. Acceleration and Collimation of Blazar Jets on Parsec Scales," *Astrophys. J.*, (2015) **798**, p. 134; https://iopscience.iop.org/article/10.1088/0004-637X/798/2/134/pdf.

1132   E.g., Y.-H. Lin et al., "Evolution and Feedback of AGN Jets of Different Cosmic-Ray Composition," *Mon. Not. R. Astron. Soc.*, (2023) **520**, p. 963; https://academic.oup.com/mnras/article-abstract/520/1/963/6993086.

1133   C. Tadhunter, quoted in editorial, "Astronomers Find Supermassive Black Hole Blasting Molecular Gas at One Million Kilometers Per Hour from a Galaxy," *Astron.*, 8 July 2014, https://www.astron.nl/astronomers-find-supermassive-black-hole-blasting-molecular-gas-at-one-million-kilometers-per-hour-from-a-galaxy/.

1134   R. Blandford et al., "Relativistic Jets in Active Galactic Nuclei," *Annu. Rev. Astron. Astrophys.*, (2019) **57**, p. 467; https://arxiv.org/pdf/1812.06025.pdf.

1135   See, e.g., Z. Savitsky, "The Tiny Physics Behind Immense Cosmic Eruptions," *Quanta Magazine*, 15 May 2023, https://www.quantamagazine.org/the-tiny-physics-behind-immense-cosmic-eruptions-20230515/.

1136   R. Antonucci, "Active Galactic Nuclei and Quasars: Why Still a Puzzle after 50 Years?," n. 1127.

1137   Ibidem.

1138   J. A. Zensus and T. J. Pearson, "Superluminal Radio Sources," in J. A. Zensus and T. J. Pearson, eds., *The Impact of VLBI on Astrophysics and Geophysics*, Proc.129th IAU Symposium, Cambridge, MA, May 1987 (Dordrecht: Kluwer Academic Publishers, 1988), p. 7.

1139   M. J. Rees, "Studies in Radio Source Structure: I. A Relativistically Expanding Model for Variable Quasi-Stellar Radio Sources," *Mon. Not. R. Astr. Soc.*, (1967) **135**, p. 345; https://adsabs.harvard.edu/full/1967MNRAS.135..345R.

1140   R. Antonucci, "Active Galactic Nuclei and Quasars: Why Still a Puzzle after 50 Years?," n. 1127.

1141   D. Clery, "X-ray Telescope to Study How Magnetic Objects Sculpt Light," n. 1122.

1142   R. Oerter, *The Theory of Almost Everything*, n. 204, p. 231.

1143   A. Guth, "The Inflationary Universe," in J. Brockman, ed., *The Universe: Leading Scientists Explore the Origin, Mysteries, and Future of the Cosmos* (New York: Harper Perennial, 2014), p. 28.

1144   F. Wilczek, *Fundamentals: Ten Keys to Reality*, n. 100, p. 193.

1145   Ibid., p. 199.

1146   D. Huterer and M. Turner, "Prospects for Probing the Dark Energy via Supernova Distance Measurements," *Phys. Rev. D*, (1999) **60**, 081371; https://arxiv.org/pdf/astro-ph/9808133.pdf.

1147   "Dark Energy," *COSMOS—The SAO Encyclopedia of Astronomy*, https://astronomy.swin.edu.au/cosmos/d/Dark+Energy.

1148   Estimates are in the order of 10^{-26} kg/m^3; see, e.g., R. Nave, "Dark Energy," *HyperPhysics*, http://hyperphysics.phy-astr.gsu.edu/hbase/Astro/dareng.html.

1149   M. S. Turner, quoted in A. Cho, "WIMPs at Last? Or More Wimpy Sightings?," *Science*, (2000) **287**, p. 1570.

1150 M. S. Turner, "Dark Matter and Dark Energy in the Universe," (1998), in B. K. Gibson et al., eds., *Astronomical Society of the Pacific Conference Series*, (1999) **165**, p. 431; https://www.aspbooks.org/a/volumes/article_details/?paper_id=17135.

1151 For a general review, see J. de Swart, "Five decades of missing matter," *Phys. Today*, (2024) **77** (8), p. 34; https://doi.org/10.1063/pt.ozhk.lfeb.

1152 A. Einstein and W. de Sitter, "On the Relation Between the Expansion and the Mean Density of the Universe," n. 491.

1153 E.g., B. Pinkerton and N. Hassenfeld, "Astronomers Were Skeptical about Dark Matter—Until Vera Rubin Came Along," *Vox*, 17 August 2021, https://www.vox.com/22576927/vera-rubin-dark-matter-astronomy-biography.

1154 See, e.g., Y. Yiu, "Remembering Vera Rubin," *Inside Science*, American Institute of Physics, 27 December 2016, https://ww2.aip.org/inside-science/remembering-vera-rubin.

1155 V. Rubin, quoted in "Vera Rubin and Dark Matter," American Museum of Natural History, https://www.amnh.org/learn-teach/curriculum-collections/cosmic-horizons-book/vera-rubin-dark-matter..

1156 E.g., R. H. Wechsler and J. L. Tinker, "The Connection Between Galaxies and Their Dark Matter Halos," *Ann. Rev. Astron. Astrophys.*, (2018) **56**, p. 435; https://arxiv.org/pdf/1804.03097.pdf.

1157 S. Hossenfelder, "The Nightmare Scenario for Dark Matter Is Inching Closer," *Science News*, 3 September 2024, https://www.patreon.com/posts/nightmare-for-is-111307303.

1158 A. Cho, "Hunt for Long-Sought Dark Matter Particle Nears a Climax," *Science*, (2022) **377**, p. 249; https://www.science.org/doi/epdf/10.1126/science.add9090.

1159 Ibidem.

1160 M. N. Ali, "The Search for Axion Dark Matter," *Science*, (2022) **375**, p. 833; https://www.science.org/doi/epdf/10.1126/science.ada1393.

1161 F. Chadha-Day et al, "Axion Dark Matter: What Is It and Why Now?," *Sci. Adv.*, (2022) **8**, p 3618.

1162 Ibidem.

1163 Y. B. Zeldovitch and I. D. Novikov, "The Hypothesis of Cores Retarded During Expansion and the Hot Cosmological Model," *Sov. Astron.*, (1967) **10**, p. 602; transl. from *Astron. Zhurn.*, **43**, p. 758.

1164 S. W. Hawking, "Gravitationally Collapsed Objects of Very Low Mass," *Mon. Not. R. Astron. Soc.*, (1971) **152**, p. 75; https://ui.adsabs.harvard.edu/abs/1971MNRAS.152...75H/abstract; see also B. J. Carr and S. W. Hawking, "Black Holes in the Early Universe," *Mon. Not. R. Astron. Soc.*, (1974) **168**, p. 399; https://ui.adsabs.harvard.edu/abs/1974MNRAS.168..399C/abstract.

1165 S. W. Hawking, "Gravitationally Collapsed Objects of Very Low Mass," ibidem.

1166 See, e.g., A. J. S. Hamilton, "Hawking Radiation," 19 April 1998, https://jila.colorado.edu/~ajsh/bh/hawk.html.

1167 For a review of some ways primordial black holes might have formed, see B. J. Carr, "Primordial Black Holes: Do They Exist and Are They Useful?," (2005), https://arxiv.org/abs/astro-ph/0511743v1.

1168 Y. B. Zeldovitch and I. D. Novikov, "The Hypothesis of Cores Retarded During Expansion and the Hot Cosmological Model," n. 1163.

1169   Physicists know the need for such new physics; see, e.g., E. Abdalla et al. (203 authors), "Cosmology Intertwined: A Review of the Particle Physics, Astrophysics, and Cosmology Associated with the Cosmological Tensions and Anomalies," *J. High En. Astrophys.*, (2022) **34**, p. 49; https://arxiv.org/abs/2203.06142.

1170   E.g., J. Sakstein et al., "Beyond the Standard Model Explanations of GW190521," *EP newsletter*, CERN, 19 September 2020, https://ep-news.web.cern.ch/content/beyond-standard-model-explanations-gw190521; and see https://arxiv.org/pdf/2009.01213.pdf.

1171   A. M. Green and B. J. Kavanagh, "Primordial Black Holes as a Dark Matter Candidate," *J. Phys. G: Nucl. Part. Phys.*, (2021) **48**, p. 3001; https://iopscience.iop.org/article/10.1088/1361-6471/abc534.

1172   E.g., P. Natarajan, "The First Monster Black Holes," *Sci. Am.*, (2022) **318**, p. 24; https://www.scientificamerican.com/article/the-puzzle-of-the-first-black-holes/.

1173   N. Cappelluti et al., "Exploring the High-Redshift PBH-CDM Universe: Early Black Hole Seeding, the First Stars and Cosmic Radiation Backgrounds," *Astrophys. J.*, (2022) **926**, p. 205; https://iopscience.iop.org/article/10.3847/1538-4357/ac332d.

1174   T. Pultarova, "Black Holes May Have Existed Since the Beginning of Time (and Could Explain Dark Matter Mystery)," *Space*, 17 December 2021, https://www.space.com/primordial-black-holes-explain-dark-matter-universe-mysteries.

1175   See also J. Sokol, "Physicists Argue That Black Holes from the Big Bang Could Be the Dark Matter," *Quanta Magazine*, 23 September 2020, https://www.quantamagazine.org/black-holes-from-the-big-bang-could-be-the-dark-matter-20200923/.

1176   K. Jedamzik, "Primordial Black Hole Formation During the QCD Epoch," *Phys. Rev. D.*, (1997) **55**, p. 5871; https://journals.aps.org/prd/abstract/10.1103/PhysRevD.55.R5871.

1177   D. Schwarz, "The First Second of the Universe," *Annalen Phys.*, (2003) **12**, p. 220; https://onlinelibrary.wiley.com/doi/10.1002/andp.200310010.

1178   See, e.g., A. M. Greene, "Astrophysical Uncertainties on Stellar Microlensing Constraints on Multisolar Mass Primordial Black Hole Dark Matter," *Phys. Rev. D*, (2017) **96**, 043020; https://journals.aps.org/prd/abstract/10.1103/PhysRevD.96.043020.

1179   K. C. Sahu et al., "An Isolated Stellar-Mass Black Hole Detected Through Astrometric Microlensing," *Astrophys. J.*, (2022) **93**, p. 83; https://iopscience.iop.org/article/10.3847/1538-4357/ac739e.

1180   As we go to press, the latest result is negative: LZ Collaboration, "Probing the scalar WIMP-pion coupling with the first LUX-ZEPLIN data," *Commun. Phys.* (2024) 7, 292; https://www.nature.com/articles/s42005-024-01774-8.

1181   A. Moberg, ed., "Written in the Stars," Nobel Prize in Physics 2011, The Royal Swedish Academy of Sciences, https://www.nobelprize.org/uploads/2018/06/popular-physicsprize2011-1.pdf.

1182   A. G. Riess and M. Turner, "From Slowdown to Speedup," *Sci. Am.*, (2008) **290**, p. 62; https://www.scientificamerican.com/article/from-slowdown-to-speedup/.

1183   O. Botner, quoted in "Discovery of Accelerating Universe Wins 2011 Nobel Prize in Physics," *Sci. Am.*, 4 October 2011, https://www.scientificamerican.com/article/the-2011-nobel-prize-in-prize-physics/.

1184   "Dark Energy, Dark Matter," *Universe*, NASA Science, https://science.nasa.gov/astrophysics/focus-areas/what-is-dark-energy.

1185   N. Aghanim et al., "Planck 2018 Results: VI. Cosmological Parameters," n. 686. But note the analysis depends upon the standard model.

1186   Appropriated from O. Wilde, *The Importance of Being Earnest*, (1865), Act 1; https://www.gutenberg.org/files/844/844-h/844-h.htm; with apologies to Oscar.

1187   N. A. Bahcall, "Hubble's Law and the Expanding Universe," *Proc. Natl. Acad. Sci. U.S.A.*, (2015) **112**, p. 3173; https://www.pnas.org/content/112/11/3173.

1188   See, e.g., "Black Holes," NASA Science, https://science.nasa.gov/astrophysics/focus-areas/black-holes.

1189   The largest known is quasar Ton 618 at some sixty-six billion $M_\odot$; C.Q. Choi, "'Stupendously Large' Black Holes Could Grow to Truly Monstrous Sizes," *Space*, 18 September 2020, https://www.space.com/black-holes-can-reach-stupendously-large-sizes.html.

1190   "Black Holes," NASA Science, https://science.nasa.gov/astrophysics/focus-areas/black-holes.

1191   In his then authoritative work on the whole field, Peebles said, "The parameter H_0 really has units of reciprocal time, but in many topics that is not the familiar way to express it." P. J. E. Peebles, *Principles of Physical Cosmology*, n. 82, p. xv.

1192   T. M. Davis and C. H. Lineweaver, "Expanding Confusion: Common Misconceptions of Cosmological Horizons and the Superluminal Expansion of the Universe," n. 444, Figure 1.

1193   For more on scientific notation see, e.g., J. M. Wenner, "Big Numbers and Scientific Notation," SERC/Carleton College, https://serc.carleton.edu/quantskills/methods/quantlit/BigNumbers.html.

1194   T. R. Lauer et al., "New Horizons Observations of the Cosmic Optical Background," *Ap. J.*, (2021) **906**, p. 77; https://iopscience.iop.org/article/10.3847/1538-4357/abc881.

1195   See, e.g., "Estimating the Size and Mass of a Black Hole," Chandra X-Ray Observatory, https://chandra.si.edu/edu/formal/math/7Page81.pdf.

1196   An evolving body of theory says the inflowing gas grows very hot, emitting intense radiation that repels further inflowing gas, tending to limit black hole growth; see, e.g., A. Levinson and E. Nakar, "Limits on the Growth Rate of Supermassive Black Holes at Early Cosmic Epochs," *Mon. Not. R. Astron. Soc.*, (2018) **473**, p. 2673.

1197   See, e.g., A. Kulier et al., "Understanding Black Hole Mass Assembly via Accretion and Mergers in Cosmological Simulations," *Astrophys. J.*, (2015) **799**, p. 178; https://arxiv.org/pdf/1307.3684.pdf.

1198   The need to rebuild the standard model is increasingly recognized: see, e.g., J. O'Callaghan, "Breaking Cosmology," *Sci. Am.*, (2022) **327**, p. 28; https://www.scientificamerican.com/article/jwsts-first-glimpses-of-early-galaxies-could-break-cosmology/.

1199   A. Einstein, "Kosmologische Betrachtungen zur allgemeinen Relativitatstheorie" (Cosmological Considerations on the Theory of General Relativity), n. 533, p. 179.

1200   This is a not entirely fanciful, or at least not new, scenario; see B. J. Carr and S. W. Hawking, "Black Holes and the Early Universe," *Mon. Not. R. Astr. Soc.*, (1974) **168**, p. 399, wherein the authors find solutions "in which the whole Universe is inside the black hole."

1201   E.g., "What Is a Gravitational Wave?," *Space Place*, NASA Science, 4 June 2020, https://spaceplace.nasa.gov/gravitational-waves/en/.

1202   See, e.g., "The Early Universe," CERN, (2024), https://home.cern/science/physics/early-universe.

1203 See, e.g., C. L. Steinhardt et al., "The Impossibly Early Galaxy Problem," *Astrophys. J.*, (2016) **824**, p. 21; https://iopscience.iop.org/article/10.3847/0004-637X/824/1/21/pdf.

1204 A. Cattaneo et al., "The Role of Black Holes in Galaxy Formation and Evolution," *Nature*, (2009) **460**, p. 213; https://www.nature.com/articles/nature08135.

1205 See, e.g., P. Natarajan, "The First Monster Black Holes," n. 1172.

1206 Quoted by science writer Jonathan Amos in "Earliest and Most Distant Galaxy Ever Observed," BBC, 31 May 2024, https://www.bbc.com/news/articles/cjeenyw8rd2o; see also S. Carniani et al., "A Shining Cosmic Dawn: Spectroscopic Confirmation of Two Luminous Galaxies at z ~ 14," https://arxiv.org/pdf/2405.18485.

1207 S. W. Hawking, "Gravitationally Collapsed Objects of Very Low Mass," n. 1164; see also B. J. Carr and S. W. Hawking, "Black Holes in the Early Universe," n. 1164.

1208 B. J. Carr et al., "New Cosmological Constraints on Primordial Black Holes," *Phys. Rev. D*, (2010) **81**, p. 1103.

1209 H. R. Quinn, "The Asymmetry Between Matter and Antimatter," *Phys. Today*, **56**, p. 30.

1210 Physics symmetries are a seemingly simple but unavoidably technical subject. For the bold, see J. Bernabeu, "Symmetries in the Standard Model," in B. G. Sidharth et al., eds., *Fundamental Physics and Physics Education Research* (Cham: Springer, 2020); https://doi.org/10.1007/978-3-030-52923-9_1.

1211 See, e.g., G. Sciolla, "The Mystery of CP Violation," *MIT Physics Annual*, (2006), p. 44; https://physics.mit.edu/wp-content/uploads/2021/01/physicsatmit_06_sciollafeature.pdf.

1212 See, e.g., "The Matter-Antimatter Asymmetry Problem," CERN, https://home.cern/science/physics/matter-antimatter-asymmetry-problem.

1213 See, e.g., I. Aitchison et al., "B-factories Confirm Matter-Antimatter Asymmetry; Leads to 2008 Nobel Prize in Physics," The BaBar Collaboration, 2008, https://www-public.slac.stanford.edu/babar/Nobel2008.htm.

1214 M. Freiberger, "Born from Broken Symmetry," *Plus*, 10 October 2008, https://plus.maths.org/content/born-broken-symmetry.

1215 See, e.g., S. Katzir, "The Emergence of the Principle of Symmetry in Physics," *Hist. Stud. Phys. Biol.*, (2004) **35**, p. 35.

1216 For a dated but comprehensive frolic in the fields of symmetries, see Martin Gardner's *The New Ambidextrous Universe: Symmetry and Asymmetry from Mirror Reflections to Superstrings*, 3rd ed. (New York: W. H. Freeman & Co., 1990).

1217 D. J. Gross, "The Role of Symmetry in Fundamental Physics," *Proc. Natl. Acad. Sci. U.S.A.*, (1996) **93**, p. 14256.

1218 L. Smolin, *Einstein's Unfinished Revolution*, n. 74, p. 264.

1219 E.g., J. Adam et al., "Measurement of e+e−Momentum and Angular Distributions from Linearly Polarized Photon Collisions," *Phys. Rev. Lett.*, (2021) **127**, 052302; https://www.bnl.gov/newsroom/news.php?a=119023.

1220 M. A. Fedderke et al., "Fireball antinucleosynthesis," *Phys. Rev. D*, (2024) **109**, 123028; https://journals.aps.org/prd/pdf/10.1103/PhysRevD.109.123028.

1221 It's a small imbalance, about one part in a billion; see, e.g., "The Matter-Antimatter Asymmetry Problem," CERN, https://home.cern/science/physics/matter-antimatter-asymmetry-problem.

1222 L. Smolin, *Einstein's Unfinished Revolution*, n. 74, p. xviii.

1223 A. Koestler, *The Act of Creation* (London: Hutchinson, 1964), p. 177.

1224 F. Bacon, *Novum Organum, sive Indicia Vera de Interpretatione Naturae*, n. 706, aph. XLIII.

1225 For a history of debate about the public (vs. private) view of language, see I. Hacking, *Historical Ontology*, n. 142, pp. 121 ff.

1226 J. Jacobs, *The Nature of Economies* (New York: Random House, 2000), p. 137.

1227 H. Reichenbach, *The Rise of Scientific Philosophy*, n. 337, p. viii.

1228 E.g., L. Boroditsky, "How Does Our Language Shape the Way We Think?," *Edge*, 11 June 2009, https://www.edge.org/conversation/lera_boroditsky-how-does-our-language-shape-the-way-we-think.

1229 J. Hintikka, *Knowledge and Belief: An Introduction to the Logic of the Two Notions* (Ithaca NY: Cornell University Press, 1962); repr., V.F. Hendricks and J. Symonds, eds., (London: King's College Publications, 2005).

1230 Ibid., p. 77 ff.

1231 P. Hut, "Seikakusa no Kontei ni aru Aimaisa (Ambiguity at the Roots of Precision)," in H. Kawai and I. Nakazawa, eds., *Aimai no Chi* (*The Wisdom of Ambiguity*) (Tokyo: Iwanami Shoten, 2003), p. 51; https://www.ias.edu/ids/~piet/publ/other/ambiguity.

1232 E.g., L. Boroditsky, "How Does Our Language Shape the Way We Think?," *Edge*, 11 June 2009, https://www.edge.org/conversation/lera_boroditsky-how-does-our-language-shape-the-way-we-think.

1233 C. Rovelli, *Quantum Gravity*, n. 41, p. 79.

1234 Plato, P. Shorey, transl., "The Allegory of the Cave," in E. Hamilton and H. Cairns, eds., *Plato: Collected Dialogues* (Princeton: Princeton University Press, 1962), p. 747; https://yale.learningu.org/download/ca778ca3-7e93-4fa6-a03f-471e6f15028f/H2664_Allegory of the Cave.pdf.

1235 R. M. Unger (as reported by Smolin) in R. M. Unger and L. Smolin, *The Singular Universe and the Reality of Time*, n. 62, p. 523; Smolin seemed to stop short of embracing this language.

1236 J. D. North, *The Measure of the Universe: A History of Modern Cosmology*, n. 1042, p. 64.

1237 G. Barros et al., "Russian Offensive Campaign Assessment, 15 April 2023," Institute for the Study of War, https://www.understandingwar.org/backgrounder/russian-offensive-campaign-assessment-april-15-2023.

1238 The ontogeny of language creation across individual to language-group to multiple-language scales depends on common processes; T. Brochhagen et al., "From Language Development to Language Evolution: A Unified View of Human Creativity," *Science*, (2023) **381**, p. 431.

1239 N. Postman and C. Weingartner, *Teaching as a Subversive Activity* (London: Penguin Books, 1969), p. 102.

1240 L. Smolin, *Einstein's Unfinished Revolution*, n. 74, p. 257.

1241 F. Markopoulou, "The Internal Description of a Causal Set: What the Universe Looks Like from the Inside," n. 560.

1242 Intuitionistic logic is concerned with "not, as in classical logic, mind-independent truth, but mental constructibility." See, e.g., M. van Atten, "The Development of

Intuitionistic Logic," *Stanford Encyclopedia of Philosophy*, n. 99, 4 May 2022, *Stanford Encyclopedia of Philosophy*, https://plato.stanford.edu/entries/intuitionistic-logic-development/.

1243   See also L. E. J. Brouwer, "Over de Grondslagen der Wiskundel," thesis, Universiteit van Amsterdam, 1907; transl. in A. Heyting, transl. and ed., "On the Foundations of Mathematics," *Collected Works. I: Philosophy and Foundations of Mathematics* (Amsterdam, North-Holland, 1975).

1244   R. M. Unger and L. Smolin, *The Singular Universe and the Reality of Time*, n. 62.

1245   F. Markopoulou, "An Insider's Guide to Quantum Causal Histories," n. 867.

1246   Physicist Rafael Sorkin has long been its principal proponent; see D. D. Reid, "Introduction to Causal Sets: An Alternate View of Spacetime Structure," *Can. J. Phys.*, (2001) **79**, p. 1; https://arxiv.org/pdf/gr-qc/9909075.pdf

1247   L. Smolin, *Time Reborn: From the Crisis in Physics to the Future of the Universe*, n. 714, p. 176.

1248   L. Smolin, *Einstein's Unfinished Revolution*, n. 74, p. 258.

1249   The new ontology requires inserting new space quanta and their links into the universe to be very rare except where space is extremely curved by extreme gravity. But in principle it happens everywhere and this gives rise to quandaries about both how it works and what this means for causality. The latter quandary is exhibited, in slightly different form, in the interface between quantum theory and general relativity; see, e.g., J. Cotler and A. Strominger, "The Universe as a Quantum Encoder," https://arxiv.org/pdf/2201.11658.pdf; for a plainer explanation see C. Wood, "Physicists Rewrite a Quantum Rule That Clashes with Our Universe," *Quanta Magazine*, 20 September 2022, https://www.quantamagazine.org/physicists-rewrite-a-quantum-rule-that-clashes-with-our-universe-20220926/.

1250   The set's evolution is an example of a quantum causal set; see F. Markopoulou, "New Directions in Background Independent Quantum Gravity," in D. Oriti, ed., *Approaches to Quantum Gravity*, n. 6, p. 129.

1251   See, e.g., R. B. Salgado, "A More Illuminating Look at the Light Cone," *The Light Cone: An Illuminating Introduction to Relativity*, 15 June 1996, https://visualrelativity.com/LIGHTCONE/lightcone.html.

1252   F. Markopoulou, "The Internal Description of a Causal Set: What the Universe Looks Like from the Inside," n. 560.

1253   Č. Brukner, "Quantum Causality," *Nature Physics*, (2014) **10**, p. 259; https://www.nature.com/articles/nphys2930.

1254   See, e.g., M. Frisch, "Causation in Physics," *Stanford Encyclopedia of Philosophy*, n. 99, 24 August 2020, https://plato.stanford.edu/entries/causation-physics/#CausQuanMech.

1255   H. Bergson, *Essai sur les données immédiates de la conscience* (Paris: Félix Alcan, 1888); F. L. Pogson, transl., *Time and Free Will: An Essay on the Immediate Data of Consciousness* (Mineola NY: Dover Publications, 2001).

1256   B. Russell, "On the Notion of Cause," presidential address to the Aristotelian Society on 4 November 1912; repr., *Proc. Aristot. Soc.*, (1912) **13**, p. 1; https://archive.org/details/1912RussellOnTheNotionOfCause.

1257   M. Planck, *Der Kausalbegriff in der Physik* (Leipzig, Barth, 1948); repr. and transl. as "The Concept of Causality in Physics," in F. Gaynor, transl., M. von Laue, ed., *Scientific Autobiography and Other Papers*, (New York: Philosophical Library, 1949), p. 121.

1258 Ibid., p. 122.

1259 A. Einstein, "Considerations Concerning the Fundaments of Theoretical Physics," n. 606.

1260 E.g., essays in P. Dowe and P. Noordhof, eds., *Cause and Chance: Causation in an Indeterministic World*, (London: Routledge, 2004); or see a review of that volume by philosopher of science Clark Glymour, https://www.cmu.edu/dietrich/philosophy/docs/glymour/glymour-dowe2004.pdf.

1261 See, e.g., Y. Ben-Menahem, "Struggling with Causality: Einstein's Case," in *Science in Context*, vol. 6, (1993), p. 291; https://www.cambridge.org/core/journals/science-in-context/article/abs/struggling-with-causality-einsteins-case/2DB574036CED542BFEA325230D217290.

1262 But this is a subject that can descend swiftly into technical debates; see, e.g., J. Earman, "Determinism: What We Have Learned and What We Still Don't Know," in J. K. Campbell, ed., *Freedom and Determinism* (Cambridge MA: MIT Press, 2004), p. 21; https://psycnet.apa.org/record/2004-18955-001.

1263 See, *e.g.*, E. Curiel, "Light Cones and Causal Structure," *Stanford Encyclopedia of Philosophy*, n. 99, (2019), https://plato.stanford.edu/entries/ spacetime-singularities/lightcone.html.

1264 Č. Brukner, "Quantum Causality," n. 1253.

1265 M. Planck, "The Concept of Causality in Physics," n. 1257, p. 125.

1266 See, e.g., "The Wavelength of an Electron," *Quantum Mechanics: A Mini-Course*, Ohio State University, https://www.asc.ohio-state.edu/mathur.16/quantummechanics27-11-17/qm1.2/qm1.2.html.

1267 A. Einstein, letter to Jerome Rothstein, 22 May 1950, *Albert Einstein Archives*, fo. 22, doc. 54; quoted in W. Isaacson, *Einstein*, n. 37.

1268 R. P. Feynman et al., *The Feynman Lectures on Physics: Volume III, Quantum Mechanics* (Reading MA: Addison-Wesley Publishing Co., 1965), p. 1-1; Feynman's emphasis.

1269 T. Young, "Bakerian Lecture: Experiments and Calculations Relative to Physical Optics," *Phil. Trans. Roy. Soc.*, (1804) **94**, p. 1; https://www.jstor.org/stable/107135.

1270 A. Becker, *What Is Real?*, n. 49, p. 28.

1271 E.g., E. Gregersen, "wave-particle duality," n. 96, https://www.britannica.com/science/wave-particle-duality.

1272 A. Einstein and L. Infeld, *The Evolution of Physics*, n. 130, p. 263.

1273 A. Einstein, lecture in Berlin, February 1927, Z. Angew. Chemie, (1927) **40**, p. 546; transl. and quoted by A. Pais, *Subtle Is the Lord*, n. 44, p. 443.

1274 P. Ball, "Quantum Physics May Be Even Spookier Than You Think," *Sci. Am.*, May 21, 2018, https://www.scientificamerican.com/article/quantum-physics-may-be-even-spookier-than-you-think/.

1275 Ibidem.

1276 For a conceptually accurate and fun demonstration, see "Dr. Quantum—Double Slit Experiment," https://www.youtube.com/watch?v=Q1YqgPAtzho.

1277 See R. Menzel et al., "Wave-Particle Dualism and Complementarity Unraveled by a Different Mode," *Proc. Nat. Acad. Sci. U.S.A.*, (2012) **109**, p. 9314; https://www.pnas.org/doi/10.1073/pnas.1201271109; and see a nontechnical commentary at https://

arstechnica.com/science/2012/05/disentangling-the-wave-particle-duality-in-the-double-slit-experiment/.

1278   M. Beller, *Quantum Dialogue*, n. 112, at pp. 265, 13, 247, 12, 263, and 265.

1279   R. P. Feynman, *The Character of Physical Law*, n. 269, p. 139.

1280   A. Einstein and L. Infeld, *The Evolution of Physics*, n. 130, p. 282.

1281   A. Becker, *What Is Real?*, n. 49, p. 17.

1282   E.g., R. de-Picciotto et al., "Direct Observation of a Fractional Charge," *Nature*, (1997) **389**, p. 162; https://www.nature.com/articles/38241; and see Z. Lu et al., "Fractional Quantum Anomalous Hall Effect in Multilayer Graphene," *Nature*, (2024) **626**, p. 759; https://www.nature.com/articles/s41586-023-07010-7.

1283   E.g., A. Augustyn, "quark: subatomic particle," n. 96, 18 May 2023, https://www.britannica.com/science/quark.

1284   E.g., C. Wood, "How Our Reality May Be a Sum of All Possible Realities," *Quanta Magazine*, 2 February 2023, https://www.quantamagazine.org/how-our-reality-may-be-a-sum-of-all-possible-realities-20230206/.

1285   A. Einstein, letter to Walter Dällenbach, after 15 February 1917, quoted by and transl. in J. Stachel, "Einstein and the Quantum," in R. G. Colodny, ed., *From Quarks to Quasars: Philosophical Problems in Modern Physics*, n. 758, p. 379; Einstein's italics. Note that A. M. Hentschel, in *The Collected Papers of Albert Einstein, Vol. 8, The Berlin Years: Correspondence, 1914–1918*, n. 757, p. 285, gave a different translation (and the correct date) but I prefer Stachel's for most of the passage; it seems attuned more closely to Einstein's concerns (though it has the wrong date).

1286   See "Walter Dällenbach 1892–1990," *ETH Zurich*, Research Collection, https://www.research-collection.ethz.ch/handle/20.500.11850/142949.

1287   A. Einstein, letter to Walter Dällenbach, n. 1285.

1288   Ibidem; Einstein's emphasis.

1289   J. J. Thomson, "XL. Cathode Rays," *Phil. Mag.*, (1897) **44**, p. 293; https://doi.org/10.1080/14786449708621070.

1290   B. Greene, *The Hidden Reality: Parallel Universes and the Deep Laws of the Cosmos*, n. 661, p. 10.

1291   M. S. Turner, "Origin of the Universe," *Sci. Am.*, (2009) **301**, p. 36; https://www.scientificamerican.com/article/origin-of-the-universe-extreme-physics-special/.

1292   P. J. E. Peebles, "Large-Scale Background Temperature and Mass Fluctuations Due to Scale-Invariant Primeval Perturbations," *Astrophys. J.*, (1986) **263**, p. L1; https://articles.adsabs.harvard.edu/pdf/1982ApJ...263L...1P.

1293   "The ΛCDM paradigm is currently the simplest and the most faithful theoretical representation of our universe . . . based on two dominant components, which are still unknown, undetected and poorly understood: dark matter and dark energy." "Beyond Planck: Precision Computational Cosmology," *Ebrary*, https://ebrary.net/77808/sociology/planck_precision_computational_cosmology.

1294   P. J. E. Peebles, *The Whole Truth*, n. 60, p. 162 ff.

1295   P. J. E. Peebles and B. Ratra, "The Cosmological Constant and Dark Energy," *Rev. Mod. Phys.*, (2003) **75**, p. 559; https://arxiv.org/pdf/astro-ph/0207347.pdf.

1296   J. Mangum, "How Many Stars Do the Andromeda and Milky Way Galaxies Have?," *Ask an Astronomer*, National Radio Astronomy Observatory, 30 July 2020, https://public.nrao.edu/ask/how-many-stars-do-the-andromeda-and-milky-way-galaxies-have/.

1297   Astronomers can see to the edge of the visible part of the universe, beyond which galaxies are (or, rather, were when light now reaching us left them) moving away from us faster than the speed of light. This does not conflict with special relativity: see chapter 65, and see Davis and Lineweaver, "Expanding Confusion: Common Misconceptions of Cosmological Horizons and the Superluminal Expansion of the Universe," n. 444.

1298   E. Siegel, "If the Universe Is 13.8 Billion Years Old, How Can We See 46 Billion Light Years Away?," *Forbes*, 23 February 2018, https://www.forbes.com/sites/startswithabang/2018/02/23/if-the-universe-is-13-8-billion-years-old-how-can-we-see-46-billion-light-years-away/.

1299   Those who want to see more of it with their own eyes but lack access to big telescopes can find many images online, including the Hubble Ultra Deep Field image of a tiny part of the sky showing some ten thousand galaxies (about a quadrillion stars) in the early universe, many of them now more than twenty billion light-years away: see https://esahubble.org/images/heic0611b/.

1300   M. Vardanyan et al., "Applications of Bayesian Model Averaging to the Curvature and Size of the Universe," *Mon. Not. R. Astron. Soc.*, (2011) **413**, p. L91; https://watermark.silverchair.com/413-1-L91.pdf.

1301   L. Smolin, *Three Roads to Quantum Gravity*, n. 61, p. 22.

1302   S. W. Hawking and G. F. R. Ellis, *The Large Scale Structure of Space-Time* (Cambridge: Cambridge University Press, 1973), pp. 348 ff.

1303   For an easy-to-read review of such small-size and high-density issues, see C. Moskowitz, "What Is the Smallest Thing in the Universe?," *Live Science*, 17 September 2012, https://www.livescience.com/23232-smallest-ingredients-universe-physics.html.

1304   S. W. Hawking and G. F. R. Ellis, *The Large Scale Structure of Space-Time*, n. 1302, p. 364.

1305   See, e.g., A. May and E. Howell, "What Is the Big Bang Theory?," *Space*, 12 May 2023, https://www.space.com/25126-big-bang-theory.html.

1306   C. Rovelli, *Reality Is Not What It Seems*, n. 7, p. 107.

1307   R. Brandenberger and C. Vafa, "Superstrings in the Early Universe," *Nucl. Phys. B*, (1989) **316**, p. 391.

1308   Ibidem.

1309   L. Smolin, *The Trouble with Physics*, n. 70.

1310   S. Weinberg, *The First Three Minutes*, n. 1018, p. 154.

1311   And, in a subsequent interview, Weinberg elaborated: "I believe that there is no point in the universe that can be discovered by the methods of science." See undated transcript, https://www.pbs.org/faithandreason/transcript/wein-frame.html.

1312   D. Harmon & J. Roiland, *Rick and Morty*, Warner Bros., (2013).

1313   M. Manson, *The Subtle Art of Not Giving a F*ck: A Counterintuitive Approach to Living a Good Life* (New York: HarperOne, 2016).

1314   J. K. Huysmans, *À rebours* (Paris: Charpentier, 1884); R. Baldick, transl., *Against Nature* (London: Penguin, 2003); and *À vau-l'eau* (Brussels: Henry Kistmaekers, 1882); R. Baldick, transl., *Downstream* (Brooklyn: Turtle Point Press, 2005).

1315   As Heidegger sought to do: M. Heidegger, J. Macquarrie, and E. Robinson, transls., *Being and Time* (New York: Harper & Row, 1962), p. 27: '"Being" has been presupposed in all ontology up till now, but not as a *concept* at one's disposal. . . . This "presupposing" of Being has rather the character of taking a look at it beforehand, so that in the light of it the entities presented to us get provisionally Articulated in their Being.'

1316   M. S. Williams, "What Is the Cosmic Microwave Background?," *Universe Today: Space & Astronomy News*, 8 September 2018, https://www.universetoday.com/135288/what-is-the-cosmic-microwave-background/.

1317   E.g., E. Yazgin, "Was Einstein Slightly Off? Cosmic-Scale Test of One of His Main Theories Throws Up Strange Result," *Cosmos*, 15 November 2022, https://cosmosmagazine.com/space/cosmic-relativity-einstein-test/.

1318   For an easier to read, fictional (but also factual) account, see Book One of this work, n. 16.

1319   D. Miller in D. D. Murphey and D. Miller, "An Intellectual Mind-Twister for Our Readers: Going to the Outer Limits of Cosmology," n. 854, p. 255.

1320   J. Baggott, *A Beginner's Guide to Reality*, n. 2, p. 228.

1321   L. Smolin, *Einstein's Unfinished Revolution*, n. 74, p. 229.

1322   Ibid., p. 232.

1323   Ibidem.

1324   Ibidem.

1325   Ibid., p. 272.

1326   Ibid., p. 279.

1327   Ibid., p. 277.

1328   R. M. Unger in R. M. Unger and L. Smolin, *The Singular Universe and the Reality of Time*, n. 62, p. 205.

1329   Ibid., p. 206.

1330   Ibid., p. 498.

1331   F. Graham, citing a survey of readers, "Reader Poll," *Nature Briefing*, 3 February 2023; https://us17.   campaign-archive.com/?e=406084aa49&u=2c6057c528fdc6f73fa196d9d&id=53693f109b.

1332   P. Maddy, *Realism in Mathematics*, n. 123, p. 13.

1333   F. H. Bradley, *Essays on Truth and Reality*, n. 1065, p. 48.

1334   A. Becker, "The Origins of Space and Time," n. 51, p. 31.

1335   Their needs differ. String theory directly predicts supersymmetry. The standard model has multiple problems supersymmetry might, or might not, resolve; see, e.g., G. L. Kane, "Supersymmetry: What? Why? When?," *Contemp. Phys.*, (2000) **41**, p. 359; https://inpp.ohio.edu/~inpp/nuclear_lunch/archive/2005/kane_supersym.pdf.

1336   P. Van Nieuwenhuizen, "Supergravity," *Phys. Rep.*, (1981) 68, p. 189.

1337   See, e.g., M. Lindner and S. Ohmer, "Emerging Internal Symmetries from Effective Spacetimes," *Phys. Lett. B*, (2017) **773**, p. 231; https://www.sciencedirect.com/science/article/pii/S0370269317306482.

1338 See, e.g., "The Search for Supersymmetry: Come Out, Come Out, Wherever You Are!," *The Economist*, 3 January 2015; https://www.businessinsider.com/the-search-for-supersymmetry-come-out-come-out-wherever-you-are-2015-1.

1339 See P. Bechtle et al., "Determination of MSSM Parameters from LHC and ILC Observables in a Global Fit," *Eur. Phys. J. C*, (2006) **46**, p. 533.

1340 "Supersymmetry Predicts a Partner Particle for Each Particle in the Standard Model, to Help Explain Why Particles Have Mass," CERN, (2022), https://home.cern/science/physics/supersymmetry.

1341 J. Lykken and M. Spiropulu, "Supersymmetry and the Crisis in Physics," *Sci. Am.*, (2014) **310**, p. 34.

1342 M. Shifman, "Reflections and Impressionistic Portrait at the Conference Frontiers Beyond the Standard Model, FTPI, Oct. 2012," https://arxiv.org/abs/1211.0004.

1343 M. Disney, 15 August 2006 interview in "Most of Our Universe Is Missing," *Horizon*, BBC, https://www.youtube.com/watch?v=agqjaX2_QVI.

1344 Ibidem.

1345 "European Scientists Reveal Plans for Next-Generation Particle Collider to Study Dark Matter," *Euronews*, 6 February 2024, https://www.euronews.com/next/2024/02/06/european-scientists-reveal-plans-for-next-generation-particle-collider-to-study-dark-matte.

1346 CERN describes, "An international collaboration of more than 150 universities, research institutes and industrial partners from all over the world.," "Future Circular Collider," https://home.cern/science/accelerators/future-circular-collider.

1347 D. Castelvecchi, "Next-Generation LHC: CERN Lays Out Plans for €21-Billion Supercollider," *Nature*, 15 January 2019, https://www.nature.com/articles/d41586-019-00173-2.

1348 S. Hossenfelder, "The World Doesn't Need a New Gigantic Particle Collider," n. 184.

1349 See further: C. J. Gillespie, "The Silence About New Fundamental Particles Sends Us a Loud and Clear Message," *Science Seen*, 21 November 2016, http://www.timeone.ca/the-silence-about-new-fundamental-particles-sends-us-a-loud-and-clear-message/.

1350 C. Rovelli, *Reality Is Not What It Seems*, n. 7, p. 259.

1351 For example, Rovelli did his best to do away with time: C. Rovelli, *The Order of Time*, n. 69; reviewed in A. Jaffe, "The Illusion of Time," *Nature*, (2018) **556**, p. 304; https://www.nature.com/articles/d41586-018-04558-7.

1352 L. Smolin, *The Trouble with Physics*, n. 70, p. 256.

1353 C. Rovelli, *The Order of Time*, n. 69, p. 2.

1354 R. A. Muller, *Now*, n. 443, p. 16.

1355 Universität Basel, "One of World's Oldest Sun Dials Dug Up in Kings' Valley, Upper Egypt," *ScienceDaily*, 14 March 2013, www.sciencedaily.com/releases/2013/03/130314085052.htm.

1356 In 1761, clockmaker John Harrison invented the nautical chronometer; see the Commissioners for the Discovery of the Longitude at Sea, *The Principles of Mr. Harrison's Time-Keeper* (London: Richardson & Clark, 1767), Board of Longitude; see also: D. Sobel, *Longitude: The True Story of a Lone Genius Who Solved the Greatest Scientific Problem of His Time* (New York: Walker & Company, 1995).

1357 This was Einstein's explicit definition; A. Einstein, *Relativity*, n. 245, p. 28.

1358 See, e.g., "Uniform Mean Time," *Sci. Am.*, (1874) **30**, p. 111: "The Pennsylvania Railroad and some of its dependencies . . . use Pittsburgh time, which is transmitted by electricity from the Allegheny observatory, an astronomical clock of the best construction. It is regulated by a telescope, which shows its return, every twenty-four hours, to the point of observation of a fixed star, so that the earth itself becomes the regulating clock of the observatory."

1359 This is an exact count; though resting on analog foundations, our definition of time is trending toward its digital roots; Bureau International des Poids et Mesures, 13th Conférence Génerale, 1967, Resolution 1; https://www.bipm.org/en/committees/cg/cgpm/13-1967/resolution-1.

1360 R. A. Muller, *Now*, n. 443, p. 119.

1361 Ibid., p. 123.

1362 Ways have been found to reintroduce time; see, e.g., A. Y. Kamenshchik et al., "Time in Quantum Theory, the Wheeler-DeWitt Equation and the Born-Oppenheimer Approximation," *Int. J. Mod. Phys. D*, (2019) **28**, 1950073; https://arxiv.org/pdf/1809.08083.pdf.

1363 L. Smolin, *Time Reborn*, n. 714, pp. 81–2.

1364 See, e.g., L. Smolin, *Three Roads to Quantum Gravity*, n. 61, p. 40.

1365 C. Rovelli, "The Strange Equation of Quantum Gravity," *Class. Quantum Grav.*, (2015) **32**, 124005; https://arxiv.org/pdf/1506.00927.pdf.

1366 The Wheeler-DeWitt equation may be fundamental where the Schrödinger equation is not; and see T. P. Shestakova, "Is the Wheeler-DeWitt Equation More Fundamental Than the Schrödinger Equation?," *Int. J. Mod. Phys. D*, (2018) **27**, 1841004; https://arxiv.org/pdf/1801.01351.pdf.

1367 Ibidem. See also T. P. Shestakova, "On the Meaning of the Wave Function of the Universe," *Int. J. Mod. Phys. D*, (2019) **28**, 1941009; https://arxiv.org/pdf/1909.05588.pdf.

1368 E.g., Agence France-Presse, "Do Not Adjust Your Clock: Scientists Call Time On the Leap Second," *Guardian*, 18 November 2022, https://www.theguardian.com/world/2022/nov/18/do-not-adjust-your-clock-scientists-call-time-on-the-leap-second; and D. C. Agnew, "A Global Timekeeping Problem Postponed by Global Warming," *Nature*, (2024), https://www.nature.com/articles/s41586-024-07170-0.

1369 N. Paquette, quoted in A. Becker, "The Origins of Space and Time," n. 51, p. 28.

1370 Ibidem.

1371 H. Reichenbach, *Philosophie der Raum-Zeit-Lehre* (Berlin: Walter de Gruyter, 1928); M. Reichenbach and J. Freund, transls., *The Philosophy of Space and Time* (New York: Dover Publications, 1957) p. 109.

1372 D. Hume, *A Treatise of Human Nature*, n. 862, Book I, Part II, Sec. VII.

1373 S. Walter, "The Historical Origins of Spacetime," in A. Ashtekar and V. Petkov, eds., *The Springer Handbook of Spacetime* (Cham: Springer, 2014), p. 27; https://halshs.archives-ouvertes.fr/halshs-01234449/document.

1374 H. Minkowski, "Von Stund' an vollen Raum, für sich und Zeit für sich völlig zu Schatten herabsinken und nur noch eine Art Union der beiden soll Selbständigkeit bewahren," address in Göttingen to the Gesellschaft Deutscher Naturforscher und Ärtze, 21 September 1908; repr. as "Raum und Zeit," *Jahresbericht der Deutschen Mathematiker-Vereinigung*, (1909) 18, p. 75; my translation. A popular translation (Perrett's and Jeffrey's) ends with ". . . preserve an independent reality." See H. Minkowski, "Space and Time," in H. A. Lorentz et al., *Das Relativitätsprinzip* (Leipzig: Teubner, 1922); W. Perrett and

G. B. Jeffery, transls., *The Principle of Relativity: A Collection of Original Memoirs on the Special and General Theory of Relativity* (New York: Dover Publications, 1952), p. 75. But, see above, Minkowski did *not* in fact speak of reality (*wirklichkeit*).

1375 H. P. Robertson, "On E. A. Milne's Theory of World Structure," Z. *Astrophys.*, (1933) 7, p. 153; https://adsabs.harvard.edu/full/1933ZA......7..153R.

1376 J. A. Wheeler, *Geons, Black Holes & Quantum Foam*, n. 78, p. 270.

1377 N. Noir, "The Last Frontier," *Nathaniel Noir's Blog*, 30 August 2020, https://www.nathanaelnoir.com/blog/the-last-frontier.

1378 M. Hersch, "The Wheeler-DeWitt Equation," *Javier Rubio blog*, 10 February 2017, https://javierrubioblog.files.wordpress.com/2016/09/notes_wheeler-dewitt_talk.pdf.

1379 P. J. E. Peebles, *The Whole Truth*, n. 60, p. 135.

1380 See R. T. W. Arthur, "Minkowski's Proper Time and the Status of the Clock Hypothesis," in V. Petkov, ed., *Space, Time, and Spacetime* (Cham: Springer, 2010), p. 159.

1381 A. Einstein, "Autobiographical Notes," n. 216, p. 59; my variation of translation.

1382 My translation; original quoted by Valentine Bargmann to, and recounted by, A. Pais in his definitive biography, *Subtle Is the Lord*, n. 44, p. 152; Pais translated it as "superfluous learnedness."

1383 A. Einstein, *Relativity*, n. 245, p. 63.

1384 A. Valentini, "Hidden Variables and the Large-Scale Structure of Space-Time," n. 1005, p. 131.

1385 Its crowning glory is that the separation of two points in spacetime is invariant in all Lorentz transformations between coordinate systems; see A. Einstein, *Relativity*, n. 245, p. 174.

1386 R. A. Muller, *Now*, n. 443, p. 7.

1387 Minkowski's address to the GDNA is often rendered in English as ascribing reality to spacetime. This is an artifact of the canonical translation; see n. 1355.

1388 H. Minkowski, "Von Stund' an vollen Raum, für sich und Zeit für sich völlig zu Schatten herabsinken und nur noch eine Art Union der beiden soll Selbständigkeit bewahren," n. 1355, my translation.

1389 This is an extensive topic; readers to whom it is new might start with, for example, "A Matter of Time," *Einstein Exhibition*, American Museum of Natural History, 15 November 2002, https://www.amnh.org/exhibitions/einstein/time/a-matter-of-time.

1390 L. Smolin, "Space and Time in the Quantum Universe," n. 613; his emphasis.

1391 F. Markopoulou, "New Directions in Background Independent Quantum Gravity," n. 1250, p. 129.

1392 Ibid., p. 142.

1393 Ibid., p. 137.

1394 Ibid., p. 143.

1395 L. Smolin, *Einstein's Unfinished Revolution*, n. 74, p. 261.

1396 J. Baez, "What Is a Background-Free Theory?," n. 161; he said, "Personally I think one can dig oneself into a hole by trying to do physics without *any* background structure."

1397 R. M. Unger and L. Smolin, *The Singular Universe and the Reality of Time: A Proposal in Natural Philosophy*, n. 62, p. 232.

1398 G. Zukav, *The Dancing Wu Li Masters* (New York: William Morrow, 1979), p. 200.

1399 N. J. Nersessian, "Faraday's Field Concept" in D. Gooding and F. A. J. L. James, eds., *Faraday Rediscovered* (Cambridge: Cambridge University Press, 1981), p. 175.

1400 J. C. Maxwell, "A Dynamical Theory of the Electromagnetic Field," *Phil. Trans. Roy. Soc. London*, (1865) **155**, p. 459; https://ia903201.us.archive.org/21/items/dynamicaltheoryo00maxw/dynamicaltheoryo00maxw.pdf.

1401 M. Strassler, *Waves in an Impossible Sea: How Everyday Life Emerges from the Cosmic Ocean* (New York: Basic Books, 2024).

1402 M. Strassler, interview by Sean Carroll, 4 March 2024; https://www.preposterousuniverse.com/podcast/2024/03/04/267-matt-strassler-on-relativity-fields-and-the-language-of-reality/.

1403 D. Tong, interviewed by S. Strogatz, "What Is Quantum Field Theory and Why Is It Incomplete?," *Quanta Magazine*, 10 August 2022, https://www.quantamagazine.org/what-is-quantum-field-theory-and-why-is-it-incomplete-20220810/.

1404 A. Einstein and L. Infeld, *The Evolution of Physics*, n. 130, p. 142.

1405 Ibid., p. 243.

1406 E.g., S. C. Ramos, "The Discovery of Calculus: Leibniz vs. Newton," StMU Research Scholars, 3 November 2017, https://stmuscholars.org/the-discovery-of-calculus-leibniz-vs-newton/.

1407 A. Hobson, "There Are No Particles, There Are Only Fields," *Am. J. Phys.*, (2013) **81**, p. 211.

1408 E. Siegel, "Are Quantum Fields Real?," *Forbes*, 17 November 2018.

1409 R. P. Feynman, *Feynman's Lectures on Physics*, Caltech (1963), vol. 1, p. 2–4.

1410 E.g., G. U. Jakobsen, "General Relativity from Quantum Field Theory," thesis, The Niels Bohr Institute, University of Copenhagen, 1 July 2020; https://arxiv.org/pdf/2010.08839.pdf.

1411 Y. V. Baryshev, "Energy-Momentum of the Gravitational Field: Crucial Point for Gravitation Physics and Cosmology," *Pr. Cosmol.*, (2008) **1**, p. 276; https://arxiv.org/abs/0809.2323.

1412 See, e.g., A. Einstein, "The Problem of Space, Ether, and the Field in Physics," n. 10.

1413 "On physical grounds this gave rise to the conviction that the metrical field was the same as the gravitational field." A. Einstein, *Relativity*, n. 245, Appendix 5, "Relativity and the Problem of Space," p. 73.

1414 A. Einstein, "On the Method of Theoretical Physics," n. 114, p. 19.

1415 A. Einstein, letter to M. Besso, 10 August 1954; *Albert Einstein Archives*, The Hebrew University of Jerusalem, doc. 7–421; quoted by and transl. in J. Stachel, "Einstein and the Quantum: Fifty Years of Struggle," n. 758, p. 380.

1416 See J. Stachel, "The Other Einstein: Einstein Contra Field Theory," *Science in Context*, (1993) **6**, p. 275; reprinted in J. Stachel, *Einstein from "B" to "Z,"* n. 594, p. 141.

1417 A. Einstein, letter to Walter Dällenbach, after 15 February 1917, in *The Collected Papers of Albert Einstein, Vol. 8*, n. 1285, p. 285.

1418 E.g., L. Bombelli et al., "Space-Time as a Causal Set," *Phys Rev. Lett.*, (1987) **59**, p. 521.

1419 A. Einstein, "The Problem of Space, Ether, and the Field in Physics," n. 10, p. 285.

1420 E. R. Harrison, *Cosmology: The Science of the Universe* (Cambridge: Cambridge University Press, 1981), p. 276.

1421 E.g., A. Augustyn, "conservation of energy," n. 96, 19 May 2023, https://www.britannica.com/science/quark.

1422 Of her, Einstein wrote: "In the judgment of the most competent living mathematicians, Fräulein Noether was the most significant creative mathematical genius thus far produced since the higher education of women began." *The New York Times*, 4 May 1935; https://mathshistory.st-andrews.ac.uk/Obituaries/Noether_Emmy_Einstein/.

1423 E. Noether, "Invariante Variationsprobleme," *Nachr. v. d. Ges. d. Wiss. zu Göttingen, (1918),* p. 235; see also, e.g., N. Byers, "E. Noether's Discovery of the Deep Connection Between Symmetries and Conservation Laws," Symposium on the Heritage of Emily Noether in Algebra, Geometry and Physics, Tel Aviv, 2 December 1996; http://cwp.library.ucla.edu/articles/noether.asg/noether.html.

1424 See, e.g., B. Crowell, "Noether's Theorem for Energy," *Chemistry LibreTexts,* 18 August 2020, https://phys.libretexts.org/Bookshelves/Classical_Mechanics/Supplemental_Modules_(Classical_Mechanics)/Conservation_of_Mass_and_Energy/Noether%27s_Theorem_for_Energy.

1425 A. Einstein, soundtrack of film *Atomic Physics,* J. Arthur Rank Organization Ltd., (1948); https://history.aip.org/exhibits/einstein/voice1.htm.

1426 E.g., S. De Haro, "Noether's Theorems and Energy in General Relativity," in J. Reid, ed., *The Philosophy and Physics of Noether's Theorems* (Cambridge: Cambridge University Press, 2022); https://philsci-archive.pitt.edu/18871/1/Noether%27s%20Theorems%20and%20Energy%20in%20GR.pdf.

1427 Efforts to rescue conservation of energy as a local law within general relativity seem mired in difficulty; see, e.g., K. Brading, "A Note on General Relativity, Energy Conservation and Noether's Theorems," in A. J. Coc and J. Eisenstaedt, eds., *The Universe of General Relativity* (Cham: Springer, 2006), p. 125; https://inspirehep.net/literature/476201.

1428 Penrose was one: "The energy—and therefore the mass—of the gravitational field is a slippery eel indeed, and refuses to be pinned down in any clear location." R. Penrose, "The Mass of the Classical Vacuum," in S. Saunders and H. R. Brown, eds., *The Philosophy of Vacuum,* n. 265, p. 21; see also R. Penrose, *The Emperor's New Mind* (Oxford: Oxford University Press, 1989).

1429 Y. V. Baryshev, "Energy-Momentum of the Gravitational Field: Crucial Point for Gravitation Physics and Cosmology," n. 1411.

1430 P. J. E. Peebles, *Principles of Physical Cosmology,* n. 82, p. 139.

1431 Y. V. Baryshev, "Paradoxes of Cosmological Physics in the Beginning of the 21st Century," in R. A. Ryutin and V. A. Petrov, eds., *30th International Workshop on High Energy Physics—Particle and Astroparticle Physics, Gravitation and Cosmology—Predictions, Observations and New Projects* (Singapore: World Scientific, 2015), p. 297; https://arxiv.org/abs/1501.01919.

1432 T. M. Davis, "Is the Universe Leaking Energy?," *Sci. Am.,* (2010) **303**, p. 38.

1433 R. P. Feynman et al., *The Feynman Lectures on Physics: Volume I, Mainly Mechanics, Radiation and Heat* (Reading MA: Addison-Wesley Publishing Co., 1965), p. 4-2; https://www.feynmanlectures.caltech.edu/I_04.html.

1434 F. Wilczek, *The Lightness of Being: Mass, Ether and the Unification of Forces*, n. 192, p. 1.

1435 R. Matarese et al., "SI Redefinition," National Institute of Standards and Technology, 4 June 2019, https://www.nist.gov/si-redefinition/kilogram.

1436 Another complication is their speed (and so their energy) depends upon the number of fleeting proton-neutron pairings, which depends on the size of the nucleus; see, e.g., B. Schmookler et al., "Modified Structure of Protons and Neutrons in Correlated Pairs," *Nature*, (2019) **566**, p. 354; https://www.nature.com/articles/s41586-019-0925-9; and see J. Chu, "Study of Quark Speeds Finds a Solution for A 35-Year Physics Mystery," *MIT News*, 20 February 2019, https://news.mit.edu/2019/quark-speed-proton-neutron-pairs-0220.

1437 F. Wilczek, *The Lightness of Being*, n. 192, p. 128 ff.

1438 Ibid., p. 74.

1439 S. Nadis, "Mass and Angular Momentum, Left Ambiguous by Einstein, Get Defined," *Quanta Magazine*, 13 July 2022, https://www.quantamagazine.org/mass-and-angular-momentum-left-ambiguous-by-einstein-get-defined-20220713/.

1440 Ibidem.

1441 E.g., P.-N. Chen et al., "Quasilocal Mass at Axially Symmetric Null Infinity," *Int. J. Mod. Phys. D*, (2019) **28**, 1930013; https://arxiv.org/pdf/1901.06948.pdf.

1442 R. P. Feynman, *The Character of Physical Law*, n. 269, p. 9.

1443 A. Pais, *Subtle Is the Lord*, n. 44, p. 288; Pais's emphasis.

1444 A. Einstein, "Die Grundlage der allgemeinen Relativitätstheorie," *Ann. Physik*, (1916) **49**; "The Foundation of the General Theory of Relativity," in H. A. Lorentz et al., W. Perrett and G. B. Jeffery, transls. (incomplete), *The Principle of Relativity*, n. 1374, p. 160; repr. and transl. in part, *The Collected Papers of Albert Einstein, Vol. 6, The Berlin Years: Writings, 1914–1917*, n. 23, p. 147. See also D. W. Sciama, "On the Origin of Inertia," *Mon. Not. Roy. Astron. Soc.*, (1952) **113**, p. 34; https://adsabs.harvard.edu/full/1953MNRAS.113...34S.

1445 See, e.g., M. Schirber, "Nobel Prize—Why Particles Have Mass," *Physics*, (2013) **6**, p. 111; https://physics.aps.org/articles/v6/111.

1446 S. Weinberg, "What Is an Elementary Particle?," *Beam Line*, Stanford Linear Accelerator Center, Spring 1997, p. 17; http://www.ub.edu/hcub/hfq/sites/default/files/27-1-weinberg.pdf.

1447 Ibid. p. 21.

1448 It ceased publication in 2001; https://www.slac.stanford.edu/pubs/beamline/pastissues.html.

1449 S. O. Bilson-Thompson et al., "Emergent Braided Matter of Quantum Geometry," n. 903.

1450 *Merriam-Webster* dictionary, https://www.merriam-webster.com/dictionary/composite.

1451 C. Rovelli, *Quantum Gravity*, n. 41, p. 19.

1452 Democritus, quoted in S. Berryman, "Democritus," *Stanford Encyclopedia of Philosophy*, n. 99, https://plato.stanford.edu/entries/democritus/; citing C. C. W. Taylor, transl., *The Atomists: Leucippus and Democritus. Fragments, A Text and Translation with Commentary* (Toronto: University of Toronto Press, 1999).

1453 R. P. Feynman, *QED: The Strange Theory of Light and Matter* (Princeton: Princeton University Press, 1985), p. 128.

1454 P. A. M. Dirac, "The Evolution of the Physicist's Picture of Nature," *Sci. Am.*, (1963) **208**, p. 45; repr. in https://blogs.scientificamerican.com/guest-blog/the-evolution-of-the-physicists-picture-of-nature/.

1455 P. A. M. Dirac, quoted in H. Kragh, *Simply Dirac* (Cambridge: Cambridge University Press, 2016), p. 184.

1456 R. P. Feynman, *QED: The Strange Theory of Light and Matter*, n. 1433.

1457 E.g., M. Constantinou and H. Panagopoulos, "Improved Renormalization Scheme for Non-Local Operators," *Phys. Rev. D*, (2023) **107**, 014503; https://arxiv.org/pdf/2207.09977.pdf.

1458 See, e.g., G. Farmelo, *The Universe Speaks in Numbers: How Modern Maths Reveals Nature's Deepest Secrets* (London: Faber and Faber, 2019), which offers a counterpoint to Hossenfelder's—and my—criticisms, n. 157.

1459 C. Rovelli, *Reality Is Not What It Seems*, n. 7, p. 148.

1460 D. Amati et al., "Can Space Time Be Probed Below the String Size?," CERN, (1988), CERN-TH.5207/88; https://cds.cern.ch/record/191788/files/198811126.pdf.

1461 See, e.g., D. Weaire, ed., *The Kelvin Problem*, special issue of *Forma*, (Tokyo: KTK Scientific Publishers, 1996), vol. 11, pp. 161–330.

1462 W. Thompson, "On the Division of Space with Minimum Partitional Area," *Phil. Mag.*, (1887) **24**, p. 503; https://zapatopi.net/kelvin/papers/on_the_division_of_space.html.

1463 D. L. Weaire and R. Phelan, "A Counter-Example to Kelvin's Conjecture on Minimal Surfaces," *Phil. Mag. Lett.*, (1994) **69**, p. 107.

1464 S. Carlip, "Hiding the Cosmological Constant," *Phys. Rev. Lett.*, (2019) **23**, 131302; https://journals.aps.org/prl/pdf/10.1103/PhysRevLett.123.131302.

1465 S. Hossenfelder, "The 'Worst Prediction' Was Never Made: The True Story," 16 April 2024, https://www.patreon.com/posts/worst-prediction-102409950.

1466 P. J. E. Peebles, *The Whole Truth*, n. 60, p. 141.

1467 E.g., T. Yoneya, "String Theory and Space-Time Uncertainty Principle," *Prog. Theor. Phys.*, (2000) **103**, p. 1081; https://core.ac.uk/download/pdf/25283203.pdf.

1468 L. Smolin, *Three Roads to Quantum Gravity*, n. 61, p. 115.

1469 Ibid., p. 11.

1470 Ibid., pp. 106 ff.

1471 G. Amelino-Camelia et al., "Tests of Quantum Gravity from Observations of Gamma-Ray Bursts," *Nature*, (1998) **393**, p. 763.; https://cds.cern.ch/record/340297/files/9712103.pdf.

1472 See also G. Amelino-Camelia, Quantum-Spacetime Phenomenology," *Living Rev. Relativity*, (2013) **16**, p. 5; https://link.springer.com/content/pdf/10.12942/lrr-2013-5.pdf.

1473 G. 't Hooft, "The Fundamental Nature of Space and Time," in D. Oriti, ed., *Approaches to Quantum Gravity*, n. 6, p. 13.

1474 E.g., H. Genz, *Die Entdeckung des Nichts*, (München: Carl Hanser Verlag, 1994); K. Heusch, transl., *Nothingness: The Science of Empty Space* (Reading MA: Perseus Books,

1999); J. D. Barrow, *The Book of Nothing: Vacuums, Voids, and the Latest Ideas about the Origins of the Universe*; F. Close, *The Void*, n. 1060.

1475 At that temperature, the only other detectable gas is hydrogen, with a vapour pressure of 10^{-7} torr; see C. Benvenuti, "Molecular Surface Pumping: Cryopumping," in S. Turner, ed., *Vacuum Technology*, CERN Accelerator School conference proceedings, Snekersten, Denmark, 28 May 1999, Table 1 at p. 52; https://www.osti.gov/etdeweb/servlets/purl/20146141.

1476 Outgassing is unavoidable even with all surfaces made of selected metals and glasses, and even after everything was baked some residual emission of adsorbed or trapped gas molecules continued; see J. L. Segovia, "Physics of Outgassing," https://cds.cern.ch/record/455557/files/p99.pdf.

1477 E.g., D. C. Webb, "Space as a Vacuum," 10 April 2014, *Science Daily*, https://sites.coloradocollege.edu/pc357ml/2014/04/10/space-as-a-vacuum/.

1478 B. Rundle, *Why There Is Something Rather Than Nothing* (Oxford: Clarendon Press, 2004), p. 116.

1479 P. Ben-Zvi, "Kant on Space," *Philosophy Now*, (2005) #49, https://philosophynow.org/issues/49/Kant_on_Space.

1480 R. Williams, "October 1644: Torricelli Demonstrates the Existence of a Vacuum," *APS News*, October 2012, https://www.aps.org/publications/apsnews/201210/physicshistory.cfm.

1481 P. Ben-Zvi, "Kant on Space," n. 1459.

1482 I. Kant, *Prolegomena zu einer jeden künftigen Metaphysik, die als Wissenschaft wird auftreten können* (Riga: Johann Friedrich Hartknoch, 1783); P. Carus, transl., *Prolegomena to Any Future Metaphysics That Will Be Able to Present Itself as a Science* (Chicago: Open Court Publishing Co., 1904), p. 16; https://www.gutenberg.org/files/52821/52821-h/52821-h.htm.

1483 See, e.g., P. Wu et al., "Space and Time Averaged Quantum Stress Tensor Fluctuations," *Phys. Rev. D*, (2121) **103**, 125014; https://arxiv.org/abs/2104.04446.

1484 See, e.g., H. S. Kragh and J. M. Overduin, *The Weight of the Vacuum: A Scientific History of Dark Energy* (Cham: Springer, 2014).

1485 W. H. Nernst, "Über einen Versuch von quantentheoretischen S. Turner, ed., *Vacuum Technology* Betrachtungen zur Annahme stetiger Energieänderungen zurückzukehren," *Verh. Dtsch. Phys. Ges.*, (1916) **18**, p. 83; and see, e.g., P. Huber and T. Jaakolla, "The Static Universe of Walther Nernst," *Semantic Scholar*, (1995), https://www.semanticscholar.org/paper/The-static-universe-of-Walther-Nernst-Huber-Jaakkola/3cea992842e3f7d5624871c694dce7127588e566.

1486 A. Almheiri, "Black Holes, Wormholes and Entanglement," *Sci. Am.*, (2022) **327**, p. 34.

1487 E.g., U. Mohideen and A. Roy, "Precision Measurement of the Casimir Force from 0.1 to 0.9µm," *Phys. Rev. Lett.*, (1998) **81**, p. 4549.

1488 M. P. Hobson et al., *General Relativity: An Introduction for Physicists* (Cambridge: Cambridge University Press, 2000), p. 187.

1489 For the full story, see H. Kragh, "Walter Nernst: Grandfather of Dark Energy?," *Astron. & Geophys.*, (2012) **53**, p. 1.24.

1490 W. V. Quine, *Methods of Logic* (rev. ed.) (New York: Holt Rinehart & Winston, 1966), p. 95.

1491 B. Russell, *Introduction to Mathematical Philosophy* (London: George Allen & Unwin, 1919), p. 252; https://people.umass.edu/klement/imp/imp.html.

1492 W. V. Quine, *From a Logical Point of View: Logico-Philosophical Essays*, ch. IX, "Meaning and Existential Inference," n. 38, p. 161.

1493 A. Almheiri, "Black Holes, Wormholes and Entanglement," n. 1466.

1494 See S. W. Hawking, "The Beginning of Time," lecture in 1996; https://www.hawking.org.uk/in-words/lectures/the-beginning-of-time; and see A. Cho, "Stephen Hawking's (Almost) Last Paper: Putting an End to the Beginning of the Universe," *Science*, 2 May 2018, https://www.science.org/content/article/stephen-hawking-s-almost-last-paper-putting-end-beginning-universe.

1495 L. Krauss, *A Universe from Nothing: Why There Is Something Rather than Nothing* (London: Simon & Schuster, 2012), p. 98.

1496 It was that rarest of reviews, one that was itself reviewed. Of the acerbic tongue, astrophysicist Adam Frank said: "It was the very title of Krauss' book that Albert picked apart. *Can physics explain how a Universe emerges from nothing?* Not surprisingly, everything depends on which 'nothing' you are talking about. That is where the knives came out." A. Frank, "Blackboard Rumble: Why Are Physicists Hating On Philosophy (and Philosophers)?," NPR, 1 May 2012, https://www.npr.org/sections/13.7/2012/05/01/151752815/blackboard-rumble-why-are-physicists-hating-on-philosophy-and-philosophers.

1497 D. Albert, "On the Origin of Everything," *The New York Times*, 23 March 2012; https://www.nytimes.com/2012/03/25/books/review/a-universe-from-nothing-by-lawrence-m-krauss.html.

1498 J.-P. Sartre, *Being and Nothingness*, n. 1031, p. 48.

1499 See, e.g., L. Smolin, *The Trouble with Physics*, n. 70.

1500 See, e.g., M. Schermer, "Wronger Than Wrong," *Sci. Am.*, (2006) **295**, p. 40; https://www.scientificamerican.com/article/wronger-than-wrong/.

1501 P. Woit, *Not Even Wrong* (New York: Basic Books, 2006).

1502 P. Woit, "Hawking Gives Up," *Not Even Wrong*, 7 September 2010, https://www.math.columbia.edu/~woit/wordpress/?p=3141.

1503 Ibidem.

1504 H. Johnston, "M-Theory, Religion and Science Funding on the BBC," *Physics World*, 8 September 2010, https://physicsworld.com/a/by-hamish-johnstonstephen-hawk/.

1505 S. W. Hawking and L. Mlodinow, *The Grand Design*, n. 430, p. 180.

1506 J. Horgan, "Cosmic Clowning: Stephen Hawking's 'New' Theory of Everything Is the Same Old CRAP," *Sci. Am.*, 13 September 2010, https://blogs.scientificamerican.com/cross-check/cosmic-clowning-stephen-hawkings-new-theory-of-everything-is-the-same-old-crap/.

1507 L. King, "Leonard Mlodinow: On Larry King," *Larry King Live*, 10 September 2010, CNN, https://transcripts.cnn.com/show/lkl/date/2010-09-10/segment/01; https://www.youtube.com/watch?v=OgQvlBrNBLk.

1508 J. W. Moffat, *Cracking the Particle Code of the Universe*, n. 198, p. 113.

1509 See G. 't Hooft, *In Search of the Ultimate Building Blocks* (Cambridge: Cambridge University Press, 1997), p. 68 ff.

1510 See, e.g., B. V. Svistunov et al., *Superfluid States of Matter* (Boca Raton: CRC Press, 2021), p. 510.

1511   J. W. Moffat, *Cracking the Particle Code of the Universe*, n. 198, p. 113.

1512   E.g., S. L. Wu, "Historic Review of Higgs Searches," Higgs Quo Vadis Conference, Aspen, 10 March 2013, slide 4; https://cds.cern.ch/record/1593025/files/ATL-PHYS-SLIDE-2013-487.pdf.

1513   J. W. Moffat, *Cracking the Particle Code of the Universe*, n. 198, p. 115.

1514   H. Johnston, "Overlooked for the Nobel: the CERN Physicists Who Discovered the Higgs Boson," *Physics World*, 29 September 2020, https://physicsworld.com/a/overlooked-for-a-nobel-the-cern-physicists-who-discovered-the-higgs-boson/.

1515   In 2018, CERN announced detection of the Higgs decaying into two bottom quarks amid an "overwhelmingly large background contribution from a number of other SM processes that can mimic its experimental signature."; https://cms.cern/news/higgs-observed-decaying-b-quarks. It is not clear these data significantly improved the evidence.

1516   Quoted in K. C. Cole, "The Unnatural Future of Physics," *Wired*, 18 October 2022, https://www.wired.com/story/unnatural-future-physics/; Cole's (and presumably Alexander's) emphasis.

1517   J. A. Wheeler, *Geons, Black Holes & Quantum Foam*, n. 78, p. 312.

1518   This spectacle occurs quite often in nearby galaxies; it is called a type 2 supernova; see, e.g., A. Briggs, "What Is a Supernova?," *EarthSky*, 12 November 2020, https://earthsky.org/astronomy-essentials/definition-what-is-a-supernova/.

1519   E. Malek, "The Singularity Theorem (Nobel Prize in Physics 2020)," Einstein Online, https://www.einstein-online.info/en/spotlight/the-singularity-theorem/.

1520   E. Curiel, "Singularities and Black Holes," *Stanford Encyclopedia of Philosophy*, (2019), n. 99, https://plato.stanford.edu/entries/spacetime-singularities/.

1521   Their formation leaves them very hot; e.g., "Astronomy Picture of the Day," 25 April 1998, NASA, https://apod.nasa.gov/apod/ap980425.html.

1522   JWST images may have revealed infrared light from a neutron star thirty-six years after a supernova in our galaxy; C. Fransson et al., "Emission Lines Due to Ionizing Radiation from a Compact Object in the Remnant of Supernova 1987A," *Science*, (2024) **383**, p.898; https://arxiv.org/abs/2403.04386.

1523   J. Hessels et al., "A Radio Pulsar Spinning at 716 Hz," *Science*, (2006) **311**, p. 5769; https://www.science.org/doi/10.1126/science.1123430.

1524   See, e.g., A. Smale, "Neutron Stars," (2017), *Imagine the Universe!*, National Aeronautics and Space Administration, https://imagine.gsfc.nasa.gov/science/objects/neutron_stars1.html.

1525   B. P. Abbott et al., "Gravitational Waves and Gamma-Rays from a Binary Neutron Star Merger: GW170817 and GRB 170817A," *Astrophys. J. Lett.*, (2017) **841**, p. L13; https://iopscience.iop.org/article/10.3847/2041-8213/aa920c/pdf.

1526   See, e.g., N. Greenfieldboyce, "Astronomers Strike Gravitational Gold in Colliding Neutron Stars," *The Two-Way*, NPR, 16 October 2017, https://www.npr.org/sections/thetwo-way/2017/10/16/557557544/astronomers-strike-gravitational-gold-in-colliding-neutron-stars.

1527   Simply, in its tweedles, not in a "frozen star," nor at its notional event horizon; see A. Feldman, "Stephen Hawking's black hole radiation paradox could finally be solved—if black holes aren't what they seem," *Live Science*, 20 September 2024, https://www.livescience.com/physics-mathematics/quantum-physics/stephen-hawking-s-black-hole-radiation-paradox-could-finally-be-solved-if-black-holes-aren-t-what-they-seem; and see J. D. Bekenstein, "Black holes and entropy," *Phys. Rev. D*, (1973) 7, p. 2333.

1528 A. Augustyn, "neutron star," n. 96, https://www.britannica.com/science/neutron-star.

1529 This kind of event is not well understood; for example, recent observations show a fireball that is spherical, a shape that is hard to explain; A. Sneppen et al., "Spherical Symmetry in the Kilonova AT2017gfo/GW170817," *Nature*, (2023) **614**, p. 436.

1530 See the CODATA internationally recommended 2018 values of the fundamental physical constants, "Planck Length," *The NIST Reference on Constants, Units and Uncertainty*, National Institute of Standards and Technology, https://physics.nist.gov/cgi-bin/cuu/Value?plkl.

1531 A. J. Ayer, *Philosophy in the Twentieth Century* (New York: Vintage Books, 1984), p. 1; https://archive.org/details/philosophyintwen00ayer/page/n9/mode/2up.

1532 R. G. Boscovich, *Philosophiae naturalis theoria redacta ad unicam legem virium in natura existentium* (The theory of natural philosophy reduced to a single law of forces existing in nature) (Vienna: Bernadi, 1758), § 539, p. 254: "Ac vanissima illorum somnia corruunt penitus, qui Mundum vel casu quodam fortuito putant, vel fatali quadam necessitate potuisse condi, vel per se ipsum existere ab aeterno suis necessariis legibus consisstentem." My translation. http://www.brera.mi.astro.it/~mario.carpino/Biblioteca_Digitale/ENB/ENB_Opere_v07.pdf.

1533 R. Leverton, "Spoon Fed Advanced Physics," *Amazon*, 17 January 2014, https://www.amazon.com/Time-One-Understanding-Colin-Gillespie-ebook/dp/B08KGRJDFR/ref=sr_1_1; and see https://www.amazon.com/gp/profile/amzn1.account.AEXCCWZW-TIY5W4IX6TCNQIQREVUQ/ref=cm_cr_dp_d_gw_tr.

1534 For a review of this chaotic "Planck time" period in context of the Wheeler-DeWitt equation, see B. L. Hu et al., "Minisuperspace as a Quantum Open System" in *Misner Festschrift*, B. L. Hu et al., eds., (Cambridge: Cambridge University Press, 1993), p. 145.

1535 M. Rees, "Pondering Astronomy in 2009," *Science*, (2009) **323**, p. 309; https://www.science.org/doi/full/10.1126/science.1170104.

1536 E.g., "Pong," *The Strong National Museum of Play*, https://www.museumofplay.org/games/pong/.

1537 E.g., E. Gregersen, "gas laws," n. 96, https://www.britannica.com/science/gas-laws.

1538 The concept of emergence extends beyond physics, for example to biology; see, e.g., D. Galas, "emergent property," n. 96, https://www.britannica.com/science/emergent-property.

1539 E.g., C. R. Nave, "Particle in a Box," *HyperPhysics*, Georgia State University, http://hyperphysics.phy-astr.gsu.edu/hbase/quantum/pbox.html.

1540 Complexity arises in emergent systems from interplay between levels in the hierarchy of emergence. See, e.g., F. E. Rosas et al., "Software in the Natural World: A Computational Approach to Hierarchical Emergence," 5 June 2024, https://arxiv.org/pdf/2402.09090.

1541 G. F. B. Riemann, "On the Hypotheses Which Lie at the Foundation of Geometry," n. 273.

1542 Ibidem.

1543 P. G. Bergmann, *The Riddle of Gravitation* (New York: Dover Publications, 1992), revised and updated ed., p. ix.

1544 Ibid., p. x.

1545 See R. D. Kaaronen et al., "Body-based Units of Measure in Cultural Evolution," *Science*, (2023) **380**, p. 948; https://www.science.org/doi/10.1126/science.adf1936.

1546   E.g., J. A. L. Rodríguez et al., "Early science and colossal stone engineering in Menga, a Neolithic dolmen (Antequera, Spain)," *Sci. Adv.*, (2024) **10**, p. 1295; https://www.science.org/doi/epdf/10.1126/sciadv.adp1295.

1547   See, e.g., "The Friedmann-Lemaitre-Robertson-Walker Metric," *LibreTexts*, 6 September 2021, https://phys.libretexts.org/Courses/Skidmore_College/Introduction_to_General_Relativity/07%3A_Cosmology/7.02%3A_The_Friedmann-Lemaitre-Robertson-Walker_Metric.

1548   F. Wilczek, *The Lightness of Being*, n. 192.

1549   G. F. B. Riemann, "On the Hypotheses Which Lie at the Foundation of Geometry," n. 273, p. 36.

1550   Ibidem.

1551   D. Overbye, "Laws of Nature, Source Unknown," *The New York Times*, 18 December 2007; https://www.nytimes.com/2007/12/18/science/18law.html.

1552   L. Lederman and D. Teresi, *The God Particle*, n. 201, p. 401.

1553   E.g., S. W. Hawking and L. Mlodinow, *The Grand Design*, n. 430; and L. M. Krauss, *A Universe from Nothing*, n. 1495; and see chapter 91.

1554   E.g., D. Albert, "On the Origin of Everything," *The New York Times*, 23 March 2012;   https://www.nytimes.com/2012/03/25/books/review/a-universe-from-nothing-by-lawrence-m-krauss.html.

1555   E.g., A. Jenkins and G. Perez, "Looking for Life in the Multiverse: Universes with Different Physical Laws Might Still Be Habitable," *Sci. Am.*, (2010) **302**, p. 42; https://www.scientificamerican.com/article/looking-for-life-in-the-multiverse/.

1556   L. Smolin, *The Life of the Cosmos*, n. 566, p. 208.

1557   F. Dowker, "Causal Sets and the Deep Structure of Spacetime," n. 716.

1558   L. Smolin, in R. M. Unger and L. Smolin, *The Singular Universe and the Reality of Time*, n. 62, p. 358.

1559   P. Atkins, *On Being: A Scientist's Exploration of the Great Questions of Existence* (Oxford: Oxford University Press, 2011), p. 12.

1560   S. Carroll, *The Biggest Ideas in the Universe: Space, Time and Motion* (New York: Dutton, 2022), p. 138.

1561   G. 't Hooft, *In Search of the Ultimate Building Blocks*, n. 1509, p. xi.

1562   R. Fitzpatrick, "Thermodynamics and Statistical Mechanics," lecture notes, University of Texas at Austin, https://farside.ph.utexas.edu/teaching/sm1/Thermalhtml/node7.html.

1563   For example, the apparently incompatible wave mechanics and matrix mechanics formulations of quantum theory are equivalent; J. von Neumann, *Mathematische Grundlagen der Quantenmechanik* (Berlin: Julius Springer, 1932), p. 10 ff.; R. T. Beyer, transl., *Mathematical Foundations of Quantum Mechanics* (Princeton: Princeton University Press, 1935).

1564   A. Einstein, "Considerations Concerning the Fundaments of Theoretical Physics," n. 606.

1565   E.g., V. J. Stenger, *The Comprehensible Cosmos: Where Do the Laws Of Physics Come From?* (New York: Prometheus Books, 2006).

1566   A. Einstein, *Physics and Reality*, n. 215, p. 353.

1567 L. Smolin, *The Life of the Cosmos*, n. 566, p. 68.

1568 A more recent estimate is 10^{500}: e.g., Y.-H. He, *The Calabi-Yau Landscape* (Cham, Springer, 2021), p. 15; see also Y.-H. He, "The Calabi-Yau Landscape: From Geometry, to Physics, to Machine-Learning," https://arxiv.org/pdf/1812.02893.pdf.

1569 S. Alexander et al., "The Autodidactic Universe," (2021), https://arxiv.org/pdf/2104.03902.pdf.

1570 Ibidem.

1571 T. Andersen, "The Universe May Be Learning," *The Infinite Universe*, 6 June 2021, https://medium.com/the-infinite-universe/the-universe-may-have-learned-the-laws-of-physics-fe2d22946432.

1572 Ibidem.

1573 J. D. Norton, "General Covariance and the Foundations of General Relativity: Eight Decades of Dispute," *Rep. Prog. Phys.*, (1993) **56**, p. 794; https://sites.pitt.edu/~jdnorton/papers/decades.pdf.

1574 A. Einstein, *Relativity*, n. 245, p. 16.

1575 Indeed, it generalizes the principle Galileo set out for low speeds; see, e.g., L. Sartori, *Understanding Relativity: A Simplified Approach to Einstein's Theories*, (Berkeley: University of California Press, 1996), ch. 1.

1576 In the end, the effects of gravity on rods and clocks became almost a footnote to the theory of general relativity; see A. Einstein, "The Foundation of the General Theory of Relativity," n. 1444, § 22.

1577 J. D. Norton, "General Covariance and the Foundations of General Relativity: Eight Decades of Dispute," n. 1573, p. 794.

1578 Mermin applied the *shut up and calculate* summary of the twentieth-century philosophy of physics to quantum mechanics in particular; N. D. Mermin, *Boojums All the Way Through: Communicating Science in a Prosaic Age*, n. 333. It applies to relativity in perhaps lesser degree.

1579 For a nontechnical review of Einstein's difficulties, see T. Andersen, "General Relativity: How Einstein's Wrong Ideas Led to His Greatest Success," *The Infinite Universe*, 5 October 2020, https://medium.com/the-infinite-universe/general-relativity-how-einsteins-wrong-ideas-led-to-his-greatest-success-dcfe085a417.

1580 Indeed, Einstein changed his mind about this repeatedly; e.g., J. D. Norton, "General Covariance and the Foundations of General Relativity: Eight Decades of Dispute," n. 1573.

1581 A. Einstein, "What Is the Theory of Relativity?," *London Times*, 28 November 1919; as retransl. and repr. in A. Einstein, *Ideas and Opinions*, n. 28, p. 227, and repr. in M. Janssen et al., eds., *The Collected Papers of Albert Einstein, Vol. 7, The Berlin Years, Writings 1918–1921* (Princeton: Princeton University Press, 2002), p. 213; the editors of the latter note the *Times* text "shows major deviations from the original German." My translation of the original agrees with the retranslated text of this passage.

1582 A. Einstein, "Über das Relativitätsprinzip und die aus demselben gezogene Folgerungen," *Jahrb. Radioact. Elektr.*, (1907) **4**, p. 411; "On the Relativity Principle and the Conclusions Drawn from It," A. Beck, transl., *The Collected Papers of Albert Einstein, Vol. 2, The Swiss Years, Correspondence 1900–1909*, n. 255, p. 252; and see, e.g., F. Fernflores, "The Equivalence of Mass and Energy," (2019), *Stanford Encyclopedia of Philosophy*, n. 99, https://plato.stanford.edu/entries/equivME/.

1583 See A. Pais, *Subtle Is the Lord*, n. 44, pp. 178 ff.

1584 A. Einstein, "Autobiographische Skizze," in C. Seelig, ed., *Helle Zeit, Dunkle Zeit* (Zürich: Europa Verlag, 1956), p. 14; my transl.

1585 J. L. Synge, "What Is Einstein's Theory of Gravitation?," in B. Hoffman, ed., *Perspectives in Geometry and Relativity* (Bloomington: Indiana University Press, 1966), p. 7.

1586 A. Einstein, "The Foundation of the General Theory of Relativity," n. 1444, p. 117.

1587 J. D. Norton, "General Covariance and the Foundations of General Relativity: Eight Decades of Dispute," n. 1573, p. 798.

1588 Ibid., p. 791.

1589 A. Pais, *Subtle Is the Lord: The Science and the Life of Albert Einstein*, n. 44, ch. 12, "The Einstein-Grossmann Collaboration."

1590 E. J. Kretschmann, "Über den physikalischen Sinn der Relativitätspostulate. A. Einsteins neue und seine ursprüngliche Relativitätstheorie," *Ann. d. Phys.*, (1917) **53**, p. 575.

1591 For example, Einstein suggested Newton's theory could not be made generally covariant; A. Einstein, "Prinzipielles zur allgemeinen Relativitatstheorie," *Annalen der Physik*, (1918) **55**, p. 241. But it has been done, see C.W. Misner et al., *Gravitation* (New York: W.H. Freeman & Co., 1970).

1592 J. D. Norton, "Did Einstein Stumble? The Debate over General Covariance," *Erkenntnis*, (1995) **42**, p. 223; https://www.jstor.org/stable/20012618.

1593 J. D. Norton, "General Covariance and the Foundations of General Relativity," n. 1573, p. 794.

1594 Ibidem.

1595 A. Einstein, letter to Felix Klein, 4 April 1917, in *The Collected Papers of Albert Einstein, Vol. 8, The Berlin Years: Correspondence*, n. 757, p. 314.

1596 A. Einstein, letter to David Bohm, 28 October 1954, repr. in J. Stachel, "Einstein and the Quantum: Fifty Years of Struggle," n. 758.

1597 See, e.g., lecture by S. Hughes, "The Einstein Field Equation," MIT OpenCourseWare, (2020); https://ocw.mit.edu/courses/8-962-general-relativity-spring-2020/resources/lecture-12-the-einstein-field-equation/.

1598 M. M. G. Ricci and T. Levi-Civita, "Méthodes de calcul différentiel absolu et leurs applications," *Math. Ann.*, (1900) **54**, p. 125; https://link.springer.com/article/10.1007/BF01454201.

1599 See, e.g., Y. Ding and C. Rovelli, "The Volume Operator in Covariant Quantum Gravity," *Class. Quantum Grav.*, (2010) **27**, 165003; https://arxiv.org/abs/0911.0543.

1600 R. P. Feynman, *The Character of Physical Law*, n. 269, p. 123.

1601 A. Einstein, letter to Hedwig and Max Born, 27 January 1920, in D. K. Buchwald et al., eds., *The Collected Papers of Albert Einstein, Vol. 9, The Berlin Years: Correspondence, January 1919–April 1920* (Princeton: Princeton University Press, 2004), p. 386, my italics; my translation of: "Das mit der Kausalität plagt auch mich viel. . . . Ich verzichte aber sehr, sehr ungern auf die vollständige Kausalität. . . . Die Frage, ob strenge Kausalität oder nicht, hat einen klaren Sinn, wenn auch wohl zu keiner Zeit eine sichere Beantwortung." In the canonical translation of this passage, *vollständige* is rendered adjectivally as *complete*, which lacks cogency here. In the *Hochdeutsch* of Einstein's day it was also used in the adverbial sense of *wholly* or *altogether* or *entirely*, corresponding to the adjectival sense of *entire* and better matching the situation of this passage; see, e.g., *Muret-Sanders Encyclopedic*

English-German and German-English Dictionary (Berlin-Schöneberg: Langenscheidtsche Verlagsbuchhandlung, 1910), vol. 2, p. 1073.

1602   J. Stachel, "Einstein and the Quantum: Fifty Years of Struggle," n. 758, p. 349.

1603   A. Einstein, letter to Michele Besso, 12 December 1951, in P. Speziali, ed., *Albert Einstein—Michele Besso Correspondence, 1903–1955* (Paris: Hermann, 1972), p. 453; cited and transl. in J. Stachel, "Einstein and the Quantum: Fifty Years of Struggle," n. 758, p. 349.

1604   J. D. Norton, "Chasing a Beam of Light: Einstein's Most Famous Thought Experiment," *Goodies*, 15 February 2005, https://sites.pitt.edu/~jdnorton/Goodies/Chasing_the_light/.

1605   Georgia State University hosts a calculator you can use to check the contraction as you approach light speed; just put lots of nines in the "v =" box; "Length Contraction," http://hyperphysics.phy-astr.gsu.edu/hbase/Relativ/tdil.html.

1606   F. Wilczek, "QCD Made Simple," *Phys. Today*, (2000) 53, p. 22; http://compsci.cas.vanderbilt.edu/~umar/p8140/section_2/Frank_Wilczek_QCD_Made_Simple_Phys_Today_2000.pdf.

1607   V. D. Burkert et al., "Determination of Shear Forces inside the Proton," 12 April 2021, https://arxiv.org/pdf/2104.02031.pdf.

1608   Well—with a tip of the hat to W. S. Gilbert—hardly ever. See, e.g., A. Hadhazy, "The Enduring Quest for Proton Decay," The Kavli Foundation, 18 March 2021, https://kavlifoundation.org/news/the-enduring-quest-for-proton-decay.

1609   See, e.g., R. Webb, "Strong Nuclear Force," *New Scientist*, https://www.newscientist.com/definition/strong-nuclear-force/.

1610   E. Siegel, "At Last, Physicists Understand Where Matter's Mass Comes From," *Forbes*, 7 November 2017, https://www.forbes.com/sites/startswithabang/2018/11/07/at-last-physicists-understand-where-matters-mass-comes-from/.

1611   E.g., C. Sutton, "quantum chromodynamics," n. 96, https://www.britannica.com/science/quantum-chromodynamics.

1612   Press release, "The Nobel Prize in Physics 2004," Nobel Prize Outreach, https://www.nobelprize.org/prizes/physics/2004/press-release/.

1613   F. Wilczek, "QCD Made Simple," n. 1606.

1614   J. Bardeen et al., "Theory of Superconductivity," *Phys. Rev.*, (1957) 108, p. 1175; https://journals.aps.org/pr/abstract/10.1103/PhysRev.108.1175.

1615   R. Penrose, "Gravitational Collapse: The Role of General Relativity," *Riv. Nuovo Cimento*, (1969) 1, p. 252.

1616   S. Chandrasekhar, "The Highly Collapsed Configurations of a Stellar Mass," *Mon. Not. Roy. Astronom. Soc.*, (1935) 95, p. 207; https://articles.adsabs.harvard.edu/pdf/1935MNRAS..95..207C.

1617   R. Oppenheimer and H. Snyder, "On Continued Gravitational Contraction," *Phys. Rev.*, (1939) 56, p. 455; https://journals.aps.org/pr/pdf/10.1103/PhysRev.56.455.

1618   A. Einstein, "On a Stationary System with Spherical Symmetry Consisting of Many Gravitating Masses," *Ann. Math.*, (1939) 40, p. 922; https://www.jstor.org/stable/1968902.

1619   See, e.g., "Roger Penrose Wins 2020 Nobel Prize in Physics for Discovery about Black Holes," Cambridge University, 6 October 2020, https://www.cam.ac.uk/research/news/roger-penrose-wins-2020-nobel-prize-in-physics-for-discovery-about-black-holes.

1620   E.g., M. Tremmel et al., "Wandering Supermassive Black Holes in Milky-Way-Mass Halos," *Astrophys. J. Lett.*, (2018) **857**, p. L22; https://iopscience.iop.org/article/10.3847/2041-8213/aabc0a/pdf.

1621   E.g., "Stars," NASA Science, https://science.nasa.gov/astrophysics/focus-areas/how-do-stars-form-and-evolve.

1622   E.g., K. Chamcham, "Stellar Nucleosynthesis," Philosophy of Cosmology, http://philosophy-of-cosmology.ox.ac.uk/stellar-nucleosynthesis.html.

1623   E.g., "Galaxies—Merging and Interacting," Center for Astrophysics, https://www.cfa.harvard.edu/research/topic/galaxies-merging-and-interacting.

1624   P. van Dokkum et al., "A Candidate Runaway Supermassive Black Hole Identified by Shocks and Star Formation in its Wake," *Astrophys. J. Lett.*, (2023) **946**, p. L50; https://iopscience.iop.org/article/10.3847/2041-8213/acba86/pdf.

1625   Most are skeptical; see, e.g., "Does CERN Create Black Holes?," *Angels & Demons: The Science behind the Story*, CERN, (2011), https://angelsanddemons.web.cern.ch/faq/black-hole.html.

1626   Ill-informed fears such a black hole could destroy Earth led to a lawsuit in the United States in 2008 (*Luis Sancho, et al. v. US Department of Energy*) that was dismissed on the ground that "speculative fear of future harm does not constitute an injury in fact sufficient to confer standing." Justia, https://law.justia.com/cases/federal/appellate-courts/ca9/08-17389/08-17389-2011-04-18.html.

1627   For a somewhat dramatic discussion, see A. D. Aczel, *Present at the Creation: The Story of CERN and the Large Hadron Collider* (New York: Crown Publishers, 2010), pp. 204 ff.

1628   S. W. Hawking, "Black Hole Explosions?," n. 1016.

1629   A 10^{-5} g black hole would evaporate in about 10^{-40} s; you can find the formula here: ttps://www.bigbadaboom.ca/docs/Hawking%20radiation.pdf; or use the calculator here: https://www.vttoth.com/CMS/physics-notes/311-hawking-radiation-calculator.

1630   S. W. Hawking, "Black Hole Explosions?," n. 1016.

1631   See, e.g., V. De Luca et al., "On the Primordial Black Hole Mass Function for Broad Spectra," *Phys. Lett. B*, (2020) **807**, 135550; https://pdf.sciencedirectassets.com/271623/1-s2.0-S0370269320X00074/1-s2.0-S0370269320303543/main.pdf.

1632   Ibidem.

1633   See, e.g., G. Gilmore, "The Short Spectacular Life of a Superstar," *Science*, (2004) **304**, p. 1915, https://www.science.org/doi/10.1126/science.1100370.

1634   "DOE Explains Supernovae," Office of Science, https://www.energy.gov/science/doe-explainssupernovae.

1635   As the number of detected mergers grows it becomes possible to draw conclusions though with low confidence: R. Abbott et al. (1634 authors), "The Population of Merging Compact Binaries Inferred Using Gravitational Waves through GWTC-3," 23 February 2022, https://arxiv.org/abs/2111.03634.

1636   Gaia Collaboration, "Discovery of a Dormant 33 Solar-Mass Black Hole in Pre-Release Gaia Astrometry," *Astron. Astrophys.*, (2024) **685**, p. L2; https://arxiv.org/abs/2404.10486.

1637   D. Natarajan, "The First Monster Black Holes," n. 1172, p. 24; and see M. C. Begelman et al., "Formation of Supermassive Black Holes by Direct Collapse in

Pregalactic Haloes," *Mon. Not. R. Astr. Soc.*, (2006) **370**, p. 289; https://articles.adsabs.harvard.edu/pdf/2006MNRAS.370..289B.

1638   Á. Bogdán et al., "Evidence for Heavy-Seed Origin of Early Supermassive Black Holes from A Z≈10 X-Ray Quasar," *Nature Astron.*, (2024) **8**, p. 126; https://www.nature.com/articles/s41550-023-02111-9.

1639   M. Häberle et al., "Fast-Moving Stars Around an Intermediate-Mass Black Hole in ω Centauri," *Nature*, (2024) **631**, p. 285; https://www.nature.com/articles/s41586-024-07511-z; and see S. Jarman, "Speedy Stars Point to Intermediate-Mass Black Hole in Globular Cluster," *Physics World*, 18 July 2024, https://physicsworld.com/a/speedy-stars-point-to-intermediate-mass-black-hole-in-globular-cluster/.

1640   E.g., J. Salazar, "First Stars and Black Holes," Texas Advanced Computing Center, 11 August 2022, https://www.tacc.utexas.edu/-/first-stars-and-black-holes.

1641   R. B. Larson and V. Bromm, "The First Stars in the Universe: Exceptionally Massive and Bright, the Earliest Stars Changed the Course of Cosmic History," *Sci. Am.*, (2002) **12**, p. 4.

1642   See, e.g., J. O'Callaghan, "How Did Supermassive Black Holes Grow So Fast?," *Horizon*, 9 December 2019, https://ec.europa.eu/research-and-innovation/en/horizon-magazine/how-did-supermassive-black-holes-grow-so-fast.

1643   D. Clery, "Astronomers Home In on Colliding Giant Black Hole Duos," *Science*, (2023) **380**, p. 328; https://www.science.org/content/article/astronomers-try-catch-titanic-black-hole-clashes-action.

1644   K. Cooper, "Pulsar Timing Irregularities Reveal Gravitational-Wave Background," *Phys. World*, (2023), 29 June 2023, https://physicsworld.com/a/pulsar-timing-irregularities-reveals-hidden-gravitational-wave-background/; D. J. Reardon et al., "Search for an Isotropic Background Gravitational-Wave Background with the Parkes Pulsar Timing Array," *Astrophys. J. Lett.*, (2023) **951**, p. L6; https://iopscience.iop.org/article/10.3847/2041-8213/acdd02/pdf.

1645   J. W. Nightingale et al., "Abell 1201: Detection of an Ultramassive Black Hole in a Strong Gravitational Lens," *Mon. Not. R. Astron. Soc.*, (2023) **521**, p. 3298; https://arxiv.org/abs/2303.15514.

1646   S. W. Hawking, "Gravitationally Collapsed Objects of Very Low Mass," n. 1207.

1647   E.g., A. Witze, "It's a Dream: JWST Spies More Black Holes Than Astronomers Predicted," *Nature*, 3 August 2023, https://www.nature.com/articles/d41586-023-02460-5.

1648   J. Silk et al., "Which Came First: Supermassive Black Holes or Galaxies? Insights from JWST," *Astrophys. J. Lett.*, (2024) **981**, p. L34.

1649   Recent James Webb observations have revealed a new anomaly, an early giant black hole with no galaxy to speak of: I. Juodžbalis *et al.*, "A dormant overmassive black hole in the early Universe," https://arxiv.org/pdf/2403.03872; and see J. O'Callaghan, "Sleeping black hole is way more massive than it should be," *New Scientist*, 13 March 2024, https://www.newscientist.com/article/2421742-sleeping-black-hole-is-way-more-massive-than-it-should-be/.

1650   G. Hasinger, quoted in "Did Black Holes Form Immediately After the Big Bang?," *ESA*, 16 December 2023, https://www.esa.int/Science_Exploration/Space_Science/Did_black_holes_form_immediately_after_the_Big_Bang.

1651   D. Castelvecchi, "Gravitational Waves from Giant Black-Hole Collision Reveal Long-Sought 'Ringing,'" *Nature*, 1 December 2021, https://www.nature.com/articles/d41586-023-03813-w.

1652   R. Boyle, "The 'Beautiful Confusion' of the First Billion Years Comes into View," *Quanta Magazine*, 9 October 2024, https://www.quantamagazine.org/the-beautiful-confusion-of-the-first-billion-years-comes-into-view-20241009/.

1653   The theorem is not rigorously proven but has observational support: M. Isi et al., "Testing the No-Hair Theorem with GW150914," *Phys. Rev. Lett.*, (2019) **123**, 111102; https://arxiv.org/pdf/1905.00869.pdf; and C. D. Capano et al, "Multimode Quasinormal Spectrum from a Perturbed Black Hole," *Phys. Rev. Lett.*, (2023) **131**, 221402; https://journals.aps.org/prl/pdf/10.1103/PhysRevLett.131.221402.

1654   E.g., J. R. Samal et al., "Experimental Test of the Quantum No-Hiding Theorem," *Phys. Rev. Lett.*, (2011) **106**, 080401; https://www.hri.res.in/~akpati/prl11.pdf.

1655   B. Cox and J. Forshaw, *Black Holes: The Key to Understanding the Universe* (New York: HarperCollins, 2023).

1656   B. Cox, interview by Bob MacDonald, *Quirks & Quarks*, CBC Radio, 29 April 2023; https://www.cbc.ca/radio/quirks/brian-cox-black-holes-qa-1.6825858.

1657   Ibidem.

1658   C. Rovelli, "Unfinished Revolution," n. 6, p. 7.

1659   C. Rovelli, "The Layers That Build Up the Notion of Time," n. 984; Rovelli's emphasis.

1660   Recent developments may bring nuclear photons to the table, with potential for clocks to be stable to one part in 10^{20}, or 10 ms in the universe's lifetime; S. Kraemer et al., "Observation of the Radiative Decay of the 229^{Th} Nuclear Clock Isomer," *Nature*, (2023) **617**, p. 706; https://www.nature.com/articles/s41586-023-05894-z.

1661   P. Maddy, *Realism in Mathematics*, n. 123, p. 1.

1662   M. L. Wilson, "The Unreasonable Uncooperativeness of Mathematics in the Natural Sciences," *Monist*, (2000) **83**, p. 296.

1663   R. A. Muller, *Now*, n. 443, p. 265.

1664   D. Hilbert, radio address in 1930, unknown transl., repr. in J. T. Smith, "David Hilbert's Radio Address," https://www.maa.org/book/export/html/326610.

1665   See also R. W. Hamming, "The unreasonable effectiveness of mathematics," *Am. Math. Mon.*, (1980) **87**, p. 81; https://math.dartmouth.edu/~matc/MathDrama/reading/Hamming.html.

1666   E. P. Wigner, "The unreasonable effectiveness of mathematics in the natural sciences," Richard Courant lecture in mathematical sciences at New York University, May 11, 1959; repr. in *Comm. Pure & Appl. Math.*, (1960) **13**, *p. 1*; https://web.archive.org/web/20210212111540/http://www.dartmouth.edu/~matc/MathDrama/reading/Wigner.html.

1667   Ibidem.

1668   A. Einstein, "Geometry and Experience," lecture to the Prussian Academy of Sciences in Berlin, 27 January 1921; https://mathshistory.st-andrews.ac.uk/Extras/Einstein_geometry/.

1669   A. Badiou, *L'être et l'évènement* (Paris: Éditions du Seuil, 1988); O. Feltham, transl., *Being and Event* (New York: Continuum, 2006), p. 4.

1670   Ibid., p. 7; his emphasis.

1671   J. J. Gray et al., "mathematics," n. 96, https://www.britannica.com/science/mathematics.

1672   P. Maddy, *Realism in Mathematics*, n. 123, p. 98; indeed, her index has sixteen distinct entries for "science/mathematics analogy."

1673   P. Kitcher, *The Nature of Mathematical Knowledge* (Oxford: Oxford University Press, 1983), p. 92.

1674   See, e.g., K. Bapat, "Is Mathematics an Invention or a Discovery?," Science ABC, 10 January 2022, https://www.scienceabc.com/eyeopeners/is-mathematics-an-invention-or-a-discovery.html.

1675   P. Maddy, *Realism in Mathematics*, n. 123, p. 21.

1676   For example, "We could say: the number 1 belongs to a concept *F*, if the proposition that *a* does not fall under *F* is not true universally, whatever *a* may be, and if from the propositions '*a* falls under *F*' and '*b* falls under *F*' it follows universally that *a* and *b* are the same." G. Frege, *Die Grundlagen der Arithmetik:Eine logisch mathematische Untersuchung über den Begriff der Zahl* (Breslau: Wilhelm Koebler, 1884); J. L. Austin, transl., *The Foundations of Arithmetic: A Logico-Mathematical Enquiry into the Concept of Number* (Evanston IL: Northwestern University Press, 1960), § 55.

1677   B. Russell, *Introduction to Mathematical Philosophy*, n. 1491, p. 13; https://people.umass.edu/klement/imp/imp.html.

1678   This (known as abstraction) is a strong thread running through the literature of mathematical realism; my comments here neglect many arguments and schools of thought. To pursue more of them, see P. Maddy, *Realism in Mathematics*, n. 123.

1679   Ibid., p. 30.

1680   B. Russell, *The Principles of Mathematics*, n. 237, vol. 1, p. 427.

1681   P. Kitcher, *The Nature of Mathematical Knowledge*, n. 1673, p. 115.

1682   See, e.g., A. P. Bird, "Realism and Antirealism," *Cantor's Paradise*, 19 May 2021, https://www.cantorsparadise.com/realism-and-antirealism-d4b94f47e9.

1683   L. Smolin, *Einstein's Unfinished Revolution*, n. 74, p. 85.

1684   B. Whittle, "Mathematical Anti-Realism and Explanatory Structure," *Synthese*, (2021) 199, p. 6203; abstract, https://philsci-archive.pitt.edu/18713/.

1685   In set theory, a set may be a conceptual collection of different mathematical objects such as symbols or even other sets; see, e.g., P. R. Halmos, *Naive Set Theory* (Princeton: D. Van Nostrand, 1960).

1686   W. V. Quine, "New Foundations for Mathematical Logic," *Am. Math. Monthly*, (1937) 44, p.70; repr. in *From a Logical Point of View: Logico-Philosophical Essays*, n. 38, p. 81.

1687   Ibid., p. 89.

1688   This aspect of the philosophy of mathematics has an extensive literature; for a representative glimpse, the bold reader might try F. MacBride, "Structuralism Reconsidered," in S. Shapiro, ed., *The Oxford Handbook of Philosophy of Mathematics and Logic* (Oxford: Oxford University Press, 2005), p. 563; https://doi.org/10.1093/oxfordhb/9780195325928.003.0018.

1689   J. S. Mill, *A System of Logic, Ratiocinative and Inductive* (New York: Harper & Brothers, 1882) 8th ed., p. 189; https://www.gutenberg.org/files/27942/27942-h/27942-h.html.

1690   J. Bigelow, *The Reality of Numbers: A Physicalist's Philosophy of Mathematics* (Oxford: Clarendon Press, 1988), p. 1.

1691   Ibid., p. 5.

1692 P. Kitcher, *The Nature of Mathematical Knowledge*, n. 1673, p. 115.

1693 K. T. Fann, *Ludwig Wittgenstein: The Man and His Philosophy* (Singapore: Partridge Publishing, 2020).

1694 See, e.g., B. Sherman and G. Harman, "Knowledge and Assumptions," *Philos. Stud.*, (2011) **156**, p. 131; https://citeseerx.ist.psu.edu/viewdoc/download?doi=10.1.1.172.7539.

1695 P. Maddy, *Realism in Mathematics*, n. 123, p. 6.

1696 Ibid., p. 8.

1697 L. Wittgenstein, A. Weiberg and S. Majetschak, eds., *Bemerkungen über die Grundlagen der Mathematik* (Oxford: B. Blackwell, 1956); G. H. von Wright and R. Rees, eds., G. E. M. Anscombe, transl., *Remarks on the Foundations of Mathematics* (Oxford: B. Blackwell, 1956), 3rd ed. 1978, vol. 1, § 37, p. 51.

1698 L. Wittgenstein, ibidem.

1699 J. Landauer and J. Rowlands, "A Priori Knowledge," *Importance of Philosophy*, (2001), http://www.importanceofphilosophy.com/Irrational_APriori.html.

1700 A. C. Paseau, *Philosophy of Mathematics* (London: Routledge, 2016); https://philpapers.org/rec/PASPOM-2.

1701 There are thousands of apple varieties; e.g., C. Kemp, "Apple Revival: How Science Is Bringing Historic Varieties Back to Life," *Nature*, (2023) **622**, p. 446; https://www.nature.com/articles/d41586-023-03229-6.

1702 G. Frege, *The Foundations of Arithmetic*, n. 1676, §§ 22 and 23.

1703 P. Maddy, *Realism in Mathematics*, n. 123, p. 60; my emphasis.

1704 Ibid., p. 83.

1705 E.g., A. Thompson, "Check Out This Picture of a Single Atom," *Popular Mechanics*, 20 January 2022, https://www.popularmechanics.com/science/a17804899/here-is-a-photo-of-a-single-atom/.

1706 W. V. Quine, "Two Dogmas of Empiricism," n. 38, p. 21.

1707 M. E. Brown, (2006), quoted in M. Inman, "Pluto Not a Planet, Astronomers Rule," *Nat. Geo.*, 23 August 2006, https://www.nationalgeographic.com/science/article/dwarf-planet-pluto-astronomy.

1708 L. Wittgenstein, *Über Gewissheit* (Frankfurt am Main: Suhrkamp Verlag, 1970); D. Paul and G. E. M. Anscombe, transls., *On Certainty* (Oxford: Basil Blackwell, 1969).

1709 P. Maddy, *Realism in Mathematics*, n. 123, p. 4.

1710 A. N. Whitehead and B. Russell, *Principia Mathematica* (Cambridge: The University Press, 1927), vol. I, p. 360.

1711 Ibid., vol. II, p. 200.

1712 See, e.g., M. Tiles, "Logical Foundations of Set Theory and Mathematics," in D. Jacquette, ed., *A Companion to Philosophical Logic* (Hoboken NJ: Blackwell Publishing Ltd., 2006), ch. 24; http://www.thatmarcusfamily.org/philosophy/Course_Websites/Readings/Tiles Logical Foundations of Math.pdf.

1713 See, e.g., A. Bellos, "Mathematics: Set Theory for Six-year-olds," *Nature*, (2014) **516**, p. 34; https://www.nature.com/articles/516034a.

1714 W. V. Quine, "New Foundations for Mathematical Logic," *Am. Math. Monthly*, (1937) **44**, p. 70.

1715 P. Benacerraf, "What Numbers Could Not Be," *Phil. Rev.*, (1965) **74**, p. 47.

1716 N. J. Wildberger, "Set Theory: Should You Believe?," UNSW, https://web.maths.unsw.edu.au/~norman/views2.htm.

1717 Ibidem.

1718 R. Kaplan, *The Nothing That Is: A Natural History of Zero* (London: Allen Lane, 1999), p. 216.

1719 The "trick" seems to have originated with John von Neumann, "Eine Axiomatisierung der Mengenlehre," *J. für die Reine und Angew. Math.*, (1925) **154**, p. 219; repr. and transl. "On the Introduction of Transfinite Numbers" in J. van Heijenoor, transl., *From Frege to Gödel: A Source Book in Mathematical Logic* (Cambridge: Harvard University Press, 1967), p. 347.

1720 P. Maddy, *Realism in Mathematics*, n. 123, p. 97.

1721 Ibid., p. 128, her n. 62.

1722 F. Dowker, "Causal Sets and the Deep Structure of Spacetime," n. 716.

1723 J. L. Bell, *The Continuous and the Infinitesimal in Mathematics and Philosophy* (Milan: Polymetrica, 2019), p. 224.

1724 P. Kitcher, *The Nature of Mathematical Knowledge*, n. 1673, p. 124.

1725 R. Dedekind, *Stetigkeit und irrationale Zahlen* (Braunschweig: Friedrich Vieweg und Sohn, 1872); "Continuity and Irrational Numbers" in, W. W. Beman, transl., *Essays on the Theory of Number* (Chicago: Open Court Publishing Co., 1901), p. 1; https://www.gutenberg.org/files/21016/21016-pdf.pdf.

1726 R. Dedekind, *Was Sind und Was Sollen die Zahlen?* (Braunschweig: Friedrich Vieweg und Sohn, 1888); "The Nature and Meaning of Numbers" in R. Dedekind, W. W. Beman, transl., *Essays on the Theory of Number* (Chicago: Open Court Publishing Co., 1901), p. 14; https://www.gutenberg.org/files/21016/21016-pdf.pdf.

1727 R. Dedekind, "Continuity and Irrational Numbers," n. 1725.

1728 P. Maddy, *Realism in Mathematics*, n. 123, p. 82.

1729 G. Frege, *The Foundations of Arithmetic*, n. 1676, p. II.

1730 J. Hadamard, *The Psychology of Invention in the Mathematical Field* (Princeton: Princeton University Press, 1945); repr. (New York: Dover Publications, 1954), p. 117, his n. 2.

1731 P. Maddy, *Realism in Mathematics*, n. 123, p. 82.

1732 Ibid., p. 83.

1733 G. Frege, *The Foundations of Arithmetic*, n. 1676, p. xiv.

1734 P. Maddy, *Realism in Mathematics*, n. 123, p. 87.

1735 Ibid., p. 82.

1736 R. Kaplan, *The Nothing That Is*, n. 1718, p. 68.

1737 J. D. Barrow, *The Book of Nothing*, n. 1474, p. 48.

1738 B. Barnett and S. Fleming, "Creating Something Out of Nothing: Symbolic and Non-Symbolic Representations of Numerical Zero in the Human Brain," 30 January 2024, https://www.biorxiv.org/content/10.1101/2024.01.30.577906v1.full.pdf.

1739 As reported by science journalist James Woodford, "Fear of Predators May Have Helped Us Conceptualise the Idea of Zero," *New Scientist*, **261**, #3481, 1 March 2024, https://www.newscientist.com/article/2419468-fear-of-predators-may-have-helped-us-conceptualise-the-idea-of-zero/.

1740 Brahmagupta, *Brahmasphutasiddhanta*; H. T. Colebrooke, transl., *Algebra, with Arithmetic and Mensuration, from the Sanscrit of Brahmagupta and Bháscara* (London: John Murray, 1817), C. 1, S. 3; https://ia600907.us.archive.org/14/items/algebrawitharith00brahuoft/algebrawitharith00brahuoft.pdf.

1741 G. Frege, *The Foundations of Arithmetic*, n. 1676.

1742 See G. Bessler, "Unity and Disunity Between Gottlob Frege and Giuseppe Peano on the Basis of Their Correspondence in the Years 1891–1903," 8th Intl. Conf. Eur. Soc. Hist. Sci., London, September 14–17, 2018; https://rebus.us.edu.pl/bitstream/20.500.12128/15173/3/Besler_Unity_and%20Disunity_between_Gottlob_Frege_and_Giuseppe_Peano.pdf.

1743 There are debates about Peano's philosophic inclinations—e.g., H. C. Kennedy, "The Mathematical Philosophy of Giuseppe Peano," *Phil. Sci.*, (1963) **30**, p. 262—but he was clearly no realist: "We think number, therefore number exists." G. Peano, *Rend. Circ. Mat. Palermo*, (1906) **21**, p. 860, transl. in H. C. Kennedy, *Peano: Life and Works of Giuseppe Peano* (Dordrecht NL: D. Reidel Publishing Co., 1980), p. 119.

1744 G. Peano, *Arithmetices principia, nova methodo exposita* (Torino: Bocca, 1889); V. Verheyen, transl., *The Principles of Arithmetic, Presented by a New Method*, April 24, 2022, https://raw.githubusercontent.com/mdnahas/Peano_Book/master/Peano.pdf.

1745 B. Russell, *The Principles of Mathematics*, n. 237, vol. 1, p. 124; Russell's italics.

1746 B. Russell, *Introduction to Mathematical Philosophy*, n. 1491, p. 4.

1747 I. Peano, *Arithmetices Principia: Nova Methodo Exposita* (Romae: Fratres Bocca, 1889), p. 1; https://github.com/mdnahas/Peano_Book.

1748 This is so in the standard rendition of his postulates; see, e.g., W. L. Hosch, "Peano axioms," n. 96, https://www.britannica.com/science/Peano-axioms.

1749 Ibidem.

1750 G. Peano, *axiomata*, repr. in J. Horne, ed., *Philosophical Perceptions on Logic and Order* (Hershey PA: IGI Global, 2018).

1751 See J. D. Barrow, *The Book of Nothing*, n. 1474, pp. 12 ff.

1752 For example, in his definitive history of the emerging foundations of mathematics in this period, historian of logic and mathematics Ivor Grattan-Guinness wrote: "Zero was allowed into the story because Peano had also proposed these definitions of 0 and 1 . . . $a + 0 = a$."; I. O. Grattan-Guinness, *The Search for Mathematical Roots 1870–1940: Logics, Set Theories and the Foundations of Mathematics from Cantor Through Russell to Gödel* (Princeton: Princeton University Press, 2000), p. 236.

1753 R. Netz, "Methods of Infinity," *The Archimedes Palimpsest*, http://archimedespalimpsest.org/about/scholarship/method-infinity.php.

1754 See, e.g., M. Goldstern and J. Kellner, "Ordnung in den Unendlichkeiten," *Spektrum der Wissenschaft*, (2021) **3**, p. 74; repr. and transl. "A Deep Math Dive into Why Some Infinities Are Bigger Than Others," *Sci. Am.*, 16 August 2021, https://www.scientificamerican.com/article/a-deep-math-dive-into-why-some-infinities-are-bigger-than-others/.

1755 Aristotle, *Physics*, n. 248, book III, part 4.

1756 I. Kant, *Critique of Pure Reason*, n. 404, Sec. VII.

1757 See, e.g., G. Cantor, "Über die verschiedenen Standpunkte in Bezug auf das aktuelle Unendliche," ("Concerning Various Perspectives on the Actual Infinite"), *Z. Philos. Philos. Krit.*, (1886) **88**, p. 224.

1758 W. Thomson, "The Wave Theory of Light" (1884), in *Popular Lectures and Addresses* (London: Macmillan and Company, 1891), Vol. 1, p. 322.

1759 F. Close, *The Infinity Puzzle: Quantum Field Theory and the Hunt for an Orderly Universe* (New York: Basic Books, 2011), p. 4.

1760 B. Russell, *Introduction to Mathematical Philosophy*, n. 1491, p. 16.

1761 D. Hilbert, "Über das Unendliche," lecture on 4 June 1925, to the Westphalian Mathematical Society in Münster, repr. in *Math. Annal.*, (1926) **95**, p. 161; E. Putnam and G. J. Massey, transl., "On the Infinite," in P. Benacerraf and H. Putnam, eds., *Philosophy of Mathematics: Selected Readings*, (Cambridge: Cambridge University Press 2012), p. 183, https://www.cambridge.org/core/books/abs/philosophy-of-mathematics/on-the-infinite/B91A325262EF69CA314F13B1B9CD98C2.

1762 Ibidem.

1763 P. Maddy, *Realism in Mathematics*, n. 123, p. 23.

1764 Ibidem.

1765 Ibidem.

1766 This asserts a factual foundation for strict finitism, a philosophic movement in aid of consistency some mathematicians have pursued, notably Hilbert; see R. Zach, "Hilbert's Program, *Stanford Encyclopedia of Philosophy*, (2019), n. 99, https://plato.stanford.edu/entries/hilbert-program/.

1767 C. H. Chihara, *Constructibility and Mathematical Existence* (New York: Oxford University Press, 1990), p.68.

1768 C. F. Gauss, quoted in W. S. von Waltershausen, *Gauss zum Gedächtnis* (Leipzig: S. Hirzel Verlag, 1856), my transl.; original text: "Die Mathematik ist die Königin der Wissenschaften und die Zahlentheorie ist die Königin der Mathematik."

1769 R. Kaplan, *The Nothing That Is*, n. 1718, p. 40.

1770 S. F. Barker, *Philosophy of Mathematics* (Englewood Cliffs, NJ: Prentice-Hall, 1964), p. 88.

1771 P. Benacerraf and H. Putnam, *Philosophy of Mathematics*, n. 1761, p. 11.

1772 W. Mueckenheim, "Physical Constraints of Numbers," A. Beckmann et al., eds., *Proc. 1st Int. Symp. Math. & Connections to Arts & Sciences* (Berlin: Franzbecker, 2005), p. 134; https://arxiv.org/abs/math/0505649.

1773 See, e.g., C. Wells, "Mathematical Objects," *Abstract Math*, 28 February 2017, https://abstractmath.org/MM/MMIntro.htm.

1774 Y. Martel, *Life of Pi* (Toronto: Alfred A. Knopf, 2001), p. 27.

1775 For a review of several methods, see J. C. Stillwell, "Pi Recipes," n. 96, 5 August 2005, https://www.britannica.com/topic/Pi-Recipes-1084437.

1776 Pi World Ranking List, https://www.pi-world-ranking-list.com/?page=lists&category=pi&sort=digits.

1777 E. H. Iwao, "Even More Pi in the Sky: Calculating 100 Trillion Digits of Pi on Google Cloud," Google Cloud, 28 June 2022, https://cloud.google.com/blog/products/compute/calculating-100-trillion-digits-of-pi-on-google-cloud/.

1778 For a neat demonstration of pi's irrationality see P. I. Dagnum, "Visualization of π Being Irrational," https://twitter.com/Dagnum_PI/status/1716338308069110114.

1779 C. Sagan, *Contact* (New York: Pocket Books, 1986).

1780 "Pioneer Plaques," NASA, 13 February 2018, https://solarsystem.nasa.gov/resources/706/pioneer-plaque/.

1781 G. F. B. Riemann, "On the Hypotheses Which Lie at the Foundation of Geometry," n. 273.

1782 And likely on papyrus; see also F. J. Swetz, "Mathematical Treasure: Euclid Proposition on Papyrus," *Mathematical Association of America*, August 2013, https://maa.org/press/periodicals/convergence/mathematical-treasure-euclid-proposition-on-papyrus.

1783 See, e.g., L. Shabel, "Kant's Philosophy of Mathematics," *Stanford Encyclopedia of Philosophy*, (2021), n. 99, https://plato.stanford.edu/entries/kant-mathematics/.

1784 E.g., J. J. Gray, "Carl Friedrich Gauss," n. 96, https://www.britannica.com/biography/Carl-Friedrich-Gauss.

1785 H. Poincaré, *Science et Methode* (Paris: Flammarion, 1920), p. 131; G. B. Halstead, transl., *The Foundations of Science: Science and Hypothesis, The Value of Science, Science and Method*, (Lancaster PA: Science Press, 1946), p. 435.

1786 E.g., M. Murzi, "Jules Henri Poincaré (1854–1912)," *Internet Encyclopedia of Philosophy*, https://iep.utm.edu/poincare/.

1787 E.g., V. Mukunth, "Beyond the Surface of Einstein's Relativity Lay a Chimerical Geometry," *Wire*, (2015), https://thewire.in/science/beyond-the-surface-of-einsteins-relativity-lay-a-chimerical-geometry.

1788 G. F. B. Riemann, "On the Hypotheses Which Lie at the Foundation of Geometry," n. 273.

1789 E.g., T. E. Collett et al., "A Precise Extragalactic Test of General Relativity," *Science*, (2018) **360**, p. 1342; https://www.science.org/doi/10.1126/science.aao2469.

1790 E.g., A. Ashtekar et al., eds., *General Relativity and Gravitation* (Cambridge: Cambridge University Press, 2015), esp. Part Three; https://www.cambridge.org/core/books/abs/general-relativity-and-gravitation/gravity-is-geometry-after-all/53C58CE9B5C00163B38D53FDB88D9B5B.

1791 For an analysis of Einstein's view about this (to which I subscribe) see D. Lehmkuhl, "Why Einstein Did Not Believe That General Relativity Geometrizes Gravity," *Stud. Hist. Phil. Sci. B*, (2014) **46**, p. 316; https://www.sciencedirect.com/science/article/pii/S1355219813000695.

1792 B. Russell, *The Autobiography of Bertrand Russell 1914–1944* (Boston: Little, Brown and Company, 1968), Vol. II, p. 341.

1793 Ibidem.

1794 It is a collection of six works; Aristotle, O. F. Owen transl., *The Organon, or Logical Treatises, of Aristotle* (London: George Bell, 1889–90); https://archive.org/details/organonorlogical02arisuoft.

1795 J. Corcoran, "George Boole," in *Encyclopedia of Philosophy*, 2nd ed., (London: Macmillan & Company, 2006).

1796 K. Gödel, "Einige mathematische Resultate über Entscheidungsdefinitheit und Widerspruchsfreiheit," *Anz. Akad. Wiss. Wien*, (1930) **67**, p. 214; and "Über formal unentscheidbare Sätze der Principia mathematica und verwandter Systeme I," *Monatsh. Math. Phys.*, (1931) **38**, p. 173; repr. and transl. in S. Feferman et al., eds., *Kurt Gödel: Collected Works, Vol. I, Publications 1929–1936* (New York: Oxford University Press, 1986), pp. 141 and 151. For a lay exposition of the method behind Gödel's proofs and of their impacts, see N. Wolchover, "How Gödel's Proof Works," *Quanta Magazine*, 14 July 2020, https://www.quantamagazine.org/how-godels-incompleteness-theorems-work-20200714/.

1797 P. Raatikainen, "Gödel's Incompleteness Theorems," *Stanford Encyclopedia of Philosophy*, 2 April 2020, n. 99, https://plato.stanford.edu/entries/goedel-incompleteness/.

1798 Ibidem.

1799 B. Russell, letter to Alice M. Hilton, 9 June 1963, quoted in *The Autobiography of Bertrand Russell 1944–1969* (New York: Simon and Schuster, 1969), Vol. III, p. 250.

1800 W. V. Quine, "Kurt Gödel," *Year Book* (Newark DE: American Philosophical Association, 1978); repr. in W. V. Quine, *Theories and Things*, n. 13, p. 144.

1801 L. Dammbeck, *Das Netz (The Net)*, (2003), https://www.imdb.com/title/tt0434231/; https://www.youtube.com/watch?v=DgHlhBceR78&t=4247s.

1802 S. F. Barker, *Philosophy of Mathematics*, n, 1770, p. 80.

1803 See, e.g., P. Raatikainen, "Gödel's Incompleteness Theorems," n. 1797, § 2.3.

1804 See, e.g., E. Nagel and J. R. Newman, *Gödel's Proof*, n. 803, p. 58: "Gödel's argument makes it unlikely that a finitistic proof of the consistency of arithmetic can be given."

1805 D. Hilbert, address to the *Gesellschaft der Deutschen Naturforscher und Ärzte in Königsberg* on 8 September 1930, quoted in C. Reid, *Hilbert-Courant* (New York: Springer Verlag, 1986), p. 196.

1806 C. Reid, ibid., p. 197.

1807 F. Kafka, letter to Milena Jesenska (a Wednesday in 1920), in W. Haas, ed., T. Stern and J. Stern, transls., *Letters to Milena* (London: Vintage Books, 1999), p. 57; my emphasis.

1808 M. Born, lecture at Massachusetts Institute of Technology, 2 January 1926; repr. in M. Born, *Problems of Atomic Dynamics*, n. 326, p. 129.

1809 J. D. Chapman and C. J. Gillespie, "Radiation-Induced Events and Their Time Scale in Mammalian Cells," n. 672.

1810 J. S. Bell, "Quantum Mechanics for Cosmologists," n. 154, p. 124.

1811 Those interested in a nontechnical but more extensive outline of quantum states might try physicist and teacher Philip Stamp, "Basic Ideas of Quantum Mechanics: I. Quantum States," University of British Columbia, https://phas.ubc.ca/~stamp/TEACHING/PHYS340/NOTES/FILES/QM-basic_1-states.pdf.

1812 E.g., P. Ball, "Experiments Spell Doom for Decades-Old Explanation of Quantum Weirdness," *Quanta Magazine*, https://www.quantamagazine.org/physics-experiments-spell-doom-for-quantum-collapse-theory-20221020/.

1813 This question is often wrongly attributed to philosopher George Berkeley, who did however express similar views; G. Berkeley, *A Treatise Concerning the Principles of Human Knowledge* (London: Jeremy Pepyat, 1710); https://www.gutenberg.org/files/4723/4723.txt.

1814 N. G. Gresnigt, "A Topological Model of Composite Preons from the Minimal Ideals of Two Clifford Algebras," *Phys. Rev. Lett. B*, (2020) 808, 135687.

1815 See also E. Siegel, "Why Is the Universe Fundamentally Left-Handed?," *Forbes*, 19 November 2020, https://www.forbes.com/sites/startswithabang/2020/11/19/why-is-the-universe-fundamentally-left-handed/.

1816 But note one can invert the spin by changing one's velocity relative to the (anti)neutrino; e.g., E. Siegel, "This Is Why the Neutrino Is the Standard Model's Greatest Puzzle," *Forbes*, 17 September 2019; https://www.forbes.com/sites/startswithabang/2019/09/17/this-is-why-neutrinos-are-the-standard-models-greatest-puzzle/.

1817 S. O. Bilson-Thompson, "A Topological Model of Composite Preons," n. 762.

1818 Ibidem.

1819 E. Majorana, "Teoria symmetrica dell'elletrone e del positrone," *Nuovo Cim.*, (1937) **14**, p. 171; L. Maiani, transl., "A Symmetric Theory of Electrons and Positrons," *Soryushiron Kenkyu*, (1981) **63**, p. 149; repr. in G. F. Bassani, ed., *Ettore Majorana Scientific Papers* (New York: Springer, 2006). p. 201; Majorana proposed an underlying theory for neutral particles—"neutrons and the hypothetical neutrinos"—the latter having then been proposed but not observed and the former's charged-quark structure being quite unknown.

1820 S. Glashow, "Resonant Scattering of Antineutrinos," *Phys. Rev.*, (1960) **118**, p. 316.

1821 With Bilson-Thompson ingredients from L. H. Kauffman, "Iterants, Majorana Fermions and the Majorana-Dirac Equation", *Symmetry*, (2021) **13**, p. 1373.

1822 The IceCube Collaboration, "Detection of a Particle Shower at the Glashow Resonance with IceCube," *Nature*, (2021) **591**, p. 220.

1823 The standard-model neutrino story leads on to a quantum number called weak isospin and to so-called sterile neutrinos, hypotheses that may do little more than prop up a once productive but now fading theory; see, e.g., R. R. Volkas, "Neutrinos in Cosmology, with Some Significant Digressions," *AIP Conf. Procs.*, (2003) **655**, p. 220; (we won't go there).

1824 The concept got a boost, but not from an "elementary" particle collision: See J. Butterworth, "Majorana Particles: Fundamentally Confusing," *Guardian*, 5 October 2014, https://www.theguardian.com/science/life-and-physics/2014/oct/05/majorana-princeton-particles-fundamentally-confusing.

1825 Multiple long-planned experiments of this kind are now underway with various isotopes that undergo double-beta decay; see, e.g., A. Garfagnini, "Neutrinoless Double Beta Decay Experiments," lecture at the Flavor Physics and CP Violation conference, Marseille, 26 May 2014; https://arxiv.org/pdf/1408.2455.pdf.

1826 See "Legend at UCL," University College London, https://www.hep.ucl.ac.uk/legend/.

1827 With Bilson-Thompson ingredients from L. H. Kauffman, n. 1821.

1828 E.g., D. Castelvecchi, "The Vanishing Neutrinos That Could Upend Fundamental Physics," *Nature*, 16 July 2021, https://www.nature.com/articles/d41586-021-01955-3.

1829 J. Evans, quoted in L. Crane, "Physicists Fail to Find Mysterious 'Sterile Neutrino' Particles," *New Scientist*, 27 October 2021; https://www.newscientist.com/article/2294958-physicists-fail-to-find-mysterious-sterile-neutrino-particles/.

1830 See, e.g., "What's a Neutrino?," All Things Neutrino, Fermi National Accelerator Laboratory, https://neutrinos.fnal.gov/whats-a-neutrino/.

1831 See, e.g., L. M. Brown, "The Idea of the Neutrino," *Phys. Today*, (1978) **31**, p. 23.

1832 E.g., C. Prescod-Weinstein, "Sterile Neutrinos Could Explain Dark Matter—If We Can Find Them," *New Scientist*, (no. 3360), 13 November 2021, https://www.newscientist.com/article/mg25233600-100-sterile-neutrinos-could-explain-dark-matter-if-we-can-find-them/.

1833 C. Moskowitz, "Can Sterile Neutrinos Exist?," *Sci. Am.*, 4 November 2021, https://www.scientificamerican.com/article/can-sterile-neutrinos-exist/.

1834 L. Smolin, *Einstein's Unfinished Revolution*, n. 74, p. xvii.

1835 Ibidem.

1836 L. Hardy, "Reformulating and Reconstructing Quantum Theory," n. 781.

1837   G. 't Hooft, *In Search of the Ultimate Building Blocks*, n. 1509, p. 165.

1838   L. Smolin, *The Life of the Cosmos*, n. 566, p. 68.

1839   T. Freeth et al., "A Model of the Cosmos in the Ancient Greek Antikythera Mechanism," *Sci. Rep.*, (2021) **11**, p. 5821; https://doi.org/10.1038/s41598-021-84310-w.

1840   N. Copernicus, *Six Books on the Revolutions of the Heavenly Spheres*, n. 1085.

1841   See, e.g., R. Macke, "Religious Scientists: Canon Nicolaus Copernicus (1473–1543); Heliocentrism," Vatican Observatory, 23 February 2020, https://www.vaticanobservatory.org/sacred-space-astronomy/religious-scientists-canon-nicolaus-copernicus-1473-1543-heliocentricism/.

1842   L. Zyga, "A Test of the Copernican Principle," Phys.org, https://phys.org/news/2008-05-copernican-principle.html.

1843   P. C. W. Davies, "Yes, the Universe Looks Like a Fix. But That Doesn't Mean That a God Fixed It," *Guardian*, 26 June 2007; https://www.theguardian.com/commentisfree/2007/jun/26/spaceexploration.comment.

1844   B. Russell, *The Scientific Outlook* (London: George Allen & Unwin Ltd., 1931), p. 126.

1845   E.g., "How Many Stars Are There in the Universe?," European Space Agency, https://www.esa.int/Science_Exploration/Space_Science/Herschel/How_many_stars_are_there_in_the_Universe; and A. Hamer, "Does Every Star Have Planets?," *Live Science*, 25 December 2021, https://www.livescience.com/does-every-star-have-planets; and, too, with so much choice, it would be sad if we succeeded in evolving on a bad one.

1846   P. C. W. Davies, *The Goldilocks Enigma: Why Is the Universe Just Right for Life?* (London: Allen Lane, 2006).

1847   L. Smolin, in R. M. Unger and L. Smolin, *The Singular Universe and the Reality of Time*, n. 62, p. 398.

1848   L. Smolin, *The Life of the Cosmos*, n. 566, p. 45.

1849   S. Weinberg, *The First Three Minutes*, n. 1018, p. 154.

1850   S. Alexander et al., "The Autodidactic Universe," n. 1569, https://arxiv.org/pdf/2104.03902.pdf.

1851   L. Susskind, *The Cosmic Landscape: String Theory and the Illusion of Intelligent Design* (New York: Little, Brown & Company, 2006), p. 20; original emphasis.

1852   B. Greene, *The Hidden Reality*, n. 661.

1853   G. F. R. Ellis, "Does the Multiverse Really Exist?," *Sci. Am.*, (2011) **305**, p. 38; https://www.scientificamerican.com/article/does-the-multiverse-really-exist/.

1854   D. Deutsch, *The Fabric of Reality: The Science of Parallel Universes* (New York: Penguin, 1997).

1855   D. Deutsch, *The Beginning of Infinity: Explanations That Transformed the World* (London: Allen Lane, 2011), pp. 96 ff.

1856   B. Rundle, *Why There Is Something Rather than Nothing*, n. 1478, p. 33.

1857   C. Rovelli, "Unfinished Revolution," n. 6, p. 3.

1858   L. Smolin, "You Think There's a Multiverse? Get Real," *New Scientist*, 14 January 2015, https://www.newscientist.com/article/mg22530040-200-you-think-theres-a-multiverse-get-real/.

1859   P. Goff, "Our Improbable Existence Is No Evidence for a Multiverse," *Sci. Am.*, 10 January 2021, https://www.scientificamerican.com/article/our-improbable-existence-is-no-evidence-for-a-multiverse/.

1860   G. F. R. Ellis, "Does the Multiverse Really Exist?," n. 1853.

1861   See, e.g., J. Howgego, "Schrödinger's Cat," *New Scientist*, https://www.newscientist.com/definition/schrodingers-cat/.

1862   Everett's concept somehow became highly influential. The interested reader might try J. A. Barrett, "Everett's Relative-State Formulation of Quantum Mechanics," *Stanford Encyclopedia of Philosophy*, 23 October 2018, n. 99, https://plato.stanford.edu/entries/goedel-incompleteness/. For more, J. A. Barrett, *The Quantum Mechanics of Minds and Worlds* (Oxford: Oxford University Press, 2001), p. 56 ff.

1863   L. Smolin, *The Life of the Cosmos*, n. 566, p. 263.

1864   E.g., A. Linde, "A Balloon Producing Balloons Producing Balloons," in J. Brockman, ed., *The Universe*, n. 1143, p. 33.

1865   B. Greene, *The Elegant Universe*, n. 932, p. 367.

1866   E.g., R. Bousso and J. Polchinski, "The String Theory Landscape," *Scientific American*, (2004) **291**, September, p. 79; http://www.scientificamerican.com/magazine/sa/2004/09-01/.

1867   L. Susskind, *The Cosmic Landscape*, n. 1851, p. 14.

1868   Ibid, p. 304.

1869   R. Penrose, interview by Alister McGrath, 25 September 2010, "Hawking, God & the Universe ," Unbelievable? (London: Premier Christian Radio, 2010).

1870   G. F. R. Ellis, "Does the Multiverse Really Exist?," n. 1853.

1871   A. J. Ayer, *Philosophy in the Twentieth Century*, n. 1531.

1872   See also, C. J. Gillespie, *This Changes Everything* (New York: Rosetta Books, 2012).

1873   M. S. Turner, "Making Sense of the New Cosmology," *Int. J. Mod. Phys.* A, (2002) **17**, p. 180; https://arxiv.org/pdf/astro-ph/0202008.pdf.

1874   See, e.g., the "Hubble tension," A. G. Riess et al., "JWST Observations Reject Unrecognized Crowding of Cepheid Photometry as an Explanation for the Hubble Tension at 8σ Confidence," *Astrophys. J. Lett.*, (2024) **962**, p. L17; https://iopscience.iop.org/article/10.3847/2041-8213/ad1ddd/pdf.

1875   G. de Vaucouleurs, "The Case for a Hierarchical Cosmology," *Science*, (1970) **167**, p. 1203; https://www.science.org/doi/epdf/10.1126/science.167.3922.1203.

1876   G. R. Burbidge, "New Vistas in Cosmology and Cosmogony," in G. Münch et al., eds., *The Universe at Large: Key Issues in Astronomy and Cosmology* (Cambridge: Cambridge University Press, 1997), p. 64.

1877   M. López-Corredoira, "Non-Standard Models and the Sociology of Cosmology," *Stud. Hist. Philos. Mod. Phys.*, (2014) **46**, p. 86; https://arxiv.org/pdf/1311.6324.pdf.

1878   Y. V. Baryshev, "Paradoxes of Cosmological Physics in the Beginning of the 21st Century," n. 1431.

1879   J. W. Moffat, *Cracking the Particle Code of the Universe*, n. 198, p. xiv.

1880   P. Volcker, in "Building a Global Community of New Economic Thinkers," Annual Report 2010–2011, Institute for New Economic Thinking, p. 5; https://files.givewell.org/files/labs/macroeconomic-policy/inet-annual-report-2010-2011.pdf.

1881 M. Belz, lecture at Melbourne University (1959). This well-worn quote is usually attributed to Mark Twain, who in turn wrongly attributed it to Disraeli (as, I think, did Belz); the earliest record I found in essentially this form cited a speech by Arthur Balfour a decade before he became prime minister of the United Kingdom, calling it "an old saying" he heard from James Munro, a professor of political economy and jurisprudence; *Manchester Guardian*, 29 June 1892, p. 5; https://www.york.ac.uk/depts/maths/histstat/balfour.pdf.

1882 Botanist William Campbell Steere said, "When plant ecology first emerged as a subscience of botany, it was largely descriptive. Today, however, it has become . . . much more quantitative." W. C. Steere, *Britannica*, https://www.britannica.com/science/botany/Areas-of-study.

1883 She measured it using cluster analysis; see J. M. Parks, "Fortran IV Program for Q-mode Cluster Analysis on Distance Function with Printed Dendogram," *Computer Contributions*, Kansas State Geological Survey, (1970), #46; https://www.kgs.ku.edu/Publications/Bulletins/CC/46/CompContr46.pdf.

1884 J. R. Dugle and D. H. Thibault, "Ecology of the Field Irradiator-Gamma Area," Atomic Energy of Canada Limited, Report 4135; https://inis.iaea.org/collection/NCLCollectionStore/_Public/05/137/5137255.pdf.

1885 S. Ouliaris, "What Is Econometrics," *Finance & Development*, (Washington, DC, International Monetary Fund, 2011) **48**, p. 38; https://www.imf.org/external/pubs/ft/fandd/2011/12/pdf/b2b.pdf.

1886 D. F. Hendry, "Econometrics—Alchemy or Science?," *Economica*, (1980) **47**, p. 387; https://www.jstor.org/stable/2553385.

1887 See "Professor Sir David Hendry," Oxford Martin School, University of Oxford, https://www.oxfordmartin.ox.ac.uk/people/professor-sir-david-hendry/.

1888 E. Tretkoff, "Economic Models for Stock Markets Should Incorporate 'Outlier' Events," *APS News*, **17**, May 2008, https://www.aps.org/publications/apsnews/200805/economicmodels.cfm.

1889 P. Krugman, "How Did Economists Get It So Wrong?," *The New York Times Magazine*, 2 September 2009; https://www.nytimes.com/2009/09/06/magazine/06Economic-t.html.

1890 E.g., F. Black and M. Scholes, "The Pricing of Options and Corporate Liabilities," *J. Political Econ.*, (1973) **81**, p. 637; https://www.cs.princeton.edu/courses/archive/fall09/cos323/papers/black_scholes73.pdf; and see an article by mathematician and catastrophe-theorist Ian Stewart, "The Mathematical Equation That Caused the Banks to Crash," *Guardian*, 12 February 2012; https://www.theguardian.com/science/2012/feb/12/black-scholes-equation-credit-crunch.

1891 E.g., the disparate worldviews underpinning John Maynard Keynes's *General Theory of Employment, Interest, and Money* (New York: The Macmillan Company, 1936) and Karl Marx's *Das Kapital: Kritik der politischen Ökonomie* (Hamburg: Otto Meissner, 1867).

1892 Its real name (if that's not a hopeless oxymoron) is Sveriges Riksbank Prize in Economic Sciences in Memory of Alfred Nobel, https://www.investopedia.com/terms/n/nobel-memorial-prize-in-economic-sciences.asp.

1893 Nagarjuna, *The Fundamental Wisdom of the Middle Way*, (c. 150 CE), quoted in Chandrakirti, G. T. Tsültrim and A. T. Jampa, transls., "Supplement to the 'Middle Way' & Explanation of the 'Supplement to the Middle Way,'" p. 51; https://media.dalailama.com/English/texts/madhyamakavatara-autocommentary-ENG.pdf.

1894 See, e.g., "Buddhism, Schools of: Mahāyāna Philosophical Schools of Buddhism," *Encyclopedia.com*, 26 August 2022, https://www.encyclopedia.com/environment/encyclopedias-almanacs-transcripts-and-maps/buddhism-schools-mahayana-philosophical-schools-buddhism.

1895 M. Siderits, *Buddhism as Philosophy: An Introduction* (Indianapolis: Hackett Publishing Company, 2007), p. 7.

1896 See, e.g., G. C. Pande, "Causality in Buddhist Philosophy," in E. Deutsch and R. Bontekoe, eds., *A Companion to World Philosophies* (Hoboken NJ: Blackwell Publishing Ltd., 2017), p. 370.

1897 E.g., "Many traditional koans don't work well in a Western context; they're too culturally inaccessible, rooted in the thought world of ancient China and Japan." C. Webb, *The Fire of the Word: Meeting God on Holy Ground* (Westmont IL: Intervarsity Press, 2011), p. 29.

1898 K. K. Inada, "The Range of Buddhist Ontology," *Phil. East & West*, (1988) **38**, p. 261; https://www.jstor.org/stable/1398866.

1899 D. T. Suzuki, "Philosophy of the Yogācāra," in M. L. Bloom and R. T. Jaffe, eds., *Selected Works of D. T. Suzuki, Volume IV Buddhist Studies* (Oakland: University of California Press, 2021), p. 46.

1900 P. G. Jones, "The Noble Nagarjuna, Logic and Non-Duality," 27 January 2019, https://philpapers.org/rec/JONTNN-3.

1901 J. Westerhoff, interview by Richard Marshall, "Emptiness and No-Self: Nāgārjuna's Madhyamaka," (2022), *3:16*, https://www.3-16am.co.uk/articles/emptiness-and-no-self-nāgārjuna-s-madhyamaka.

1902 J. Westerhoff, "An argument for ontological nihilism," *Inquiry*, 17 June 2021, https://www.tandfonline.com/doi/pdf/10.1080/0020174X.2021.1934268.

1903 Nagarjuna, *Madhyamakakarika* (c. 150 CE); J. L. Garfield, transl., *The Fundamental Wisdom of the Middle Way* (New York: Oxford University Press, 1995).

1904 Ibid., p. 90.

1905 Students of Western philosophy will likely be unsurprised to find there are disputes about even such central propositions in Eastern philosophy; see J. W. Smith, "'Snakes and Ladders'—'Therapy' as Liberation in Nagarjuna and Wittgenstein's Tractatus," *Sophia*, (2021) **60**, p. 411; https://www.researchgate.net/publication/348752841_'Snakes_and_Ladders'_-_'Therapy'_as_Liberation_in_Nagarjuna_and_Wittgenstein's_Tractatus.

1906 J. L. Garfield and G. Priest, "Nagarjuna and the Limits of Thought," *Phil. East & West*, (2003) **53**, p. 1; https://scholarworks.smith.edu/cgi/viewcontent.cgi?article=1021&context=phi_facpubs.

1907 E.g., J. B. Glattfelder, *Information–Consciousness–Reality* (Cham: Springer Nature, 2019), pp. 515 ff.

1908 See, e.g., W. Oh, "Understanding of Self: Buddhism and Psychoanalysis," *J. Relig. & Health*, (2022) **61**, p. 4696.

1909 K. K. Inada, "The Range of Buddhist Ontology," n. 1898, p. 266.

1910 See, e.g., G. C. Pande, "Causality in Buddhist Philosophy," in E. Deutsch and R. Bontecoe, eds., *A Companion to World Philosophies* (Hoboken NJ: Blackwell Publishing Ltd., 2017), p. 370.

1911 The Dalai Lama, *The Universe in a Single Atom: How Science and Spirituality Can Serve Our World* (London: Abacus, 2006), p. 41.

1912   T. Hoover, *The Zen Experience* (New York: New American Library, 1980), p. 2; my emphasis; http://www.gutenberg.org/files/34325/34325-h/34325-h.htm.

1913   See, e.g., "History of Zen Buddhism," Association Zen Internationale, https://www.zen-azi.org/en/history-zen-buddhism.

1914   S. O. Bilson-Thompson, "A Topological Model of Composite Preons," n. 762.

1915   R. v. Wray, [1971] S.C.R. 272, per Judson, J., at p. 299; https://scc-csc.lexum.com/scc-csc/scc-csc/en/item/2700/index.do.

1916   S. Turow, *The Laws of Our Fathers* (New York: Warner Books, 1996), p. 595.

1917   R. v. Wray, n. 1915, per Martland, J., at p. 273.

1918   Direct or eyewitness evidence often has serious weaknesses; see, e.g., C. Laney and E. F. Loftus, "Eyewitness Testimony and Memory Biases," in R. Biswas-Diener and E. Diener (eds.), Noba textbook series: *Psychology* (Champaign IL: DEF Publishers, 2023); https://nobaproject.com/modules/eyewitness-testimony-and-memory-biases.

1919   E.g., L. Booth, "The Utter Unreliability of Eyewitness Testimony," *Plaintiff*, September 2011; https://plaintiffmagazine.com/recent-issues/item/the-utter-unreliability-of-eyewitness-testimony; and the famous Brian Williams recall incident, note c at page 653; see also H. Arkowitz and S. O. Lilienfeld, "Do the 'Eyes' Have It?," *Sci. Am. Mind*, (2010) **20**, p. 68.

1920   See, e.g., W. A. Dunning, "Truth in History," *Am. Hist. Rev.*, (1914) **19**, p. 217.

1921   G. W. Elton, *The Practice of History* (Sydney: Sydney University Press, 1967), repr. (Glasgow: Collins, 1982), p. 70.

1922   C. Russell, "Obituaries: Professor Sir Geoffrey Elton," *Independent*, 19 December 1994,   https://www.independent.co.uk/news/people/obituaries-professor-sir-geoffrey-elton-1388031.html.

1923   A. Einstein, "Autobiographical Notes," n. 216, p. 339.

1924   M. Robert, *Seul comme Franz Kafka* (Paris: Calmann-Lévy, 1994); R. Manheim, transl., *As Lonely as Franz Kafka* (New York: Harcourt Brace Jovanovich, 1982), p. 176.

1925   J. Spencer, "Hidden Influences," *Science*, (2022) **376**, p. 587; https://www.science.org/doi/pdf/10.1126/science.abo6935.

1926   M. D. Sahlins, "The Original Political Society," *HAU: J. Ethnogr. Theory*, (2017) 7, p. 91; https://www.journals.uchicago.edu/doi/pdf/10.14318/hau7.2.014.

1927   J. Spencer, "Hidden Influences," n. 1925.

1928   E.g., P. E. Smaldino, "Social Identity and Cooperation in Cultural Evolution," *Behav. Process.*, (2019) **161**, p. 108.

1929   E.g., M. Douglas, *How Institutions Think* (Syracuse NY: Syracuse University Press, 1986), p. 18.

1930   W. B. Ashworth Jr., "Alexander Koyré," Linda Hall Library, 29 August 2019, https://www.lindahall.org/about/news/scientist-of-the-day/alexandre-koyre.

1931   A. Koyré, lecture at the Johns Hopkins Institute of the History of Medicine in Baltimore, 15 December 1953; repr. in A. Koyré, *From the Closed World to the Infinite Universe* (Baltimore: Johns Hopkins University Press, 1957), p. 1.

1932   Ibid., p. 2.

1933   Ibidem.

1934   M. Somerville, *On the Connexion of the Physical Sciences* (London: John Murray, 1834), p. 2.

1935   J. Carville, (1992), see, e.g., "It's the Economy, Stupid," Political Dictionary, https://politicaldictionary.com/words/its-the-economy-stupid/.

1936   A. Freeman, "A Penny for Your Thoughts: A Note on the Impact of Ideas," Academia, (2016), https://www.academia.edu/30225601/A_Penny_for_Your_Thoughts_a_Note_on_the_Impact_of_Ideas.

1937   See, e.g., J. Reddekop, "Thinking Across Worlds: Indigenous Thought, Relational Ontology and the Politics of Nature," thesis (2014), University of Western Ontario; https://ir.lib.uwo.ca/cgi/viewcontent.cgi?article=3410&context=etd.

1938   Philosopher Nicholas Maxwell urged the need for its revival: N. Maxwell, *In Praise of Natural Philosophy: A Revolution for Thought and Life* (Montreal: McGill-Queen's University Press, 2017).

1939   M. Robert, *As Lonely as Franz Kafka*, n. 1924, p. 164.

1940   W. Shakespeare, *Henry IV, Part I* (London: Andrew Wise, 1598).

1941   A. Fletcher, *Time, Space and Motion in the Age of Shakespeare* (Cambridge: Harvard University Press, 2006).

1942   C. Rovelli, "Quantum Spacetime: What Do We Know?," in C. Callender and N. Huggett, eds., *Physics Meets Philosophy at the Planck Scale*, n. 366, p. 102.

1943   B. M. Sattler, *The Concept of Motion in Ancient Greek Thought*, n. 239.

1944   Aristotle, *Physics*, n. 248, book IV, part 10.

1945   D. Bohm, *Quantum Theory* (Edgewood Cliffs NJ: Prentice-Hall, 1951), p. 145; his emphasis.

1946   Ibid., p. 100.

1947   Einstein famously said, "In any case, I am convinced He does not play dice." My translation. Original text: "Jedenfalls bin ich überzeugt, daß Der nicht würfelt." Letter to Max Born, 4 December 1926, repr. in: *Einstein/Born Briefwechsel 1916–1955* (Münich: Nymphenburger, 1969); I. Born, transl., *The Born-Einstein Letters* (New York: Walker and Company, 1971), p. 90.

1948   B. M. Sattler, *The Concept of Motion in Ancient Greek Thought*, n. 239, p. 21.

1949   L. Smolin, *Einstein's Unfinished Revolution*, n. 74, p. 264.

1950   E.g., Y. Yang et al., "Non-Abelian physics in light and sound," *Science*, (2024) **383**, p. 844; https://www.science.org/doi/epdf/10.1126/science.adf9621.

1951   C. Rovelli, *Reality Is Not What It Seems*, n. 7, p. 263.

1952   G. Galilei, *Sidereus Nuncius* (Venezia: Tommaso Baglioni, 1610); W. R. Shea, transl., *Sidereus Nuncius or A Sidereal Message* (Sagamore Beach MA: Science History Publications, 2009), p. 73.

1953   This was first proposed by Thomas Wright in *An Original Theory or New Hypothesis of the Universe* (London: H. Chapelle, 1750); for a discourse on nebulae as known around the time of Einstein's annus mirabilis, see astronomer Agnes Mary Clerke's *The System of the Stars*, 2nd ed. (London: Adam & Charles Black, 1905), pp. 242 ff.

1954   E. A. Poe, *The Murders in the Rue Morgue*, in *Graham's Lady's and Gentleman's Magazine*, Philadelphia, April 1841; https://gutenberg.org/cache/epub/2147/pg2147.txt.

1955   See, e.g., A. Cappi, "The Cosmology of Edgar Allan Poe," in D. Valls-Gabaud and A. Boksenberg, eds., *The Role of Astronomy in Society and Culture*, Proc. IAU Symp., (2009) **260**, p. 315.

1956   E. A. Poe, *Eureka*, n. 85.

1957   S. Weinberg, *The First Three Minutes*, n. 1018, p. 132, his emphasis.

1958   The IceCube Neutrino Observatory at the South Pole, which has been operating for a decade; https://icecube.wisc.edu/.

1959   The Laser Interferometer Gravitational-Wave Observatory or LIGO, the European Gravitational Observatory or Virgo, and the Kamioka Gravitational Wave Detector or KAGRA, operating collaboratively; https://www.ligo.caltech.edu/, https://www.virgo-gw.eu/, and https://gwcenter.icrr.u-tokyo.ac.jp/en/.

1960   The James Webb Space Telescope now in operation (https://webbtelescope.org/webb-science), with the Euclid wide-angle telescope just coming on line (https://www.nasa.gov/missions/euclid/new-images-from-euclid-mission-reveal-wide-view-of-the-dark-universe/).

1961   R. Abasi et al., (the IceCube Collaboration), "Evidence for Neutrino Emission from the Nearby Active Galaxy NGC 1068," *Science*, (2022) **378**, p. 538; https://www.science.org/doi/10.1126/science.abg3395.

1962   A major upgrade is planned; "Beyond IceCube," IceCube Neutrino Observatory, https://icecube.wisc.edu/science/beyond/.

1963   Very-high-energy neutrinos are predicted and measured to have higher rates of collision; The IceCube Collaboration, "Measurement of the Multi-TeV Neutrino Interaction Cross-section with IceCube Using Earth Absorption," *Nature*, (2017) **551**, 596.

1964   Ibidem; and see K. Cooper, "IceCube Detects High-Energy Neutrinos from an Active Galactic Nucleus," *Phys. World*, 8 November 2022, https://physicsworld.com/a/icecube-detects-high-energy-neutrinos-from-an-active-galactic-nucleus/.

1965   P. Plait, "Neutrinos Reveal Black Hole Secrets," *Sci. Am.*, (2023) **328**, p. 72.

1966   K. Murase et al., "Hidden Cores of Active Galactic Nuclei as the Origin of Medium-Energy Neutrinos: Critical Tests with the MeV Gamma-Ray Connection," *Phys. Rev. Lett.*, (2020) **125**, 011101; https://arxiv.org/pdf/1904.04226.pdf.

1967   As recently happened; e.g., K. Patel, "The Power of a Volcanic Eruption: This One Was Bigger Than Any U.S. Nuclear Blast," *Washington Post*, 14 April 2023, https://www.washingtonpost.com/climate-environment/2023/04/14/tonga-volcano-explosion-destruction-study/.

1968   E.g., S. Maital, "How 'You're Out of Your Mind' Won a Nobel Prize," TIMnovate, 4 October 2017, https://timnovate.com/tag/thorne/.

1969   B. P. Abbott et al., (LIGO Scientific Collaboration & Virgo Collaboration), "GWTC-1: A Gravitational-Wave Transient Catalog of Compact Binary Mergers Observed by LIGO and Virgo During the First and Second Observing Runs," *Phys. Rev. X*, (2019) **9**, 031040.

1970   A merger of two neutron stars was caught in the act of manufacturing, among other things, gold, so solving the long-standing problem of its origins in the universe's matter; see C. J. Gillespie, "There's Gold in Them Thar Hills; And There's Much More Where It Came From," *Science Seen*, 4 January 2018, http://www.timeone.ca/theres-gold-thar-hills-theres-much-came/.

1971   See "Early Universe," James Webb Space Telescope, NASA: Goddard Space Flight Center, https://webb.nasa.gov/content/science/firstLight.html.

1972   See the figure showing typical spectrum at Sloan Digital Sky Survey (note: 10 Angstroms = 1 nanometer), https://cas.sdss.org/dr6/en/proj/basic/spectraltypes/stellarspectra.asp.

1973   J. O'Callaghan, "JWST's First Glimpses of Early Galaxies Could Break Cosmology," n. 1198.

1974   P. A. Oesch et al., "A Remarkably Luminous Galaxy at Z = 11.1 Measured with Hubble Space Telescope Grism Spectroscopy," *Astrophys. J.*, (2016) **819**, p. 129; https://iopscience.iop.org/article/10.3847/0004-637X/819/2/129.

1975   T. Bakx et al., "Deep ALMA Redshift Search of a z 12 GLASS-JWST Galaxy Candidate," *Mon. Not. R. Astr. Soc.*, (2023) **519**, p. 5076; https://arxiv.org/pdf/2208.13642.pdf.

1976   Royal Astronomical Society, "Astronomers Confirm Age of Most Distant Galaxy Using Oxygen," 25 January 2023, https://phys.org/news/2023-01-astronomers-age-distant-galaxy-oxygen.html.

1977   B. E. Robertson et al., "Identification and Properties of Intense Star-Forming Galaxies at Redshifts z > 10," *Nature Astron.*, (2023) 7, p. 611; https://www.nature.com/articles/s41550-023-01921-1; https://arxiv.org/ftp/arxiv/papers/2212/2212.04480.pdf.

1978   Quoted in J. O'Callaghan, "JWST's First Glimpses of Early Galaxies Could Break Cosmology," n. 1198.

1979   Ibidem.

1980   A. Frank and M. Gleiser, "The Story of Our Universe May Be Starting to Unravel," *The New York Times*, 2 September 2023; https://www.nytimes.com/2023/09/02/opinion/cosmology-crisis-webb-telescope.html.

1981   J. R. Rigby, 12 July 2022, quoted in F. Manjoo, "The Webb Telescope Restored (Some of) My Faith in Humanity," *The New York Times*, 14 July 2022, https://www.nytimes.com/2022/07/14/opinion/webb-telescope-triumph.html.

1982   Aristotle, *Metaphysics* (350 BCE); W. D. Ross, transl., *Metaphysics*, Book V, Part 17; http://classics.mit.edu/Aristotle/metaphysics.html.

1983   R. P. Feynman et al., "Atoms in Motion," *The Feynman Lectures on Physics: Vol. I* (Reading MA: Addison-Wesley Publishing Co., 1965), s. 1–4.

1984   Telomeres are short DNA stretches that protect the ends of chromosomes from fraying. Shortening of telomeres leads to cell senescence; see, e.g., M. A. Shamas, "Telomeres, Lifestyle, Cancer, and Aging," *Curr. Opin. Clin. Nutr. Metab. Care*, (2011) **14**, p. 28; https://www.ncbi.nlm.nih.gov/pmc/articles/PMC3370421/; for a description of telomeres see, e.g., L. H. Chadwick, "Telomere," 26 January 2023, National Human Genome Research Institute, https://www.genome.gov/genetics-glossary/Telomere.

1985   See, e.g., J. Emsley, *The Elements*, 3rd ed. (Oxford: Clarendon Press, 1998).

1986   See, e.g., D. R. Lide (ed.) *Handbook of Chemistry and Physics*, 95th ed. (Boca Raton: CRC, 1996), p. 14.

1987   See, for example, stable levels of xenon gas at Oahu in weeks following the Fukushima reactor release; N. I. Kristiansen et al., "Atmospheric Removal Times of the Aerosol-Bound Radionuclides ^{137}Cs and ^{131}I Measured After the Fukushima Dai-ichi Nuclear Accident—A Constraint for Air Quality and Climate Models," *Atmos. Chem. Phys.*, (2012) **12**, p. 10759; https://acp.copernicus.org/articles/12/10759/2012/acp-12-10759-2012.pdf.

1988   Good data on the exchange of various atomic species from the body are hard to find. The oft-cited 1953 "report" of P. C. Aebersold is not a reliable source.

1989   E.g., H. González-Pardo and M. P. Álvarez, "Epigenetics and Its Implications for Psychology," *Psicothema*, (2013) **25**, p. 3; http://www.psicothema.com/pmidlookup.asp?pmid=23336536. As well, with new CRISPR gene-editing technology we are now able to intentionally modify our genetic information; M. Jinek et al., "A Programmable Dual-RNA-Guided DNA Endonuclease in Adaptive Bacterial Immunity," *Science*, (2012) **337**, p. 816; https://www.science.org/doi/full/10.1126/science.1225829; for a review, see C. Zimmer, "CRISPR, 10 Years On: Learning to Rewrite the Code of Life," *The New York Times*, 27 June 2022, https://www.nytimes.com/2022/06/27/science/crispr-gene-editing-10-years.html.

1990   K. Walsh, (2023), quoted in M. Leslie, "NIH Project Probes the Human Body's Multitude of Genomes," *Science*, (2023) **381**, p. 719; https://www.science.org/doi/epdf/10.1126/science.add3503.

1991   S. Gopalan, quoted by K. Harris, 10 May 2023, The White House, "Remarks by Vice President Harris at Swearing-In Ceremony of Commissioners for the White House Initiative on Advancing Educational Equity, Excellence, and Economic Opportunity for Hispanics"; https://www.whitehouse.gov/briefing-room/speeches-remarks/2023/05/10/remarks-by-vice-president-harris-at-swearing-in-ceremony-of-commissioners-for-the-white-house-initiative-on-advancing-educational-equity-excellence-and-economic-opportunity-for-hispanics/.

1992   D. Hume, *A Treatise of Human Nature*, n. 862, Book I, Part II, Sec. VII.

1993   M. Planck, 25 January 1931, interview in the *Observer*, p. 17.

1994   F. Dyson, "In Praise of Diversity," lecture in Aberdeen, 1985; rev'd and repr. in *Infinite in All Directions* (New York: Harper & Row, 1988), p. 7.

1995   See, e.g., "Realism," n. 96, https://www.britannica.com/topic/realism-philosophy.

1996   Ibidem.

1997   A. Fine, "The Natural Ontological Attitude," in J. Leplin, ed., *Scientific Realism*, n. 12, p. 93.

1998   Ibid., p. 83.

1999   Ibid., p. 97.

2000   Ibidem.

2001   Ibid., p. 98.

2002   W. S. Churchill, *The Second World War*, n. 682, Vol. III, p. 520.

2003   There is an extensive literature espousing varied versions of the scientific method of which Popper's, n. 211, pp. 49 ff., is but one of many; I concur with Smolin and Feyerabend that the concept has no substance; see P. Feyerabend, *Against Method* (London: Verso, 1993), 3rd ed. and L. Smolin, "There Is No Scientific Method," Big Think, 1 May 2013, https://bigthink.com/articles/there-is-no-scientific-method/.

2004   A. Einstein, "On the Method of Theoretical Physics," n. 114, p. 18.

2005   P. J. E. Peebles, *The Whole Truth*, n. 60, p. 85.

2006   B. Russell, *The Problems of Philosophy*, n. 96, p. 156.

2007   A. Cusmariu, "'Subsistence Demystified,'" *Auslegung*, (1978) **6**, p. 24; https://www.researchgate.net/profile/Arnold-Cusmariu/publication/273178057_Subsistence_Demystified/links/593406d9a6fdcc89e7df5a76/Subsistence-Demystified.pdf; and see R.G. Cronin, "Subsistence Demystified?! A Note on Cusmariu's "Subsistence Demystified,"" *Auslegung*, (1979) **6**, p. 192; https://journals.ku.edu/auslegung/article/view/12794.

2008 Plato held pure forms to be the ultimate reality; see, e.g., D. Banach, "Plato's Theory of Forms," (2006), https://olli.gmu.edu/docstore/600docs/0903-602-Plato-Theory of Forms.pdf.

2009 W. V. Quine, "Russell's Ontological Development," *J. Phil.*, (1966) **63**, p. 657; repr. in W. V. Quine, *Theories and Things*, n. 13, p. 73.

2010 But here I trench upon the mind-brain problem, one aspect of which may have the mind extend beyond the brain or even the body; see, e.g., R. A. Wilson, *Boundaries of the Mind: The Individual in the Fragile Sciences—Cognition* (New York: Cambridge University Press, 2004).

2011 E.g., T. Welsh, "Exactly How Long Does It Take to Think a Thought?," *Christian Science Monitor*, 1 July 2015, https://www.csmonitor.com/Technology/Breakthroughs-Voices/2015/0701/Exactly-how-long-does-it-take-to-think-a-thought.

2012 M. Reicher, "Nonexistent Objects," *Stanford Encyclopedia of Philosophy*, n. 99, 7 December 2022, https://plato.stanford.edu/entries/nonexistent-objects/.

2013 See P. M. S. Hacker, "The Sad and Sorry History of Consciousness: Being, Among Other Things, a Challenge to the 'Consciousness-Studies Community,'" *R. Inst. Philos. suppl.*, (2012) **70**, p. 149; https://doi.org/10.1017/S1358246112000082.

2014 E.g., K. H. Pribram, "The Implicate Brain," n. 601, p. 365; D. Chalmers, *The Conscious Mind: In Search of a Fundamental Theory* (Oxford: Oxford University Press, 1996); F. Wilczek, "The Function of the Universe and Origin of Consciousness," The Karlfeldt Center, https://www.thekarlfeldtcenter.com/the-function-of-the-universe-and-origin-of-consciousness-with-frank-wilczek/; and see note 1964.

2015 A. Einstein and L. Infeld, *The Evolution of Physics*, n. 130, p. 277.

2016 However, philosopher of science Michal Tempczyk said, "The book was in fact written by Infeld; Einstein only added some remarks to his manuscript." M. Tempczyk, "Leopold Infeld: The Problem of Matter and Field," *Poznan Studies in the Philosophy of the Science, and the Humanities*, (2001) **74**, p. 207; repr. in W. Krajewski, ed., *Polish Philosophers of Science and Nature in the 20th Century* (Leiden: Brill, 2001), p. 202.

2017 Several hundred years BCE, Hindu philosopher Kaṇāda sought science based on observation and reason (and arrived at atomism independently of the Greeks); see, e.g., J. D. Fowler, *Perspectives of Reality: An Introduction to the Philosophy of Hinduism* (Liverpool: Liverpool University Press, 2002), pp. 98 ff.

2018 A. Pope, *Essay on Man*, Epistle I, (London: J. Wilford, 1733); https://www.poetryfoundation.org/poems/44899/an-essay-on-man-epistle-i.

2019 M. Planck, "A Scientific Autobiography," n. 281, p. 33.

2020 See also T. Kuhn, *The Structure of Scientific Revolutions* (Chicago: University of Chicago Press, 1962), p. 151. There are data in support of Planck's view: P. Azoulay et al., "Does Science Advance One Funeral at a Time?," *Am. Econ. Rev.*, (2019) **109**, p. 2889.

2021 Daniel [cohort2], quoted in M. Inglis, thesis, "Making Sense of Physics: Student Teachers' Experiences of a Physics Subject Knowledge Enhancement Course," The Open University, (2015), p. 94; https://oro.open.ac.uk/63631/1/13889388.pdf.

2022 Ibidem.

2023 L. Smolin, *Einstein's Unfinished Revolution*, n. 74, p. 277.

2024 C. Rovelli, *Quantum Gravity*, n. 41, p. 415.

2025 The scientific method as we know it has a long history, dating a thousand years back to Ibn al-Haytham and in some degree to Aristotle, whose works influenced him; see, e.g., B. Steffens, *Ibn al-Haytham: First Scientist* (Greensboro NC: Morgan Reynolds, 2006).

2026 G. Amelino-Camelia, "Quantum Gravity Phenomenology," in D. Oriti, ed., *Approaches to Quantum Gravity*, n. 6, p. 427.

2027 Philosopher Richard Dawid is a leading exponent; R. Dawid, *String Theory and the Scientific Method*, n. 365.

2028 A. Einstein, "Remarks Concerning the Essays Brought Together in This Cooperative Volume," in P. A. Schilpp, transl. and ed., *Albert Einstein: Philosopher-Scientist*, n. 216, p. 665.

2029 L. Smolin, *Einstein's Unfinished Revolution*, n. 74, p. 11.

2030 L. Smolin, "The Place of Qualia in a Relational Universe," n. 71.

2031 E.g., S. Vosoughi et al., "The Spread of True and False News Online," *Science*, (2018) **359**, p. 1146; https://www.science.org/doi/10.1126/science.aap9559.

2032 B. Glassner, *The Culture of Fear: Why Americans Are Afraid of the Wrong Things* (New York: Basic Books, 1999); https://news.usc.edu/8743/A-Sociologist-Explores-the-Culture-of-Fear/.

2033 See, e.g., T. Nichols, *The Death of Expertise: The Campaign Against Established Knowledge and Why it Matters* (New York: Oxford University Press, 2017); and M. Kakutani, "The Death of Truth: How We Gave Up on Facts and Ended Up with Trump," *Guardian*, 14 July 2018; https://www.theguardian.com/books/2018/jul/14/the-death-of-truth-how-we-gave-up-on-facts-and-ended-up-with-trump.

2034 M. Planck, "The Meaning and Limits of Exact Science," lecture in Zagreb, 15 September 1942; repr., *Science*, (1949) **110**, p. 319; https://www.science.org/doi/10.1126/science.110.2857.319.

2035 R. M. Unger and L. Smolin, *The Singular Universe and the Reality of Time*, n. 62, p. 500.

2036 L. Smolin, in *The Singular Universe and the Reality of Time*, ibid., pp. 484 ff.

2037 T. de Chardin, *The Phenomenon of Man*, n. 81, p. 216.

2038 J.-P. Sartre, *Being and Nothingness*, n. 1031, p. 559 ff.

2039 M. O'Bonsawin, J., speech in Ottawa, 28 November 2022, quoted in A. Francis, "Odanak Justice Michelle O'Bonsawin Officially Welcomed to the Supreme Court," *APTN National News*, https://www.aptnnews.ca/national-news/odanak-justice-michelle-obonsawin-officially-welcomed-to-the-supreme-court/.

2040 Image provided by Dr. Viren Jain; see V. Jain, "Ten years of neuroscience at Google yields maps of human brain," *Google Research*, 2 May 2024, https://research.google/blog/ten-years-of-neuroscience-at-google-yields-maps-of-human-brain/.

2041 C. S. von Bartheld et al., "The Search for True Numbers of Neurons and Glial Cells in the Human Brain: A Review of 150 Years of Cell Counting," *J. Comp. Neurol.*, (2016) **524**, p. 3865; https://www.ncbi.nlm.nih.gov/pmc/articles/PMC5063692/.

2042 See, e.g., M. H. Grider, *Physiology, Action Potentials* (Tampa: StatPearls Publishing, 2022), https://www.ncbi.nlm.nih.gov/books/NBK538143/.

2043 G. Conroy, "This Is the Largest Map of the Human Brain Ever Made," *Nature*, 12 October 2023, https://www.nature.com/articles/d41586-023-03192-2; and A. Weninger and P. Arlotta, "A Family Portrait of Human Brain Cells," *Science*, (2023) **382**, p. 168; https://www.science.org/doi/epdf/10.1126/science.adk4857.

2044 C. Wong, "Cubic Millimetre of Brain Mapped in Spectacular Detail," *Nature*, 9 May 2024, https://www.nature.com/articles/d41586-024-01387-9; A. Shapson-Coe et al., "A Petavoxel Fragment of Human Cerebral Cortex Reconstructed at Nanoscale Resolution," *Science*, (2024) **384**, p. 635; https://www.science.org/doi/epdf/10.1126/science.adk4858.

2045 M. Axer and K. Amunts, "Scale Matters: The Nested Human Connectome," *Science*, (2022) **378**, p. 500; https://www.science.org/doi/epdf/10.1126/science.abq2599.

2046 This was intensely studied by multiple research institutions under the Human Connectome Project; see A. W. Toga et al., "Mapping the Human Connectome," *Neurosurgery*, (2012) **71**, p. 1; https://www.ncbi.nlm.nih.gov/pmc/articles/PMC3555558/.

2047 S. R. y Cajal, ink drawing (1899) in *Instituto Cajal*, Madrid; repr. in L. Swanson, *The Beautiful Brain: The Drawings of Santiago Ramón y Cajal* (Minneapolis: Weisman Art Museum, 2017), p. 15; http://brainu.org/sites/brainu.org/files/lessons/tg-the-beautiful-brain-do-you-see-what-i-see.pdf.

2048 T. Lewton, "Is the Human Brain Really the Most Complex Object in the Universe?," *New Scientist*, 21 February 2024, https://www.newscientist.com/article/mg26134792-100-is-the-human-brain-really-the-most-complex-object-in-the-universe/.

2049 J. D. Watson, foreword in Institute of Medicine and National Academy of Sciences, *Discovering the Brain* (Washington, DC: The National Academies Press, 1992); and see "Discovering the Brain," National Center for Biotechnology Information, (1992), https://www.ncbi.nlm.nih.gov/books/NBK234155/.

2050 "What Is Synaptic Plasticity?," Queensland Brain Institute, https://qbi.uq.edu.au/brain-basics/brain/brain-physiology/what-synaptic-plasticity.

2051 Yet neuronal involvement can be detected at the individual-neuron level; M. Jamali et al., "Semantic Encoding During Language Comprehension at Single-Cell Resolution," *Nature*, (2024), https://www.nature.com/articles/s41586-024-07643-2.

2052 E.g., P. Brodal, "Restitution of Function After Brain Damage," *The Central Nervous System*, (2016), ch. 11, https://doi.org/10.1093/med/9780190228958.003.0011.

2053 J. M. Harlow, "Recovery from the Passage of an Iron Bar Through the Head," *Bull. Mass. Med. Soc.*, (1868) **2**, 327; https://collections.nlm.nih.gov/ext/kirtasbse/66210360R/PDF/66210360R.pdf.

2054 K. Pribram, "Localization and Distribution of Function in the Brain," in J. Orbach (ed.), *Neuropsychology after Lashley* (New York: Erlbaum, 1982), p. 273; http://karlpribram.com/wp-content/uploads/pdf/theory/T-136.pdf.

2055 S. Reardon, "Ultra-Detailed Brain Map Shows Neurons That Encode Words' Meaning," *Nature*, 3 July 2024, https://www.nature.com/articles/d41586-024-02146-6.

2056 See, e.g., E. Camina and F. Güell, "The Neuroanatomical, Neurophysiological and Psychological Basis of Memory: Current Models and Their Origins," *Front. Pharmacol. Sec. Neuropharmacol.*, 30 June 2017, https://www.frontiersin.org/articles/10.3389/fphar.2017.00438/full.

2057 E.g., D. S. Roy et al., "Brain-wide Mapping Reveals That Engrams for a Single Memory Are Distributed Across Multiple Brain Regions," *Nature Comms.*, (2022) **13**, no. 1799; https://www.nature.com/articles/s41467-022-29384-4.

2058 Something like this concept (minus its crucial time element) has been starting to circulate; see, e.g., C. Savodova, "Brain as a Hologram: Exploring Vision at ITMO's Museum of Optics," *iTMO/NEWS*, 17 November 2020, https://news.itmo.ru/en/news/9905/.

2059 J. Ghomeshi, *Grammar Matters: The Social Significance of How We Use Language* (Winnipeg: Arbeiter Ring Publishing, 2010), p. 65.

2060 And see M. Merleau-Ponty, *Phenomenology of Perception*, n. 1049.

2061 Augustine of Hippo, *De Trinitate*, c. 420 CE; S. McKenna, transl., *The Trinity* (Washington, DC: Catholic University of America Press, 1963), Vol. 45, *The Fathers of the Church*, Book XI, p. 333.

2062 This is not to say there are no others; even slime molds show signs of intelligence: C. R. Reid et al., "Slime Mold Uses an Externalized Spatial 'Memory' to Navigate in Complex Environments," *Proc. Nat. Acad. Sci. USA*, (2012) **109**, p. 17490.

2063 I. Kant, *Prolegomena zu einer jeden künftigen Metaphysik, die als Wissenschaft wird auftreten können* (Riga: Johann Friedrich Hartknoch, 1783); P. Carus, transl., *Prolegomena to Any Future Metaphysics That Will Be Able to Present Itself as a Science*, n. 1482, Appendix: *On What Can be Done to Make Metaphysics Actual as a Science*, p. 163.

2064 A. Einstein, letter to George Jaffe, 19 January 1954, item 13-405 in the Control Index to the Collected Papers of Albert Einstein; transl. and quoted by J. Stachel, *Einstein from "B" to "Z,"* n. 588, p. 390.

2065 B. Russell, letter to Gilbert Murray, 21 March 1903, repr. in *The Autobiography of Bertrand Russell: 1872–1914* (London: George Allen and Unwin, 1967), p. 154.

2066 S. W. Hawking, *A Brief History of Time*, n. 530, p. 191.

2067 M. F. Rehbock, personal communication, 1976.

2068 A. Pope, *Essay on Man*, Epistle II, (London: J. Wilford, 1733); https://www.poetryfoundation.org/poems/44900/an-essay-on-man-epistle-ii.

2069 R. Robertson, quoted by Jim Farber, "Band," https://citizenfreak.com/artists/91075-band, from *The New York Times*, 9 August 2023.

2070 A. Einstein, (1955), advice to Pat Miller, quoted in W. Miller, "Death of a Genius," n. 209.

2071 See, e.g., Scientific Foresight Unit, "Automated Tackling of Disinformation," European Parliament, https://www.europarl.europa.eu/RegData/etudes/STUD/2019/624278/EPRS_STU(2019)624278_EN.pdf.

2072 "a. That cannot be believed; (colloq.) hard to believe, surprising." *The Concise Oxford Dictionary of Current English* (Oxford: Clarendon Press, 1968).

2073 "a. Fancied (rare); extravagantly fanciful, capricious, eccentric; grotesque or quaint in design etc."; ibidem.

2074 "*adj.* 1. Too dubious or improbable to be believed: *an unbelievable excuse.* 2. So remarkable as to strain credulity; extraordinary: *the unbelievable fury of the storm; an unbelievable athlete.*"; *Webster's Encyclopedic Unabridged Dictionary of the English Language* (New York: Gramercy Books, 1994).

2075 L. Charteris, *The Saint Overboard* (New York: Charter, 1935), author's note.

2076 E.g., J. Hoerger, "Missing the Night Sky: On Light Pollution, Enlightenment, and Our Sense of Finitude," *New Atlantis*, (2016), https://www.thenewatlantis.com/publications/missing-the-night-sky.

2077 T. Nowakowski, "Merriam-Webster's 2023 Word of the Year Is 'Authentic,'" *Smithsonian Magazine*, 29 November 2023, https://www.smithsonianmag.com/smart-news/why-merriam-websters-2023-word-of-the-year-is-authentic-180983329/.

2078   See, e.g., S. Nee, "The Great Chain of Being," *Nature*, (2005) **435**, p. 429; https://www.nature.com/articles/435429a.

2079   A. O. Lovejoy, lectures at Harvard University, (1933); repr. as *The Great Chain of Being* (Cambridge MA: Harvard University Press, 1936), p. 242.

2080   S. Nee, "The Great Chain of Being," n. 2078, p. 429.

2081   H. Arendt, *The Human Condition*, n. 102, p. 176.

2082   K. J. Mitchell, *Free Agents: How Evolution Gave Us Free Will* (Princeton: Princeton University Press, 2023) and R. M. Sapolsky, *A Science of Life Without Free Will* (New York: Penguin Press, 2023).

2083   See, e.g., G. Strawson, "Nietzsche's Metaphysics?," in M. Dries and P. J. E. Kail, eds., *Nietzsche on Mind and Nature* (Oxford: Oxford University Press, 2015), p. 10.

2084   This is often depicted as the block universe; see, e.g., D. Falk, "A Debate Over the Physics of Time," *Quanta Magazine*, 19 July 2016, https://www.quantamagazine.org/a-debate-over-the-physics-of-time-20160719/#.

2085   A. Einstein, "Über den freien Willen," in R. Chattergee, ed., *The Golden Book of Tagore: A Homage to Rabindranath Tagore from India and the World in Celebration of his Seventieth Birthday* (Calcutta: The Golden Book Committee, 1931), p. 11, my translation; https://ia902901.us.archive.org/10/items/dli.csl.3598/3598.pdf.

2086   F. Herbert, *God Emperor of Dune* (New York: G.P. Putnam's Sons, 1981), p. 403.

2087   H. Clay, speech in the Senate, 2 February 1832, *Gales and Seaton's Register*, p. 275; https://www.senate.gov/artandhistory/history/resources/pdf/AmericanSystem.pdf..

2088   E. R. R. Moody et al., "The Nature of the Last Universal Common Ancestor and Its Impact on the Early Earth System," *Nat. Ecol. Evol.*, (2024) DOI: 10.1038/s41559-024-02461-1. More generally, see D. A. King, "Origins of Life," Lunar and Planetary Institute, https://www.lpi.usra.edu/science/kring/epo_web/impact_cratering/origin_of_life/.

2089   J. W. Johnson, "Deep in the Quiet Wood," in *Fifty Years & Other Poems* (Boston: The Cornhill Company, 1917); https://poets.org/poem/deep-quiet-wood.

2090   L. Besner, "A Cosmic and Human Event," *Globe and Mail*, 6 April 2024, p. O8.

2091   J. E. Carter, Jr., statement accompanying the *Voyager 1* satellite into interstellar space, (1977); https://www.nasa.gov/missions/deepspace/MI_CM_Feature_02.html.

2092   The skeptic sketch was inspired by Dave Coverly, *Speed Bump*, 5 March 2021; https://www.gocomics.com/speedbump/2022/03/05.

SUBJECT INDEX

1
1 → 2
   as source of physics laws, 483
   first law
      of nature, 480
      of the universe, 118, 121, 223-7, 645
   law created matter, 368
   reified succession and addition, 537
   the only law of nature, 223

A

absolute frame of reference
   and simultaneity, 327
   special relativity and the, 160
absolute velocity, 305
accretion disk, *meaning*, 346
action
   at a distance, problem of, 69
   potential
      is the element of thought, 635
      meaning, 648
      seemingly unphysical, 648
   the, *meaning*, 92
   the quantum of the, 82–3
action quantum
   message of the, 92–3
   Planck published nothing further on the, 82
   Planck's derivation of the, 93
   plays a fundamental part in atomic physics, 83
   the, and Planck, 76-94
   tiny size of the, 87-8
active galactic nucleus, 619
   meaning, 343
   review of, 346
Aczel, Amir, 145-6
addition
   is an object of arithmetic, 549
   reified, 521
æther, 66–7, 166
aether
   wind, *meaning*, 71
Albert, David, 454
Alexander, Stephon, 459
al-Haytham, Ibn, 779
Al-Khalili, Jim, 60
Almheiri, Ahmed
   on something out of nothing, 454
   on the emptiness of vacuum, 451
ambiguity
   language is never free of, 653

   of language, 540
   of language, inherent, 374
   of math definitions, 211
Amelino-Camelia, Giovanni, 446, 642, 722
Amoroso, Richard, 99
analog
   change seen as, is digital, 311
   clock is, 510
   neuron is, to digital converter, 218
   physics laws are, approximations, 384
   quantum field particle is, 428
Anaxagoras, 614
Anaximander, 23–24, 51–2, 221, 598, 614
Anaximenes, 614
Andersen, Tim, 484
anthropic principle, 456
   meaning, 20
antigravity
   cosmological constant as, 292, 436
   dark energy is a label for the, effect, 413
   effect causes expansion, 357
   is not, but solid space, 436
   "pressure", *meaning*, 170
   unexplained, 170, 349
Antikythera, 580
antimatter
   excess of matter over, 365
   the problem of, 367–8
antineutrino, 251–2, 572–4
antirealism, 629
   meaning, 519
   reconciliation with realism, 549
anti-realist position, Bohr's, 519
apeiron, 51
apples
   are unsuitable as counters, 525
   as counters, 523–7
Aquinas, Thomas, 374, 450
Archytas, 133
area
   links between manifolds have, 236
      that is also a real quantum, 264
   links represent, 249-50
   metric, is one Planck unit, 198
   Planck, is real, 444
   quanta of, 558
      Rovelli's grains of space are, 264, 440
   smallest surface, of bubbles, 445
Arendt, Hannah
   and demonstration of truth, 645
   and Descartes' doubt, 26–7
   and doing, 110

and intertwined strands of thought, 11
and Kafka's parable of past and future, 156
and *Now*, 301
and the absence of thinking, 638
and thought vs action, 638
worldview of, 660
Aristotle, 74, 174, 217, 374, 614, 779
   and logic, 559
   and space, 67–73
      is never empty, 69
      is something, 73
   invented ontology, 38
   on causes, 312
   on infinity and physics, 543
   on limits, 624–6
   on void, 67
   on Zeno's flying arrow paradox, 310–1
   problems raised by, 465
   problems raised by, are unsolved, 590
   saw the universe as finite, 133
   the time of, 285
   thought space is something, 221
arithmetic
   consistent system of, must be inconsistent, 212
   essential primitive objects of, 549
   exact, of fractional charges, 254, 269
   is not complete and consistent, 174
   logical foundation for, 211, 528
   natural numbers and, 518
   the universe's, 521
   truths are useful, 522
   universe's steps reified, 516
   we discover not invent, Russell claimed, 519
Arkani-Hamed, Nima, 585
arrow of time, 322
   meaning, 77
Aspect, Alain, 29, 147, 680
assumptions
   about ether, Lorentz and, 68
   and our concept of space and time, 32, 34
   and the standard model, 225, 412
   as the weak spot of ontologies, 224
   either ontology depends on bizarre, 664
   fewest and simplest possible, 403
   hidden, and the role of philosophy, 523
   inherent in mathematics, 209, 211
   input, and logic, 637
   more, would be needed for a different ontology, 662
   no new, to explain dark energy, 349
   opposite, of physics' two theories, 152
   Planck kept his distance from unneeded, 90
   simplicity begins with one's, 57
   Smolin said physics is making wrong, 32, 34, 40

so many, for simulations, 504
some, are necessary, 57, 523
astronomers
   and supermassive black holes, 359
   expected gravity to slow expansion, 357
   observe vast jets, 345
   see a big universe, 398
Atkins, Peter, 479
atom
   mass of, 435
   maybe only one kind of final, 51
   means no-cut, 47
atomic scale, 577
a-tomos or no-cut, 47
Augustine of Hippo, 301, 374, 653
Avogadro's number, 573, 625
axiom
   meaning, 181
   of infinity, 548
   Spinoza's, 181
axioms
   Euclidean, 212
   Peano's, 540–1
axion, *meaning*, 352
Ayer, A.J.
   allusion in Greene's novel, 214
   and empiricism, 213–4
   and logical positivism, 214
   and the Copleston debate, 313
   on ancient problems with no solutions, 465, 590
   on reason as a guide to reality, 35
   wrote of time and motion, 309

B

Back, universe does not go, 321
background independence, 178, 181, 283, 369, 404
   and the relational approach, 44
   and Smolin, 19
   and the Wheeler-DeWitt equation, 419
   Markopoulou's definition, 425
   meaning, 44–6
background-free, 240
Bacon, Francis, 176, 373
Badiou, Alain, 516–7
Baez, John, 15-6, 223, 662
   background independence as an objective, 425–26
   inconsistency of arithmetic, 212
   we are part of the world being studied, 143
Baggott, Jim, 28, 403, 671
Bahcall, Neta, 358
Bakx, Tom, 622
Balfour, Arthur, 771
Ball, Philip, 148, 387-8
Banksy, 455
Barbour, Julian, 324-5

Barker, Stephen, 548, 561
Barnes, Hazel, 301
Barnett, Benjy, 539
barrier
    shape, 295, 309-10
        changes, 121
    tunneling, 119–21, 295
        meaning, 119
Barrow, John, 35, 118-9, 130, 137, 205, 212, 271, 539
baryon
    mass of, based on energy, 436
    meaning, 436
Baryshev, Yurij, 291, 433–4, 592
be, to, 647
beables, *meaning*, 110
Becker, Adam, 16
    and BBC interview, 303
    on Einstein and photons, 386–7
    on not finding supersymmetry, 411
    on not seeing half-electrons, 390
    on Paquette and dimensions, 237
    on reality of time and space, 420
beginning of the universe, 105–11, 182, 236–41
    according to general relativity, 399
    according to the standard model, 297
    and exponential growth, 222
    and low entropy,139–40
    and simplicity, 62
    and string theory, 400
    chain of causation back to the, 312
    first steps at the, were mathematical, 516
    Hawking's summary of the, 7
    is obscured from view in standard model, 396
    Lemaître's, 106–11, 126, 139–40, 229
    meaning, 467–70
    needed something like tunneling, 224
    no laws of physics at the, 432, 482
    physics should focus on the, 184–5
    the closed cosmos came into existence at the, 238
    there was only one space quantum in the, 42
    was when the first manifold tunneled into two, 468
    were laws of physics inplace at the, 478
being and *doing*
    interplay as a universal theme, 4
    is the nature of the world, 664
    laws as descriptions of, 480
    Sartre and the interplay between, 647
    the universe's, 322
        are real ontologic objects, 661
Bekenstein, Jacob, 507
Bell inequalities,147
Bell, John, 40, 142, 146–7, 190, 310, 568
    and local beables, 110

Bell, John Lane, 532
Bell's theorem, 147, 190
Beller, Mara, 29
    on complementarity and Bohr, 388–9
Belz, Maurice, 593
Benacerraf, Paul, 548
    numbers can't be sets, 529
Ben-Zvi, Pinhas, 450
Bergmann, Peter, 88, 330–1, 418, 475
Bergson, Henri, 301–2
    on causation and the nature of time, 381
Berkeley, George, 767
Besso, Michele, 493
beta decay, 251, 573, 768
Big Bang
    and general relativity, 185
    and ordered nascent universe, 139
    beginning of the, portrayed by Hawking, 7
    galaxy seen just 370 million years after the, 622
    general relativity and the, Epoch, 354
    Greene's blunt critique of the, 17
    Guth's blunt critique of the, 184
    had no explosion, 298
    happened with the universe well under-way, 184
    Hoyle naming the, 107
    Lemaître as father of the, 106–7, 140
    name is misleading, 107
    presents a profound puzzle, 137
    separated from the beginning, 108
    is a source of physical confusion, 112
    theory begins with a point universe, 399–400
    working on what happened before the, 197
Big Crunch, 140
Big Fizz, 139, 293, 296, 377, 407
    continues today, 361
    fast then slow, 241
    getting fizzier, 361
    meaning, 293
Big Flash, 129
    meaning, 327
Bigelow, John, 521
Bilson-Thompson, Sundance, 240, 432, 643
    and neutrino-antineutrino annihilation, 572–3
    and the electron, 269
    and the photon, 273–5
        and the Z^0, 274
    and the proton, 269
    assumed twist is digital, 393
    came up with concepts Einstein lacked, 193
    depicted a twisted-ribbon worldview, 250
    finding, was a coincidence, 244
    gave a lecture at CCM, 249–54

has confusing diagrams, 233
Markopoulou mentored, 18
saw matter as braided twisted ribbons, 221
model
   and links, 228–35
   and the two-slit experiment, 387–91
   explains standard model particles, 196, 571
   has no Higgs boson, 459
   has only left-handed neutrinos, 571-2
   has no room for sterile neutrino, 576
   helons are bound in triplets, 279–83
   neutrino is same as antineutrino, 572
   ribbons, *meaning*, 231
   shows extreme economy, 193
   offers an understanding of fractional charges, 269
   on nodes and links - volume and area, 249–54
ribbons
   lie flat, 393–4
said locality can be tricky in quantum gravity, 332
saw a braid as connected to "the universe", 281
search for, 198–9
set me on the path to causal sets, 243
shed a whole new light on quantum physics, 256
showed particles as composite, 439
Binnig, Gerd, 121
black bodies
   Planck studied, 77
   radiation of, without regard for sources, 80
black hole
   initial universe as a very big, 185, 343
   meaning, 125, 359
   size of the non-singular center of a, 125
   the initial universe as a big, 185, 363
   Wheeler on the fascination of the, 460
black holes, 499–507
   and quasars, 503
   and Robert Oppenheimer, 499
   and Roger Penrose, 499
   as dark matter, 353–6
   as sources of vast jets, 343–7
   big
      are very cold, 354
      as galaxy organizers, 366
      as space factories, 344, 359, 361
      at centers of galaxies, 359–62
      primordial, 353, 622
      seen too early, 353
   colliding, 620
   direct-collapse, 503
   extreme curvature of space in, 359
   from first stars, 622
   have singularities, Penrose showed, 460–3
   holographic principle and, 338
   masses of colliding, 621
   organizing early galaxies, 622
   primordial, 364–5, 469, 501–7
      meaning, 353–6
   probed by new instruments, 619
   sizes of, 359
   supermassive, 500–7
      at centers of large galaxies, 359
      in the early universe, 355
      jets from, and a new telescope, 347
      origin of, 355–6
   temperature of, 354
block universe, 39, 303, 782
   meaning, 155
blunder, Einstein's biggest (or best), 165-72
Bohm, David, 89, 192
   deprecated the two theories, 151–2
   on discontinuous motion, 615
   worked on a new ontology, 35
Bohr, Niels, 31, 678
   and complementarity, 388-9
   obscurity of writings of, 678
   worldview of, 519
Boltzmann, Ludwig, 78, 84, 640, 648, 688
   and entropy, 317–9
Bondar, Roberta, 66, 685
Boole, George, 559
Born, Max, 246
   on meaning and quantum systems, 565
   the special character is whole numbers, 88
Boscovich, Rogerio, 467-8
Bose, Satyendranath, 90
Bostrom, Nick, 104
Botner, Olga, 357
Boulding, Kenneth, 222, 227
Boylan-Kolchin, Michael, 622–3
Bradley, Francis, 409
Brahmagupta, 540
braid, 195
   end connected to "the Universe", 281
   trivial, 272-3, 276, 279
braids, 240, 250, 279
   may interchange with twists, 281
   meaning, 233
   new, from new space, 267
Brandenberger, Robert, 400
Broglie, Louie-Victor de, 257
Bronstein, Matvei, 321
Brown, Mike, 526
Brukner, Časlav, 381, 383
Bruno, Giordiano, 134
brute fact, *meaning*, 314, 728
Buckyballs and quantum interference, 255
Buddha, Gautama, 598
Buddhism, 597–600
Bunge, Mario, 215
Buniy, Roman, 149
Burbidge, Geoffrey, 591

Butterfield, Jeremy, 97

C
Cajal, Ramón y, 650–1
Calabi, Eugenio, 98
Calabi-Yau threefold, 101, 264, 692
calculus, 308, 416
    infinitesimal, 317, 428, 557
        meaning, 93
cancer treatment
    and tunneling, 120
    more effective with good theory, 170
Canovan, Margaret, 11
Cantor, Georg, 543
Cappelluti, Nico, 355
Cappi, Alberto, 694
Carlip, Steven, 445
Carnap, Rudolf, 598
Carniani, Stefano, 366
Carr, Bernard, 366, 735
Carroll, Lewis, 195, 601–2
Carroll, Sean, 70, 138, 322, 480
Carter, Jimmy, 664
Carus, Titus, 181
Carville, James, 612
castle in the air
    Einstein's, 103, 429, 493
causal completeness, 245, 405
    guaranteed, 182
    meaning, 182
causal set, 378–9, 532, 538
    an element that is key to the new worldview, 180
    partially ordered
        finite, 378
        the universe is a, 163, 245
        real numbers are not real in a, 532
    theory, 378
        and Markopoulou, 160
        and matter, 179
        and simultaneity, 179
        is a candidate for quantum gravity, 178–80
        Markopoulou led me to, 244–5
    tracking events as a, 163
    universe as a, 380
    universe's, 405
causal sets
    meaning, 378–80
    partially ordered
        Miller's thesis on, 245
causality. *See also* causation
    Buddhism and, 597
    Einstein's concern about, 385, 492, 656
    new view of strict, based on random moves, 382
    Planck and the meaning of the word, 381
    strict, 248, 492
    uncertainty, reconciling with, 247

causation, 142–3, 243–8
    deterministic, meaning, 243
    Einstein's view of strict, 661
    experience does not provide a reason for, 244
    probabilistic, meaning, 243
    the great chain of, 313
    the real nature of, 663
central processing unit, universe has no, 285
Chadha-Day, Francesca, 352–3
Chadwick, Sir James, 48
Chalmers, David, 598
Chandrasekhar, Subrahmanyan, 499
change
    a photon into a Z^0 particle, 274
    an inherent quantum's magnitude cannot, 260
    and motion, 614–5
    Aristotle said time is not a kind of, 285
    nature of, puzzled thinkers, 307–9
    seems smooth due to tiny increments, 325
    the primitive element of, is digital, 310–1
    the problem of, 103–4, 308–11
    the seminal idea of digital, 104
Chapman, Don, 565, 625–6
Chardin, Teilhard de, 51, 183, 647
    on these complicated and fragile edifices, 676
charge
    conjugation symmetry, 367
    fractional, 269, 723
    net total, on proton, 269
    why the proton's, is equal to the electron's, 268–9
Charteris, Leslie, 659
chemistry, 52-3, 169, 471, 625
Chihara, Charles, 547
China
    early concept of cosmos in, 22
    picked up where Perimeter Institute left off, 254
chirality
    how particles have both left and right, 253
    meaning, 571
    why neutrinos have only one, 253
Cho, Adrian, 352
Chown, Marcus, 112
Christie, Agatha, 188
Churchill, Winston, 171
Clarke, Arthur C., 228
classical
    continuum is thousands of years old, 30
    meaning, 119
    metric, 198
    part of a quantum experiment is necessary, 141
    physics, 118–9
        foundations of, Planck's results conflict with, 88

the action in, 92
system
    measuring apparatus must be a, 569
    quantum system needs a classical, 141–2
    theory, 274–5
        quantizing a, not good strategy, 179, 259
    view, *meaning*, 472
Clausius, Rudolf, 316–7
Clay, Henry, 663
Clery, Daniel, 343
clock, 158
    motion of molecules in aging people as a, 287
    the system, 103
    there is no, outside the universe, 418
    time
        Einstein used, 423
        has conceptual problems, 287
        is a asource of endless confusion, 419
        is a concept we devise, 155
        meaning, 159
clocks
    accelerating, and general relativity, 486
    and Minkowski's first attempt to define time, 423
    atomic, 160, 290
    do not measure true time, 164
    inside the universe and universal iterations, 278
    measure motion, 289–90
    motion
        in, 157, 473
        of, 157, 473
    should emerge from the theory, 423
    special relativity and, 287, 485
    sundials as, 288, 417
    tell common time, 163
    we should reassess the role of, 321
Close, Frank, 312–3, 544–5
closed universe, 168, 295, 326–9, 422
    and the concept of the Cosmos, 611
    and the curvature of space, 544
    as a causal set, 380
    depicted in a NASA image, 396
    Einstein's, 133–6
    meaning, 122–3, 136
    origin of the, 238, 468
    seminal idea of a, 133–6
coincidence
    and Miller's thesis, 244
    in the quantum story, 87
    of *be*, *do* and *go*, 647
    of the proton and electron charges, 722
    of the universe's tiny curvature, 130–1, 170
    of the viable universe, 584
    unbelievable

of the knife edge, 131. 170-1, 452
of the Λ value, 170
Vonnegut on, 245–6
common time. *See* time
compactified dimensions
    are vanishingly small in string theory, 177
    enormous number of possible sets of, 101
    meaning, 100
    there must be six for string theory, 99–101
    were compact when the universe began, 400
complementarity
    Bohr's attachment to, 388–9
    meaning, 388
complexity
    and conundrums of physics, 61
    of the brain, 650–3
    of the human body, Feynman and the, 624
    of the standard model, 292
    origin of, in emergent systems 753
    we are at the far end of the, scale, 624
composite
    counters trouble philosophers of math, 526
    hadrons are, 55
    helons regarded as, 249
    meaning, 440
    particles
        all elementary, are, 197
        and their charge combinations, 272, 439
        electrons are, 254, 269, 393
        standard model, are, 261
    preons, 193
    proton is known to be, 268
    standard model particles may be called, 440
    structure and Hackett, 49
    substance is assembly of monads, 51–2
computer
    bit is physically analog, 393
    hacker in *Time One*, 402–3
    language is not ambiguous, 653
    programmable digital, and Zuse, 103–4
    reality as a, simulation, 104, 503
    simulation of primordial black holes, 503
    universe is like a, 285
Comte, Auguste, 59, 561
concept
    of a cosmos, 22
    of a natural number, 536
    of a number, 632
    of *chair*, 652
    of infinity, 532–3
    of locality in new ontology, 382
    of locality will change in quantum gravity, 330–1
    of mass, 435

of measurement, 475
of nonlocal memory, 651
of real quanta suggests experiments, 639
of reality, 27, 492, 644
of space as a real entity, 487
of space expanding is hard to understand, 291
of the field, Faraday developed the, 427
of the idea, Plato and the, 25
of time, 511
of time is a mental construct, 287
of vacuum, 450-3
of validity of logical statements, 452
of zero, 539–40
when a, takes on form in thought, 637
concepts
and Plato's forms, 633
as simple and as few as possible, 57
as unalterable givens, Einstein's warning, 205-6
assumptions include wrong, of space and time, 40
clash between the, of position and motion, 615
different, all called entropy, 318
distilling, from physics, 209
elementary, and Bilson-Thompson, 193
embodied in our language, Ayer and, 215
fundamental
lack of understanding of, 536
new, need new langauge, 376
starting afresh with quite different, 40
hierarchy of, and backbone of an ontology, 37
many of physics', lack evidence, 28
observer-centric, of quantum theory, 569
of a number and of its square root, 633
of space and time, wrong assumptions about, 40
six key, underlying the new ontology, 41
the aim of science is a minimum of primary, 58
Unger and Smolin came close but lacked two, 407
unreal, used as if they describe reality, 375
connectome, *meaning*, 652
consciousness, 653
and external reality, 599–600
and the Moon per Einstein, 661
emergent, 662
Planck (foolishly) said matter derives from, 628
the nature of, 637
the universe now has, 638
conservation laws
and Noether, 432
general relativity abandoned, 433
there were no, in the beginning, 482
conservation of energy, 77, 140, 431–4

an approximate and local law, 432
rescue of, seems mired in difficulty, 747
consistency
and finitism, 765
and Gödel, 613
Greeks demanded logical, 539
is a key virtue of mathematics, 211
is a quality to seek in fundamental physics, 408
is lacking in today's physics, 27
of math is mostly an illusion, 211
requires that space is real, 163
thought experiments can test ideas for, 174
we should lean on, of the whole picture, 219
constant
arbitrary, added when integrating, 317
Boltzmann, 317, 648
cosmological or cosmic. *See* cosmological constant
Hubble. See Hubble constant
Planck. See Planck constant
speed of light. See speed of light
the most fundamental, of the universe, 80
continuity
and Zeno, 64
as a source of confusion, 578
challenge of the granular space objective, 179
fields as an expression of the, compulsion, 428
of space, the assumed, 74
of time, Smolin's commitment to, 162
"real" numbers and the imagined, of space, 532–3
the quantum dis-, and Kuhn, 78
continuous
assumption true time is, 160
creation, 107, 291, 299, 694
line, 532–3
time
is not, 417-8
or space in differential equations, 74
view of Planck-scale space, 91
continuous space. *See* space; *see also* continuum
continuum, 192
and the uncertainty principle, 446
atomicity challenges the notion of a, 179
basis of Einstein's castle in the air, 493
black hole theory uses a, 500
contradiction of formulating a quantum in a, 93
contradictions of motion in a, 615
could be based on faulty presuppositions, 531
dis-
Einstein considering a, 392

Einstein favored a, 430
Einstein toiled on the form of a, 430
without resorting to continuum, difficulty of, 192
drawback of the, Einstein noted, 392
Einstein
   and Infeld with trains, quantizing a, 639
   plowing barren fields in the, 482
eliminating the, and Smolin, 262
Hawking saw no reason to abandon the, 93
is unphysical, Rovelli said, 92
math of the, is hard to give up, 93
must avoid the, Einstein said, 192, 259, 490
old ontological commitment to the, 31, 583
pi would have problems even in a, 533
possibility leads to renunciation of the, 192
quantum foam conceived in a, 445
quantum mechanics needs a background, 569
real numbers and the, 534
reason energy conserved in a, Noether and, 432
Riemann went on to exploit the convenient, 687
singularity is a consequence of the, 460, 500
Einstein did not argue space is a, 29
first response is try to discretize the, 179
string theory
   manifolds in a, 177, 223
   needs Calabi-Yau manifold at every point in a, 99
strings wiggle in a, 100
time is not a, 156, 406
first response is try to discretize the, 179
vs granular space, 473
contradiction
   apparent, of universe's preferred frame, 304
   in basing laws on nothing, 478–9
   real numners are permeated by, Weyl said, 532
   light vs wave, disappears in quantum space, 493
   no, of ftl motion ouside local inertial frame, 340
   of quantum theory in continuous space, 93
   sinkholes of, in science, 10
contradictions
   how, arose in Frege's arithmetic, 537
   in science, 10
   and truth, 630
   in the standard model of cosmology, 108

new ontology to settle all, in one fell swoop, 186
of a two-theory situation and Einstein, 152
of motion in continuous space, 615
of physics
   and philosophy offer clues, 186–7
   offer clues, 205
of relativity and quantum theory, 206
of the universe both being and doing, 307
our best theory of the world is bogged down in, 534
reconciling the seeming, of two theories, 385
scientists are aware of, in the old ontology, 21
solving all, is easier than solving only one, 35
the old ontology is filled with, 34
zero is chief source of, 539
Coope, Ursula, 285
Copenhagen group, hegemony of the, 246
Copernican principle, the, 580
Copernicus, Nicolaus, 272, 326, 580
Copleston, Frederick, 313
Corcoran, John, 559
Cosmic Clock, 284-90, 553
   in *Time One*, 322
   is simultaneous everywhere, 288
   meaning, 284, 288
   universe as a, 284
   universe, is the, 321
cosmic inflation, 587
cosmic microwave background, 39, 129, 396–7, 618
   and the absolute frame of reference, 160
   cooling plasma and the, 470
   meaning, 108–9
   radiation is uniform, 304-5
cosmic time, what clock can measure, 321
cosmogony
   and Buddhist ontology, 598
   Lemaître's, was similar to Poe's, 81
   meaning, 105
   nebular, and Poe 617-8
   Penrose's far-fetched, 140
   Poe
      and Eureka's, 22, 81, 618, 694
      was fascinated by, 618
cosmological
   constant
      and the calculated energy density, 445
      and the vacuum energy, 451
      as an arbitrary afterthought, 490
      as an antigravity effect, 165, 292
      as dark energy, 225
      as space, 169
      as a thorn in the side of physicists, 168
      assumed to be zero, 167

Carlip on the problem of the, 445
confusion about the, 433
Einstein's idea, 165–72
is measured to be non-zero, 168
may not be constant, 436
meaning, 167–72
rebranded as dark energy, 349
event horizon, meaning, 328
expansion, *meaning*, 113
fallacy, *meaning*, 644
inflation, 126, 186, 226, 292, 296, 354
and exponential growth, 227
Guth invented, 128
is old ontology, 169
many versions of, 128
meaning, 126–32
observations don't support, 131
solves two big problems, 128, 226
redshift, *meaning*, 621
cosmology
a real theory of, 142
and expansion of the universe, 113
assumes initial conditions, 109
conformism in, 591
deep questions about, 19
Einstein launched modern, 114
inflation theory is at the heart of, 131
is based on observing subsystems, 183
low initial entropy is an unresolved issue of, 137
meaning, 39
mysteries of modern, 355
observer not outside the system in quantum, 141
physical, 74, 397
revolutionary insights for, 171
standard model of. *See* standard model of cosmology
cosmos, the, 107
meaning, 23
count
each counter only once, 536
integer in step sequence, 419
iterations, 286
living things as a metric in biology, 594
not to, a set twice, 530
paces of an infinite walk, 544
physical objects, 518
Planck needed distinct energy packets to, 82
points in lines, 548
quanta, 447, 527, 552
set that holds empty set as one, 530
sets, 530
space granules, 82
the metric, 594
things to, 523–7
true time, 164
universe's, began with one, 540, 542

volumes and areas, 477
counters
apples as, 524
body parts as, 526
composite, 526
eggs as, 526
fingers as, 525-6
medium-sized physical objects as, 523
planets as, 526-7
sets as, 527, 529–31
space quanta as, 527
the problem of defining, 527
counting
and Newtonian time, 509
and thought experiments, 522
as an essential element of mathematics, 516-7
as the basis for Planck's insight, 88
as the only metric, 230
begins with one, 542
Gödel needed infinitely many numbers for, 613
grains of space, and Rovelli, 293
granules *(quanta)*, and Riemann, 75, 475-6
medium sized objects, 525
needs distinct counters, 531
no medium-sized object for, 527
Planck required, to get a probability, 82
planets, 527
points on a continuous line, 544
quanta, 522, 527, 552
the universe does real, 522
the, rule against recycling objects, 549
to get the natural numbers, 521
vs measurement, 533
at Planck scale, 230
Wittgenstein and, 523-4
Cox, Brian, 507
Crane, Louis, 30
Crease, Robert, 110
creation myths, 137
crime fiction
connection with fundamental physics, 186–8
solution strategy, 188-9
crisis
in physics as opportunity, 216
in cosmology, 585
in fundamental physics, Smolin on the, 150
in particle physics, 412
in philosophy of science, 213
of modern physics, 630
the Chinese word for, 216
Critchley, Simon, 38
crystal structures, analagous, 334-5
Curie, Pierre, 368
curvature of space, 295, 403

assumed to be zero, 124
barrier shape varies with, 309
decreased as the universe grew, 365
extreme, 296, 358-61, 468-9, 556
infinite, 125
is gravity, 295
meaning, 123
near the beginning of the universe, 238
negative or positive, 130
of the universe today is small but nonzero, 122
replication rate of space quanta depends on, 247
where, is small, 362
curve fitting
and expansion of the universe, 170
and general relativity, 490
and Lorentz, 85
and Planck, 79, 80, 84
and the standard model of cosmology, 590-1
cell survival data, 169
Cushing, James, 246
Cusmariu, Arnold, 633

D
Dalai Lama, 600
Dällenbach, Walter, 392
Dammbeck, Lutz, 560
Daniel, science student, 641
dark energy. *See* energy, dark
Davies, Paul, 104, 136, 324, 582, 585
Davis, Tamara, 113, 340, 434
on expansion of the universe, 340-2
Dawid, Richard, 779
Dedekind, Richard, 533
Democritus, 47, 440, 450, 614
density
infinite, 198
nonuniformities of, 365
Descartes, René, 26, 74, 645
on space, 67-9
Desjarlais, Blake, 150
determinate number property, 522, 526
Deutsch, David, 585
DeWitt, Bryce, 177, 418, 585
Wheeler-DeWitt equation, 177, 184, 263, 418-9, 753
Dickens, Charles, 243
differential equations
constant added when integrating, 317
in calculus, 317
in the standard model of cosmology, 74
meaning, 74
digital, 93, 558
arithmetic, 269
change is, 311
math, 443

metric, 379
reality, approximations to a, 384
state, universe in a, 103
the math of the universe is, 75
the universe is, 103, 310
tweedle's motion is, 282
twist is, 269, 393
universe, *meaning*, 325
watch, electrons moving in a, 289
dimension
new, cannot create a, 236
there is no real entity with only one, 231
dimensionality of space emerges, 265
dimensions
a point has no, 400
any number of, 74
as loops, 98
curvatures seen in additional, 123
each manifold's links use all six, 240
externalized by replication, 239
four, of Minkowski spacetime, 421
four, of the mind, 652
links have two, 231-41, 236
loops, as, 136
many ways to curl up six, 315, 483, 580
no unlinked, by step three, 334
origin of a link's two, 237
origin of manifold's external, 237
pairs of, 226, 235
physics invents, 325
rings and, 239
six compactified, 99-101
six, of Calabi-Yau manifolds, 191
spacetime is a mix of emergent and fictional, 377
spatial, origin of, 231
string theory's need for six extra, 97-101, 177, 400
tangled, 223, 280
the universe may have only two, 336
theories make sense in a different number of, 264
three, are emergent, 265
two, of a ribbon, 197
unseen, of string theory, 99
volume
has three, 264
of quanta gives space three, 404
why there are three, 181, 263
dipole anisotropy, *meaning*, 304
Dirac, Paul, 260, 442-3
disaggregation, 150-1
discovery
a new, Lazaridis sought, 196
Einstein's, of the photon, 272
failure to grasp the new, 254
Lazaridis's, left on a bicycle, 202
of a new particle, 457
of the cosmic microwave background, 618

of the Higgs boson, 457
Planck's, 403
vs invention of numbers, 519
Disney, Mike, 413
distributed processing, 285
do, to, 647,
Dokkum, Pieter van, 500
Dowker, Fay
  and causal set theory, 178
  and granular space, 179
  and Planck-scale, 264
  and real numbers, 532
  laws arise from the universe, 479–80
Doyle, Arthur Conan, 188, 618
dream
  grasp of reality has been a, 609
  of reality, Kafka's, 609
  reality as a, 26-7
  to unify matter with geometry and gravity, 283
duality, 338
  and the two-slit experiment, 387
  beauty and, 213
  holographic, 339
  understanding wave-particle, 275
  wave-particle, *meaning*, 387
Duck, Ian, 78, 83, 101
Dugle, Janet, 594
dynamics, the, 282
  meaning, 281
Dyson, Freeman, 629

E
$E = mc^2$, 250, 432
econometrics, *meaning*, 595
economics
  and statistics, 593
  is there a real metric for, 595-6
econophysics, 595
Eddington, Arthur, 320
Einstein, Albert, 2, 8, 21, 36-7, 47, 57, 66, 75, 77, 100, 103, 106, 122, 145, 146, 152, 185, 192, 205-6, 228, 236, 246, 258-9, 265-7, 278, 286, 331-2, 344, 349, 363, 384, 392, 482, 516, 544, 607, 615, 632, 642, 645
  adopted general covariance, 433, 487-8
  and 5-D space, 98
  and a discontinuum, 192, 392
  and a finite universe, 134-6, 326
  and a leading role for mathematics, 210
  and a new age of cosmology, 134
  and a softening view of the photon, 272-3
  and background-independence, 44
  and Bose's derivation of the Planck formula, 91
  and causality vs realism, 385
  and dark matter, 351
  and elementary foundations for physical theory, 45
  and expansion of the universe, 114-5
  and fields as the only reality, 428-30
  adopted general covariance, 433, 487-8
  and geometry of the universe, 556
  and language slips, 375
  and mass-energy equivalence, 432
  and no firm foundation in Planck's result, 87–94
  and origins of string theory, 96
  and photons that behave as particles or waves, 386
  and Poe's *Eureka*, 694
  and Poincaré, 555-6
  and realism, 9, 629
  and Ricci curvature, 491
  and the cosmological constant, 165-72, 292
  and the new ontology, 36–40
  and the old ontology, 29–31
  and the problem of inertia, 616
  and the quantum, 31, 88, 492–5
  and the source of laws of science, 482
  and thought experiments, 173
  and time, 154-5, 419
  and Unified Field Theory, 187
  as a realist, 9, 29
  as essential guide, 656
  as philosopher-physicist, 210
  black holes and, 500
  said clocks define time, 287
  criticized Friedmann paper, 115
  curvature of space, 295
    can be determined, 699
  found a special frame to be unnecessary, 304
  free
    inventions and, 26, 33, 36
    will and, 661
  had weekly chats with Russell and Gödel, 559
  had wrong ideas about space, 490
  mass and, 435-6
  mathematics and, 632
  Minkowski and, 421-4
  most consequential insight of, canceled, 73
  motion and, 308
  on crime fiction, 187
  on Emmy Noether, 747
  on essential role of fundamental ideas, 639
  on Planck's imperfect derivation, 79
  on simplicity, 60
  on speed of light as a fundamental constant, 277-8
  on fundamental physics strategy, 179
  on the ether, 68–73
  on the fictitious character of theory, 33
  on wave-particle duality, 387

problem of quanta in a continuous universe, 31
reason for his Nobel Prize, 272
said clocks define time, 287
said objective reality exists, 628
said there is nothing special about *Now*, 302-6
said theory decides what we can observe, 169
saw no place for free will, 381
saw his physics as a castle in the air, 103, 429
seminal ideas of, 221
sought math for granular space, 430
thought space is granular, 486
thought the continuum idea is wrong, 29
translations of, 11
used clock time, 423
Eisenstaedt, Jean, 60
electric charge
   is quantized, why, 253, 268
   quanta of, 259
electromagnetic waves, 273
electron, 54, 85, 196
   assumed to be elementary, 390
   Bilson-Thompson's, 269
   charge on the, 393
   classical view of the, 384
   is real, 634
   new ontology view of the, 384
   quantum theory view of the, 384
   tunneling, 121
electrons, 48
   and Lorentz, 68
   as points, 392
   fractional, 390
   Wheeler's reason all, are identical, 390
electro-weak theory, 274
elementary
   entities
      consist entirely in relationships, 45
      what are the, 45
   structures of physics, 439
elementary particles. *See* particles
elephant in the room
   our suitable universe is the, 483
   the scientific method is the, 642
   the supervisor is the, 484
Eliot, T.S., 197
Ellis, George, 96, 399, 584–586, 589
Elton, Sir Geoffrey, 605
emergence, 471-4
   hierarchy of, 753
   meaning, 471
emergent
   approximations, equations reflect, 282
   black holes, 499
   causation is, 383
   dimensions of space are, 265

entities are all essential for us, 662
laws of physics are, 384
mass is, 438
properties are inexact, 377
Smolin on, laws of physics, 479
space is, 236
Emerson, Ralph, 43
Emerson's transparent eyeball, 96, 331, 447, 552, 579
Empedocles, 614
empirical incoherence, *meaning*, 46
empiricism
   and sense perception, 213
   and what is real, 216
   meaning, 213–7
empty set, the, 528-31
Emspak, Jesse, 145
energy
   conservation of, 432–4
   dark, 214
      and Guth proposing a negative pressure, 348
      and superconductivity, 120
      and Wilczek, 349
      as a label for a mystery, 413
      as the density of space, 349
      careful calculations of, 171
      concepts related to Λ, 168
      density of, is nearly zero, 348
      universe's expansion is attributed to, 357
      is a challenge for physics, 357
      is a mystery, 262, 349, 358
      is like a cosmological constant, 225
      is like missing mass, 171, 358
      is space, 349, 362
      is two-thirds of universe's mass, 358, 436
      mass of, 185
      may be seen as everything, 350
      meaning, 348
      is one of the greatest unsolved problems, 348
      Planck satellite data on, 358
      Turner on, and fundamental physics, 590
      vague descriptions of, 349
   of the universe is undefinable, 434
   ontological commitment to conserved, 31
   physics has great difficulty with what, is, 31
   the world does not run on, 323
      needs real physics of quantum space, 438
engram
   for the concept of "chair", 652
   for the concept of zero, 539
   is not localized in the brain, 652
   meaning, 652

remembered, has been processed, 653
entanglement. *See* quantum entanglement
entropy, 316-9
    absolute, *meaning*, 319
    analogous concept of, 78
    and Parisi, 318
    high, 138
    low, 138
        initial, 137, 322
        meaning, 137-40
    machine, universe as an, 139, 323
    meaning, 18, 316
    of radiation, 79-81
    Planck
        and the idea of, 77
        derived an, 84
    zero initial, 319
epistemology
    meaning, 38
    physics is now about, 679
EPR, 146–7
equations
    do not reflect reality, 320
    physics', work both ways in time, 138-9
    Planck's problem in deriving his, 79
    time in physics', 154
ether, 84
    was not needed for relativity, 69-70
    Einstein shows, is real, 70
    Einstein's, 70-1
    grasp of motion may not be applied to, 171
    Lorentz's, 70
    meaning, 166
    space as the new, 66
    space without, is unthinkable, 70-1
    the story of the, 66-71
Euclid, 98, 124, 426, 475, 555
    assumed space is flat, 130
    wide-angle telescope, 775
Euclidean
    space, 473, 534
        meaning, 692
        non-, needs a non-Euclidean metric, 475
        quantum theory in, 473
Euler, Leonhard, 92
Eureka, Poe's prose poem, 81, 108, 618
Evans, Justin, 575
event horizon, the, 344, 360, 507
    meaning, 338
event is prior to all events, what, 313
events
    in causal set theory, 160
        meaning, 379
    instantly linked distant, 145
    new, steadily created from present events, 162
    quanta of space as, 161

simultaneity of far-flung, 39
space quanta and links as, 246
Everett, Hugh, 20, 586-7
evidence, 603-4
    direct vs circumstantial, 604
    energy of the universe is not conserved, 431
    for dark matter, 351
    for elementary particles, 55
    for finite size of the universe, 136
    for space expanding, 114
    for the universe's beginning, 109
    lack of experimental, at Planck scale, 333
    legal, is admissible if logically probative, 603
    loss of contact with empirical, 591
    no, for proposed new laws of physics, 129
    no, for strings, 400
    of a million pairs of supermassive black holes, 504
    of common and true times, 510
    of discreteness at Planck scale, 264
    of early big black holes, 353
    of neutron stars, 461
    of Newton's true time, 510
    of three supermassive black holes in a galaxy, 501
    or clues, an embarrassment of,188
    seminal idea of explaining all the, 221
    that all math is translatable into logic, 520
    the photon is not elementary, 389
existence
    and Sartre, 110
    building better language for, 631
    Dalai Lama and the nature of human, 600
    fleeting, of exotic particles, 447
    how physics laws came into, 583
    how space came into, 197
    mathematical, of Calabi-Yau manifold, 98
    new view of our, 660
    Next has no, 647
    numbers get meaning from, of apples, 525
    of atoms, 83
    of black holes, 125
    of dark matter, 352
    of extra dimensions, 100
    of God, Stephen Hawking and the, 455
    of infinitely many entities, 548
    of mathematical objects, Maddy and, 519
    of medium-sized physical objects, 523
    of the Planck-scale world, 77
    of real mathematical objects, 521
    of space and time, 44
    of space is a structural quality of the field, 429
    of the ether, Einstein and, 68-70
    of the true atom, 48-50
    of thoughts, 635–7

ontology as the study of, 38
  our own, and the universe, 401
  pending non-, and Holt, 313
  physical, tied to mathematical, 515
  pointless, 401
  points of space have no, 399
  real math objects and operations came
    into, 527
  reason for, of quarks and leptons, 252
  the idea of, and Hume, 628
  two kinds of, 633-7
  virtual particles come into, 332, 451
expanding or collapsing universe, balancing,
  131
expansion
  and dark energy, 358
  exponential, 127
    meaning, 222
  faster-than-the-speed-of-light, 293
  rewind, to a point at the beginning, 399
  universe's, projected backwards, 185
experience
  and logical positivism, 59
  Einstein
    and the world of, 9
    and thought distilled from, 36
  sensory. See sensory experience
experiment
  classical part of a quantum, 141
  gedanken, 173-5
  no, can determine absolute rest, 61
  of counter-traveling atomic clocks, 160
  testing theory with, 4
  the Michelson-Morley, Einstein
    explained, 71
  the two-slit, explained, 386-91
experimental
  data constrain wild theorizing, 642
  evidence is lacking at Planck scale, 333
  evidence is lacking for physics concepts,
    28
  little, evidence for M-theory, 455
  no, test of string theory, 96
experiments
  and the history of our universe, 585
  Bell-type, 149
  cannot foresee theories and, to come, 639
  Land's, 217
  Newton and, with light, 67
  on color vision, 217
  particle-collider, are the most expensive,
    414
  test theories not ontology, 42
  the new ontology suggests, 639
  the universe lays on, in space for free, 414
explanation
  a single self-consistent, is needed, 153
  beauty of a simple, 60
  consistent, 153, 196

dark energy is like magic as an, 358
fields as physics', for everything, 427
for color charge, 252
for $E = mc^2$, 250
for equal charges on proton and electron,
  253, 269
for the fractional charges of colored
  entities, 252
for the loose localization of a particle, 333
for other generations of particles, 253
for our perception of Now, 306
for the Heisenberg uncertainty principle,
  241
for the universe's three dimensions, 265
for the two-slit experiment, 389
needed for fractional charges, 56
needed for motions of Sun and planets,
  580
needed for quantum theory, 9
needed for string theory's extra dimensions, 99
needed for the whole universe, 183
no
  for quantum tunneling, 119
  in standard model for how expansion
    began, 292
of continuing expansion, 344
of down-quark decay, 251
of electron-positron annihilation, 279
of electroweak interactions, 253
of expansion speeding up, 361
of how galaxies came to exist so soon, 366
of motion, 615
of quark and lepton charge ratios, 252
of the arrow of time, 322
of the origin of matter, 191
of the speed of light, 276
of the standard model's particles, 571
particles are made of relationships, 253
why all neutrinos are left-handed, 253
why charge is quantized, 269
why mass and energy are interconvertible,
  253
why quarks have six flavors, 253
why the universe's charges are balanced,
  270
why there are six leptons, 253
why there are sixteen particles, 253
explanatory power, 27, 408, 584
  justifies a new assumption, 224
  meaning, 219
  of a consistent picture, 27
exponential
  awesome power of an, 293
  epoch
    black holes arising in the, 353, 355
    consumed no resources, 227
    ending, 292, 295
    extremely dense matter in the, 354

meaning, 226
  space emerged in the, 265
  growth, 126, 226, 293
    an idea to solve problems, 126
    and Guth, 128
    came to an end, 127, 294-6
    differences of, from cosmological inflation, 227
    in a finite world, Boulding and, 222
    Lemaître's, 108
    new ontology's, consumed no resources, 222
    slowed, 227
    the power of, 226, 469
    unexplained end of, in the standard model, 396
extreme economy
  of the Bilson-Thompson model, 252
  of the new ontology, 361

F
Fann, Kuni, 523
Faraday, Michael, 427, 642
Farrell, John, 115–6, 139–40
faster than light
  apparent motion of AGN jets, 343-7
  cosmological inflation moved, 128
  huge masses moving apart, 294
  in relativity frame matter cannot move, 343
  infant universe expanding, 127, 241
  inflation, 293
  matter
    in space that is leaving, 340-1
    moving, but not in a local frame, 340
  space moving, 341
  two cases where matter may go, 341
Fermi, Enrico, 717
Feyerabend, Paul, 631, 777
Feynman, Richard
  and curvature of space, 122
  and difficulty of a new idea, 192
  and identical electrons, 271
  and path integral, 92, 391
  and the two-slit experiment, 386–91
  did not believe space is continuous, 74
  on conservation of energy, 434
  on ideas in another field, 10
  on inertia, 437
  on nobody understanding quantum mechanics, 492
  on our atomic complexity, 624
  on renormalization as a dippy process, 442
  on the reality of fields, 429
fiction
  atoms were long seen as, 53
  fundamental physics and detective, 186-9
  hard for physicists to see their theories as, 152
  Hume saw causation as, 244
  math, 209-12, 400, 403, 522
  meaning, 26
  multiple universes are, 586
  objects severable from the universe are, 151, 183
  observer outside universe is, 286
  physicists seek beauty in, 209
  Planck's oscillators were, 83
  reality and, 26
  some math is, and some is not, 517
  spacetime is, 289
  the block universe is, 39
  Time One is, 4
  useful, 325
    also entertains, 613
    cosmology based on subsystems is, 183
    embraces almost all of physics, 219
    Euclidean geometry is, 212
    fundamentals of scientific theory are, 33
    may be essential, 26
    pi is, 551
    real numbers are, 534
    science's story of the world is extremely, 566
    time is treasured, 511
fields
  and continuous space, Einstein's view of, 429
  and Einstein, 153
  and inertia, 437
  and reality, 427
  are not real, 153
  Einstein and Infeld on the role in physics of, 428
  have values to infinity, 333
  the reality of, and Feynman, 429
final atom, the, 50, 200, 440
Fine, Arthur
  and natural ontological attitude, 629
  and rational belief, 630–1
finite
  atoms are, in size and number, 47
  causal set, the universe is a, 245, 380
  density in a black hole, 462
  set, each iteration is a, 246
  universe
    according to Plato and Russell, 133
    assumed to be, 133
    could have no edge, 134
    Einstein's, 133-6, 326
    holds only a finite number of things, 547
    in the new ontology, 326
    Kant and a, 543
    measurement revealed a, 168
    Russell and a, 545

Weinberg on a, 297
Firestein, Stuart, 16
first law
   meaning, 645
   Newton's, 616
   of nature, 480
Fitzpatrick, Richard, 480-1
flat space, 123
   meaning, 123
   the nonexistent, of Euclidean geometry, 533
flatness
   deviation from, 123
   problem, *meaning*, 130
Fleming, Stephen, 539
Fletcher, Angus, 614
foliation, 289, 305-6, 422, 726
   meaning, 289
Foot, Robert, 268
Forrest, Peter, 288
foundations, structures built from elementary, 45
Four Rings, the, *meaning*, 240-1
frame of reference
   absolute
      acceleration requires an, 135
      and absolute velocity are both impossible, 61
      cosmic microwave background provides an, 305
      Mach and the universe as an, 135
   local
      faster than light jets are not in a, 343
      matter cannot move faster than light in a, 340
      meaning, 343
   special, *meaning*, 39
   that there is no absolute, is old ontology, 169
Frank, Adam, 28, 623, 657, 751
free invention, physics as, 26, 33
   Einstein and, 36
free will
   Bergson defended, 381
   Einstein's view of, 381-2, 661
   taking a position on, 660-1
Freeman, Alan, 612, 643, 665
Frege, Gottlob, 561
   and arithmetic, 536-7, 540
   and composite counters, 526
   and modern logic, 559
   and Peano, 540
   on the nature of numbers, 518
   on nonexistent objects, 636-7
   on zero, 540
   said counting requires distinct objects, 535-6, 548
   set out to derive math from geometry, 548
Freiberger, Marianne, 368

Friedman, Michael, 59
Friedmann, Aleksandr, 115, 294, 476
Frisch, Mathias, 212
Fuchs, Guido, 79
fundamental physics, 15-6, 19
   and reality, 197, 205
   approach to achieving a breakthrough in, 162
   beauty as a guide to, 213
   composite particles are a revolution in, 197
   connection of, with detective fiction, 186-9
   crisis in, 150, 216, 585, 630
   dark energy's connections to, 590
   holographic principle is a new frontier in, 337
   lack of progress in, 405
   proposed principles for, 181, 404
   prospects for, at Planck scale, 333
   reason for turning to, in quest for what is real, 205
   spacetime is a roadblock for, 420-6
   the central goal of, 206
   the kind of math needed for, 244
   three bold programs for, 176
   we should aim money at the frontier of, 415
Furley, David, 133
Future Circular Collider, 414

G
Gage, Phineas, 652
galaxies
   black holes as organizers of, 366
   curve space, 124
   dark matter controls, 351
   distant, are moving faster, 116
   formation of, in the standard model, 365
   kinetic energy of moving, 433
   light from distant, will never get to us, 327-8, 398
   large-scale motion of, is due to new space, 299
   moving away from us, 106, 113-4
   new matter said to fill gaps between, 299
   new space enlarges gaps between, 299
   quasars at the centers of, 344
   receding from us faster than light, 340
   riding along as space expands, 343
   spiral, and dark matter, 351
   stars in, are gravitationally bound, 113
   supermassive black holes in, 359
   two hundred billion, in the visible universe, 360
Galileo, 173, 617, 755
Gamow, Barbara, 112
Gamow, George, 112, 165
Gardner, Martin, 736

Garfield, Jay, 599
Gates, Evalyn, 114
Gauss, Carl, 547, 555
Gedankenexperiment, 173, 418
Gefter, Amanda, 142
Gell-Mann, Murray, 100, 269, 682
general covariance, 433, 487
    Einstein's view of, 488
    is only a formality, 489–90
    meaning, 488
    Norton on, 488–9
general relativity, *See* relativity, general
geometry
    and Calabi-Yau space, 223
    and Euclid, 130, 554
    and Gauss, 555
    and gravity, 75, 557
    and Kant, 555
    and measurements of motion, 490
    and Poincaré, 555
    and the spatialization of time, 426
    and the metric relations of space, 477
    and the universe's mean density, 130
    dynamical, 283
    Einstein used Riemann's, to describe gravity, 75
    Euclidean, 74, 212, 533
    for measuring Earth, 554
    Frege deriving mathematics from, 548
    in imaginary spaces, 426
    is approximate, 425
    objects needed for, 554
    Poincaré's non-Euclidean, 556
    Riemann on doing, in spaces, 490, 554-7
    Riemannian, 74
    Smolin on discrete quantum, 230
    time evolution of
        3-D, 421
        space-time as, 289
    with no points or lines, 554
Ghomeshi, Jila, 653
Giaever, Ivar, 121
Gibbs, Willard, 316-7
Gilder, Louisa, 141, 155
Gimbel, Richard, 694
Gleiser, Marcelo, 28, 104, 144, 623
global state, 285
    meaning, 725
gluons, *meaning*, 435
Gödel, Kurt, 613, 715
    and logic, 559-62
    and math's basic object, 537
    and unprovable arithmetic, 211
    and weekly chats with Einstein, 559
    Quine's view on, 520
    upended programs seeking math foundations, 520
Goff, Philip, 586
Goldberg, Rube, 325, 589

Gorgias, 614
Graham, Flora, 408
granular space, 74, 76, 92, 157, 161, 178-9, 223, 283, 308, 332, 379, 399, 430, 443-5, 473, 483, 490, 534
granularity
    and nonlocality, 145
    and Riemann, 101
    Einstein and, in a continuous universe, 493
    elementary, 493
    new math needed for, 93
    of space and precision of pi, 552
granules, space quanta are the, 235
Grattan-Guinness, Ivor, 764
gravitational-wave observatories
    and black holes, 502-3
    LIGO, 619
        and neutron stars, 621
        found colliding black holes, 355
        found colliding neutron stars, 461
        most precise position measurements are in, 330
    other, 621
gravity, 69, 292
    extreme
        and tunneling barrier shape, 295
        must be space with extreme curvature, 358–61
    space production in, 300
    strong near the beginning, 238
    universe's, became less intense, 295
Great Attractor and absolute rest, 305
great chain of being, the, 659
Greece, ancient, 24, 710
Greene, Brian
    as science communicator, 16
    on multiverses, 584-8
    on order in the early universe, 139
    on physics in other universes, 588
    on the Big Bang, 17
    on the fleeting *now*, 289
    on the possibly holographic universe, 338
    on the size of the universe, 395
    on understanding quantum mechanics, 255
Greene, Graham, 214
Gresnigt, Niels, 254, 571
Gribbin, John, 104, 128
Gross, David, 34, 98, 228, 368
Grossman, Marcel, 488, 491
Gruber, Thomas, 36
Guarino, Nicola, 37
Gutfreund, Hanoch, 72
Guth, Alan, 128-9, 184, 221, 226, 348, 585

H
Hackett, Jonathan, 49, 199, 202, 281, 595
Hacking, Ian, 37

Hadamard, Jacques, 535–6
hadrons
 are composite, 55
 made of quarks, 54
 meaning, 53
 quarks coalesced as, 365
Hainline, Kevin, 366
Hammerstein, Oscar, 100
Hanson, Andrew, 667
Harari, Yuval, 1
Hardy, Lucien, 203, 578
Harrison, Edward, 431
Hasinger, Günther, 505
Hawking, Stephen, 100, 585
 and his eponymous radiation, 354, 453-4, 501
 and M-theory as source of the universe, 455
 and philosophy, 174-5
 and primordial black holes, 354, 366, 501-5
 and quantum tunneling, 119
 and scientific questions, 201
 and the beginning of time, 109-10
 and the radius of space, 295
 and the single-quantum beginning, 110
 and the universe beginning as a point, 399
 and turtles and the infinite regress, 140
 as a positivist, 672-3
 as comic performance artist, 109, 455
 believed in nothing, 454-6
 did not avoid the beginning, 109
 on a complete theory, 656-7
 said Newton needed infinite space, 134
 the continuum has been successful, 93
 version of the Big Bang, 7
Hegel, Georg, 65
Heidegger, Martin, 38
Hein, Piet, 1, 20, 375, 664
Heisenberg, Werner, 103, 259, 240, 642
 and his eponymous uncertainty principle, 241, 615
 on a universal substance, 52
 on the early view of atoms and space, 47
helon model
 Bilson-Thompson's, 195
  as a theory of spacetime and matter, 439-40
  has a single fundamental object, 249
helons, 276
 Bilson-Thompson's meaning of, 249
 meaning, 196
 as pairs of tweedles, 269
 particles are braids of three, 195
Hendry, David, 595
Heraclitus, 614
Herbert, Frank, 7, 15
Hersch, Matías, 422

Hertog, Thomas, 110
hidden variables
 Bell called, an absurd name, 704
 Bell's prescription for, 310
 meaning, 146
 twist states as, 310
hierarchy
 levels in the, of emergence, 753
 of concepts in the new ontology, 37
Higgs
 boson
  and Moffat, 458
  and the standard model, 55
  meaning, 457–9
  timing of discovery of the, 457
 field, 438
Hilbert, David, 515, 545, 561–2, 692
Hiley, Basil, 35
Hill, J. Colin, 32
Hintikka, Jaakko, 374
Hinton, Geoffrey, 36
history
 and metapersons, 609-10
 and the dream of a real worldview, 609
 as a creative enterprise, 654
 as an aspect of the present, 604-5
 causal set as the universe's complete, 179
 cosciousness in the, of philosophy, 637
 distant stars as ancient, 327
 Elton and, 605
 intellectual, and Koyré, 610
 of atomicity, the, 179
 of dark energy and dark matter, the, 348
 of entanglement, the, 145
 of explaining the world, the, 4
 of more powerful accelerators, the, 49
 of Newton's two times, the, 160
 of non-existent new particles, the, 457
 of quantum mechanics, the, 246
 of science and ontological programs, the, 33
 of space expanding, studying the, 168
 of the foundations of mathematics, the, 764
 of the scientific method, the, 779
 of the universe and dark matter, the, 353
 of the universe in emergent terms, the, 474
 of the universe is finite, the, 546
 of theories of the ether, the, 72-3
 of Zen, the, 601
 Planck-scale discrete causal, 379
 quantum, a quick tour of, 87
 truth in, 605
Hobbs, Bernie, 49
Hobson, Art, 428
Hobson, Michael, 451
Hoffman, Donald, 46
Hoffmann, Banesh, 87

Hogan, Craig, 332
holistic
  Smolin's, perspective on quantum gravity, 153
  view
    indigenous peoples', 151
    the new ontology's, is real, 151
hologram
  we live in a 2-D, 337
  Susskind's meaning of, 337-8
holographic principle, 336
  and quantum gravity, 337
  at first was centered on black holes, 338
  is cradled in a highly artificial worldview, 339
  meaning, 337
  more on versions of the, 730
  strong, *meaning*, 337
  weak, *meaning*, 337
Holt, Jim, 313
Holton, Gerald, 8, 29–30, 57, 156–7
Hooft, Gerard 't, 257
  and the holographic principle, 337
  and the vacuum state, 448
  on the need to define exactly what space is, 93
  on the search for simple physics, 54
  said all we know will be invalid at Planck-scale, 92
Hoover, Thomas, 601–2
Horgan, John, on Hawking, 109, 456
Horizon Problem, the, 129
Hossenfelder, Sabine, 452, 681
  math's link to reality, 209
  on particle colliders, 414
  on physics being lost in math, 208
  on the vacuum catastrophe, 446
  on the value of string theory, 101
Hoyle, Fred, 107
Hsu, Stephen, 149
Hubble
  constant, 300
    and the recession speed of galaxies, 340
  Edwin, 106, 116, 221, 266
  may not be constant, 697
  meaning, 106
  obscure units of the, 696
  related to new space quanta, 359
  sphere, *meaning*, 327
  telescope, 622
Hübsch, Tristan, 315
Huggett, Nick, 46
Hume, David, 480
  on causation, 244
  on metaphysics, 59
  on nonexistent objects, 636
  on space and time, 420
  on the idea of existence, 628

Hund, Friedrich, 119, 221
Hut, Piet, 374
Huterer, Dragan, 349
Huysmans, Joris-Karl, 401
hypersphere
  Einstein's universe is a, 135
  halfway through a, 329
  meaning, 328
  the universe as a, 363, 702
hypersurface of the present, *meaning*, 289

I
IceCube Collaboration, 619
IceCube Neutrino Observatory, 572, 619
idea
  a new, is difficult to think of, 192
  of a finite universe with no edges, 134
  of a ridiculously tiny fundamental constant, 84
  of a universal substance, 51
  of compactified dimensions, 100
  of curvature of space, 122
  of extra dimensions, 100
  of infinite space, 133
  of particles as modifications of ether, 68-9
  of six compactified dimensions, 101
  of something that's both one and many, 63
  of the Calabi-Yau manifold, 99
  seminal, *See* seminal idea 101
  that the world can be understood, 23
  the missing, 194
  the unifying, 44
ideas
  Anaximander's, about nature as a cosmos, 23
  language to communicate, 23
  lasting, need the right kind of society, 24
  physics was slow to grasp Planck-scale, 89
  Quine's, about ontology, 60
  seminal, 4, 13
  six key, 41-2
  key, come together as a whole, 41
Inada, Kenneth, 598, 600
inconsistency
  in logic, 536
  in physics and math, 174
  is accepted in math and physics, 212
  of relativity and quantum theory, 206
indiscernibles, the identity of, 182
inertia
  institutional, 408
  origin of, is the most obscure subject 437
  the relationship of, to mass of the universe, 437
  the nature of, 92, 616
  the problem of, 282
  the relationship of, to mass of the universe, 437

the search for the source of, 282
Infeld, Leopold, 273, 387-9, 493, 639
   and crime fiction, 187
   and the free inventions of physics, 33
   co-authored a book with Einstein, 428, 778
   erred about Einstein and the ether, 70
   on Einstein's special relativity paper, 156
infinite
   density, 399, 402, 460-1, 463
      of the initial universe, 198
   extension of the digits of π, 551
   number of universes, 584
   or finite space
      hardly matters at all Weinberg said, 297
      matters greatly, 298
   regress
      and Hawking and turtles, 140
      logical trap called an, 312
   space
      a quantum system in, 568
      physics assumes, 544
      has no edge, 133
      is, Newton reasoned, 134
      may not be, 134
   the, is nowhere to be found in reality, 545
   universe
      Archytas reasoned it's an, 133
      physics tended to assume an, 106
   velocity of waves in incompressible fluid, 115
infinitesimal
   calculus, 93
   initial universe, 298
   pocket of a megaverse, universe as an, 584
infinitesimals do not exist in granular space, 93
infinities
   fundamental problems with, 529
   renormalization eliminates, 442
infinity
   and Cantor, 543
   and Kelvin, 544
   and Maddy, 545
   and multiverses, 586
   and number theory, 547-8
   axiom of, the, 548
      is wrong, 548
   cannot be actualized, 544
   fallacy in the argument for, 544
   fields extend to, 333
   inanities and absurdities of, 545
   is a code word for disaster for physicists, 545
   nature of, needs careful study, 543-6
   need to appeal to the hypothesis of, 547
   of pocket universes, an, 589
   of points
      between any two points, 533
      in continuous space, 29
      on a line, 544
   the reality of, has always been in question, 31
   the role of, in Gödel's proofs, 561
   there is no, 546
inflation, cosmological. *See* cosmological inflation
inflationary epoch, 354
   Sutter on shutting off the, 127
inflaton, *meaning*, 127
initial conditions
   Gaussian, 225
   Hawking and, 109
   meaning, 109
   must have been very special, the universe's, 138
intelligence
   artificial
      and how the brain works, 649
      cannot discern its own nonsense, 631
      struggling, and Teilhard de Chardin, 647
      the universe has evolved, 654
interference
   of buckyballs, 255
   pattern, 386-8
   single-particle, 388-91
intuitionistic logic, *meaning*, 378
Ionian Enchantment, *meaning*, 8
irony
   in language, 376
   of best-developed English character in French, 188
   of 'compactified' dimensions expanding, 239
   of how 'atomic' physics undercut the a-tom, 48
   of space and the quantum being the same, 188
   of Feynman and the path integral method, 391
   of foliating spacetime, 289
   of magic illusions also needing an observer, 703
   of being unable to see real counting objects, 522
   of pi driving the PI name, 714
   of Rovelli's dismissal of the cosmic clock, 288
   of Russell's dissing poverty of imagination, 105
   of using relativity to support a global time, 288-9
   of the basis of reality being seen as fiction, 102
Isham, Christopher, 97
iterations
   are not measurable, 510

Subject Index

each of the, is the whole universe, 179
first two, and math objects, 537
give us an "absolute time", 164
in lockstep, 303
meaning, 104
sequential, 284
successive, 42
time-like quality of, 164
IXPE X-ray telescope, 347
Iyengar, Sheena, 2

J

Jacobs, Jane, 373
Jacobson, Ted, 178, 419
James Webb space telescope, 116, 506, 621
James, William, 63
Jeans, Sir James, 8, 183
Jedamzik, Karsten, 356
jets
  AGN axial, 343-7
  axial, of space, 344
  faster-than-light, 345
  of molecular hydrogen, 345-6
  stunning energies required to set, in motion, 344
jigsaw, 41, 631
Johnston, Hamish, 455
Jones, Peter, 598
Judson, Justice, 603

K

Kafka, Franz
  and parable of battling past and future, 156, 301-2
  as observer of reality, 613
  dreamed of reality, 609-13
  said truth is alive and changing, 563
Kahn, Charles, 23
Kaku, Michio, 585
Kaluza, Theodor, 99–100
Kant, Immanuel, 598, 655
  and the size of the universe, 543
  assumed the universe had no beginning, 105
  on logic of nonexistent objects, 636
  said geometry is a creation of the mind, 555
  said space is not something real, 68-9
  said there is no nothing, 450
Kaplan, Robert, 530, 539, 547
Kaufman, Bruria, 91
Kelvin problem, the, 445
Kelvin, Lord, 445, 544
Kepler, Johannes, 578
kinematics, *meaning*, 281
kinetics, *meaning*, 281
King, Larry, 456
Kirchhoff, Gustav, 84
Kirchhoff's radiation law, 84

KISS acronym, 61, 684
Kitcher, Philip, 517, 519, 522
Klein, Felix, 100
Klein, Oscar, 100
knife edge, 171, 452
  balance of the universe, *meaning*, 131
  there is no, balance, 171
knowledge, theory of, 38
Kobayashi, Makoto, 368
Koestler, Arthur, 373
Koyré, Alexandre, 610
Kragh, Helge, 79, 88
Kramm, Gerhard, 90
Krauss, Lawrence, 454
Kretschmann, Erich, 488
Kribs, David, 283
Krugman, Paul, 595
Kuhn, Thomas, 78, 778
Kumar, Manjit, 93

L

lambda
  adding, to get the standard model, 395–6
  and Peebles, 395
  Einstein invented, 166
  we must learn to live with, 169
Land, Edwin, 217
Landauer, Jeff, 524
language. *See* also linguistics
  a subject is its, 376
  ambiguity of ordinary, 540
  and disaggregation, 150
  and meaning of the word quantum, 256-7
  and the changing meanings of adjectives, 658-9
  and the meaning of causation, 381
  and word meaning extension, 718
  and worldview, 1
  and zero, 541
  as a screen between thinker and reality, 373
  as the operating system of our civilization, 1
  assumption could have been enshrined in our, 268
  atom smashers changed our, 48
  Big Bang is misleading, 107
  biliteral verbs and the English, 647
  building better, 631
  building common, 41
  calculus as the preferred, of space, 93
  centrality of, and Wittgenstein, 523
  clockiness of time is buried in usage of, 286-7
  fact and the concepts embodied in our, 215
  for a number concept, 632
  forms foundations for thought, 659
  German

Einstein was a highly gifted stylist of
the, 11
is best for philosophy, 674
inexact, contrasted with that of computers, 653
infants learn, without being taught, 218
is collectively produced, 373
links to intuitionistic logic and causal sets, 378
loose, of the quantum theory story, 388
math is only, 211
mathematical, is the correct language, 516
matters for our understanding, 480
is needed to speak of a worldview, 36
new, and transcending the accepted worldview, 376
there is no such thing as ambiguity-free, 653
of a new idea, 26
of causal set theory, 378
of causation, 243
of fields, 153
of logic, 636
of much English common law is French, 524
of ontological commitment, 518
of physics, 40
of reality, 429
of the universe, 647
reassignments of, 407
Riemann and the term quanta, 256
shapes thought, 374-7
Smolin's, writing about time, 162
the value of novel, 375
the view of reality embedded in, 1
translation. See translation & translations
ultramatter, *meaning* 461-3
used for modeling the world, 36
we need, to conceive and communicate ideas, 23
languages
are the mediators between holograms, 653
growth of, 25-6
Laplace, Pierre-Simon, 307, 309
Laplace's Demon, 309
Large Hadron Collider, 49
and the Higgs boson, 457-8
and the search for supersymmetry, 412
and the standard model, 414
black holes imagined to be made in the, 501
macroscopic quantum states in the, 255-6
Laue, Max von, 77
Laughlin, Robert, 276
law
and the nature of reality, 603-4
English common, 524

ideal gas, the, 472
of conservation of energy, the, 431-4
of inertia, the, 437
of light propagation, the, 277-8
of nature, the first, 480
of number theory, a, 548
the simple, of the universe, 118
the statistical root of every, of physics, 383
laws of physics
a simple law of nature gives rise to all, 121
and chirality of the neutrino, 571
are approximations to reality, 377
are emergent, 384, 432, 482
are not laws of nature, 583
are not real, 481
are the same
in all coordinate systems, 485
in all inertial frames, 304
in both chiralities, 571
over time, 432
changes of coordinates don't change, 487
hold at the beginning of the universe, 109
include matter-antimatter symmetry, 367
reason why the, are the way they are, 483
supposedly vary in the string-theory landscape, 588
the point of creation and, 399
ultimate cause of the, 314
Lazaridis, Mike, 196, 202
Lea, Rob, 168
least action, the principle of, 92
Leavitt, Henrietta, 168
Lederman, Leon, 54, 478
Leibniz, Gottfried, 51, 240, 450, 674
and calculus, 428
and sufficient reason, 181
on nonexistent objects, 636
on time and causal order, 246
on time and the ordering of events, 285
Leiden, Einstein's lecture on the ether at, 70, 72
Lemaître, Georges, 107, 167, 226
and conservation of mass, 140
and exponential growth, 108
and the FLRW metric, 476
and the origin of matter, 266
and the single-quantum beginning, 118, 126, 221, 222, 229, 234, 240
and the space quantum, 105, 257, 322, 618
as father of the Big Bang, 106, 140
best understood general relativity, 106
found no boundary for space, 133
religious affiliation of, the, 702
solved the general relativity equations, 115
length
Planck, is not a fundamental unit, 80
there is no real property of, 231
leptons
charge ratios of, 252

meaning, 252
properties of, deduced from tweedles, 602
the reason for six, 253
Leucippus, 47, 614
Leverton, Robert, 467
LHC. *See* Large Hadron Collider
life
  chances in the chain of, 663
  is the result of many emergences, 471-2
  lifeless atoms led to, 471
  purpose of, the, 661
  reason the universe suits, the, 20
  which slips gently toward death, 662
light cone
  and spacetime, 382
  past and future, 382
  meaning, 382
  is not needed in the new ontology, 382
light years, *meaning*, 129
LIGO
  detected large black holes colliding, 461, 506
  gravitational-wave observatory, *meaning*, 619, 621
Linde, Andrei, 129, 585, 670
Lineweaver, Charles, 113, 340
  low initial entropy, and, 137
  universe, expansion of the, 340-2
linguist
  Arthur Koestler, 373
  Gottfried Leibniz, 51, 181
  Jila Gomeshi, 653
linguistics, 376, 674
link
  has the Planck area, 231
  is a connection between space quanta, 376
  meaning, 230
links
  Bilson-Thompson's
    represent areas, 249
    ribbons are also termed, 232
    use of the term, 197
  three kinds of, there are, 231
  twisted, 125, 235, 271, 332, 382, 404
  untwisted, 247, 276, 393, 494
literature
  best developed characters in English, 188
  Nobel Prize for, 63, 171
  on causation, 244, 382
  on entropy, 318
  on fallacies in theoretical physics, 27
  on foams and bubbles, 445
  on indigenous worldviews, 705
  on infinity in mathematics, 545
  on learning, 218
  on mathematical realism, 761
  on real vs invented math, 715
  on the nature of consciousness, 637

on the reality of numbers, 761
on the scientific method, 777
on the suitability of the universe, 582
on the vacuum, 450
on thought vs action, 638
Lloyd, Seth, 104
Lobanov, Andrei, 345
logic
  an empty universe in, 452
  and deduction in science, 2
  and development of physics, 36
  and laws of the mathematics of number, 561
  and ontological commitment, 453
  and Peano's axioms, 540-1
  and set theory, 361
  and the Axiom of Infinity, 548
  classical, and quantum theory, 143
  condemns arithmetic to inconsistency, 211-2
  as a foundation for arithmetic, 211, 528, 630
  Gödel's
    achievement in, 211
    incompleteness theorems and, 560
  has no solid link to arithmetic, 560
  has problems with the concept of nothing, 452
  in arithmetic, 452
  in language, 374
  intuitionistic, 378
  language of, the, 636
  mathematics is all translatable into, 520
  MOSFET switches and, 104
  of logical positivism, 216
  tangle that math, is today, 518
logical positivism
  and Ayer, 213-6
  and Heisenberg, 672
  and Reichenbach, 686
  and sense perception, 216, 653
  and the empiricist principle, 216
  and the future of physics, 567
  and the philosophy of science, 174
  and understanding estranged from physics, 631
  ascendancy of, 213
  is dead, 59
  is the tacit philosophy of many scientists, 716
  meaning, 59
loop quantum gravity, 19, 419
  and links, 334
  and timeless space, 178-80
  as a candidate for quantum gravity, 176-8
  meaning, 178
loop, a pair of dimensions forms a closed, 232
López-Corredoira, Martin, 591
Lorentz, Hendrik

and Einstein's ether, 73
and matter made of space, 68–70, 198, 221
and the basis for Planck's constant, 84–5
and the seminal idea space is something, 221
brilliant insight of, 68-9, 73
invariance and nonlocality, 145
lecture on electrons, 68
space is something, 221
the electrodynamics of, 487
Lovejoy, Arthur, 659
Lundmark, Knut, 115

M
Mach, Ernst, 135, 173, 304, 437
MacMillan, William, 107
Maddy, Penelope, 31
and counters, 523-6
and existence of unobservable entities, 408
and Frege's being scandalized, 536
and infinity, 545–6
and realism in mathematics, 408
on composite counters, 526
on determinate number property as key, 522, 526
on math and physical sciences, 517-9
on medium-sized physical objects, 523-4
on scientific ontological commitment, 453
on set theory, 528-34
  and mathematical things, 31, 528-31
on what makes mathematical truths true, 513
Maigret's strategy, 188
Majorana particle, *meaning*, 572–4
Maldacena, Juan, 338
manifold
  a single replicating, as source of physics laws, 483
  Calabi-Yau, 177, 191, 221, 223, 264
    an element that is key to the new worldview, 180
    and string theory, a real, 403
    existence of the, 98
    has six tangled dimensions, 97-9
    in the beginning there was one, 468
    investigations of the, 483
    is a math abstraction in string theory, 98, 339
    is real in the new ontology, 101, 400
    is the final atom, 440
    is the ultimate uncuttable entity, 440
    is the key to understanding space and time, 98
    may be the most-studied geometric object, 315
    meaning, 97–9

online images of a, 369
ribbons run right through each, 280
string theories need a, 99-101
that replicated, a, 229, 314
with a tiny volume, 241
continuous
  needs an external metric, 75
  vs discrete, and Riemann, 473
  is a space with any number of dimensions, 98
  meaning, 75
Riemannian, foliated into space-like slices, 422
space as an unbounded, 134
manifolds, Calabi-Yau
  are naturals for forming ribbons and links, 232
  the first two, begin to make space, 236
  in space, the fiction of, 375
Manson, Mark, 401
Markopoulou, Fotini, 283, 284, 408
  and a coincidence, 244-6
  and causal sets, 160, 243, 378-80, 725, 738
  and doing physics inside the universe, 141-2, 244
  and Smolin, as Bilson-Thompson collaborators, 196
  and spacetime, 425
  co-author with Bilson-Thompson, 231
  deepest fundamental-physics thinker, 17-20
  ditched physics, 202
Martel, Yann, 551
Martland, Justice, 603
Maskawa, Toshihide, 368
mass, 435-8
  and energy are the same, 250, 436
  conservation of, as a mindset, 266
  quasilocal, *meaning*, 437
  question of what, is, 376, 432, 435-8
mathematics
  as language, 211
  brevity of, may be expensive, 211
  consistency of, is mostly an illusion, 211
  creative principle and, 210, 632
  defined with words, 211
  digital, for granular space, 75
  Einstein and, 210, 516, 632
  Gödel cut to the roots of, 212
  halls of imaginary, are vast, 210
  Hossenfelder assailed abuse of, 208
  imaginary, 188
  laws of the, of number reduced to logic, 561
  logical foundations for, 520
  Lucasian Chair of, the, 109
  number theory in, 560
  obscures the creative process, 561
  of causal set theory, 244

of quantum field theory is not rigorous, 428
of string theory is deeply rooted in nature, 101
ontologic issues of, 30
ontology of, 31
real. See real math
search for successful, 210
source of, the, 517
string theory's beautiful new world of, 97
usefulness of, 516
virtues of, 211
matrix mechanics and wave mechanics, 568, 754
matter
 and antimatter, 367-70
 and the ether, 66-9
 and the steady-state model, 299
 dark, 52, 171, 214, 351-6
  a year after the Big Bang, 352-3
  and dark energy are aspects of one thing, 361
  and the sterile neutrino, 575
  as a label for a mystery, 413
  black holes as, 355
  careful calculations of, 171-2
  is inferred from gravitational effects, 352
  is missing, 262
  mass of, as percent of total mass, 185
  meaning, 351
  primordial black holes as, 506
  supersymmetry and, 412
 in a primeval atom when universe began, 107
 in axial jets, 343-7
 made from braided twisted ribbons, 221
 made of atoms, 47-9
 made of many kinds of atoms, 53-6
 made of monads, 51-2
 made of space, 198, 234
 made of twists between space quanta, 193-201
 making new space made new, 240
 new, and conservation laws, 431
 ordinary, 52, 171, 435, 462
  as a fiction we don't understand, 413
 ponderable, Einstein's concept of, 68, 71, 85, 686
 radiating, in relation to Planck's constant, 79
 distinction from spacetime likely ill-founded, 192
 the origin of, 191, 197-8, 266-7
 the problem of no source of, 229
Maudlin, Tim, 306
Maxwell, James, 317, 427, 487
Maxwell, Nicholas, 774
Mead, Alden, 89, 330

measuring rods and clocks, 154-5, 485, 755
 accelerated, 486
 should emerge from the theory, 423
Meinong, Alexius, 637
memory
 and synapses, 651
 digits of pi from, 423
 engrams are elements of, 652
 engrams could not be localized, 652
 is highly processed, 653
 is relational, 652
 the basis of, 651-2
Merleau-Ponty, Maurice, 308
metaphysics, 38
 meaning, 59
metric
 a, or measurable time as Rovelli redefined it, 508-9
 arbitrary, 211, 379, 476, 491
 Aristotle's time has no, 285
 artificial, from outside the space, 476
 classical, woven out of quantum loops, 198
 clock time as a, 309
 counted granules of space are a natural, 476
 digital, a, 379, 491
 discrete, a, 75
 ecological, an, 594
 field as the primary ingredient of reality, 476
 FLRW, the, 476
 locality is tricky with no background, 332
 meaning, 211, 475-7
 new ontology has no need to assume a, 328
 Newton's true time has no, 285, 511
metrics
 common, defined by body parts, 475
 econo-, called alchemy, 595
 economic, 612
 fuzzy, of economics, 594
Meyer, David, 707
Michell, John, 125
Michelson, Albert, 71
Miletus, 23
Milky Way, 343, 360
 Galileo saw the, is made of stars, 617
Mill, John Stuart, 520
Miller, Amber, 130
Miller, David, 245, 665
 and axial jets, 343-7
Millikan, Robert, 268
Milne, Alan, 105
Minkowski, Hermann, 421-4, 432-3, 488, 744
Mlodinow, Leonard, 455-6, 529
Moberg, Annika, 357
Möbius strip, 231-2

Moffat, John
   letter from Einstein, 258–9
   on non-existent new particles, 457–9
   on the particle zoo, 54
   said physicists become emotionally invested, 592
Mölders, Nicole, 90
monads, matter made of, 51–2
monism, *meaning*, 51
Moon, the
   and action at a distance, 144
   and Einstein's view of free will, 661
   and its relation to the Earth, 69
   as a counting object, and Frege, 535
Morley, Edward, 71–2
Moskowitz, Clara, 575
motion, 281-2
   accelerated, 487
   aether assumed to have a state of, 71
   and clocks, 155-7
   and position. *See* position, and motion
   and the ether, 70
   common time is measured by, 159
   Cosmic Clock is not based on, the, 321
   Descartes and, 67
   digital, resolves paradoxes, 310
   grasp of, the, 71, 73, 171
   inertial, *meaning*, 487
   light-speed, and time, 164
   mystery of, 75, 614-6
   Newton and, 67
   Newton's laws of, 134, 308
   of ponderable bodies, and Einstein, 68
   paradoxes of, 64
   power of the idea of, 614
   put in at the beginning, 616
   relativity as background-free physics of, 44
   smooth, due to tiny scale of increments, 325
   Zeno's view of, 64
M-theory
   and Hawking, 454–5
   little evidence for, 455
   meaning, 454
Mückenheim, Wolfgang, 548
Muller, Richard
   on mathematics and reality, 515
   on matter in expanding space, 113
   on the fundamentals of time, 417
   on the meaning of now, 423
   on what makes time move on, 417
multiplication
   is an object of arithmetic, 549
   of the world, infinite, 587
   reified, 521
multiverses
   and Hawking's beginning of the universe, 456
   and Susskind *et al.*, 584–9
   and the suitability of the universe, 582
   and Wheeler, 20
Murphey, Dwight, 245

N
Nadis, Steve, 436
Nagarjuna, 124, 598–9
Natarajan, Priyamvada, 503
Nature's knitting, knots in, 369
Nauenberg, Michael, 190
Neamati, Saeed, 317
nebulas, 617
Nee, Sean, 659
Nernst, Walther, 451
networks
   neural, 484, 631, 649
      and Hinton, 36
   ribbon, and Bilson-Thompson model 281
   spin, *meaning*, 230
Netz, Reviel, 543
Neumann, John von, 32, 243, 318
neurons
   and elements of memory, 652
   and the action potential, 648
   and the connectome, 652
   linked by about a quadrillion connections, 649
   meaning, 649
neutrino, the sterile, 575-6, 768
neutrinos
   as Majorana particles, 572–4
   why, have only one chirality, 253
neutron star, 463
   collapse to a, 502
   colliding, 125, 621, 775
   pair collapse into a black hole, 125, 461-3
neutronium, *meaning*, 461
neutrons, 125
   are made of three quarks, 462
   Chadwick discovered, 48
   crushed in a black hole, 460–3
   from crushed protons and electrons, 125
Newsum, Joe, 151
Newton, Isaac, 316, 418, 642
   and calculus, 317, 428
   and common time, 416
   and Descartes' vortices, 685
   and light particles, 277
   and locality, 144
   and motion, 67-9, 290, 308-11
   and two kinds of time, 158-64
   Hawking as successor to, 109
   needed an infinite universe, 134
   three-color theory of, 217
Newton's
   first law, 282
   laws, 486
Next
   meaning, 286

program, 285-6
next iteration, 246, 282, 285
Ng, Yee Jack, 84
Nobel Prize, 200, 459, 498, 596, 723, 736
   Adam Riess, 168, 357
   Albert Einstein, 88, 152, 206, 272, 277
   Albert Michelson, 71
   Bertrand Russell, 63
   David Gross, 34
   Enrico Fermi, 717
   Ernest Rutherford, 48
   Eugene Wigner, 516
   for economics, 594
   Frank Wilczek, 20, 55, 497
   Gerard 't Hooft, 54
   Gerd Binnig, 121
   Giorgio Parisi, 318
   Heinrich Röhrer, 120-1
   Hendrik Lorentz, 85
   Ivar Giaever, 121
   James Chadwick, 48
   James Watson, 651
   Jim Peebles, 27
   Joseph (J.J.) Thompson, 48
   Leon Lederman, 54
   Makoto Kobayashi, 368
   Martin Perl, 55
   Max Born, 88
   Max Planck, 84
   Max von Laue, 77
   Murray Gell-Mann, 269
   Niels Bohr, 31
   Paul Dirac, 260
   Paul Krugman, 595
   Pierre Curie, 368
   Rabindranath Tagore, 678
   Ramón y Cajal, 650
   Richard Feynman, 10
   Richard Smalley, 255
   Robert Laughlin, 276
   Robert Millikan, 268
   Roger Penrose, 18, 460, 500
   Steven Weinberg, 4, 274, 297
   Subrahmanyan Chandrasekhar, 499
   Thomas (T.S.) Eliot, 197
   Toshihide Maskawa, 368
   Werner Heisenberg, 47
   Winston Churchill, 171
   Wolfgang Pauli, 455
nodes
   Bilson-Thompson's, 249
      and volume, 250
   meaning, 250
Noether, Emmy
   and conservation law, 432
   and Minkowski space, 432
   Einstein's comment on, 747
nonexistence, logic of, 636
nonlocal

functioning of the brain, 652
memory is, 651
Planck scale must be radically, 145
universe, 144
   and spatial simultaneity, 145
   must be, 147
   was, from the beginning, 148
nonlocality, 151, 289
   and our twists, 626
   Einstein and, 145
   meaning, 144
   seminal idea of, 148
North, John (J. D.), 375
   and Lemaître's letter to Nature, 694
Norton, John (J. D.)
   and debate on Einstein's theories, 485
   on curvature of space near the Sun, 122
   on foundations of relativity, 489
   on general covariance, 488
   on Newton's laws of motion, 134
nothing, 448-53; *see also* void
   a universe from almost, 118
   an absurdity, 109
   expanding is said to make no sense, 113
   Hawking believed in, 454
   the universe in, fiction of, 375
nothingness
   and Jim Holt, 313
   and Sartre, 301
   including absence of space, 450
Novikov, Igor, 353-4
Now
   becomes the next *Now*, 308
   is all that is real, 156
   is different from the last *Now*, 308
   is timeless, 157
Nows
   Barbour's, 324
   the world is made of, 324-5
number theory, 547-8, 560-1
   and no largest natural number, 548
   meaning, 560
numbers, 529-31
   and Frege, 540
   and Mill, 520
   and real math, 209
   complex, 692
   elementary quanta reified, 516
   imaginary, *meaning*, 533
   irrational, 551
   natural
      and set theory, 528-31
      are based on counting objects, 524
      are properties of the universe, 518
      cannot be based on set theory, 531
      infinitely many, Gödel needed, 613
      largest of the, 547-50
      Peano's, 541
      the universe constructs the, 537-8

Whitehead and Russell defined, 528
of particles and antiparticles must be equal, 367
rational, *meaning*, 552
real. *See also* real numbers
  (really), 522
  are misleadingly named, 533
  are not real, 532-4
  Dedekind and, 533
  definition relies on infinity, 532
  meaning, 532
what are natural, 523-7
whole, and Born, 88
Nutt, Amy Ellis, 653

O

O'Bonsawin, Michelle, 648
O'Callaghan, Jonathan, 622–3
object of arithmetic, *meaning*, 549
objects
  are real, the notion that, 518
  mathematical, 516-22
  motions of, 473
  of reality, 516
observation
  a reality independent of, 516
  entropy was low at the beginning, a given that, 139
  in quantum theory, 40
  Planck scale is beyond reach of direct, 219
  quantum theory is fundamentally about, 142, 568
  realms of reality beyond all possibility of, 174, 213
  wave function collapses instantly upon, 569
observer
  a reality independent of any, of experiments 28
  and observed are one, 602
  is also required for a magic trick, 703
  Kafka as, 613
  local, and speed of light, 340
  motion relative to an, 473
  must be a classical system, 569
  must be outside the quantum system, 141
  need not be a person, 141
  no need for an, in the new ontology, 570
  of color scene, 217
  of quantum system, 142, 569-70, 586
  of relativistic jet, 346
  of relativistic system, 154-5, 473
  outside the universe, 285
    is unphysical, 141
    Markopoulou on, 725
  Pauli on the, 672
  problem for quantum mechanics, 569
  quantum theory requires an external, 141, 600

Occam's Razor, 58-62, 237, 240, 247
Ockham, William of, 58-62, 221, 280, 349, 393
Oersted, Hans, 173
Oerter, Robert, 55, 348
One and the Many
  paradox of the, 63
  paradox of the, and Ayer, 214
  paradox of the, and Riemann, 75
  paradox of the, and Rovelli, 614
ontological
  assumption, space quanta tunneling, 224
  attitude, natural, 629
  commitment, 454, 477, 645
    change in, 191
    most troubled, is the quantum of energy, 31
    to elementary particles, 390
    to granular space, 283
    to locality, 147
    to the block universe, 155
    to the continuity of time, 162
    to zero and infinity, 30-1
    universal, to continuous space, 267
    unshakeable, borders on belief, 31
    without a coherent, 283
  commitments, 291
    and Quine, 30
    failing, 356
    for physicists, a set of, 30
    lost in old, 583
    of a theory, 30
    of economics, 596
    of ether-seeking experiments, 71
    of physicists, 642
    Sartre's, 301
    Smolin's and Unger's suggested, 406
    string theoey's, 99
  problem, the, 32, 60
ontology, 140
  an extremely simple, 62
  Ayer and any possible Planck-scale, 215
  Aristotle invented, 38
  backbone of an, 37, 221
  Bilson-Thompson's, 201
  conventional scientific, 68
  credible, as a guide to real physics, 567
  Einstein sought to build a real, 211
  idea of, 36
  meaning, 4
  of quantum physics, 40, 577
  physics imprisoned by, 60
  pointers to a new, 134
  Quine dominated modern study of, 30
  real, 36, 41, 163, 209
  Riemannian-Einsteinian, 33
  the new, 168, 320, 392, 396, 402, 418, 492
    absence of symmetries, 369
    and curvature of space, 124

and emergent space, 236
and infinity, 544-6
and inherent vs conventional quanta, 231, 259
and localization of a particle, 333
and Lorentz and Einstein, 73
and mathematics, 516
and *Now*, 303
and *Nows*, 324
and Occam's Razor, 247
and quantum tunneling, 118
and real math vs math fiction, 209
and spacetime, 422–6
and special relativity, 157
and the Big Fizz, 241
and the exact number of space quanta, 291
and the growth of space, 299
shows the holistic view is real, 151
and the infant universe, 363
and the nature of time and space, 565
and the nonlocal universe, 148
and the source of laws of physics, 482-3
and the structure of space, 445
and the uncertainty principle, 615
and time, 160-3
and understanding dark energy, 349
and understanding dark matter, 353, 358-61
and what change really is, 310
adding a new assumption to, 278
causation in, 309, 382, 383, 647
compared with some Buddhist ontology, 597
elements from quantum gravity in, 179
explains all kinds of matter and energy, 413
has a closed universe, 135
has new roles for black holes, 501
has no room for supersymmetry, 411
illustrious sources of, 41
meaning, 7
no points in, 400-1
says the electron is divided, 390
says the universe is finite, 326
shows the holistic view is real, 151
shows the universe as the creative source, 212
time is not a dimension in, 263
Zeno and, 64
the old, 7, 29, 277, 321, 569
causation in, 383
Einstein and, 171
false assumptions about ether in, 71
math and, 633
meaning, 32
problems and contradictions of, 186
the true physicist seeks the correct, 203
value of a real, 631

working with an, that's obviously wrong, 640
order
a state of perfect, 140, 319, 468
universe began in a state of, 137-40
Organon, The, 559
Oriti, Daniele, 95
oscillators
consensus seized on, 90
Planck's
assumption about, 82
imagined, 82
invented, 84
not assumed to emit radiation, 80
were a pure artifice, 83
Ouliaris, Sam, 595
Overbye, Dennis, 478

P
Pais, Abraham, 11, 87, 437
Paquette, Natalie, 237, 420
paradox
of creation from the void, 312
of set of sets that do not contain themselves, 520
of the need for classical systems, 141
of the One and the Many, 63-5, 75, 214, 614
of two theories, 152
of Zeno's flying arrow
and Aristotle, 310–1
and Wilczek, 311
paradoxes
of cosmological physics, 747
of motion, 64, 310
of special relativity, 157
of theoretical cosmology, 27, 706
Zeno's, simple to solve, 310
parameters
adjustable
of physics, 582
of the standard model, 225
assume no new, 239
of the standard model, 583
prop up the standard models, 21
standard model needs twenty new, 55
supersymmetry needs more than a hundred, 412
Parisi, Giorgio, 318
parity, 571
meaning, 367
Parmenides, 64, 614
particle, 376
accelerators, 235
found no supersymmetry, 411
new, for higher energies, 49-50
unstable fermions made in, 197
-antiparticle annihilation, 251, 367
elementary

electron seen as an, 384
    is imaginary, 385
    is vast at Planck scale, 20
    only one truly, 48
    photon as an, 276-7
    there is only one, 261
    useful concept of the, 31
horizon, *meaning*, 327
meaning of, is not well understood, 673
physics
    and spin, *meaning*, 571
    and symmetries, 369
    crisis in, 412
    essential method of, 705
    problems in, 590
    standard model of. *See* standard model of particle physics
    wordplay in, 195
Poe's single, 257, 618
particles
    and antiparticles in the early universe, 365
    composite, charges on, 272
    concealing real identities, 43
    elementary
        abandonment of, 191
        all reduced to some universal substance, 52
        all sixteen, taken to be elementary, 49
        are assumed to be, 269
        are composite, 261
        are not elementary, 377
        being identical is most remarkable property, 271
        electric charges of, 55
        found to be composite, 49
        ontological commitment to, 390
        physics of, at Planck scale, 91
        supersymmetric partners of, 411
        supposedly, 49
        there are no, 440
        Weinberg on, 439
        yet another model of, 439
    examine what Bilson-Thompson's, do, 279
    light waves as, 9
    new, of energy and matter, 431
    none of the standard model, are elementary, 197
    subatomic, are real, 271
Paseau, Alexander, 524
Passmore, John, 59
Pauli, Wolfgang, 100, 455, 559
Peano, Giuseppe
    and Gödel's theorems, 560-1
    and zero, 540
    and zero according to Russell, 541
    did not start with zero, 531
    set out to base arithmetic on axioms, 540-1
    was not a realist, 764

Peebles, Jim, 108
    and accelerating expansion, 114
    and cold dark matter, 395
    and energy conservation, 433
    and expansion of the universe, 397
    and learning to live with lambda, 169
    and Lemaître, 106
    and space sections, 422
    on continuous space, 74
    on cosmological inflation theory, 128
    on expansion as a baroque arrangement, 292
    on general relativity as a social construction, 206
    on match of general relativity to data, 490
    on physics' changing story, 27
    on physics theories as approximations, 207
    on the cosmological constant problem, 446
    on the roles of pure thought, 632
    on the theory of everything, 675
    on units for the Hubble constant, 735
    on usefulness of physics theories, 33
Penrose, Roger
    and simplicity, 57-61
    and spin networks, 230
    and the problem of low initial entropy, 137-40
    black holes have singularities, 460
    on low initial entropy, 317-9
    on multiverses, 589
    on reality, 18
    plea for black holes, 499-507
Perimeter Institute, 18-9, 196, 198, 249, 714
Perkowitz, Sidney, 173
Perl, Martin, 55
PET scanner, 279
Phelan, Robert, 445
Phillips, Philip, 339
philosophy, 38
    and logical positivism, 59
    and physics at odds about reality, 3
    and the flow of time, 324
    and thought experiments, 174
    calls to abandon, 216
    claimed irrelevance of, 210
    concerns the whole universe, 183
    Einstein's, 9
    fundamental questions of, 2
    Hegel and modern, 65
    is concerned with motion, 63
    natural, 613, 715, 753
        and Boscovich, 467
        and Newton, 58
        and thought experiments, 173
        and Unger, 18
        Unger and Smolin on, 321
    of existentialism, 647

Subject Index

of mathematics, 517, 518, 539, 560
of science, crisis in, 213
Parmenides', and Zeno's paradoxes, 64
physics turned its back on, 38
physics' disdain for, 210
skepticism of the senses in, 26
western, 23, 51, 598
photon
   and calculating true time, 164
   and how it manifests an e-m wave, 273-5
   as six untwisted links, 276
   changing charge in a, 275
   Einstein soft on particle character of the, 273
   evidence that the, isn't elementary, 389
   has no definable location, 494
   is the null braid, 272
   may behave like moving charges, 273-4
   sees no path, 495
   transforming into into Z^{0}, 274
photons
   clocks follow, 286
   Einstein's discovery of, 272
   from particle-antiparticle collisions, 253, 365
   from positron-electron collisions, 279
   in atomic clocks, 289-90
   made of links between three pairs of quanta, 340
   made of quantum space, 493
   yellow, yielding full color vision, 217
physical world, no consistent picture of the, 192
physics
   a future for, 639-46
   and philosophy, 3
   as a worldview, 27
   basis for causal laws of, 383
   Einstein sought, without the continuum, 192
   fundamental. *See* fundamental physics
   handmaiden of epistemology, 38
   is unconcerned with reality, 28
   may not be based on continuous structures, 259
   posture of, 38
   pursuing, with no space and time, 43
   real
      few physicists do, 640
      fresh fields for, 407
      must explore universe as an entire entity, 379
      ontology as a guide to, 567
      real math may blaze a better trail to, 209
      requires a real metric, 477
   search for new, 443
   test of truth for, 3
PI, *meaning*, 196

Plait, Philip, 619
Planck area
   given no metric meaning, 476
   pixels of black hole's information, 507
   size of each link is the, 468
   square root of, is Planck length, 444
   window or link, 230
Planck constant
   comedy of, 79-81
   hidden implications of the, 88
   made to measure for world of atoms, 89
   meaning, 81
   ruled the roost for the last hundred years, 476
Planck length, 20, 88, 231, 278, 295, 448, 462
   is not a fundamental unit, 80
   is not real, 444
Planck satellite
   data on dark energy, 358
   data show a positive curvature of universe, 556
Planck scale, 179, *See also* Planck-scale
   and the rise of post-empiricism, 642
   and quantum gravity, 96
   and Rovelli on time, 508
   concepts that are beyond the pale at, 393
   evidence of discreteness at, 264
   foundations of physics are at, 485
   fundamental physics at, 333
   geometries emerging beyond, 557
   impossibility of observing events at, 489
   intuition is likely no guide at, 224
   is an existential level, 406
   is where the action is, 385
   length is undefined at, 533
   meaning, 7
   metric problem at, 477
   need a consistent understanding at the, 183
   no symmetries at, 369
   observing motion at, for universe's clock, 285
   origin of the light-speed limit at, 473
   particles are composite at, 261
   Planck's number pointed to the, 257
   quantizing relativity fails at, 228
   quantum gravity at, 475
   Smolin's definite picture of the world at, 446
   space at, 177, 446, 447
Planck size sets a limit on location, 330
Planck time
   is the smallest time with physical meaning, 104
   new space per, 360
   one, per iteration, 278, 286
   universe doubling each, 364
Planck volume
   cube root of the, gives the Planck length,

444
  is far too small to be detected, 293
  is given no metric meaning, 476
  is real, 444
  is the most fundamental constant, 80
  manifold has the, 230
  meaning, 230
Planck, Erwin, 82
Planck, Max, 8, 101, 221
  an act of desperation by, 79-82
  and change in physics, 639-45
  and entropy, 77, 81, 84, 323
  and facts continuous space can't explain, 477
  and granular space, 103
  and loose talk about his work, 90
  and size of Riemann's granules, 75, 223
  and the action quantum, 87-94
  and the Cosmic Clock, 290
  found a hidden window into granular space, 76-86
  message of
    about the real nature of energy, 92
    for quantum theorists, 579
    yet to be driven home, 330
  on size and speed of change, 311
  on speed of light and the quantum of action, 277
  on statistical laws of physics, 383
  space quanta at the heart of his discovery, 403
  stumbled across the quantum, 93
  understanding has not changed since, 566
Planck's
  constant, 79
    2π as divisor of, 534
    and Einstein, 88
    and the photon let loose the dogs of war, 492
    meaning of, confronts us with deep problems, 83
  quantum, 92, 229, 257
    confused message from, 556
    not a quantum of energy, 31
Planck-scale, 20, 91, 198, *See also* Planck scale
  behavior of light, 490
  bias needs new physics, 247
  changes accumulate to change we see, 310
  cosmological constant, 445
  meaning of Zeno's view of motion, 615
  physics, 282, 645
    't Hooft on, 92
    action quantum pointed to, 84
    doing, without turning their minds to it, 491
    is an almost empty field, 579
    it is foolhardy to mention, 84
  observations at, 446
  revolution, economic value of a, 612
  weaknesses of, 691
  probing the structure of space at, 446
  property of space, light speed is a, 473
  space at, 444-7
  systems, Markopoulou and, 379
  topological rules, 281
  was unreal to physicists, 84
  world, navigating the, 310
Plato, 374, 450, 598, 614
  and introspection, 174
  and realism, 629
  and Russell's subsistence, 633
  and seminal idea of a real ontology, 42
  and the Cave, 375, 657, 664
  and the first principle of all that exists, 7
  and the idea of the idea, 25
  and the theory of forms, 633
  assumed the universe is finite, 133
  problems raised by, are unsolved, 465, 590
Podolsky, Boris, 146
Poe, Edgar Allan, 188
  and *Eureka*, 694
  and *The Murders in the Rue Morgue*, 617
  on the term *universe*, 22
  usage of *quantum*, 257
Poincaré, Henri, 555–6
Polchinski, Joseph, and string theory, 213-5
Pope, Alexander, 640, 658
position
  and motion
    clash between concepts of, 615
    dual relationship in quantum theory, 615
    source of quantum theory's battle between, 241
  definite, in classical view, 384
  flying arrow hops from one, to another, 311
  limit on measuring, 330
  no definite, in quantum theory view, 385
  our most precise measurements of, 330
positron as an electron going back in time, 271
post-empiricism
  and Planck scale, 642
  meaning, 96
preferred frame, 304, 305, *See also* spécial frame
preon models, variant of, 254
preons
  a topological model of composite, 193, 713
  meaning, 250
presentism, *meaning*, 706
pressure, example of emergence, 472
Pribram, Karl, 652
primeval atom, Lemaître's 107-11, 618, 701

Principia Mathematica, 158–59, 453, 520, 528, 548, 560
principles
   and Aristotle, 312
   and Gibbs, 316
   and Smolin, 61, 181, 373, 404
   as sources of potential errors, 61
   inapt aim does not invalidate Smolin's, 404
   incompatible, Einstein proposed for relativity, 486
   Planck sought physical, 77
   universal, and 't Hooft, 53
proton
   assumed to be elementary, 269
   Bilson-Thompson's, 269
   composed of three quarks, 435
   could have been named positive electron, 268
   -electron plasma, 470
   is stable, 496
   why, charge is equal to electron's, 269
psychology
   gestalt, not invoked, 218
   role of ontology in, 679
Putnam, Hillary, 548

Q

quanta, 78, 89, 95, 106-7
   conventional, *meaning*, 258-60
   counting, 527
   inherent
      meaning, 259
      three kinds of, 231
   light, making sense of, 493
   of energy, 31
   of space. *See* space quanta
   relinking, 334
   replicating, 362
   two kinds of, 258
quantizing
   classical theory, Einstein doubted, 179, 259
   Einstein and Infeld, using trains, 639
   general relativity, Rovelli and Smolin, 230
   approach of, has more fans, 557
   meaning, 258
quantum
   chromodynamics, *meaning*, 497
   coherence, when buckyballs lose, 255
   computer and Lloyd, 693
   cosmology, 141, 178
   entanglement
      direct connection could give rise to, 149
      EPR and, 147
      meaning, 144–9
      separation and, 148
   entanglement, a high degree of, 149

field theory, 15, 261, 333, 427
   and string theory, 101
fluctuations, 452
   in the early universe, 129
foam
   and loop quantum gravity, 178
   not like Wheeler's, 445
   Wheeler's, *meaning*, 91
gravity, 153, 247, 419, 446
   and Markopoulou, 17
   and quantum nature of space and time, 97
   and Rovelli, 18
   and Smolin, 19
   and string theory, 95
   and understanding inflation, 126
   candidates for, 179
   causal set theory as candidate for, 379
   causal sets in, 246
   loop. *See* loop quantum gravity
   math of early, 418
   meaning, 95
   mix-and-match goal, 206
   string theory and, 264
   three programs for, 176-80
   topological defects in geometry, and, 283
   why it doesn't work, 153
hard to construe as a real, 257
link is also a real, 264
meaning of the word, 256
mechanics, 92, 568-9
   and absurdity, 454
   and change in the worldview of physics, 29
   and Greene, 16
   and realism, 28
   change in understanding of, is needed, 577
   Einstein sought the deeper theory beyond, 145
   establishing the hegemony of, 246
   implies continuous space is unphysical, 92
   is ad hoc and abstract, 577
   is fundamentally about 'observations', 142
   Krauss and where laws of, came from, 454
   nobody understands, 492
   of universe has no place for time, 184
   particles can behave like waves in, 387-8
   was a unique intellectual adventure, 29
multiples of a real, 494
new, concept has two kinds of quanta, 258
of action, 82-93, 101, 277
   a closer look at the, 92

reality of the, 101
physics, paper shed new light on, 256
Planck did not use the term, 82
state of a system, 142
system
   entire universe as a, 141
   large-scale, 256
systems, 255, 275
   are quantum-entangled forever, 147
   need classical systems, 141
the real, 577
theory, 95, 151-3, 206
   and large systems, 255
   Dalai Lama and, 600
   many worlds interpretation, *meaning*, 587
   must be incomplete, *EPR said*, 146
   needs simultaneity, 39
   path for a real, 262
   reconciling, with relativity, 247
universe began as a single, 111, 139
quantum tunneling. See tunneling, quantum
quark
   decay of the down, 251
   decay of the top, 565
   never seen an isolated, 269
quarks
   antiquarks, no reason to pair with, 369
   Bilson-Thompson model and, 251
   explanation for existence of, 252
   fractional charges on, 55, 269, 390
   gluons binding, 435
   meaning, 54
   motion of, 436
   neutron made of three, 462
   proton and, 254
   reason for charge ratios of, 252
   reason for color charge on, 252
   reason for six flavors of, 253
   seem confined to composite particles, 55
   up and down, in protons and neutrons, 251
quasars
   already in the early universe, 355
   meaning, 344
Quine, Willard, 5, 383, 431
   and changing ontological commitments, 191
   and logic, 452-3
   and ontological commitments, 30
   and the ontological problem, 59-61
   on counting planets, 526
   on logic and mathematics, 520
   said numbers are sets, 529
Quinn, Helen, 367
quintessence, 168

R
R. v. Wray, 603
Raatikainen, Panu, 560
Rachinger, William, 685
radiation
   assumed sources of, 90
   blackbody, law, 87
   damage threshold is not a property of cells, 169
   energy loss of, due to expansion, 433
   entropy of, in Planck's equation 81
   extremely energetic, in the infant universe, 364
   gravitational capture of, in Hawking's picture, 354
   study of cells surviving, 169
radioactive decay, tunneling and, 119
Randall, Lisa, 675
random
   and causal events, 385
   aspects of human reproduction, 663
   biased Planck-scale changes, change made of, 310
   boreal forest squares, 594
   causation, 243-8
   choice
      of futures, 304
      of the universe's parameters, 582-3
   clumps in the early universe, 469
   coin tosses
      10^{152}, all being heads, 509
      and motion stopping, 509
      standard error of the mean of, 383
   density variations, 366
   element and time's direction, 320
   fluctuations in the early universe, 505
   meaning, 247
   molecular impacts and pressure, 472
   origin in prebiotic photochemistry, 663
   tweedle moves, 282
   tweedle pathways, 385
   twist-state relocations, 247, 473, 509
   walk of twist states in one-link steps, 447
randomness, and causality, 492
Rankine, William, 316
Ratra, Bahrat, 397
real
   but unreal numbers
      as points on a line, 532-3
      infinitely many, 533
      pi for example, 551-3
   language, need for, 373
   line, *meaning*, 544
   math
      and black holes, 507
      and Maddy, 545
      and unreal math, 209
      blazing a trail for physics, 209

has no infinity, 443, 545-6
includes some natural numbers, 549
is not the math of real numbers, 209
is unexplored, 209
meaning, 209, 517-22
objects, 518-21
operations, 527
vs math fiction, 522, 545
will be digital, 443
realism
a new foundation for, 628-31
and set theory, 531
axiom of infinity and, 548
Einstein and, 629
in mathematics, 517
Maddy and, 408, 518, 523
meaning, 9
reconciling with antirealism, 549
Smolin and, 643
vs quantum mechanics, 28
realist approaches, 3
reality
and entanglement, 145-6
and Plato, 174
Bell's theorem and, 147
epic struggle over the nature of, 146
hard to conceive of atom-scale, 83
hidden variables and, 146
idea of, may not match what is, 26
of *Now*, 306
Planck could not imagine Planck-scale, 83
two contradictory pictures of, 387, 493
unity of, 152
we all work with a mix of fiction and, 26
reciprocity, 182
Rees, Martin, 471, 584
Rehbock, Manfred, 657
Reichenbach, Hans, 9, 90, 320
on lacking logical means to solve problems, 190
on language, 373
on time and causal order, 246
on treating time like space, 420
Reicher, Maria, 636
Reid, Constance, 562
relational
approach, *meaning*, 44-6
causal set theory is completely, 378
mass must be, 438
space and time are, Smolin said, 182, 405
universe, motion in the space of a, 425
relationship
between two space quanta, 230, 256
close, between math and reality, 515
in definition of relational, 405
manifold-to-manifold, 230
topological, 447
twist is a, 194

relationships
a minimum of fundamental, 482
all particles are made of, 253
among brain's neurons, 652
and motion, 425
as a seminal idea, 221
between elementary entities, 45
between space quanta, 283, 334, 600
causal, in a Buddhist worldview, 600
entities emerging from, 480
particle's larger-scale, 494
physical theory must rest upon, 405
the mind's information is in, 652
the world is all made of, 612
universe as a causal network of, 161
relativity, 95, 151-3, 206, 277, 382
Dalai Lama and, 600
Einstein's work on, 44
foundations of, 491
general, 192, 211, 264-5, 344, 365
and accelerated motion, 488-90
and cosmology, 39
and dynamical geometry, 283
and Einstein's extra constant, 165-70
and ether, 70-1
and gravitational waves, 621
and non-Euclidean geometry, 555-6
and string theory, 101
and three- (vs four-) dimensional geometry, 421
assumed correct on cosmological scales, 225
de Sitter solved equations of, 106
Einsein's main paper on, 114
Einstein's manuscript on, 72
Friedmann solved equations of, 115
in support of simultaneity, 288-9
insights behind, 15
is approximate, 265
is best seen as a statistical theory of space, 473
is the basis of the Big Bang model, 185
Lemaître on, and the Big Bang, 140
Lemaître solved equations of, 107-9
showed space is something, 122-4
granular math for, 178
had no need for an ether, 69
meaning of theories of, 485
misconception of, 302
principle of, 278, 304
reconciling, with quantum theory, 247
special, 39, 68-70, 124, 397, 417, 485
1905 paper on, 156
and clocks, 287, 424
and common time, 286-90
and energy conservation, 432
and Minkowski, 421
and the Cosmic Clock, 288
and the no-special-frame problem, 340

and the present of the universe, 154-7
did not show there is no ether, 69
does not apply to the universe, 288
equations do not describe reality, 303
is a statistical theory of motion, 473
misapplied, and Newton's true time, 159-64
sets a speed-of-light limit for matter, 343
shows mass and energy are the same, 432
uses flat space, 124
was a respite for Einstein, 31
religion terminates a physics conversation, 110
renormalization, 458
is not mathematically legitimate, 442
meaning, 442
replication
by quantum tunneling, 224
of a space quantum, 247
slowing, 295-6
step one, 223
step two, 225
repulsion, as a cosmic blunder, 165
revolution
17th century scientific and philosophical, 611
a new math, 550
a new, from new kinds of instruments, 619
Boole initated a, in the domain of logic, 559
convulsion in the standard model's worldview, 197
in the economy, 612
in view of the early universe, 621
of which modern science is root and fruit, 610
prophet of the second quantum, 40
quantum mechanics wrought a, 29
Heisenberg and the quantum, 52
to rethink the standard model of cosmology, 28
ribbon
forms a ring, 237, 270
meaning, 376
twist moves along a, 309-10
ribbons, 236
and helons, 195
Bilson-Thompson's, 194, 229-35
as links, 197, 247
three, 250
meaning, 231
run right through the manifold, 280
three, through each manifold, 240
twists in, 194, 250
twists in, making matter, 267
Ricci curvature, *meaning*, 491
Ricci-Curbestrato, Gregorio, 491

Riemann, Bernhard, 74-5, 93, 101, 134, 379
a quantum of Riemann's granular space, 223
and causal sets, 707
and geometry, 554-7
coined the term space quanta, 256
continuous space needs a metric, 211, 475-7
Einstein's view of, 74-5
genius of, 236
on accuracy in continuous space, 473
on geometry in imaginary spaces, 426
on granular space, 103, 490
on quanta of space, 256, 493
said space may be granular, 74
seminal ideas of, 221
thesis, 74
Riemannian
-Einsteinian, ontology, 33, 426
geometry, 74
manifold foliated, 422
space
and general relativity, 473
and pi, 534
and Ricci curvature, 491
granular, 491
Riess, Adam, 168, 357
Rigby, Jane, 623
ring or loop, dimensions form a, 237
Ring, the One
is vast, 239-40
meaning, 239
traverses every manifold exactly once, 240, 241
Rings
no net twist in any of the, 270
seven in total, 240
the Four, 240-1
the Two, 239-41
three, traverse each manifold, 241
road to reality, 33, 57
Penrose's book, 675
Robert, Marthe, 609
Robertson, Howard, 421, 476
Robertson, Robbie, 658
Rodych, Victor, 561
Röhrer, Heinrich, 120–1
Rosen, Nathan, 146
Rovelli, Carlo, 154
and reality is not what it seems, 220
and spacetime, 155
and Wheeler-DeWitt equation, 419
biographical note on, 18
correcting our worldview, 25
dismissed the Cosmic Clock, 288
on a metric made of quantum loops, 198
on Anaximander's grammar for the world, 23
on background independence, 45

on change, 308
on counting grains of space, 293
on dialog with philosophy, 641
on eight notions of time, 287
on freeing ourselves from wrong ideas, 416
on giving up the notion of time, 508
on low-entropy beginning, 138
on multiple universes, 585
on Newtonian time, 508-11
on the value of a novel language in physics, 375
on Planck scale discreteness, 178
on pragmatic scientists, 10
on grains of space separated by area, 264, 440
on small-scale space, 444
on ten notions of time, 707
on the beginning of the universe, 399
on the continuing relevance of philosophy, 210
on the idea of an elementary substance, 47
on the meaning of *moving*, 614
on the nature of time, 416-9
on the need for explanation of quantum theory, 9
on the present of the universe, 39, 154, 301
on the problem of evidence at Planck scale, 333
on the purpose of scientific research, 210
on the question of reality, 2
on trying to understand, 617
on understanding the world, 2
on vision in science, 13
pointed in what proved to be the right direction, 192
quantizing general relativity, 230
quantum mechanics is unphysical, 92
something still eludes us, 152
space quanta *are* space, 258
Rowlands, Joseph, 524
Rube Goldberg machine
 and the string theory landscape, 589
 Barbour's pile of *Nows* as a, 325
Rubin, Vera, and dark matter, 351–52
Rundle, Bede, 449, 585
Russell, Bertrand, 215, 598, 656
 and existence, 633-7
 and set of sets that don't contain themselves, 520
 and subsistence, 633
 and Whitehead, 528
 and Zeno's paradox, 63-5
 at Princeton with Einstein et al., 595
 needed the axiom of infinity, 548
 on an empty universe, 452-3
 on Frege and number, 518
on Gödel, 559-61
on nonexistent objects, 636-7
on Peano and arithmetic, 540-1
on the meaning of causation, 381
said a finite world seems unlikely, 545
said few stars have planets, 582
said numbers are discovered, 519
said the world had no beginning, 105
tried to found arithmetic on logic, 211
Rutherford, Ernest, 48

S
Sagan, Carl, 112, 553
Sahlins, Marshall, 609–12
San Francisco, meeting in, 200
Sartre, Jean-Paul, 301
 and a worldview, 301
 and being, 110
 and non-existence, 313
 and the interplay of being and doing, 647
 and the notion of *Now*, 301
 said nothingness is not an origin, 454-5
satellite
 Euclid, 621
 Gaia, 503
 JWST, 621
 Planck, 122, 358, 436, 556
 Voyager 1, 782
 WMAP, 130
Sattler, Barbara
 on paradoxes of motion, 64
 on Zeno's view of motion, 615
scale. *See also* Planck-scale
 atomic vs quantum, 578
 hidden machinery at small, 192
 macroscopic, 282
 of very early expanding universe, 328
 of whole universe, 434
 time-, 76
  in biophysics, 565
  in physics, 566
  millisecond, of neural events, 218
  of real events, 566
  the universe's, 114
 tunneling and, 224
 understanding the universe at every, 220
Schopenhauer, Arthur, 374
Schrödinger equation, 418–9
Schrödinger, Erwin, 257
Schrödinger's
 views on consciousness, 638
 cat, 586
scientific method, 631
 and the future of physics, 642
 has no substance, 777
 myth of, 3, 23, 631
scientific notation, 360, 462
 example of, 509
 meaning, 360

Scott, Douglas, 128, 224
Second Law of Thermodynamics, 77, 137, 320
Selety, Franz, 701
seminal idea, 220, 221
    a consistent understanding, 27, 37, 41, 65
    a finite universe with no edge, 133-6
    a real ontology, 36-41
    Calabi-Yau manifold is real, 101-2
    digital change, 103-4
    exponential expansion, 126-32
    granular space, 74-5
    it is all relational and background-free, 44-6
    low entropy beginning of the universe, 137-40
    matter made from space, 190-8
    new space, 112-6
    Newton's two times, 158-64
    now, 154-7
    of the atom, stultified, 48-50
    oneness of the universe, 150-3
    physics from inside the universe, 141-3
    Planck's quantum of action, 82-4
    Planck-scale reality, 77-86
    probabilistic causation, 246-8
    quantum tunneling, 118-21
    real space, 166-72
    ribbons, 229-34
    simplicity, 58-61
    space is curved, 124-5
    space is something, 66-73
    sufficient reason, 181-2
    the a-tom, 47-50
    the causal set, 243-6
    the cosmos, 23-4
    the one and the many, 63-5
    the quantum of space, 97-102
    the single ingredient, 51-2
    the single solution, 186-9
    the thought experiment, 173-5
    the universe at first was very ordered, 137-40
    the universe began as one quantum, 105-11
    the universe is expanding, 112-7
    the universe is nonlocal, 144-8
    the whole universe, 183-5
    twists make matter, 193-8
    Zeno's episodic motion, 64-5
seminal ideas
    and Newton's times, 159
    are in Part I, 41
    disparate, fit together, 41, 196
    provide the ontology's hierarchy of concepts, 37
    three, from quantum gravity, 176-80
sense perception
    and empiricism, 213-4
    and limitations of logical positivism, 653
sensory experience, 213, 216
    and Einstein, 58
    as arbiter of what is real, 59
    now largely cut off from, 659
    skepticism about, 218
    useful fiction based on, 219
    visual, limitations of, 217
sequence
    a step or iteration of 3-D universes in the, 377
    and the Planck time, 104
    and Zeno's paradox, 615
    as opposed to continual action, 162
    as the source of causation, 647
    causal set's, absolute time emerges from, 380
    causal, *meaning*, 121
    instead of time, 383, 404
    leads to a universal kind of time, 42, 246
    of impressions from the world, 77
    of separated states, and Smolin, 307
    of spaces, and Barbour, 324
    registers the universe's age, 510
    time of any kind derives from, 406
    universal, as a guide to Rovelli's times, 508
set
    another meaning for, 761
    meaning, 529
    the universe is a finite, 121
    theory
        and Cantor, 543
        and counters, 528-31
        and natural numbers, 529-31
        and {} or Ø, 531
        and something from nothing, 529
        and the new ontology, 530
        and Wildberger's objections, 529
        Badiou and, 516–7
        causal. *See* causal set theory
        fallibility of, 531
        in grade school, 529
        Maddy on, 517
        Maddy on, and arithmetic, 530-1
        Maddy on, and mathematical analysis, 533–4
        neglected problems of, 529-30
        reliance on, 31
        Zermelo–Fraenkel, 520
sets as counters, 527, 528-31
Shaghoulian, Edgar, 337
Shakespeare, William, 43, 614
Shaknov, Irving, 146
Sharma, Suresh, 551–2
sheets, folded, 335
Shestakova, Tatyana, 419
Shifman, Mikhail, 412
shut up and calculate, 89, 486

Siderits, Mark, 597
Siegel, Ethan, 428-9, 497
Silk, Joseph, 505
Simenon, Georges, 188-9, 221
simplest
   assumption
      and explains a lot, 238
      is all quanta have six links, 334
      is lockstep iterations, 303
      is matter stops collapsing, 461
      replication is the, 239
   conceptual scheme, Quine on the, 60
   form of laws, Einstein showed, 555-6
   multiverse, 586
   particle, the photon is the, 272
   possibilities, Einstein on the, 2
   single-manifold origin is, 241
   story would assume nothing, 57
   way, Planck's, 689
   when the universe was, 184
   ΛCDM paradigm is the, representation, 740
simplicity
   and Maigret's strategy, 713
   and Penrose, 57-61
   and the ontological problem, 60
   and William of Ockham, 58-62
   astonishing, of the one law, 8
   extreme, Poe sought, 694
   grounds for seeking, 61
   impressive, of a theory's premises, 59
   meaning, intuitive, 683
   of causal set theory, 160-1, 380
   of physics, 60
   Planck's concept of, 83
   seminal idea of, 58-61
simultaneity, 39
   and causal set theory, 179
   Einstein and, 154
   inherent, in new ontology, 161
   nonlocal universe requires, 145
   perfect expression of, 417
   problem in causal sets, 161
   quantum theory needs, 39-40, 288
   relativity purported to eliminate, 286
singularities
   black holes in theory are, 125
   strings resolve, 400
singularity
   black hole is not a, 463
   depends on continuous space, 460, 500
   general relativity and, 460
   initial, not totally believable, 400
   initial, outside scope of physics Hawking said, 399
   vs multiplicity, 53
   where time has a beginning, Hawking and, 108
Sitter, Willem de
   curvature of space can be determined, 699
   on dark matter, 351
   on Lemaître's solution of Einstein's equations, 106
   solved relativity equations for an empty space, 714
   universe found by, had constant size, 116
Slipher, Vesto, 114
Smalley, Richard, 255
Smolin, Lee, 92, 157, 231, 321
   2005 paper on quantum gravity, 249
   and causal set theory, 378
   and Markopoulou, 17, 379
   and principles based on sufficient reason, 181-2
   and simplicity, 61
   and spacetime, 18, 425
   and the cosmic clock, 284, 288
   and the future of physics, 643
   and the old ontology, 27-32
   and the Wheeler-DeWitt equation, 178
   as champion of time, 160-3
   biographical note on, 19
   discrete lattice view of space, 446
   getting close, 157, 163, 407
   laws arise from the universe, 479-80
   mathematics obscures the creative process, 208
   no background, 44-6
   no symmetries with background independence, 369
   on a metric made of quantum loops, 194
   on a purely quantum structure, 240-1
   on a relational approach, 181, 221
   on a theory of a whole universe, 184
   on Barbour's time, 324
   on Bilson-Thompson's paper, 194
   on causation, 245
   on causal completeness in the universe, 312
   on cosmology in crisis, 585-7
   on crisis in fundamental physics, 150
   on detective work, 187
   on disagreement about the quantum world, 577
   on Einstein's search for deeper theory, 145
   on elementary particles, 197
   on eliminating the continuum, 261-2
   on essential role of the observer, 142
   on motion, 616
   on new research directions, 641-5
   on physics inside the universe, 141-3
   on Planck's derivation as comedy, 79
   on points of space having no exisstence, 399
   on principles for fundamental physics, 403-5
   on quantizing general relativity, 230

on the need for radical change, 371
  on realism, 3
  on string theory, 100-1
  on the dimensionality of space, 264
  on the multiverse, 587
  on the reason for laws of physics, 483-4
  on time, 158-63, 416-8
  on two theories, 152
  on ways to curl up six dimensions, 483, 582-3
  on wrong assumptions, 32, 34, 40
  primary goal of, 196
  primary goal vanished, 202
  principles for fundamental physics, 404-6
  reaction to Bilson-Thompson paper, 714
  said space is emergent, 236
  said the two theories are wrong about time, 416
  said time is fundamental, 307
Smuts, Jan, 186
social media
  and geopolotical intrusions, 594
  and metapersons, 610
  enhance untruth, 644
Socrates, 174
Solvay Conference, 76
Somerville, Mary, 611
Sorkin, Rafael, 178
  and causal sets, 161
  on granularity and nonlocality, 145
Southey, Robert, 51
space
  and time, the deep roots of, 242
  as a fluid, 170
  ascribing physical reality to, 66
  beyond the visible universe, 328
  calculating, 104
  compressibility of, 170
  comprises real quanta, 489
  concept of moving, 341
  conceptualizations of, 170
  continuous, 74, 427, 444-6
    assumed, 392, 442
    calculus needs, 93
    disconnect between quanta and, 259
    Euclidean, 74, 473, 534, 692
    fields depend on, 429
    in the old ontology, 29-31
    is infinite, in quantum mechanics, 568
    is and old-ontological commitment, 169
    is the foundation of physics, 493
    light-cone equivalent in, 380
    no natural metric for, 211
    Planck's derivation used, 93
    quantum foam in, 445
    quantum theory in, 473
    real numbers in, 533
    string theory in, 177, 483
  curvature of, 122, 135
  de Sitter, *meaning*, 106
  deep misunderstanding of, 437
  Descartes thought, is discrete, 74
  discontinuous, 265, 430
  discrete, *meaning*, 74
  empty, 47, 278, 448-52, 587
    and vacuum, 166
    attempt to breathe in, Einstein on, 259
    does not exist, 377
    Einstein on ascribing physical reality to, 66
    is confused old ontology, 169
    light propagating in, 278
    no such thing as, Einstein said, 429
  Euclidian, 534
  expanding, 112-7, 291-3
    Hubble showed, 106
    Lemaître showed, 106
  finite, 134-5
  grains of
    Rovelli counting, 293
    Rovelli's, 260
    separated by quanta of area, 440
  granular, 179, 223
    digital metric inherent in, 379
    history of, 74-5
    made of tiny volumes, 491
    may be, Riemann said, 74
  how, came to be closed, 238
  is a real object, 376
  is closed, 122
    Einstein deduced, 134-5
  is not continuous, 489
  is something, 42, 66-73, 117, 221, 341, 403
    Einstein said, 267
    Lorentz and Einstein discerned, 73
    made of quanta, 403
  mass of, 169, 299, 431
  math for granular, 430
  matter is a property of, 489
  Minkowski
    and Noether's theorem, 432
    general relativity and, 433
    rejecting, leads to difficulties, 433
  new, 116, 126, 171
  no theory for granular, 483
  Planck-scale makeup of, 214
  quanta. *See* space quanta
  quantum. *See* space quantum
  Riemannian, 491, 534
    granular, 491
  starting to emerge, 240
  structure of, 334-5, 444-7
  the substantiality of, 294
  -time. *See* spacetime
  timeless, seminal idea of, 180
space quanta, 104, 230, 246, 393
  all things arose from vast numbers of, 471

and structure of space, 444-7
and tweedles, 440
are all that exists, 102
are identical, 272
are not in space, 18, 42, 258
as events of the causal set, 380
existence is the totality of, 634
have a fixed volume, 42, 80, 230
jets of, 344
largest natural number and, 538, 549
laws of physics emerge from many, 432
links between, are basis of causal set, 161
may be linked in a 2-D sheet, 338
meaning, 42
merging, universe would not be affected by, 322
number of, 286, 291-3, 349, 383, 549
particles made of twist relations between, 437
relationships between, 161
replicating, 223-7
matter made of tiny linked, 148
static and dynamic ensemble of, 600
tangles in links between, 369
two, at step one, 468
universe's mathematic objects arise from, 542
vs quanta of quantum theory, 494
why did doubling of, stop, 294
space quantum
has never been observed, 494
is an indivisible Planck-volume of space, 376
is the final atom, 261
naming the, 314
nature of the, 97
one, tunneled to two, 293
volume of the, 462
spacetime, 377, 382, 420-6
and causal sets, 178
and Minkowski, 421
and familiar usage of fictional concepts, 375
an entire, is unphysical 725
background is difficult to let go, 45
foliated, 289, 422
has no physical reality, 424
hints that, is not fundamental, 420
imaginary continuous, 379
is a source of confusion, 578
is confusing old ontology, 169
may be discrete, Hawking said, 93
meaning, 421
Minkowski's, replaced Newton's, 423
Newtonian, 423
problem, the, 18, 40, 211, 289
Smolin seemed to walk away from, 425
stretching, 358
structure of, at Planck scale, 96

stultifying progress, 374
Sparrow, Giles, 9
special frame. *See also* absolute frame
Einstein did not need a, 304
for acceleration, 304
meaning, 39
that no, exists was an assumption, 304
the universe is a, 304-6
speed of light
as a fundamental constant, 69-70
axial jets may exceed the, 343-7
fundamental mystery of the, 276
galaxies receding faster than the, 327, 340
is a digital property of space, 164
is a fundamental property of space, 341
is a property of space at Planck scale, 382-3
is constant because it just is, Laughlin said, 276
is one link per tock, 278
limit on Big Bang explosion wavefront, 298
measurement in different directions, 71
universe grew faster than, 127, 226, 241
Spinoza, Baruch, 181
Spinoza, Benedict, 374
spooky action, 146–7
Stachel, John, 178, 493
standard model of cosmology, 185
and beginning with a point universe, 197-8
and changing modes of growth, 292
and primordial black holes, 354
and the imagined knife edge, 170
and the old ontology, 20-1
and the problem of expanding space, 292
and unanswered fundamental questions, 590
and Weinberg, 297-8
how deeply the, assumes space is continuous, 74
cannot explain
black hole sizes, 355
early galaxies, 622
space expanding, 358
changing story of the world's beginning in the, 299
contradictions of the, 108
meaning, 365
primordial black holes and the, 470
problems of the, 413
rests on general relativity, 39, 170
requires revision, 362, 623
says early space expanded faster than light, 293
story keeps changing, Peebles said, 27
supported Lemaître's assumption, 140, 222
truism of why space is expanding in the,

112
standard model of particle physics
  is based on simple symmetries, 367
  is completely flawed, 49
  matter and antimatter in the, 367
  meaning, 20-1, 49
  needs supersymmetric particles, 411-2
  particles
    are assumed to be elementary, 439
    are made of six paired and braided
      links, 404
    are made of three pairs of links, 235
    strategy to move beyond the, isn't working, 414
    why there are sixteen particles, 253
standard muddle. *See* standard model of
  cosmology
stars
  big black holes consume, 361
  black hole is the ultimate fate of many
    large, 359
  merging neutron, 621
  need to annihilate millions of, to power
    jet, 344-5
  some, become supernovas, 361
  the first, were huge 622
statistical mechanics
  and Bose on Planck's formula, 90-1
  and entropy, 316-9
  and thermodynamics, 316
statistics
  and Bell's theorem, 146-7
  of 10^{24} particles in a thermodynamic
    system, 481
  of energy of invented oscillators, 84
  quantum, and Bose, 90-1
  social, 593
steady-state model, 107, 298, 299
  meaning, 300
Steinhardt, Paul, 129-31
step four, seven Rings, 240
step one, 468
  made two quanta, 225
  meaning, 223
step ten was expanding faster than light, 241,
  469
step three
  and granular space, 468-9
  eight quanta, 240
step twenty, proto-space, 226
steps, *meaning*, 238
Strassler, Matt, 427-8
strategic
  choices, 209
  single seminal, framework, 46
  issue for philosophy of physics, 210
  question of the creative source, 212
  question, Muller's, 417
strategy

detective fiction, of Maigret, 188, 713
Einstein on, 179
  for fundamental physics, 259
  not a good, for finding a real answer, 186
string theory, 19, 180, 400
  beautiful new math of, 97
  continuous space and, 177, 446
  holographic principle adapted to, 337-8
  is a candidate for quantum gravity, 176-7
  is (literally) pointless, 400
  is the lead candidate for quantum gravity,
    177
  landscape of Rube Gooldberg machines,
    589
  math needs six extra dimensions, 264
  meaning, 96
  mistake, people make, 201
  needs supersymmetry, 411
  the, landscape, 588
strings
  are math abstractions, 99
  replace point particles, 96, 99
  resolve singularities, 400
  there is no evidence for, 400
  wiggle in continuous space, 100, 177
Strogatz, Steven, 428
strong nuclear force, *meaning*, 497
structure
  cubic honeycomb, *meaning*, 334
  hexagonal close-packed, *meaning*, 335
succession
  and Newton's true time, 311
  endless, of cutting atoms, 48
  is a math object, 537
  is an object of arithmetic, 549
  motion as a, 615
  of 2-D universes, 653
  of events, time as the, 508, 511
  of space is perceived as time, 325
  of synchronous spatial states, 104
  reified, 521
successor
  one of Peano's indefinables, 540
  to 9, 537
  to zero, one as Peano's supposed, 541
Sudarshan, George, 78, 83, 101
sufficient reason
  Leibniz and, 181
  principle of, 403
    meaning, 181
  Smolin's principles based on, 181
śūnyatā and zero, 598
superconductivity
  and color charge of quarks, 496
  and holographic duality, 339
  and QCD, 497
  and my study of quantum tunneling, 120
  and the Higgs, 457
  and the structure of space, 446

high-temperature, 120
macroscopic quantum state of, 255
meaning, 120
superfluidity, 339, 457
supernovas, 697, 752
as standard candles, 168
can leave black holes, 622
light from a neutron star after a, 752
meaning, 502
stars can become, 361
supersymmetry
evidence for, was not found, 411-2
meaning, 411
Surya, Sumati, 246
Susskind, Leonard
as multiverse proponent, 584
on Calabi-Yau manifolds, 99-100
on meaning of the Planck length, 77
on hidden machinery of particles, 191-2
on string theory and pocket universes, 588-9
on the holographic principle, 337
on why quantum gravity never worked well, 153
on wormholes in nonlocality, 148
Sutter, Paul, 118–9
Suzuki, Daisetsu, 598
Swift, Jonathan, 48
symmetries
and deducing laws of nature, 368
and Martin Gardner, 736
and supersymmetry, 411-2
are properties of fixed backgrounds, 369
depend on continuous space and time, 432
in Baez's language, 16
physics', emerged with large-scale space, 369
standard model is based on assumed simple, 367
synapses
and memory, 651
and thought, 637
meaning, 649
Synge, John, 487

T
Tadhunter, Clive, 345
Tagore, Rabindranath, 678
teachers
math, could do better, 536
my Hochdeutsch, 11
physics, 73
teaching
beans for, sums, 524
elementary, 535
math, 529
Tegmark, Max, 104, 585
Tempczyk, Michal, 778

Thales of Miletus, 8, 23, 614
theorem
Bell's, 147, 190, 680
causation, 664
Hackett's relocation, 281
Noether's, for energy conservation, 432
no-hair, 506, 760
of number theory, 548
theorems
Gödel's, 520, 559-62
proofs may fail in reality, 561
theory
decides what we can observe, 169
interpretation of, 97
of everything, 18, 95, 215
thermodynamics, 316
meaning, 316
Second Law of, 77, 137, 319, 320
thesis
Badiou's, 516
Miller's, on causal sets, 245
Planck's doctoral, 77
Riemann's habilitation, 74, 134, 236, 686-7
Thompson, Joseph (J.J.), 48
thought experiment, 133, 284, 363, 525
meaning, 173
thought experiments, 173-5, 356
and inconsistency, 174
impacts on science of, 173
time
absolute zero of, 312
and Einstein, 287
and Rovelli, 416-9, 508
and Smolin, 158, 307
and space are not both real, Smolin said, 19
and Zeno, 64
Aristotle's
has no metric, 285
is a universal order, 285
background
quantum theory needs, 383
quantum theory assumes, 288
common, 416
is a measure of motion, 159, 405
is defined by clocks, 417
is not a property of the universe, 155
meaning, 289-90
Newton's, 158-64
confronting deep questions about, 19
continuity of, is an ontological commitment, 162
eight kinds of, 287
Einstein's definition of, 160
emerges from the causal set, 380
global conception of, 288
Heidegger's holistic view of, 38
how can cosmic, be measured, 284

is approximate but may be precise, 511
is not a dimension, 155, 164, 325, 489
is not a property of the universe, 287, 325
limitations of the term, 377
Minkowski's definitions of, 423
Newton's absolute, 158-9
Newton's, is assumed to be continuous, 160
Newtonian as defined by Rovelli, 508-11
Newton's true, has no artificial metric, 511
our true, is digital and absolute, 164
paradoxes of motion and, 64, 310
projected back to the beginning, 106
proper, *meaning*, 423
quantum nature of, and space, 97
reassess the role of, 321
relativity has troubles with, 39
spatialization of, 18, 375, 426
ten notions of, Rovelli on, 707
the arrow of
    and Carroll, 322
    and Eddington, 330
    meaning, 77
the idea of, 416
the problem of, 9
the universe has no property of, 404
the Wheeler-DeWitt equation has no, 177
true, 163
    is discontinuous, 163, 417
    meaning, 160
    Newton's, 158-64, 510-1
        has no metric, 285
    something close to, is real, 311
two kinds of, 158-64, 404
universe has no property of, 417
what makes, move on, 417
times
    Newton's two, 508
        are both the succession of events, 511
        meaning, 159
        the difference between, is fundamental, 511
    Nexts drive all our, 418
    Rovelli's two, 508-9
tock, 377
    meaning, 285
    no way to determine duration of each, 285
    number of space quanta doubled each, 293
    one link per, 278, 286, 383
tocks
    connecting, to our clocks, 510
    of the Cosmic Clock, 553
Tolkien, J. R. R., 240, 705, 718
Tong, David, 428
topology
    meaning, 231
    of linking, 469

Topper, David, 135, 665
Torricelli, Evangelina, 450
translation
    artifact of, 307, 745
    Diels', of Zeno's paradox, 63
    Motte's early, of Newton's *Principia*, 683
    of Boscovich, 753
    of Einstein, 71, 91, 678, 686, 740, 745, 755, 756, 774, 782
    of Hegel, 684-5
    of Minkowski, 744
    of Newton's Principia, 159
    of Planck, 688
    of Rovelli, 260
    of Simenon's works, 188
    Saunders', 91
    where, is mine, 11
triangles
    angles of, to measure curvature, 123
    largest, angles add to 180°, 123
true time. *See* time
truth
    -seeking value of science, 645
    a new scientific, and Planck, 640
    and history, 605
    and social media, 644
    and the law of evidence, 603
    as correspondence with the world, 629
    difficult to speak the, 563
    is a whole, Smuts said, 186
    of $2 + 3 = 5$, 523
    of myths, Kafka setting out to test the, 613
    science as a stand-in for, 630
    Smolin and how to search for scientific, 641
    the idea observation would reveal, 640
    two indicia of, 645
tunneling, quantum, 118-21, 334, 300, 344, 361, 403
    and superconductivity, 120
    barrier, 118
    shape, 120
    explains the universe expanding, 361
    happens faster than light, 119
    only to available states, 570
    routinely observed, 119
    space quanta replicating by, 224
    was the beginning of the universe, 468
Turner, Michael, 349
    and WIMPs, 352
    on our changing understanding of the universe, 395
    on dark energy, 590
    on gravity slowing expansion, 357
    on the detective story and dark matter, 351
Turok, Neil, 18-9
Turow, Scott, 603
Twain, Mark, 771
tweedle, 376

Subject Index

charge on the, 254
is the model's single fundamental object, 249
meaning, 195
Smolin and the lost excitement of the, 254
tweedles
are half twists, 194
Bilson-Thompson backed away from, 439
detector causes all six, to pick one slit, 391
identified as twisted links, 234
only parenthetic mention of, 249
seminal idea of, 190, 257
six, per standard model particle, 261
two per helon, 233
twist
may be digital by default, 394
meaning, 195-6
motion of, 310
twists
and Bilson-Thompson's paper, 193
are all the same, 393
are created in matched pairs, 270
are "far" apart, 461
are irreducible, 440
are one-sixth of the electron's charge, 254
as hidden variables, 310
braided pairs of, 197
do a kind of random walk, 447
dum and dee have opposite half-, 195
from chaotic replication make particles, 469
in links, matter is made of, 125, 267
may interchange with braids, 281
move at most one link per step, 404
new space makes new, 267
are not shown in most diagrams, 195
opposite, can cancel, 270
represent the model's fundamental objects, 231
settled into particles, 470
unimaginable, inside Calabi-Yau manifolds, 99
up to three per space quantum, 462
we are poorpl defined clouds of, 626
Two Rings, the, *meaning*, 239-41
two-slit experiment, 387-91

U
ultimate cause, 312-5, 405, 663
ultramatter, 461-3
uncertainty
and language, 374
ontological, 164
principle, 332
and location at Planck scale, 332
meaning, 615
Planck-scale basis for the, 615
was set in a continuum, 446
reconciling causality with, 247

understanding
a different story of the beginning, 402
a good theory helps our, 170
a Humean, of the laws of physics, 480
a single, of the whole, 189
a way to make direct distant connections, 149
abandon hope of, inertia 616
all we see, 235
and a holistic view, 151
atom-scale, of quantum theory, 35
became foreign to physics, 631
behavior of particles, 271
more, might enable better theory to emerge, 220
black holes, 500-7
without singularities, 462
causal connections, 244
causation, 244-8
causation of a particle arriving, 385
change, 310, 325
charge conservation, 270
cosmologists', of the special frame, 305
cosmology and the expansion of the universe, 113
dark energy, 349
and dark matter as two sides of one coin, 361
deep difficulty in, continuity and motion, 63
discovery of the tweedle is key to, 196
nobody has an, of duality, 387
Einstein sought an, 210
Einstein's, of causality has no place, 381-5
Einstein's definition of time, 155
everything is made of one thing, 52, 240
evolution of the universe, 618
expanding confusion, 340-1
fractional charges, 56, 254, 269
Frege and mathematicians' lack of, 536
granular space, 76
how mass and energy are the same, 250
how the photon can be electromagnetic, 275
how the universe began, 109
humankind's long climb to, 611
implications of the Planck constant, 88-93
language for fundamental concepts, 376
limitations of our inside view, 143
mass
and energy, 438
and inertia, 491
problems of, 435-8
entropy and, microstates of quanta and links, 319
more particles than antiparticles, 370
must rest on unobservables, 408
needs language for ideas, 23
new ontology may offer more and better,

        662
not, inertia, 437
numbers, what the universe can tell us
    about, 515
of the physical world, 2
our experience as a whole, 218
our lifelong search for, 218
ourselves, 600, 613
over timescale ranges, 565-6
paradoxes of motion, 310-1
philosophy may help physics to find a
    path to, 408
physics does not do, 4
planetary motions, 580
property of replication offers, of ultimate
    cause, 314
quantum
    entanglement, 149
    making sense of theory of space and
        time, 419
    tiny quantum workers, 578
    tunneling, 121, 223
quest for, 2
quests for personal, 25
radical change in our, is required, 371
real, 567
    black holes with real math, 507
    ideas, language for, 41
    space, 172, 231, 421, 477
reality, 20, 25-6, 35, 77, 191, 401, 493, 490,
    567
    and the universe as a causal set, 248
    beyond all possibility of observation,
        174
reasons for everything, 181
relativity, 485-91
sense perception, 216
Smolin sought an, of background inde-
    pendence, 44
space, 73, 98
    expanding, 291
    growing, 299
    is something, 117
spacetime, 420-1
the Big Bang, 184
the Calabi-Yau manifold is the key to,
    space, 198
the closed universe, 326
the cosmos, 24
the difference between Newton's times,
    511
the direction of time, 321-3
the entire (early) universe, 634
the fallacy of multiverses, 586
the final atom, 50, 64
the finite closed universe, 135-6
the flow of time, 324
the foundations of relativity theory, 486
the great chain of being imposed an, 659

the holographic principle, 339
the low entropy of the early universe, 137,
    319
the meaning of general relativity, 490
the meaning of the Planck constant, 83
the Planck quantum, 101
the real space quantum, 94, 235, 257-8
the relational approach, 45
the search for, that drove physics, 4
the shape of the universe, 136
the size of the whole universe, 398
the source of new space, 116, 358-62
the source of the laws of physics, 483
the structure of space, 444
the two-slit experiment, 389-91
the universe
    as a causal set, 248, 380
    as an entropy machine, 323
    as the creative source, 212
    at every scale and every time, 220
    black holes are central to, 507
    has no outside, 329
    may have only two dimensions, 336
the usefulness of mathematics, 516
the wave function, 40
the whole universe, 64, 153, 183
the world, 44
    great minds struggled with, 41
    rather than the metric, 309
three imagined elements as real, 180
through disaggregation, 150-1
time, 158-64, 164
    and motion, 418
    paradoxes, 157
wave-particle duality, 275
we have a growing, 658
what is real, 1-4, 11
    key to, 98
what reality is like, 203
what was needed to create space, 241
whether the laws or the universe came
    first, 478
why all electrons have the same charge,
    393
why electric charge is balanced, 270
why exponential growth stopped, 294
why fundamental physics is the way it is,
    205
why it is always now, 303
why space has three dimensions, 263-5
why the Planck time is the smallest, 103
why the speed of light is the speed limit,
    278
why these are the laws of physics, 583
without recourse to the supernatural, 23
Zeno, 98
Zeno's paradox, 64, 75
Unger, Roberto Mangabeira, 32, 143, 162,
    406

and causation, 244
and the cosmic clock, 288
and the cosmological fallacy, 586, 644-6
and the spatialization of time, 18, 375
on measuring cosmic time, 284
on time in the whole universe, 184
units
  absurd, of the Hubble constant, 359
  clock-time, 164
  digital, of twist, 393
  or percent per billion years, 359
  physics' arbitrary, 278
  Planck's
    fundamental or natural, 81
    proposed system of, 86
  universal, for the speed of light, 278
universals
  meaning and the concept of, 633
  Plato called, ideas, 25
  subsistence of, and Russell, 633-5
universe
  a new physics of the, 419
  a preposterous picture of our, 585-6
  a spherical, 135-6
  as a manifold, 98
  as a whole is a very different kind of system, 183
  assumed to be fluid-filled, 115
  closed. *See* closed universe
  Einstein saw the, as static, 106
  entropy of the early, 137-40
  equation for the entire, 418
  expanding, 112-7
    Einstein's initial rejection of, 112
    or contracting, 115
  expansion rate of the, 116
  explanatorily closed theory of the whole, 143
  growing to galaxy sized, 296
  how to get a complete description of the, 141
  infant, when Solar-System sized, 295
  infinite, 106
    has no limit on numbers, 547
  infinitely small, theory does not apply to, 402
  inflation makes a, that is close to flat, 131
  inflationary, *meaning*, 587
  is a causal set
    in the new ontology, 163
    Markopoulou may have suspected the, 378
    Smolin said the, 163, 379
  is a perpetual motion machine, 434
  is a whole, 183-5
  is digital, 310, 325
  is finite and closed, 135-6
  is nonlocal, 144
  is spherical, 326

laws at the beginning of the, 109
low entropy at the beginning of the, 137-40
must be explanatorily closed, 143
there was no matter in the first instant of the, 431
observers inside the, 141
one
  understanding ourselves and, 600
  Smolin said there is, 19
  vs many, explanatory power of, 584
other half of the, 329
  data inconsistent with Copernican principle, 581
in its entirety, as an existential level, 406
the missing piece of a consistent picture of, 191
the steady-state, *meaning*, 107
the story of the, 647
total energy of the, 434
visible
  astronomers explore the entire, 397
  galaxies in the, 360
  the whole universe is bigger than the, 398
was unchanging in Einstein's view, 112
wave function collapsing throughout the, 40
we are inside the, 17, 104, 141-3, 163, 228, 244, 278, 284, 290, 378-80, 479
workings expected to be simple, 61
universes, multiple, assume everything, 585
unobservable entities, 146, 408
Updike, John, 126-7, 190

V

vacuum, 113, 166
  case, *meaning*, 451
  catastrophe, *meaning*, 451
  concept of the, 450
  energy
    clash of values for the, 451
    drives cosmological inflation, 127-9
  ether in the, 68
  false, *meaning*, 587
  implies a nonexistent property, 377
  is a fiction, 448
  is an old ontological commitment, 169
  radiation state of the, 80
  state requires special attention, 448
Vafa, Cumrun, 400
Valentini, Antony, 289, 423
Vardanyan, Mihran, 398
Vaucouleurs, Gérard de, 591
Vilenkin, Alexander, 585
virtual particles, *meaning*, 332
void
  and Aristotle, 67
  and Buddhist thought, 598

and Copernicus, 326
and Democritus, 440–1, 450
and Descartes, 67
and ether, 68
and fleeting particles, 166
and Kant, 450–1
and Newton, 53
and Sartre, 313
and space expanding, 113
creation from the, 312–3
does not exist, 539–40
is inconceivable, Kant said, 450–1
no role for, in the new ontology, 314
space is not, 122
space of Democritus is, 441
Volcker, Paul, 593–4
volume
  doubling every Planck time, 364
  is a property of a Calabi-Yau threefold, 264, 468
  is how we see 3D, 265
  nodes represent, in Bilson-Thompson model, 249
  of extreme-gravity regions, 361
  quanta of, 558
  smallest-possible, 260
  the universe's, 291
volumes, Rovelli's quantized, 260
Vonnegut, Kurt, 245, 297

W

W particle, antineutrino and electron make a, 572–3
Walker, Arthur, 476
Walsh, Kenneth, 626
Watson, James, 651
wave function, 568
  Bell on the, 40
  collapsing, 39-40, 384, 568
  meaning, 40
waves, electromagnetic, 166
  neutral photons as, 273
Weaire, Denis, 445
weak force, 274, 575
  meaning, 274
Webb, Richard, 148, 220
Weinberg mixing, 274
Weinberg, Steven, 4, 100, 585
  and a hostile universe, 583
  and absolute zero of time, 312
  and elementary particles, 439
  and space expanding, 291
  and the standard model of cosmology, 297-9
  and the Z^0 *particle*, 274
  as multiverse proponent, 584
  said the universe seems pointless, 400-1
Weinstein, Steven, 176
Westerhoff, Jan, 598–9

Weyl, Hermann, 532
Wheeler, John, 20, 91, 103, 178, 421, 585
  and quantum foam, 91, 445
  and the importance of black holes, 460
  and the Wheeler-DeWitt equation, 177, 418
  asked why there are three dimensions, 263-5
  on why electrons are identical, 271
Wheeler-DeWitt equation, 177, 184, 263, 419, 753
Whitehead, Alfred North, 158, 520, 528, 560-1
  and the axiom of infinity, 548
Whittaker, Sir Edmund, 72-3
Whittle, Bruno, 519–20
Wien's law, 77–8
Wigner, Eugene, 516
Wilczek, Frank, 20, 26, 272, 497
  and quarks, 55
  dark energy as density of space, 349
  on cosmological inflation, 127–8
  on mass as energy, 435
  on Planck's system of units, 85–6
  on the ingredient of space and time, 52
  on the metric field, 476
  Zeno's arrow could hop each tick of time, 311
Wildberger, Norman, 529-30
Wilde, Oscar, 358, 735
Williams, Matthew, 402
Wilson, Edward O., 10
Wilson, Mark, 515, 522
WIMPs
  meaning, 352
  search for, 352, 356
windows
  between quanta, 236
  meaning, 230
  found, on the tiny world of granular space, 76
Winnipeg, meeting Bilson-Thompson in, 201
witchcraft, 27
Wittgenstein, Ludwig, 523-7, 598
  and the centrality of language, 523–7
WMAP satellite, 130
Woit, Peter, 455
Wollack, Edward, 131
worldview, 27
  a Buddhist, 600
  a coherent, 599
  a common, 610
  a communal, 25-6
  a new, 93, 275, 613, 660
    consistent, 403
    physical, 407
  and turtles all the way down, 140
  a real, 18, 33-4, 97, 609, 643

a simple, 62
change of cosmic situation in our, 617
disaggregation as the basis for our, 150-1
Einstein's, was local, 146
ether and the then new, 68
great chain of being was central in the Western, 659
holographic principle in an artificial, 339
individual, and Planck, 644
language needed for a new, 36
of a civilization is consequential, 609
of indigenous peoples, 150-1, 600, 626
of physics, 40
    is not grounded in reality, 29
of scientists influences how they work, 645
origin of causal thinking in our, 381
physics precepts amount to a, 27
positivism's influence on physics', 59
Sartre's, 301
the canonical, 557
    has consequences, 27
the continuum embedded in our, 30
the long evolution of our, 611
Weltanschauung, 173
Western philosophic, 598
Wheeler's, 20
without physics' useful fictions, 36
worldviews
    Hawking and how science changes, 201
    inconsistent, of physics' two theories, 206
    language may spawn unreal, 375
wormhole, *meaning*, 148
Wu, Chien-Shiung, 146
Wüthrich, Christian, 46

Y
Yau, Shing-Tung, 98
yin and yang
    existential, of the universe, 661
Young, Monica, 345
Young, Thomas, 386

Z
Zeilinger, Anton, 255
Zeldovitch, Yakov, 353–4
Zen
    and living in the moment, 325
    in pseudo-, fashion, 657
    meaning, 601
    the end or objective of, 601
Zeno of Alea, 162, 156, 221
    and motion, 307-11, 614-5
    and problems of continuity and motion, 63–5
    and successive places or beings, 601
    and the nature of time, 64
    key to understanding, 98
zero

and śūnyatā, 598
as an assumed math object, 561
as place holder, 537
curvature of space assumed to be, 74
energy of the initial universe, 431
entropy of the initial universe, 317-9
entropy of one microstate is, 317
Frege and, 540
Gödel used and maybe needed, 560-1
in Peano's and Godel's basic objects, 537
is not a natural number, 542
length of path in photon's frame is, 495
logarithm of one is, 317
of time, 312
ontological commitment to, 30-1
origin of the engram for, 539
Peano and, 540
role of, in Gödel's proofs, 561
Russell and Peano's use of, 541
set-theory equivalent of, 530
size
    initial universe as a point of, 7
    points of, 31, 198, 400
why dark energy (density) is so nearly, 348
zeroes, 539
    and problems based in logic, 452
    there are no, in the math of quanta, 557
Zukav, Gary, 427
Zuse, Konrad, 103–4
Zwicky, Fritz, 351
Zyga, Lisa, 580

Λ
ΛCDM
    paradigm is the simplest representation, 740
    -standard model, *meaning*, 395-6